SONAR and Anti-Submarine Warfare

May 25, 00
O. Ann Grace

SONAR and

Anti-Submarine Warfare

O. Donn Grace

SAGE Publishing Company Escondido, California

SONAR and Anti-Submarine Warfare
First Edition September, 1999

Published in the United States of America by:

SAGE Publishing Company
P.O. Box 460394
Escondido, California 92046

Library of Congress Catalog Card Number 99-71089
International Standard Book Number 0-9671414-0-0

Manufactured in the United States of America by:

Publisher's Press
P.O. Box 27408
Salt Lake City, Utah 84127

10 9 8 7 6 5 4 3 2 1

Ordering Information: pg. 451

ACKNOWLEDGEMENTS

If I could give only one acknowledgement it would be to Patrick "Pat" J. Donahoe of the Naval Command, Control and Ocean Surveillance Center (NCCOSC), in San Diego, CA. Pat has been a friend and colleague through much of my career. Our work together and, most notably, the opportunity he provided of being his Principal Investigator for Reverberation during the early research phase of the Low Frequency Aperture Program has had a substantial impact on my understanding of sonar and on the outcome of this book.

Many others in the ASW community have also contributed to my intellectual growth and, thereby, to whatever success that this book will achieve in ways that even they may not be aware. I am especially indebted to Al Aburto, Larry McKinley, Cliff Meland and Drs. Bob Smith and Bill Marsh, all of NCCOSC, for introducing me to the world of at-sea test and evaluation. Dr. Bill Ellison and Pat Skarp, of Marine Acoustics Inc., were most helpful in conversations concerning naval acoustics and anti-submarine warfare. My work with Tom Cantwell, of Hughes Ground Systems, was largely responsible for the formulation of my views on nonstationary noise. Professor Fred Harris, of San Diego State University, was a major contributor to my understanding of signal processing. The opportunities and conversations provided by Drs. Dan Alspach and Dale Klamer, of the ORINCON Corporation, formed the starting point for my work on multi-target tracking. My intellectual debt to these friends and acquaintances and to others is great. To those who didn't make the list I apologize for my oversight. Of course, I hold each and all blameless for my transcription of their wisdom into text and expect that some will disagree with portions of the final product.

Beyond the ASW community, my special thanks go to Dr. George H. Curtis; historian, Director at the Truman Presidential Library (retired), and, previously, Supervisory Archivist at the Eisenhower Presidential Library. George has been a long-time friend and graciously agreed to review Chapter 1. His recommendations are especially valued and have substantially improved the historical content of this book.

Last, but certainly not least, my debt to Alice L. Courteau for indulging me in this project and for her continual support during the long and sometimes trying periods of its development has earned her my undying gratitude. Thanks, Alice!

PREFACE

The East-West cold war of the late twentieth century is history and, to the credit and good fortune of the Western Allied Nations, the West prevailed. Many factors contributed to the winning of the cold war but one of the more decisive was the anti-submarine warfare (ASW) campaign that was waged for 50 years by western navies in all oceans of the world, by government scientists and engineers and by their counterparts in private industry. For most private citizens, this war was out of sight and out of mind because, by its very nature, it was conducted quietly within secured military facilities and civilian laboratories and, especially, on the high seas but its effectiveness is uncontested. It owes its success to the efforts of a dedicated community of professionals and to substantial allocations of national treasure to the development of advanced ASW systems and to the research and development which raised the state of ASW art and science to its present sophisticated levels.

With the winning of the cold war, national interest in defense and, in particular, ASW has been circumscribed by budget reductions. Under such conditions it is reasonable to ask, "Why another book on ASW sonar when several outstanding texts already exist?" My answer is a personal one. In brief terms, I had motive, borne out of intellectual momentum, and opportunity, afforded by a measure of independence. Actually, the writing of this book was set in motion early in my career as a sonar scientist. However, as my career progressed, the demands of the work increasingly diverted energy away from book writing and it became apparent that completing the book was to become a project for later years. Now, with the termination of the cold war, the later years have arrived. The downsizing of the ASW community has afforded myself the double luxury of a more leisurely lifestyle and the opportunity of bringing this book to fruition under the guidance of still functioning mental and physical faculties.

As the project began, my intent was to write two books on ASW sonar. The first would have been concerned with state-of-the-art ASW sonar basics, and the second would have been concerned with proposed advances in ASW sonar. I viewed these projects as being statements of, respectively, ASW sonar - as it is practiced - and ASW sonar - as it might be practiced. The first book would have been the book that I should have liked to have read at the beginning of my career. The second would have discussed concepts and developments that I believe should be considered for inclusion in modern ASW sonar systems. In the course of the writing, though, I discovered that I was not able to keep the two viewpoints neatly separated and it appeared that the effort required to write two complimentary books would be substantially greater than the effort required to write one inclusive book. As a result, I capitulated to expediency and have incorporated both sets of ideas in one book. In so doing, though, I have not bothered to explicitly call out whether material is 'sonar - as it is practiced' or 'sonar - as it might be practiced'. This should pose no problem to most readers since, on the one hand, the book deals primarily with sonar concepts and not with actual sonar systems and, on the other hand, most readers will bring some background to the study and will be able to recognize what is new.

The resulting book is essentially a treatise, in the sense that it develops in logical form the tale of what ASW sonar is about based upon first principles. It is structured to address the most important aspects of ASW sonar and, hopefully, to do so at a depth and breadth that most practitioners will find both interesting and useful. The reader, of course, will be the final judge of the success of the venture but I will forewarn that, to provide a balance between reach and grasp, the book was allowed to fall short of the intended goal in certain areas. Notable among the omissions is the abbreviated discussion of acoustic propagation and the lack of a discussion of fluctuations on ocean propagated signals. Much is known about these topics but the whole story is yet to be told. To formulate a comprehensive treatment of these concepts, at a level comparable to the other topics in the book, we will require additional material sufficient for the writing of another book. These are not the only objections that might be raised and I am confident that readers will find that other of their favorite topics have been either omitted or given short shrift. Since the book was necessarily shaped and limited by my interests and past experiences, I look forward with anticipation to reader's comments and suggestions for how the book may be expanded or improved.

A word about mathematics. For a full understanding of the material, the reader should be well grounded in basic calculus and bring to the study an acquaintanceship with complex analysis, vector analysis and probability theory. Sonar is a very technical subject and its proper explanation requires a measure of mathematical development. Such developments all follow a similar pattern: an initial mathematical statement of a physical principle is given; it is mathematically manipulated; resulting in a final mathematical statement of a physical principle. The initial statement is self-evident, or is obtained from a previous development, and the final statement is, hopefully, believable and useful but is generally not self-evident (otherwise we wouldn't need the mathematical derivation) and therein lies the value of the mathematical development. However, mathematical statements are information dense and, since most readers, the author included, are looking for quick comprehension, we sometimes shy away from math based discussions. Because of this, I considered skimping on the mathematical portions, presenting only final results in some cases, to attract readership and to minimize the cost of the book. I rejected this approach because it is a style that does not appeal to me personally and it does not appear to provide true understanding or insight. In fact, for serious students, such a presentation would have substantially increased the reading effort required for true comprehension. Accordingly, I elected to present full mathematical developments including a sufficient number of trail marker equations in each development such that the reader should have no difficulty in navigating from marker-to-marker using the mind's eye as a compass. I hope that in the process I have hit a balance between clarity and redundancy. In the age of desk top publishing and equation editing software, I have tried to err on the side of redundancy but my efforts, I fear, may have produced mixed results.

The proposed advances to ASW sonar that are discussed are believed to be new or relatively unknown to the ASW community. At least I have not seen this material elsewhere. However, much of the other material is well known and the value added to be gleaned from the following pages is in the selection and presentation of the material. My debt to other authors and researchers for this established material is enormous and I have tried to make this apparent through footnotes and a bibliography. However, I have not been compulsive in citing references and my citations are not exhaustive. The reader should view them as a starting point for further searches.

The book is composed of chapters with supporting appendices. Each chapter and appendix, for the most part, can be read independently of the rest with cross links called out where possible. The appendices can be omitted on a first reading but they contain some of the more interesting and important material and are not simply addenda of mathematical derivations. The chapters are numbered consecutively and supporting appendices carry the chapter number followed by a letter. For example: Chapter 2 is supported by Appendices 2-A and 2-B. Each chapter and appendix generally contains equations, figures and tables that are each numbered consecutively within the chapter and appendix. Within a chapter these entities are referred to as, respectively, (1), Figure 2 and Table 3. When referencing a chapter object from another section, we precede with the chapter number. For example: (2.1) is Equation 1 in Chapter 2. When referencing an appendix object from another section, we append the appendix letter. For example: Figure 2.1-B is Figure 1 in Appendix 2-B. All mathematical symbols are defined and explained as they are introduced while vectors and matrices are always represented in bold face type. References to standard mathematical results are usually to the text by Abramowitz and Stegun or to the text by Gradshteyn and Ryzhik. Full citations for each are given in the bibliography.

O. Donn Grace

Elfin Forest, CA.
April, 1999

Contents

Contents

CHAPTER 1
THE SUBMARINE THREAT AND ANTI-SUBMARINE WARFARE

Modern submarines routinely embark on extended patrols lasting several months that can penetrate any of the world's oceans. During these patrols they are largely undetectable and generally perform their missions unimpeded. These missions can include searches for opposing submarines, interdiction of opposing forces, protection of friendly forces, surveillance or simply remaining covert until required to launch weapons. The threat posed by these vessels is due in part to the lethality of their weapons and in part to their stealth. Their weaponry can include torpedoes and missiles, tipped with either nuclear or conventional warheads. Their missiles can be cruise missiles or nuclear-tipped ballistic missiles. Moreover, their stealth gives them an advantage of surprise which effectively magnifies the force of their already powerful weapons.

Anti-submarine warfare (ASW) is the countervailing force against the submarine threat. ASW doctrine is concerned with strategy and tactics that can be effected by available fleet and land based assets and is also concerned with affecting future asset development for the purpose of implementing more effective strategies and tactics. Consequently, the specifics of ASW are continually developing. However, in general, the first order of business for ASW forces is the determination of the presence and whereabouts of submarines with the second order of business being the follow-on neutralization of these vessels. Neutralization can be accomplished either by destroying the submarine or by causing it to abort its mission. Clearly, the term signifies more than a polite evasion of a harsh reality. Since ASW has evolved over the years, a brief examination of its history is in order before discussing its current state.

1.1 World Wars I and II

Naval forces, historically, have been organized around battle groups or task forces that usually consisted of one or more capital ships with support vessels. Capital ships are the principal fighters that can establish control of the sea and have the staying power to maintain sea control throughout the conflict. At the beginning of the 20th century and during WWI, battleships were the premier capital ships, with battle cruisers and cruisers sometimes filling that role. The heavy guns of the battleship could dominate coastal cities and opposing fleets by hurling explosive shells of several tons onto targets out to about 40 km and the armor of the battleships could withstand similar retaliatory barrages. Battleships were supported by cruisers, destroyers, frigates and other smaller ships. During WWII aircraft carriers became the premier capital ship because their aircraft could carry bombs and other firepower out to targets about 400 km distant and these aircraft could also provide protection to their carriers and accompanying escort ships. Battleships continued to function as capital ships, providing coastal offensives and coverage for beach landings, but relied upon aircraft carriers for their defense since they were vulnerable to opposing airpower and were rarely able to directly attack the opposing aircraft carriers.

The role of submarines, during the first half of the 20th century and the two world wars, evolved dramatically. During WWI, submarines developed from the experimental devices of the early 1900's into submersible diesel-electric fighting machines. Through their use by Germany in the blockade of England, they were nearly the decisive force in both world wars. Armed with torpedoes, deck guns and sometimes mines, they proved their effectiveness against opposing naval forces, including capital ships, and against merchant shipping. The principal defenses against the submarine at the start of WWI were mines and barrier nets. Cannon fire or ramming by surface ships sometimes proved to be effective against submarines found on the surface or diving. To detect and destroy submerged submarines, passive listening devices and depth charges were developed by the Allied forces in 1916. The passive listening devices could detect sounds radiated by submarines which were then prosecuted using depth charges. The listening devices consisted of arrays of carbon button microphones that were mounted on surface ship hulls or were towed by surface ships and these arrays were sufficiently directional to estimate the bearing to the target. Lacking range and depth information, surface ships did not have precise information on target location but, since a near miss by a depth charge could be lethal for the submarine, this did not always matter. In the early stages of the war, aircraft were found to be very effective in locating submarines on the surface and even sometimes when they were submerged. Once located, the aircraft crew could immobilize the submarine if they were fortunate enough to make a direct bomb hit. Depth charges modified for aerial use improved the effectiveness of air attacks since a direct hit by a depth charge was not required. To reduce seriously mounting Allied merchant shipping losses to submarine attacks, convoy operations were introduced in 1917 which included military escorts, routine evasive zigzag maneuvers and camouflage paint all of which proved to be very effective. In the on-going Allied effort to neutralize the submarine threat, a group of scientists were secretly brought together in 1918 to formulate methods for defeating submarines. They produced an underwater sound projector of sufficient power to achieve detection and ranging echoes from submarine hulls at distances out to 1500 m. This system was referred to as "Asdic" which are the initials of the name of the group that developed the system[1]. Since their work was performed during the last year of the war, Asdic was not completed and operationally deployed in time to help in the war effort but it was subsequently developed during the post-war era and played a major role in WWII.

In the period between the two world wars, all major nations embarked on rearmament programs which, recognizing the important and effective role that diesel-electric submarines played during the war, included submarine building and modernization efforts as parts of those programs. Consequently, at the outbreak of WWII all belligerents had fleets of modern well equipped diesel-electric submarines. In contrast, ASW technology was very nearly where it was at the end of WWI, with

---

[1] The group responsible for developing Asdic has been variously reported as the Allied Submarine Detection Investigation Committee, Anti-Submarine Devices Investigation Committee, Allied Submarine Devices Investigation Committee or Anti-Submarine Division. In the later case the acronym Asdic adds the "ic" suffix to mean "having to do" with the ASD group.

the exception of Asdic, which was more fully developed and widely used on Allied vessels. On U.S. vessels, Asdic was referred to as sonar (*so*und *n*avigation *a*nd *r*anging) which is an acronym patterned after radar. During the opening days of WWII the submarine threat was regarded with complacency at the command levels of the Allied forces. However, the German submarine commanders' discovery that night surface attacks could circumvent most convoy defenses, including sonar, served to focus Allied Command attentions and efforts. Moreover, the German submarine commanders also developed the highly effective "wolf pack" strategy concentrating about 5 to 15 submarines (and in some cases many more) to overwhelm convoy defenses with coordinated attacks. During this period it appeared that the Axis submarine forces would prevail. Allied forces initially countered with illuminating flares, termed Snowflakes, that could reveal the presence of the submarines on the surface. However, the real solution to the German submarine threat was surface search radar and high frequency direction finding (HFDF) for localizing submarine radio transmissions. Both were deployed in Allied fleets in 1941. Although German developed radar impulse detectors countered meter band surface search radar, the Allied follow-on developments of centimeter band radar and the breaking of German codes, neither of which were apparently realized by the Germans, were never countered. These developments, coupled with the increased use of aircraft in anti-submarine warfare, marked the turning point in the submarine war. As in WWI, aircraft had proven, once again, to be effective in finding and immobilizing submarines. Convoy protection was significantly improved through the use of escort aircraft carriers and long-range land-based bombers. The carrier based aircraft maintained continuous air coverage over the convoys and the long-range land-based bombers would assist and also perform continuous pursuit of submarine targets after the convoy passed.

During WWI and WWII, the effectiveness of submarines on both sides of the conflict was repeatedly demonstrated by their destructive attacks on naval forces and merchant convoys. In fact, the impact of German submarine forces on the course of both wars was nearly decisive, although, in the end, they were overwhelmed by Allied ASW forces. In contrast, American submarine forces during WWII were highly successful in the Pacific Theater of operations, especially in the destruction of Japan's merchant fleet. Their success was due, in part, to the lack of an aggressive or adequate ASW effort. All submarines were generally devastating in offensive, but, when their presence was revealed, they were aggressively hunted and frequently neutralized. They performed as essentially guerrilla fighters in both wars and, as is the practice of guerrilla fighters, when confronted by a large coordinated counterattack, they retreated. They suffered from the handicaps of maximum submerged speeds of about 10 knots, restrictions to depths of less than about 100 m and their need to return to the surface every few hours for air and to recharge batteries. Moreover, they lacked information regarding the goings-on around them and especially information as to what their hunters were doing. Consequently, under attack by ASW forces, submariners could only wait it out while guessing at appropriate evasive maneuvers and hope to either slip away or to simply endure. At these times all hands, no doubt, were mindful that only a near miss by a depth charge was all that was required to destroy their boat. During the cold war of the second half of the 20th century, this all changed.

1.2 Cold War Period

Under the impetus of the east-west cold war that followed WWII, submarine and anti-submarine warfare proceeded apace. The early work integrated lessons learned from WWII and was evolutionary in nature. However, much of the work during the cold war period can only be described as revolutionary. In broad terms it led to enhanced tactical submarines, on the one hand, and to strategically capable submarines, on the other, leading to the general classification of submarines as being either tactical or strategic. Tactical submarines include attack, guided missile and, for a brief period, hunter-killer submarines, while strategic submarines are the nuclear-tipped ballistic missile submarines. In broad terms, the mission of attack submarines is interdiction of enemy fleets including enemy submarines. The mission of guided missile submarines is similar but includes power projection against shore installations. Hunter-killer submarines were an early submarine development that were outfitted with advanced sonar systems for the purpose of anti-submarine warfare. As all submarines became similarly outfitted the hunter-killer designation was dropped. U.S. designations for the nuclear powered versions of these boats are, respectively, SSN, SSGN, SSKN and SSBN. The designations for the corresponding diesel-electric boats are similar but without the suffix "N". The principal cold war submarine developments were achieved by the U.S. and U.S.S.R. but other major powers also contributed. A brief history of those efforts follow and a summary of submarine developments for the U.S. and U.S.S.R. is presented in Figure 1.

U.S. Tactical Submarine Development

In the U.S. the designs for the diesel-electric *Gato*, *Balao* and *Tench*-class submarines, that were instrumental in the winning of WWII, gave way to a series of advanced submarine concepts during the post-war period. The streamlined tear-drop shaped hull of the experimental submarine *Albacore* (completed in 1953) increased submerged speeds to 33 knots, improved depth capability and maneuverability, enabling future submarines to outrun surface craft and torpedoes. Moreover, the new hull shape augmented the interior space and improved crew habitability. The production model of the *Albacore* was the diesel-electric *Barbel* class attack submarines which were armed with torpedoes.

The USS *Nautilus* (launched in 1954) was the world's first nuclear powered submarine. Its power plant and air purifying carbon dioxide scrubbers eliminated the submarine's need for frequent visits to the surface and created the first true submarine, as opposed to a submersible ship. The *Nautilus*, at 4000 tons submerged displacement, was roughly twice as large as modern diesel-electric submarines, such as the USS *Barbel*. The size of its nuclear power plant and, more importantly, the habitability needs of the crew, for extended underwater cruises, necessitated the larger size. The *Skate* class submarine (initiated in 1957) was the production model of the *Nautilus* which, in order to minimize development risk and following the

Nautilus design, retained the conventional submarine hull shape. The *Sea Wolf* (also initiated in 1957) was similar to the *Nautilus* but was powered by a liquid sodium cooled reactor in contrast to the pressurized water cooled reactor of the *Nautilus* and *Skate* class. The problems encountered made the pressurized water cooled reactor the fleet standard for follow-on designs.

The combining of nuclear power with the *Albacore*-type hull was first achieved in the *Skipjack* class attack submarines, starting in 1959. They set the pattern for all future U.S. submarines and, at 3500 tons submerged displacement and armed with torpedoes, could reach about 300m in depth and 30 knots submerged speed. The conventional hulled USS *Tullibee* was built as a hunter-killer submarine to complement the fleet work of the *Skipjacks*. It was slower than the *Skipjacks* but was equipped with the AN/BQQ-1 enhanced sonar suite that featured a large spherical active-passive bow array which was the predecessor to the modern U.S. submarine sonar suites. The *Tullibee* design was not pursued but, when the *Skipjacks* were upgraded by the larger *Permit* and *Sturgeon* classes in 1961 and 1967, respectively, the ASW function was retained and these classes were outfitted with the improved AN/BQQ-2 sonar suite which subsequently acquired a towed array capability. Having stronger hulls these boats can dive to about 400m in depth and, in addition to torpedoes, carry Sub-Harpoon anti-ship missiles, having ranges of about 150 km, and SUBROC anti-submarine missiles, which are missile-borne nuclear depth charges with a range of about 50 km.

The *Los Angeles* class (started in 1976) is the most recent attack class submarine and, at 31 knots submerged speed and 6900 tons submerged displacement, is as fast or faster than its predecessors. It is also much larger (twice as large as the *Skipjacks*). High speeds are achieved even though its more economical cylindrical hull deviates somewhat from the *Albacore* streamlined hull. Its purpose was to counter the U.S.S.R. threat to U.S. battle groups by the *Charlie* class guided missile and other submarines. Armaments include Tomahawk missiles, with ranges of about 460 km for the anti-ship version and with ranges of about 2600 km for the land-attack version. Additionally, the *Los Angeles* class is armed with Sub-Harpoon anti-ship missiles and SUBROC anti-submarine missiles and torpedoes. It was initially fitted with the AN/BQQ-5 sonar suite which was subsequently upgraded to the Submarine Advanced Combat System (SUBACS) which is an integrated weapons and sonar control system.

British and French Tactical Submarine Development

Britain and France also fielded SSN submarines. The first United Kingdom nuclear submarine was the *Dreadnought* class (started in 1963) which was succeeded, in turn, by the *Valiant* class in 1966, the *Swiftsure* class in 1973 and the *Trafalgar* class in 1983. These boats are armed with torpedoes and Sub-Harpoon anti-ship missiles and provide close support to the fleet task force and to convoys in contrast with the independent operations of the U.S. submarine forces. The French *Rubis* class (started in 1982) is a small but efficient attack submarine armed with torpedoes and Exocet anti-ship missiles.

U.S.S.R. Tactical Submarine Development

The U.S.S.R. pursued an aggressive program of attack submarine development after WWII. Their initial efforts were a series of boats modeled on the successful German WWII Type XXI diesel-electric submarines with follow-on classes evolving from those early designs. The early designs included the *Zulu*[2] class, the *Whiskey* class, and its successor the *Romeo* class, all built in the 1950s. They were used for coastal defense and long range reconnaissance. All were armed with torpedoes.

The need for submarines with heavier torpedoes to counter forward positioned U.S. battle groups and to attack port installations led to the larger *Foxtrot* class in 1958 and to the *November* class, which is the first U.S.S.R. nuclear-powered submarine, both of which carried nuclear-tipped torpedoes. Since the *Foxtrots* and *Novembers* were required to close to within sonar range of their adversaries, a risky tactic, the nuclear powered *Echo* class cruise missile submarines and its diesel-electric counterpart the *Juliet* class (both started in 1962) were specially built for attacking from long range standoff positions. They succeeded an early line of converted diesel-electric cruise missile boats. Armed with SS-N-12 cruise missiles, they could launch from maximum ranges of approximately 450 km. Because they were required to be on the surface for launch and for providing midcourse guidance to the missile, from friendly reconnaissance aircraft data, they suffered their own vulnerabilities.

An alternative system for countering the risk to the *Foxtrot* and *November* classes was the *Charlie* class cruise missile submarine which was started in 1967. Its attack mode was close-in with SS-N-7 anti-ship missiles, having ranges of about 60 km, which could be launched while submerged using passive sonar targeting data while positioned beyond the target's reliable active sonar range. The *Charlies* were additionally armed with torpedoes and SS-N-15 anti-submarine missiles, similar to SUBROC missiles, having a maximum range of approximately 40 km.

After the U.S. deployment of the USS *George Washington* and follow-on SSBNs, U.S.S.R. strategy added ASW activities with U.S. ballistic missile submarines being the primary targets. The first U.S.S.R. ASW nuclear-powered hunter-killer submarine was the *Victor* class (started in 1968). Due to production problems, it was supplemented by the diesel-electric *Tango* class (started in 1972). Both boats were outfitted with active-passive bow arrays and were initially armed only with torpedoes, but upgrades to the *Victor* class were armed with SS-N-15 and SS-N-16 anti-submarine missiles which were missile borne torpedoes having ranges of about 100 km. The ASW strategy was to intercept U.S. SSBNs as they egressed from port, trail them while they patrolled and be prepared to attack as required. The plan was not successful, in part due to the intervention of U.S. SSNs, and in the early 1970s the *Victors* and *Tangos* were reassigned to the defense of Soviet SSBNs within Soviet bastion areas.

The *Alpha* class, were small but fast deep diving titanium hulled submarines (started in 1972). They were outfitted with active-passive bow arrays

---

[2] The designations for the U.S.S.R. submarines and missiles are NATO designations. U.S.S.R. designations for these ships are unknown.

and armed with torpedoes and SS-N-15 anti-submarine missiles. They could transit at 45 knots to interdict hostile surface and submarine forces and could evade counterattacks from Western ASW forces by diving beneath those boats to 900m. The *Mike* class (started in 1984), appears to have been a larger improved version of the *Alpha* class. The *Oscar* class cruise missile submarines (started in 1980) were replacements for the aging *Echoes* and had similar capabilities, but were equipped with the SS-N-19 anti-ship cruise missile that could be launched from underwater. Aging *Whiskeys* and *Romeos* were updated by the modern diesel-electric *Kilo* class, (started in 1982) for coastal defense and shallow water work and were equipped with active-passive bow arrays and a passive hull mounted conformal array.

The successors to the *Victor* class are the *Sierra* class (started in 1984) and *Akula* class (started in 1986). Both are similar to the *Victors* and, in addition to torpedoes, carry the SS-N-15 and SS-N-16 anti-submarine missiles and the SS-NX-21 missiles which are Tomahawk-like land attack missiles having ranges of about 2000 km. Their sonar systems are similar to the *Victors* but they also carry a passive towed array.

U.S. Strategic Submarine Development

Paralleling the development of the tactical submarines, the first ballistic missile submarine, capable of submerged missile launch, was the USS *George Washington* (completed in 1959). It was created by inserting a ballistic missile section into a lengthened *Skipjack* hull. This class of fleet ballistic missile submarines was made possible by the successful completion of the Polaris missile in 1957 which had a range of about 2800 km. The first specially designed SSBNs were members of the *Ethan Allen* class followed by the *Lafayette* class in 1963. Compared to the *George Washington* class, the *Ethan Allen* and *Lafayette* classes were larger, at 8250 tons submerged displacement, quieter and deeper diving and could accommodate larger MRV'd (multiple reentry vehicles) missiles, with ranges of 4600 km.

The *Ohio* class of ballistic missile submarines, at 18,750 tons submerged displacement, is the largest of U.S. SSBNs. It was launched in 1981 to accommodate the MIRV'd (multiple independently targetable reentry vehicles) Poseidon and Trident missiles with ranges of 11,000 km. The nominal loads for these missiles are 3 warheads for the Poseidon and 10 warheads for the larger Trident. Some of the later boats are fitted with MaRV'd (maneuverable reentry vehicles) Poseidon and Trident missiles. Many of the earlier SSBN classes were subsequently retrofitted to carry the Poseidon and Trident missiles. All of the U.S. SSBNs are armed with heavy torpedoes for defense.

British and French Strategic Submarine Development

The British and French also produced classes of SSBNs. The British initiated the *Resolution* class SSBNs in 1967 armed with U.S. Polaris missiles and the French initiated the *Le Redoubtable* class in 1971 armed with French M-1 missiles. Both classes are similar to the U.S. Lafayette class.

U.S.S.R. Strategic Submarine Development

The U.S.S.R. produced the world's first ballistic missile submarines, launched in 1958, which were *Zulu's* modified to accommodate two SS-N-4 missiles standing in the aft portion of the fin and having a range of about 700 km. However, they could only be launched from the surface. These boats were soon augmented by the *Golf* class (built in 1959) which were modified *Foxtrots* and the first specially built ballistic missile boats. They were armed with three SS-N-5 missiles, mounted similarly to the *Zulu* class missiles, with ranges of about 1400 km and which could be launched from underwater. All of these boats were diesel-electric powered but were followed by the nuclear powered *Hotel* class in 1960 which was an updated version of the *Golf* class.

The *Yankee* class (started in 1967) was a major departure from the older SSBNs and were similar in design and size, at 9600 tons submerged displacement, to early western SSBNs. They were armed with the compact SS-N-6 strategic missiles having ranges of about 3000 km. They were equipped with torpedoes and an active-passive bow sonar array. The *Yankees* were a major improvement over the older SSBNs but, even with the longer range strategic missiles, the *Yankees* were still required to patrol in the vicinity of the continental U.S. in order to target U.S. cities. Since the *Yankees* were noisy submarines, being close to U.S. continental surveillance assets made them vulnerable to detection and prosecution while a long way from home.

The *Delta* class was initiated in 1972 and carried the longer range SS-N-8 strategic missiles, which had ranges of about 9100 km, and the somewhat shorter range SS-N-18 MIRV'd version. With these missiles, the *Deltas* could target western cities from waters neighboring the U.S.S.R., including the Arctic Ocean and the Sea of Okhotsk. The *Deltas* were equipped with sonar systems comparable to the *Yankees* and were armed with torpedoes.

The desirability of operating from relatively safe waters nearby the U.S.S.R. led to the development of the *Typhoon* class in 1981 which, at 30,000 tons displacement, are the world's largest submarines. The Typhoons are armed with twenty SS-N-20 strategic missiles, having ranges of about 8300 km, and can launch several missiles simultaneously in contrast to all other SSBNs. The Typhoons are armed with torpedoes and also SS-N-15 and SS-N-16 anti-submarine missiles. They are equipped with active-passive bow sonar arrays and possibly passive hull mounted conformal arrays for detection and targeting of anti-submarine weapons.

The Submarine Threat in the Late 20th Century

Modern submarines of the major nations have developed during the cold war into formidable weapon systems that may be proven, in some coming major conflict, to be the next generation of premier capital ships. SSNs can attack fleet ships with anti-ship missiles, anti-submarine missiles and homing torpedoes, and can attack land facilities with land attack missiles. SSNs, operating in close proximity to their adversaries, are vulnerable to retaliatory ASW defenses but, because of the modern SSN's speed, depth, armaments and sensors, they are not

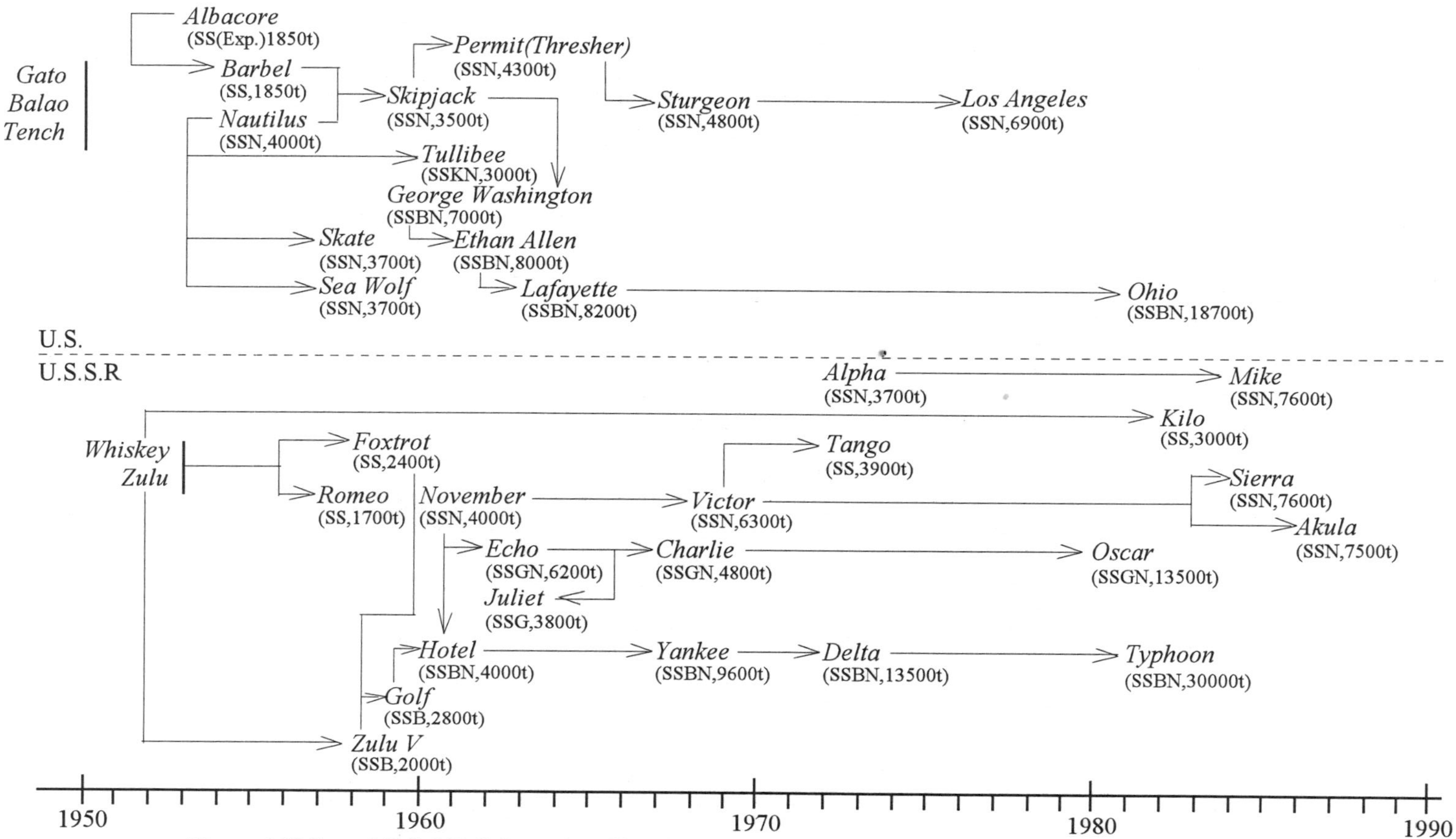

Figure 1 U.S. and U.S.S.R Submarine Class Developments (Type,Approximate Submerged Displacement (tons))

expected to be the hunted prey of surface ASW forces, as were their diesel-electric ancestors, but rather they are expected to have the potential of being the aggressive and intimidating hunters. In fact, it has been argued that submarine-against-submarine conflicts may be more evenly matched than submarine against surface conflicts but this has yet to be proven in real life combat. SSBNs are less aggressive than SSNs but are, arguably, the most lethal weapons in the world. They have the capability of launching missile borne nuclear warheads onto opposing nations and each missile has the power and accuracy sufficient to destroy large portions of any enemy nation's infrastructure. The SSBNs are virtually invulnerable to detection, localization and attack prior to missile launch and are expected to be exceedingly difficult to locate and attack after missile launch. Moreover, if detected, the SSBNs can mount aggressive defenses against any adversary. In summary, SSNs and/or SSBNs may be paramount threats in future conflicts and whether or not ASW forces will be able to neutralize them is an open question.

1.3 Modern Anti-Submarine Warfare

Anti-submarine warfare in the late twentieth century has been a dominant preoccupation of western naval forces. It is pursued by naval surface forces, land-based ASW airborne forces and submarine forces. All are supported by surveillance assets, including the SOund SUrveillance System (SOSUS) and the mobile SURveillance Towed Array SenSors (SURTASS) systems. Surface and submarine forces perform ASW as part of their larger mission requirements but, whereas surface forces, land-based airborne forces and surveillance assets generally perform highly coordinated ASW activities, submarines tend to wage independent and solitary battles in the ASW effort. Land-based ASW airborne forces and surveillance assets normally perform on-going broad area searches under the control of the land based ASW forces but, when required by the fleet, mobile surveillance assets, such as SURTASS, can come under the control of the battle group commander.

ASW doctrine for surface, land-based airborne and submarine forces has evolved during the twentieth century. The common thread in this evolution are the five phases: search, contact, approach, engage, disengage. In the search phase, the ASW forces monitor sensor systems for submarine detections and are guided in their efforts by previous contacts, contact reports from surveillance assets and other intelligence. The contact phase is initiated by one or more detections from which estimates of the target position, its movements and type are obtained. In the approach phase, ASW forces advance on the target to confirm the contact and to position themselves for engagement. The engagement phase is initiated either by an attack on the target or by defensive maneuvers in response to an attack by the target. Combat is pursued until the target is neutralized. In the disengagement phase, target disposition is confirmed and the ASW forces return to the search phase.

ASW for Surface Forces

Generally speaking, surface forces, in the performance of their larger missions, are organized around one or more capital ships with supporting vessels, including an ASW complement, and are clustered for efficient communications and mutual protection. However, they can be dispersed as much as 20 nautical miles between ships to minimize the effects of a nuclear attack. Depending upon the mission, ASW zones are established around the group, which generally include an inner zone extending out to about 150 km, a midzone from 150 km to 300 km and an outer zone from 300 km to 450 km. ASW activities are zone-dependent as well as being mission-dependent. Typical mission scenarios would include the formation and staging of a battle group or convoy, a battle group performing forward force projection or a convoy escort shepherding a convoy through its area. The primary mission in each scenario is necessarily uppermost in the group commander's mind and ASW activities, while essential and a large consumer of time and effort, are peripheral to the larger mission.

ASW activities in the outer zone, which extends from about 300 km to 450 km, are performed by aircraft and by surveillance assets against cruise missile submarines such as the *Echo*, *Juliet* or *Oscars*. The aircraft can be organic to the battle group, such as the U.S. S-3's, or can be land-based long range maritime patrol (LRMP) aircraft, such as the U.S. P-3 Orion or the British Nimrod. These aircraft can be vectored to a detection site, localized by one of the surveillance assets, such as SOSUS or SURTASS, and can search suspected target positions or transit lanes using radar and sonobuoys. Radar contacts are infrequent and are expected to become less frequent during times of conflict but radar contacts on periscopes, snorkels and on the occasional surfaced submarine can be made. Sonobuoy fields or barriers are the more successful methods of search and are laid by aircraft in the hope of detecting a patrolling or transiting submarine. Since each sonobuoy in the field can be localized by means of radio transponders on the sonobuoys, the position of all sonobuoy contacts can thereby be determined. The sonobuoys can be simple passive omni-directional receivers, such as the basic AN/SSQ-41B, or can be passive directional DIFAR receivers, such as the AN/SSQ-53. Passive receivers acquire no target range information but target position can sometimes be determined by correlating the received acoustic signals from different passive sonobuoys. A more positive approach to localizing the target is through the use of active sonobuoys. Modern active sonobuoys transmit omnidirectionally but they can receive either omnidirectionally or directionally. Omnidirectional active sonobuoys can be triggered automatically upon entry into the water for continuous operation or can be CASS (commanded active sonobuoy system) sonobuoys which are keyed by the aircraft crew. Directional active sonobuoys are DICASS buoys. In a typical scenario, passive DIFAR buoys would produce a target contact without alerting the submarine. After attack preparations are completed, the aircrew confirms the contact presence and updates its localization estimate by means of active CASS or DICASS buoys and, if required, dispatches the target via lightweight torpedo strikes.

ASW activities in the midzone, extending from about 150 km to 300 km, are similar to the outer zone with the addition of surface craft. Targets of interest in the midzone include *Foxtrot*, *November* and *Charlie* class submarines. The *Foxtrots* and *Novembers* would need to close within torpedo range of the battle group but the *Charlies* would be able to launch missiles while submerged and while outside active sonar range of the main battle group. The surface ASW vessels might include destroyers and/or frigates, which might be equipped with the passive AN/SQR-19 TACTASS tactical towed array sensor system. TACTASS detection ranges extend reliably to about 50 km and sometimes to 150 km and it receives data from preformed beams which are processed to produce frequency and azimuth (FRAZ) displays of the underwater environment. If a target is detected, it might be confirmed via AN/SQS-26 or AN/SQS-53 active sonar transmissions, which can echo range out to about 50 km and sometimes to 100 km, and S-3's or helicopters can be vectored to the target position. Helicopters can transit to targets out to about 100 km which they can detect with their AN/AQS-13 active dunking sonar systems or with sonobuoys. The detection range of the dunking sonar is only a few thousand meters but, since it can alert a target at much greater ranges, it is used only against alerted targets or when expediency counts or when preceded by passive sonobuoy detections. After a target is confirmed, the helicopters can prosecute it with lightweight torpedoes or surface ships can prosecute it using ASROC missiles, torpedoes and depth charges. The ASROCs are Mk-44 or Mk-46 lightweight torpedoes that are rocketed out to ranges of about 16 km to the target position. The torpedo is parachuted into the water and it searches for the target using active homing sonar. For shorter-range targets surface ships can tube launch lightweight torpedoes. For targets within about 3 km, depth charges can be deployed. Depth charges are not expected to be an effective tactic against an aggressive nuclear attack submarine and every attempt is made to keep the target at longer ranges.

ASW activities in the inner zone, extending out to about 150 km, are similar to midzone ASW activities but are performed by guided-missile cruisers as well as destroyers and frigates and their helicopter complements. In contrast to midzone ASW, these vessels rely on their AN/SQS-26 and AN/SQS-53 active sonars to search the inner zone because the underwater noise level around the battle group seriously degrades the detection performance of the towed passive array systems. In search mode these active sonar systems transmit into selected sectors and receive data on preformed beams which are processed to produce bearing-range maps of the underwater environment. Similarly, helicopters rely more heavily on active dunking sonar and less on sonobuoys.

ASW for Land Based Airborne Forces

Land based airborne forces can support battle group or convoy ASW efforts, as specified above in the section on "ASW for Surface Forces," or they can be vectored to a suspected submarine position by SOSUS or SURTASS surveillance assets as part of the on-going broad area search performed by those assets. Since the surveillance detections have an associated target position uncertainty of about 50 square nautical miles, which can expand while the aircraft is transiting to the area, the aircraft is required to perform an area search upon arriving on station. Position uncertainty can grow by an order of magnitude during a long transit but its rate of growth is constrained by later surveillance detections or by projecting target track estimates obtained from previous detections. Once on station the aircraft deploys sonobuoy barriers across suspected target tracks, which can be done from high altitudes to avoid alerting the target. After detections have been acquired and attack preparations completed, the aircraft can confirm target detection using additional, perhaps active, sonobuoys or by magnetic anomaly detection techniques, which require a low pass over the suspected target position. The alerted target can be immediately prosecuted with lightweight torpedoes before it can escape.

ASW for Submarine Forces

SSN submarines can perform ASW activities beyond the outer zone but they can also perform ASW within the zones if coordinated with battle group ASW activities. In such cases, coordination may be to simply assign a zone sector exclusively to one submarine for sanitization. During ASW searches, *Los Angeles* class boats rely on SUBACS, which is an integrated weapons/sonar control suite consisting of the AN/BQS-13 active-passive spherical bow array, the AN/BQR-7 conformal passive bow array, the AN/BQR-23 passive submarine towed array sensor system (STASS) and the passive underwater firecontrol feasibility system (PUFFS). Older SSNs rely on the AN/BQQ-2 or AN/BQQ-5 sonar suites, which are similar to the sonar suite portion of SUBACS. After a target is detected, the SSN submarine can attack with heavy torpedoes or with SUBROC. SUBROC is a rocket-borne depth charge that is launched out of an SSN torpedo tube. SUBROC follows a ballistic trajectory to the target area, falls into the sea and a nuclear warhead detonates at a preset depth.

SSN submarines can coordinate with surface forces and aircraft but, during ASW, and most other activities, they generally remain covert and communicate only occasionally. Their communication options include line-of-sight UHF or SHF links to surface, air or satellite receivers, or the more detectable and localizable HF burst links, all via sail mounted antennae, or an AN/BRT-1 transmitting buoy, which repeats a prerecorded message delayed, up to one hour, to allow the submarine to depart, or ELF or VLF links by means of trailing-wire antennae. When pursuing a target, submarines are especially noncommunicative since they cannot break from their efforts to report or to receive direction or information without compromising their positions or losing the contact. However, preselected messages can be acoustically sent to the submarine without compromising its position via the air or ship dropped AN/SSQ-86 down link communications (DLC) buoy.

Nonacoustic ASW Methods

The search and contact phases of ASW are performed predominately by either active or passive sonar which are in operational use by all navies. Both are subject to the laws of ocean acoustic propagation and the vicissitudes of oceanic conditions and, while it is found that the detection of submarines by sonar can be operationally effective, it is also found to be very difficult. Because of this, alternative methods are always sought with the objective of finding better techniques or of finding techniques for corroborating sonar contacts. Some of the better known methods are:

1. Magnetic anomaly detection (MAD): Submarine hulls made of ferrous materials cause the earth's magnetic field to constrict in the region of the hull. This anomaly in the earth's magnetic field can be detected by a magnetometer at ranges out to about 300 m and is routinely used on ASW airborne patrols.
2. Electrical fields: Electrical fields are established around all ships for a variety of reasons. For example, bimetal construction causes electrical potentials and the consequent conduction of currents through the surrounding sea water. These fields can be detected out to ranges of hundreds of meters depending upon the target and sea conditions.
3. Wake detection: When underway, submarines produce several wakes that can be detected on the surface even when the submarine is submerged. Cavitation bubbles from propellers can be detected by active sonar, at depth, for some time after the passage of the submarine. These same bubbles, after rising to the surface, cause a disturbance to the pattern of surface waves which can also be detected by radar. During the bubble assent, upwelling of colder/warmer water and thermal discharges by the submarine produce thermal gradients on the surface that can be detected with infra-red detectors. Internal hydrodynamic waves generated by an underway submerged submarine cause a wake on the surface above the submarine that can also be detected by radar.

All of these nonacoustic techniques have been proven to be effective under appropriate conditions. However, they generally require a detector positioned in the proximity of the submarine position, usually on an airborne platform, and this restricts their usefulness to the corroboration of sonar contacts. Thus far, only MAD has been operationally implemented for this purpose. Until longer range, broader area coverage implementations, possibly spaced based, or alternative techniques are developed, sonar will continue to be the primary ASW detector.

1.4 Entering the 21st Century

As the 20th century closes, the allied nations have entered a new era in which credible military threats are not apparent and major retrenchments in military strength from cold war levels, including ASW, are in order. For the demise of the cold war we owe much to the efforts of the world's leadership and to the effectiveness of the military of the allied nations. The relief from cold war tensions purchased by their efforts is apparent in the dual effects of opening lines of communication throughout the world and the release of a ground swell of logic for indulging in the long awaited peace dividend of permanent force reductions. As appealing as both trends are, though, we note that the opening lines of communication can also increase the tempo of international conflict which, in turn, makes the call for force reductions somewhat like the siren's song. Even though the threat of global war is apparently over, the stability of its constrictions has given way to the fact of small hot skirmishes between developed and developing nations and tribal wars in third world nations. Moreover, these hot conflicts are embedded within a climate of trade warfare in which battlefields are the market place; businesses are force projectors; strategy is a business plan; tactics are marketing ploys; and weaponry has become information and consumer products. Given that all nations are concerned with security and prosperity, since natural resources are limited, competition and aggression among nations can be expected to increase in tomorrow's more open world. The battle for human physical needs, i.e. food, energy, minerals, living space, etc., is never ending and, whether the battlefield for these needs is the marketplace or a place of physical combat, claims on natural resources and the supply lines to those resources will be contested and will require protection. Much of this protection is expected to come from international law but, when dealing with intransigent nations, since international law will likely breach national sovereignty primarily through the use of force, a measure of military and, especially, naval strength will be required to ensure that allied national interests are protected. Conflict resolutions at those times is expected to be more satisfactory for strong nations. Accordingly, the need for military strength, including an ASW contingent, remains an important requirement as we enter the 21st century. Consequently, it's useful to speculate on what those requirements for the 21st century ASW forces may be - all the while acknowledging that surprises are bound to occur.

Looking to the future, the growth of information technologies will have a major impact on all military developments including ASW. In our new era of small hot wars and trade wars, it's fashionable to talk about future battlefields moving to cyberspace. However, it's more correct to observe that the space of the human mind, i.e. perceptions, mindsets, decision processes, etc., has always been the principal battlefield in wars of all stripes - hot, cold, trade, etc. Information flows in cyberspace are, in actuality, a subset of the human mind space and cyberspace technology is a force multiplier in the battle for the human mind. Accordingly, the battlefield is not so much moving to cyberspace as it is expanding into cyberspace. If we date the staging of the cyberspace battlefield with the ARPANET, the expansion was started in the 1970s and will continue in response to, but regardless of the nature of, future conflicts. ASW forces will certainly be in the vanguard of this expansion in an attempt to better control information flows concerning world submarine forces and their activities.

The coming world's submarine capabilities are a matter of speculation but, at the close of the 20th century, most developed nations and some of the developing nations have fleets of modern nuclear and diesel-electric submarines. The U.S. is the most advanced in this area with approximately 100 nuclear attack boats, 40 nuclear ballistic missile boats and 4 diesel-electric boats. The submarine fleet of the former Soviet Union is approximately two and half times as large but with reduced capabilities and includes numerous diesel-electric boats. Many of these boats are currently being sold to client states. The United Kingdom has about 20 nuclear attack boats, 8 nuclear ballistic missile boats and 15 diesel-electric boats while France has about 5 nuclear attack boats, 7 nuclear ballistic missile boats and 15 diesel-electric boats. Among the developing nations, China has the largest submarine fleet of approximately 5 nuclear submarines and 100 diesel-electric submarines. The combined fleets of the rest of the world, including third world nations, number approximately 300 which are mostly diesel-electric powered attack boats. In sum, as a rough estimate, there are on the order of 1000 operationally effective submarines in the world at the close of the 20th century.

Since future conflicts are mostly expected to occur between developed and developing nations and also between developing nations, it's realistic to expect that, as the world economy grows and the latter accumulate wealth, they will augment their fleets to include nuclear and enhanced capability non-nuclear submarines. The enhanced capability boats will have submerged speeds of about 20 knots and submerged endurances of up to a month. They will be extremely quiet and maneuverable. Such boats will be readily purchased from suppliers in western Europe and in the newly formed countries from the former Soviet Union since these states have surplus submarine and armament production capabilities. As the developed nations reduce their fleet sizes, to accommodate the ending of the cold war, a time will possibly come when a perceived semblance of parity in submarines is achieved between the developed and developing nations which may seduce some developing nations into pursuing adventurous projects. Additionally, private organizations, including organized crime, may possibly acquire submarine capabilities for the purpose of furthering their, possibly nefarious, activities.

The missions of submarines in the future can be expected to include the traditional missions of: interdiction of sea lines of communications (SLOC) and battle groups; surveillance; special forces injection; bastion defense; and ballistic missile patrols. For rogue nations and criminal elements, missions can also be expected to include: terrorist operations; espionage; and smuggling. To counter these activities, allied ASW forces will pursue the traditional missions of: sea use protection for friendly battle groups and convoys; and sea denial to enemy forces. Sea use protection can be conducted as an area defense or as a moving box defense. Sea denial strategies will include: monitoring of egress/ingress from ports and other regions; establishment of barriers at choke points; and area sanitization.

Future submarine missions are expected to emphasize shallow water littoral operations, rather than the long range deep ocean operations of the cold war period, since this is where the submarines of developing nations will be found. Typical operational areas would be the Persian Gulf, Eastern Mediterranean, Sea of Japan, Yellow Sea and the South China Sea. These waters are deep enough to afford ample coverage for submarine activities but are shallow enough and have enough surface traffic that target detections will be difficult to discriminate from reverberation and background noise. Acoustic detections are expected to be, generally, relatively short range and MAD detections will be unreliable because of the accumulation of wrecks on the sea bottom. Western navies will likely augment their ASW fleets to include smaller enhanced capability non-nuclear submarines to match their future opponent's advantages of quietness and maneuverability. Surface ASW platforms will be required to be fast and maneuverable, in order to be able to reposition near new contacts, and the ASW fleet may be augmented by additional SWATH hulled ships, air cushion ships or hydrofoils. Rapidly deployable sensor systems configured for bottom mounting or suspension within the water column will likely be developed for both air and ship deployment.

From Allied experiences in WWI and WWII, a major tactic for defeating the enhanced capability non-nuclear boats will likely be to maintain overhead coverage until they are, eventually, required to surface. However, since these boats will be very quiet and can remain submerged for extended periods, this tactic will be successful only if the boats can be prevented from slipping away. To do so will require continuous overhead coverage so that monitoring can be maintained for long periods of time. Aircraft can do this but, for long duration monitoring, dirigibles may be a better solution. Space based ASW surveillance systems that scan the ocean for surface wakes or visual contacts may also be developed.

From the preceding descriptions of ASW, as it is practiced in the late twentieth century, it would be correct to surmise that ASW technology has grown to include a wide variety of naval operational practitioners and civilian science and engineering specialists. Each plays a necessary part in countering the submarine threat and all contribute singular expertise to their respective roles in bringing sonar to bear on the submarine problem. Brief introductions to critical portions of this expertise are presented in the following chapters.

CHAPTER 2
SONAR FUNDAMENTALS

ASW sonar practitioners draw eclectically where they must for technology but they draw most heavily from the sciences of naval operations analysis, ocean acoustics and sonar system engineering. Naval operations analysis specifies the objectives and functions of ASW sonar systems; ocean acoustics describes the physics of acoustic phenomenon in the oceans; and sonar system engineering bridges the gap between the two disciplines by providing ASW sonar systems that achieve the required objectives by means of the specified functionality within specified ocean environments.

In broad terms, the objectives of ASW sonar systems are to acquire operationally useful information from within the undersea environment and the functions of ASW sonar systems are the collection, processing and dissemination of that information in a timely manner. Operational information is carried by acoustic waveforms and is extracted by means of signal and data processing algorithms. These algorithms are developed from sonar system engineering principles, e.g. detection, localization, tracking and classification theory, and are based upon models of ocean propagated acoustic signals.

2.1 Sonar System Concepts

Sonar systems are either active, passive or both. Active sonar systems transmit acoustic energy into the medium which can produce echoes from submarine targets that can reveal their presence and much about their operations. Submarines can also reveal their presence and operations through the acoustic energy that they radiate into the ocean and target information can be gleaned passively from intercepted radiated waveforms. Active sonar bistatic operations are performed when transmissions from an active platform are received by other active platforms, or by passive platforms, after the signals reflect from submarine targets. The additional "look" at targets from another angle provides data that complements the information obtained from the more normal monostatic operations performed by active systems when acting independently. Signal, data and information processing techniques are the key elements that enable modern ASW sonar systems to extract target information from the intercepted signals.

It is useful to distinguish between signal, data and information processing based upon their underlying logics. Signal processing techniques are signal transformations in which one or more input signals lead to one or more unique outputs, e.g. Fourier transformations and correlations are signal processing algorithms. Data processing techniques are data transformations in which one or more input data sets leads to one or more of several possible outputs which are selected by engineering decision operations, e.g. detectors and classifiers are data processing algorithms. Information processing techniques are transformations of information in which input information leads to one or more of several possible operational decisions that cause or constrain actions. From this viewpoint, the distinguishing characteristic of signals, data and information is the processing they undergo and is not an absolute property[1] that marks each entity. ASW sonar systems are the embodiment of signal, data and information processing algorithms in seaworthy platforms in which, generally speaking, signals are transformed into data which are, in turn, transformed into information which are, in turn, transformed into operational decisions. The operational decisions made on the ASW platform can lead to target prosecutions or to information being passed to a higher authority for further processing that can lead to other operational decisions.

While only active systems are effective against quiet submarines that cannot be detected by passive means, for all other targets, active and passive systems produce complementary information regarding the targets and their operations. Accordingly, whereas older sonar systems tended to be either active or passive, modern systems tend to be both active and passive but differ in the methods by which active and passive data are combined. Modern systems use more computer automation for data integration while older systems rely more on operator integration. This trend is driven by the need to accommodate increasingly higher data throughputs while reducing the demands on operators. In modern systems, the operator focuses more on information processing leading to decisions concerning target contact data and focuses less on the "number crunching" inherent in signal and data processing.

Target contact data are based upon detection data, from which estimates of target parameters are obtained, i.e. detections reveal both the presence of targets and target parameter estimates concerning the location, track and classification of the target. Since there are always errors associated with these estimates, sonar operators perform assessments of the analyses and decisions made by the ASW system. Portions of the errors in these estimates are due to the physics of ocean acoustics and the engineering embedded in the system hardware design and are unavoidable. Other portions of the errors, though, are due to environmental and operational factors over which the operator has some control. The operator's ability to "make sense" of the data and to recognize the presence and source of extraordinary errors that are correctable is part of the on-going evaluation of contact data in determining whether or not a contact is a valid submarine target. This work is generally aided by computer generated displays that report upon computer assessments of the sonar data. The computer assessments are determined programmatically from the received acoustic data, sonar system parameters and the sonar equations. The sonar equations are one of the principal tools by which both the operator and the system evaluate the received ocean borne signals and determine how they are impacted by the ASW system. These equations were a WWII development that serve to simplify complicated system and ocean acoustic problems. The importance of the sonar equations cannot be overstated. They are key to the operational evaluation of contact

[1] Processing is always performed on either a physical quantity, such as acoustic pressure or voltage, or a set of alpha numeric characters. Describing the processed entity as signal, data or information does not describe one of its properties but rather a property of the processing performed, e.g. sunlight is a signal when spectrally decomposed by a prism; is data when it triggers a photocell; and is information when the attack is to start at dawn. Moreover, information is a general term that can be applied to virtually any processed entity, including signals and data, if the processing evokes an operational decision by the observer.

data. Because knowledge of the sonar equations and their meaning is so important, they are introduced in the following sections and much of the remaining chapters can be viewed as an elaboration of this introduction.

Active Sonar Systems and Sonar Equations

The general concept for active operations is depicted in Figure 1. An active transmit signal is generated by the system signal control function which sets the transmit power level, i.e. the source level (SL) of the transmission. The signal is modified for transmission by the transmit beamformer such that the transmit beam pattern and direction of propagation, which are characterized by the directivity index (DI), puts the maximum energy onto the suspected target location or into the suspected target sector. The transmitting response (α) determines the power radiated into the medium. The transmitted signal propagates to the target, with a source-to-target transmission loss (TL_{ST}), reflects from the target, with target strength gain (TS), and propagates to the receiver, with a target-to-receiver transmission loss (T_{TR}), where it is received by the hydrophone elements of the receive array. For monostatic transmissions, the transmission loss to-and-from the target are the same but, for bistatic transmissions, this in not the case. In the absence of interference, the received signal would always be detected but, with interference, the receive array also acquires unwanted signals, characterized by the background noise level (NL) and reverberation level (RL). One or the other will dominate in any given situation and either can seriously degrade the detectability of the signal. The in-water received acoustic data are transducted into electrical data and processed by the receive beamformer and by signal conditioning circuitry. The receiving response (β) determines the transducer output voltage and, in principle, determines the transducer sensitivity, i.e. the minimum acoustic pressure required to exceed electronic noise. However, since the background acoustic noise pressure is always much greater than this minimum, the lower limit is academic. The beamformer performs directional noise suppression on the data, in which the beamformer array gain (AG) enhances the signal-to-noise ratio, and, preceding the beamformer, the signal conditioning circuitry performs background noise suppression by filtering the data to the frequency band of interest. The resulting mix of signal with additive noise and reverberation is beam-band data and corresponds to the in-water signal plus reverberation and noise coming from the beam direction of interest and in the frequency band of interest. The beam-band data is passed to the detection system which processes the data for the presence of targets. The minimum pre-detector signal-to-noise ratio that achieves the required probability of detection is the detection threshold (DT). To clarify terminology, we note that the pre-detector signal is the input to the detector and is also the same as the in-water beam-band signal. Also, for active systems, we shall sometimes refer to noise when we mean noise plus reverberation. The in-detector signal-to-noise ratio that achieves the required probability of detection is the detection index (d) and this quantity corresponds to the detection threshold.

The received signal-to-noise ratio will generally be greater than that required for detection of targets within the area of coverage of the system. The signal excess is the exceedance of the received in-water beam-band signal-to-noise ratio SNR_R over the detection threshold which is given by

$$SE = SNR_R - DT, \tag{1a}$$

$$SNR_R = SL - 2TL + TS - (NL - AG)\text{: Noise Limited}, \tag{1b}$$

$$SNR_R = SL - 2TL + TS - RL_{BB}\text{: Reverberation Limited}. \tag{1c}$$

SE = Signal Excess
DT = Detection Threshold
SNR_R = SNR received at detector
SL = Source Level
TL = Transmission Loss
NL = Noise Level
AG = Array Gain
RL_{BB} = Beam-Band Reverberation Level

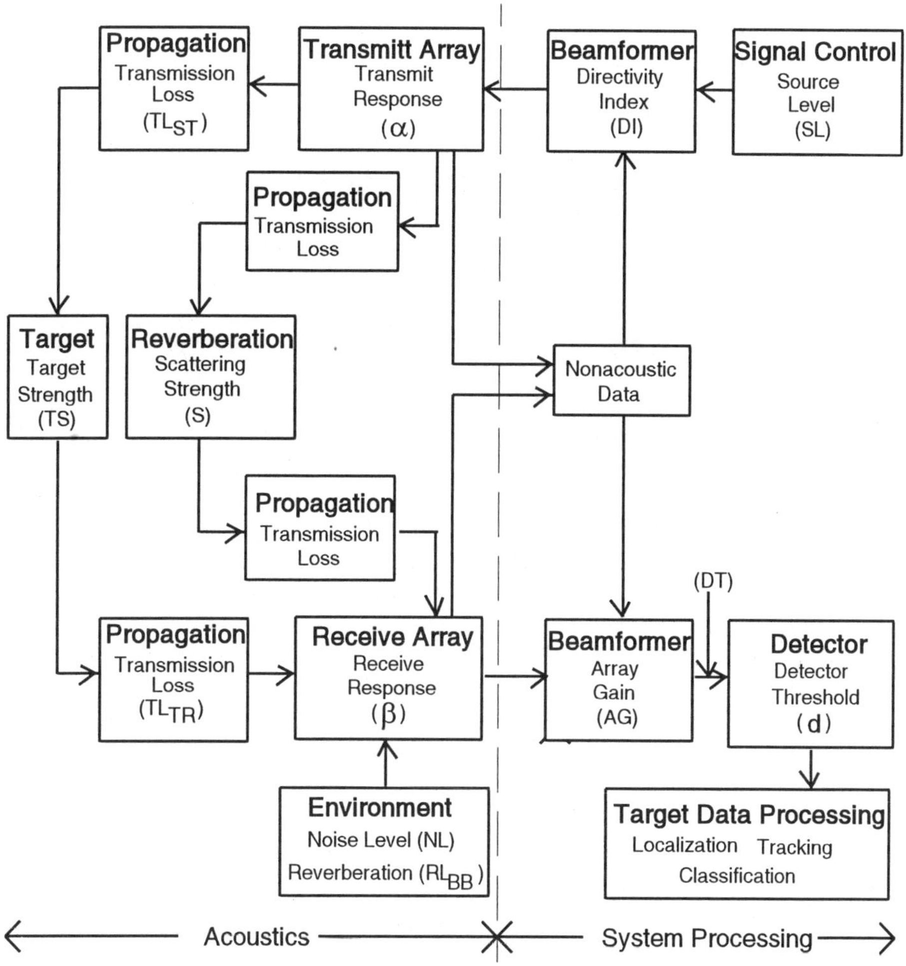

Figure 1 Active Sonar Concept

The quantity SNR_R represents the received in-water beam-band signal-to-noise ratio in which the quantity SL–2TL+TS represents the beam-band signal level at the receive array hydrophone elements. (For simplicity, this relation assumes that the transmission loss is the same for the source-to-target and target-to-receiver paths.) For the noise limited case, the quantity NL–AG represents the suppressed noise level, i.e. the beam-band noise level out of the beamformer, and for the reverberation limited case the quantity RL_{BB} represents the beam-band reverberation out of the beamformer. If the beamformed transmission was focused solely in the direction of the received echoes, RL_{BB} would be the total received reverberation level (RL). If the beamformed transmission was omnidirectional into all azimuths, RL_{BB} would be the total received reverberation level suppressed by the array gain, RL–AG. For the case of a transmission into a sector, RL_{BB} would be the total received reverberation with partial suppression by the array gain.

Passive Sonar Systems and Sonar Equations

The general concept for the passive sonar operations is depicted in Figure 2. Target radiation, characterized by its source level strength (SL), propagates through the medium, in which it is attenuated by the transmission loss (TL), and is received by the transducer elements of the sonar platform receive array. Similarly to active sonar, in the absence of interference, the received signal would always be detected but the receive array also receives background noise, characterized by the noise level (NL), which can seriously degrade the detectability of the target radiations. The in-water received acoustic data are transducted into electrical data and processed by the beamformer and signal conditioning circuitry. The transducer receiving response (β) determines the transducer output voltage as for the active receive system. The beamformer performs directional noise suppression on this data, in which the beamformer array gain (AG) enhances the signal-to-noise ratio, and, preceding the beamformer, the signal conditioning circuitry performs background noise suppression by filtering the data to the frequency band of interest. The resulting mix of signal with additive noise is beam-band data and corresponds to the in-water signal plus noise coming from the beam direction of interest and in the band of interest. The beam-band data is passed to the detection system which processes the data for the presence of targets. The minimum pre-detector signal-to-noise ratio that achieves the required probability of detection is the detection threshold (DT). To clarify terminology, we note that the pre-detector signal is the input to the detector and is also the same as the in-water beam-band signal. As for active sonar, the in-detector signal-to-noise ratio that achieves the required probability of detection is the detection index (d) which corresponds to the detection threshold.

The received signal-to-noise ratio will generally be greater than that required for detection of targets within the area of coverage of the system. The signal excess is the exceedance of the signal-to-noise ratio SNR_R over the detection threshold which is given by

$$SE = SNR_R - DT, \tag{2a}$$

$$SNR_R = SL - TL - (NL - AG). \tag{2b}$$

$$
\begin{aligned}
SE &= \text{Signal Excess}\\
DT &= \text{Detection Threshold}\\
SNR_R &= \text{SNR received at detector}\\
SL &= \text{Source Level}\\
TL &= \text{Transmission Loss}\\
NL &= \text{Noise Level}\\
AG &= \text{Array Gain}
\end{aligned}
$$

The quantity SNR_R represents the received beam-band signal-to-noise ratio in which the quantity SL–TL represents the beam-band signal level at the receive array hydrophone elements and the quantity NL–AG represents the suppressed noise level out of the beamformer, i.e. the beam-band noise level out of the beamformer.

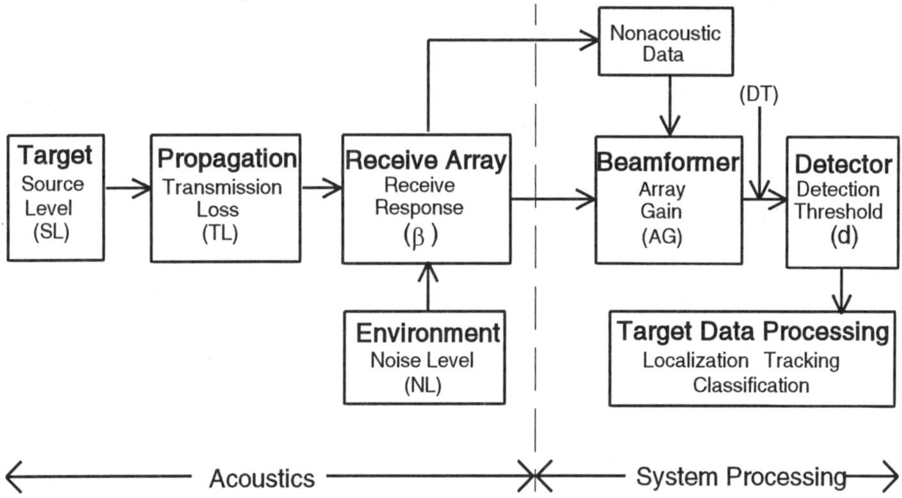

Figure 2 Passive Sonar Concept

Detection and Figure of Merit

For both active and passive systems the in-detector background noise level is estimated and from this a threshold is set at a high enough level to ensure that the noise crossing rate is within acceptable bounds, i.e. the false alarm rate is bounded from above. If the in-detector signal level exceeds the detector threshold a detection is called. In this case the detection data are passed to the target data processing functions for localization, tracking and classification estimates. Otherwise, if the in-detector signal level is less than or equal to the detector threshold no target decision is called. In brief, the detector performs the function of determining the detector threshold and the outcome of the detection test by comparing the in-detector signal level to the detector threshold. Note that the in-detector detector threshold differs from the in-water detection threshold

The performance of the detector is dependent upon the rate at which it makes correct detections and correct dismissals. Note that the adjective "correct" implies that there will be incorrect decisions and, of course, such decisions are unavoidable. In fact, a detector threshold crossing does not mean, necessarily, that a target is present. The crossing might have been caused by a noise burst which produced a false alarm. Similarly, the lack of a detector threshold crossing does not mean, necessarily, that no target is present. It may have been masked by noise which produced a false dismissal. The rates of false alarm and false dismissal are determined by the system design and are set to achieve a balance among system and operational requirements. Moreover, the setting of allowable error rates automatically sets the probabilities for correct detection and correct dismissal and, consequently, any two of these rates defines the detector's performance.

The detection function is performed by means of signal "compression" techniques. More about this later but, in simple terms, the detection function gathers the energy of the received in-water beam-band sum of signal and noise and compares the result to the detector threshold for an exceedance. The detector threshold is set to ensure that noise-only signals will exceed the threshold at a specified maximum probability of false alarm rate and, therefore, received target signals in noise will exceed the threshold at some greater rate depending on the signal-to-noise ratio. A specified minimum rate of exceedance for target signals in noise, in effect, specifies the minimum in-detector signal-to-noise ratio that will ensure a minimum probability of detection. This minimum in-detector signal-to-noise ratio is the detection index (d) and it corresponds to the in-water beam-band detection threshold (DT). Consequently, the detector threshold and detection index jointly specify the detection performance of the system and the detection index specifies the detection threshold that is required by the sonar equations. The specified minimum probability of detection corresponds to targets of interest that are at the extremes of the operations area. For other targets that are well within the operations area, including weaker that expected targets, detection can occur at rates that exceed the minimum required probability of detection. These concepts are elaborated in Chapters 9-11 but of importance here is the correspondence between the in-water and in-detector signal-to-noise ratios and their relationship to the detector threshold.

For a given target and environment, the minimum required received signal-to-noise ratio, i.e. the detection threshold, for the required probability of detection determines the maximum transmission loss that the signal can undergo and still be detectable. This maximum transmission loss is called the figure-of-merit and is defined by

$$\mathrm{FOM} = \mathrm{TL}_{\mathrm{max}}(\mathrm{P_D}), \tag{3}$$

where the dependence of the figure-of-merit on the probability of detection is made explicit. Figure-of-merit is a measure of the detection performance of the system that is independent of the propagation conditions in the medium but is dependent upon everything else. It is a useful system parameter because, once known, for a given environment, it can be translated into a maximum detection range. Since at the maximum detection range the signal excess is zero[2], the figure-of-merit expressions for active and passive systems, obtained from (1) and (2), are given by

[2] It is sometimes asserted that at zero signal excess the received signal-plus-noise has a 50% chance of being above or below the detection threshold and, therefore, the probability of detection is 50%. This notion is incorrect. It appears to confuse detection threshold with

Active Sonar

$$2FOM = SL + TS - (NL - AG) - DT: \quad \text{Noise Limited}, \tag{4a}$$

$$2FOM = SL + TS - RL_{BB} - DT: \quad \text{Reverberation Limited}, \tag{4b}$$

Passive Sonar

$$FOM = SL - (NL - AG) - DT. \tag{4c}$$

For the active case, the figure-of-merit is against a target having target strength TS which is usually chosen to be the target strength for the target of interest but which is sometimes standardized at zero. For the passive case, the figure-of-merit is against a target having source level SL which is chosen to be representative of the target class of interest. For both active and passive cases, the figure-of-merit is against an ambient noise level of NL which is chosen to be a nominal operational value for the region of interest and, of course, figure-of-merit is dependent upon the source-receiver system parameters: active projector SL, AG and DT. Even though the figure-of-merit is dependent upon target class and environmental noise, it is still an exceptionally useful measure of system performance.

Target Data Processing

Target detections are passed to the target data processing subsystem for localization, tracking and classification processing. These functions are depicted in Figure 3. The detections from the current time period are associated with detections obtained during previous time periods and those that associate are linked together as part of a contact data package. The associations are based upon localization estimates which are performed on the detections to determine bearing, range, depth, range rate, etc. of the target. The contact data package is passed to the system target tracking function and target tracks are estimated for the associated detections. The processing stream is generally left-to-right in Figure 3 but, as indicated in the figure, feedback from each of the functions, sometimes in the form of operator judgments, can modify previous results. Based upon all available information, a target classification is determined. This information includes the detection, localization and tracking parameters and, also, information from previous contacts and intelligence reports, which include contacts obtained by other platforms, that are disseminated by the battle group and the ASW surveillance community. In fact, the trend of the future is to treat the larger ASW community as one system in which computer driven information sharing is performed electronically and automatically. This promises to substantially reduce delays in circulating contact data that can occur due to slow operator throughput rates. Such slow rates are, of course, normal for human driven information channels, in part, because of operator reluctance to report an ambiguous contact. Automated computer channels not only pass information swiftly but can associate a probability of correct assessment with the

---

detector threshold. In fact, at zero signal excess the probability of detection is always the design probability of detection. Of course, the design probability is frequently set to be 50%.

information which weights it in any final compilation. Of course, the operator is still required to play a key role in evaluating all contacts.

The principal classification decisions are target, unknown and nontarget. If sufficient information is available, these classifications can be further refined, as indicated in the Figure 3, by specifying the type of contact. Classifications are based, in part, on estimated target motion coupled with information regarding previous contacts and intelligence information. For active system data, classifications are also based upon the strength of the received echo and target features that may have been impressed upon the signal envelope. For passive system data, classifications are also based upon the target frequency signature.

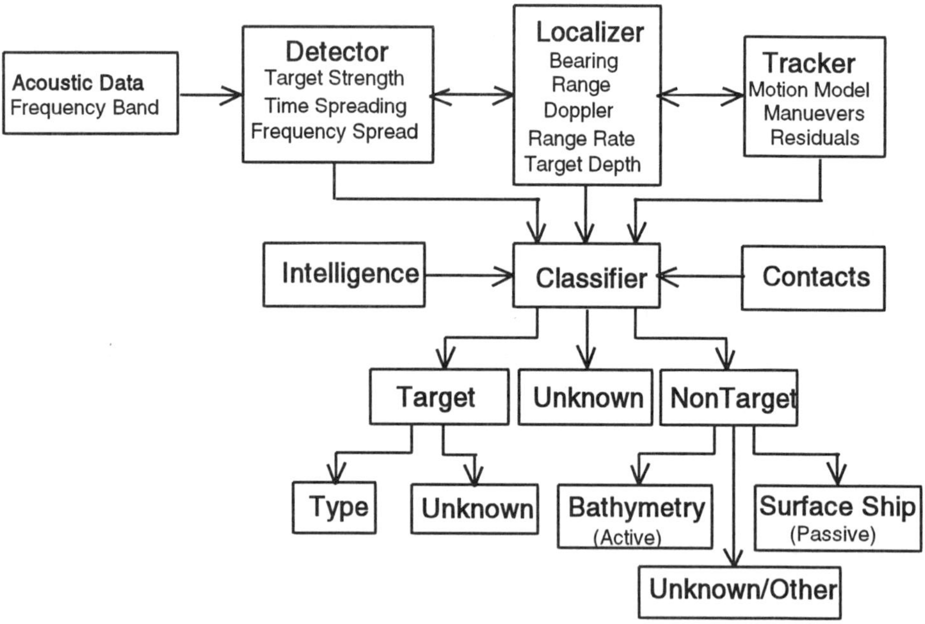

Figure 3 Target Data Processing

As can be inferred from the preceding discussions, the sonar equations and the sonar parameters are the basis of much of the work concerning sonar which includes naval operations, test and evaluation, as well as, research and development. The strength of the sonar equations lies in their simplicity and the ease of obtaining useful assessments of sonar performance. Their weakness lies in their simplicity and the ease of obtaining misleading assessments of sonar performance. Crucial to the successful application of the sonar equations is, of course, knowing precisely what the terms in the sonar equations represent. In a large sense, all that follows is focused on just the task of determining precisely what those terms do represent. As we progress through the material, it will be found that an extensive set of concepts is required for a full understanding of the sonar equations and that the apparent simplicity of these equations can be deceptive.

2.2 Signal Theory for Ocean Acoustics √

Acoustic waves that reflect from targets and acoustic waves radiated by targets have information imposed upon them that can be extracted and used to characterize target presence and modes of target operations. The methods for extracting this information are dependent upon the theories of detection, localization, tracking and classification and are all dependent upon mathematical models of ocean borne signals. In the remainder of this chapter we are concerned with elementary mathematical methods of describing how acoustic waves in the ocean carry information and how signal decibel levels are computed which is fundamental to the understanding of the sonar equations.

Ocean acoustic waves cannot be arbitrarily specified. If they could, ASW work would be far less difficult. They must conform to physical laws that govern their processes of generation, propagation and extinction. These laws are discussed more fully in Appendix 2-A where it is shown that they lead to the basic wave equation

$$\nabla^2 p - \frac{1}{c^2}\frac{\partial^2 p}{\partial^2 t} = 0 , \tag{5}$$

where p is the acoustic pressure and c is the sound speed. The importance of the wave equation cannot be overstated. It is the necessary condition that acoustic waves must satisfy which, when coupled with boundary conditions, specifies their structure and mode of carrying information as they propagate through the ocean.

All oceanic acoustic waves propagate and the simplest propagating wave is the plane wave. A thorough understanding of these waves is an important step toward the understanding of all wave propagation in the ocean. The plane wave has the general mathematical form

$$p(t, x) = p(t \mp x / c) , \tag{6a}$$

where p denotes the acoustic pressure in the medium and is a function of time and one space coordinate oriented along the direction of propagation. It causes fluid particle motion along the axis of propagation and imaginary surfaces perpendicular to the fluid particle motion are called wave fronts. Imaginary lines perpendicular to the wave fronts and parallel to the particle motion are called rays. The propagating plane wave can easily be shown to be a solution of the wave equation.

An amplitude point of the plane wave, call it point A, will propagate in the +x or −x directions, depending upon the chosen sign in the argument of (6a). The argument corresponding to point A is

$$t \mp \frac{x}{c} = \text{constant(A)} . \tag{6b}$$

For increasing t, the choice of upper sign requires x to increase, i.e. the wave moves in the +x direction, and the choice of lower sign requires x to decrease, i.e. the wave moves in the −x direction.

If at time t_1 the point A is at position x_1 and at later time t_2 it is at position x_2, the corresponding arguments are equal

$$t_1 \mp \frac{x_1}{c} = t_2 \mp \frac{x_2}{c}. \tag{6c}$$

Solving for c

$$c = \pm \frac{(x_2 - x_1)}{(t_2 - t_1)} = \frac{|x_2 - x_1|}{|t_2 - t_1|}, \tag{6d}$$

gives the speed of propagation of the wave. The sound speed is always positive because, for the case of $t_2 > t_1$, if $x_2 > x_1$ the wave is moving to the right and the upper sign holds and conversely if $x_2 < x_1$ the wave is moving to the left and the lower sign holds. A similar result is obtained for the case of $t_2 < t_1$, i.e. the case for which the wave is moving backward in time - an interesting but purely theoretical exercise for ASW sonar.

An example of a plane wave, having a contrived waveform and propagating in the positive x direction, is shown in Figure 1. It is represented, in Figure 1a, as pressure measured over time as received by a probe at position x = 0 and, in Figure 1b, as a snapshot of the wave extended along the x spatial coordinate at time t = 0. In both figures it is represented by an amplitude plot, i.e. an A-scan plot. The waveform has duration T and spatial extent cT.

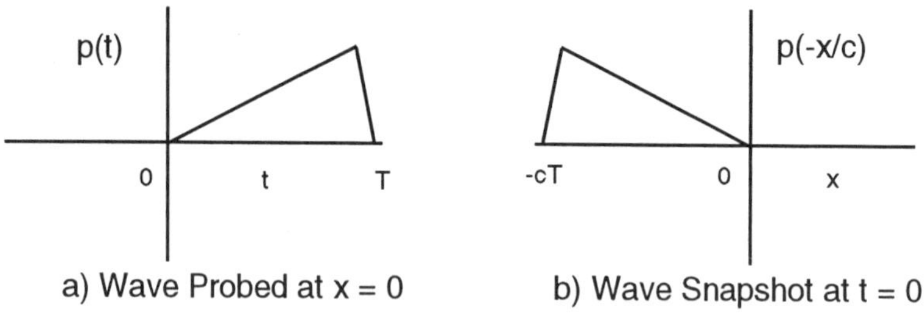

Figure 4 Plane Wave Propagating in the +x Direction

Sinusoidal Waveforms

ASW waveforms are diverse but all can be represented in terms of sinusoidal waves. Because of the endlessly repeating structure of sinusoids they are referred to as continuous waves (CW). A two cycle segment of a sinusoidal wave is depicted in Figure 5 in which, viewed as a time series plot from a sensor probe at x = 0 in Figure 5a, the wave period is denoted τ, or, viewed as a snapshot of the wave extended in space at t = 0 in Figure 5b, the wave length is denoted λ. In Figure 5 the period of the wave τ and the speed of sound c were chosen, somewhat arbitrarily, to be, respectively, 1 msec and 1500 m/sec which is the approximate sound speed in the ocean. Since in one period, the sine wave moves one wave length, the relation between these two quantities is

$$\lambda = c\tau\,, \tag{7a}$$

and, for the chosen values, λ = 1.5 m. The reciprocal of the wave period is the wave frequency

$$f = \frac{1}{\tau}\,. \tag{7b}$$

If the period is measured in seconds, the frequency is the number of cycles per second and is denoted Hertz[3] (Hz). For the chosen values, f = 1000 Hz. An alternative propagation relation that is frequently used is given by

$$\lambda f = c\,, \tag{7c}$$

and is obtained by combining (7a) and (7b).

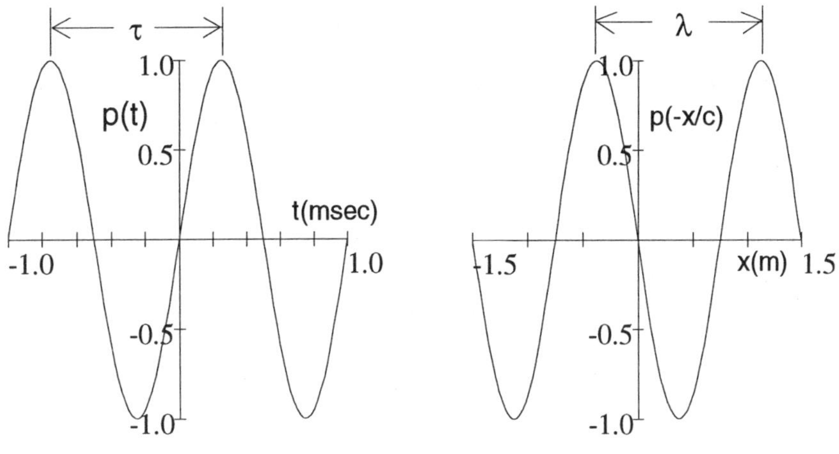

a) Sine Wave Probed at x = 0 b) Sine Wave Snapshot at t = 0

Figure 5 Sine Wave Propagating in the +x Direction

General sinusoidal plane waves can be represented by the mathematical expression

$$p(t,x) = A\cos\left(\omega\left(t \mp \frac{x}{c}\right) + \theta\right), \tag{8a}$$

where the quantity A is the amplitude of the wave which, here, is chosen to be pressure but could be velocity or density etc. The quantity ω is called the angular frequency and is introduced to make the sine function argument dimensionless. Its unit of measure is radians per unit time.

---

[3] After Heinrich Hertz, the discoverer of radio waves.

The first term in the sinusoidal waveform argument ωt must satisfy the relation $\omega\tau = 2\pi$, since the sine wave repeats every period. This gives

$$\omega = 2\pi f . \tag{8b}$$

The quantity ωt can also be expressed as $2\pi t/\tau$ which is 2π times the number of cycle periods occurring in the time interval t or, equivalently, is the number of radians in the cycle periods occurring in the time interval t.

The second term in the sinusoidal waveform argument $\omega x/c$ must also satisfy the relation $\omega\lambda/c = 2\pi$ because of the periodicity of the sine wave. This gives the relation

$$k = \frac{\omega}{c} = \frac{2\pi}{\lambda}, \tag{8c}$$

where k is called the angular wave number. Its unit of measure is radians per unit length. The quantity kx can be expressed as $2\pi x/\lambda$ which is 2π times the number of wavelengths occurring in the range interval x or, equivalently, is the number of radians in the wavelength periods occurring in the range interval x.

The last term in the sinusoidal waveform argument θ is a phase shift of the waveform and is measured in radians. It shifts the waveform $\theta/2\pi$ cycles i.e. a fixed number of cycles. This differs from a time shift or range shift in which the number of cycles shifted depends, respectively, upon the frequency or wavelength.

The time dependent pressure measured by a single probe in an acoustic field or the snapshot of pressures measured at one time from a distribution of probes in an acoustic field give no clues regarding the movement of the medium or the direction of propagation of the wave in the medium. However, the pressure field coupled with the particle velocity field does specify the medium motion and the direction of propagation. When a sinusoidal acoustic wave passes through a medium, the particles of the medium move back and forth parallel to the direction of wave propagation creating alternating zones of compression and rarefaction. The net force per unit area on a slab of the medium, that is perpendicular to the particle motion, is the negative of the pressure difference across the slab. This quantity is denoted $-\Delta p = p(t,x_2)-p(t,x_1)$. By Newton's second law this force is equal to the mass per unit area of the slab times the rate of change of velocity of the slab

$$-\Delta p = \rho_0 \Delta x \frac{du}{dt}, \tag{9a}$$

where ρ_0 is the fluid density, $\Delta x = x_2-x_1$ is the slab width, which is taken to be vanishing small, and u is the slab velocity. Integrating over time gives

$$-\int \frac{\partial p}{\partial x} dt + \text{constant} = \rho_0 u . \tag{9b}$$

The integration constant is necessarily zero because there is no net velocity of the medium due to the acoustic wave. Substituting p from (8a) and integrating gives the relation between pressure and particle velocity

$$p = \pm \rho_0 c u, \tag{10}$$

where the sign ± denotes the direction of wave propagation. Note that, if the pressure and direction of propagation are specified, the particle velocity is also specified and that, in general, specifying any two of these three variables necessarily specifies the third. In analogy with Ohm's law, with p acting as electrical potential and u acting as electrical current, ρc is called the acoustic impedance of the medium and represents the resistance of the medium to motion that opposes the impressed pressure. The pressure and particle velocity are seen to be in phase, i.e. when the excess pressure is positive, the fluid moves in the direction of wave propagation and, when the excess pressure turns negative, the fluid rebounds in preparation for the next pressure push.

Propagating waves carry information between points in the ocean. However, the usefulness of this information is dependent upon the detectability of the wave which, in turn, is dependent upon the waveform intensity. The intensity of an acoustic wave is the propagated power per unit area of the wave. This is the work performed by the wave on the fluid per unit time and per unit area which is given by

$$I(t) = pu. \tag{11a}$$

The time argument is shown to emphasize that this is the instantaneous intensity. Averaged over one cycle of duration τ, the average intensity of a sinusoidal plane wave is given by

$$I = \frac{A^2}{\tau \rho_0 c} \int_0^{\tau} \cos^2(\omega(t - x/c) + \theta)dt = \frac{A^2}{2\rho_0 c}, \tag{11b}$$

where p from (8a) and u from (10) were substituted. Since the quantity

$$P = \frac{A^2}{2}, \tag{11c}$$

is the mean square amplitude of the wave, the intensity is sometimes written

$$I = \frac{P}{\rho_0 c}. \tag{11d}$$

In either form the average intensity of the sinusoidal plane wave is constant.

Nominal values for environmental parameters are given in Table 1 for sea water and air in both centimeter-gram-second (cgs) and meter-kilogram-second (mks) units. The values for air are included for comparative purposes and because aural detection is still practiced in some modern sonar systems. In fact, the human ear is sometimes a better detector than automated detection systems. The problem with human detection systems is that they require special talents and a state of heightened alertness that cannot be maintained for long periods of time. In addition to the parameters in the table, sound speed is often expressed in British units as: $c \cong 4920$ ft/sec $\cong 0.8$ nm/sec but density and acoustic impedance are hardly ever expressed in British units.

Table 1 Environmental Parameters

Quantity	Sea Water (cgs)	Sea Water (mks)	Air (cgs)	Air (mks)
sound speed (c)	$1.5 \times 10^5 \frac{\text{cm}}{\text{sec}}$	$1.5 \times 10^3 \frac{\text{m}}{\text{sec}}$	$3.43 \times 10^4 \frac{\text{cm}}{\text{sec}}$	$343 \frac{\text{m}}{\text{sec}}$
density (ρ_0)	$1 \frac{\text{gm}}{\text{cm}^3}$	$10^3 \frac{\text{kg}}{\text{m}^3}$	$1.21 \times 10^{-3} \frac{\text{gm}}{\text{cm}}$	$1.21 \frac{\text{kg}}{\text{m}^3}$
acoustic impedance ($\rho_0 c$)	$1.5 \times 10^5 \frac{\text{gm}}{\text{cm}^2 - \text{sec}}$	$1.5 \times 10^6 \frac{\text{kg}}{\text{m}^2 - \text{sec}}$	$41.5 \frac{\text{gm}}{\text{cm}^2 - \text{sec}}$	$415 \frac{\text{kg}}{\text{m}^2 - \text{sec}}$

Narrow Band Waveforms

CW signals are easy to conceptualize and manipulate mathematically but, being infinite in duration and varying at only single frequencies, they represent an ideal that is never realized in practical ocean acoustic applications except approximately. Their importance to ASW sonar arises because all ocean acoustic waveforms can be constructed from sums of sinusoidal waveforms. One generalization of sinusoidal waveforms is the narrow band waveform

$$p(t,x) = A(t)\cos(\omega(t \mp x/c) + \theta(t)) . \tag{12a}$$

This is just the sinusoidal waveform in (8a) with a time varying amplitude and phase. The time varying amplitude is referred to as the envelope of the waveform and, because it can be zero before and after certain times, the waveform can be finite in duration. The time derivative of the phase

$$\omega(t) = \frac{d}{dt}\left(\omega(t \mp x/c) + \theta(t)\right) = \omega + \frac{d\theta(t)}{dt} , \tag{12b}$$

is referred to as the instantaneous angular frequency and, because it can vary in time, the waveform varies at different frequencies at different times. The bulk of the information in the waveform is in the amplitude and phase and the sinusoidal portion of the signal is the carrier signal for which only the frequency is in question.

The variations in the amplitude and phase cause spreading of the frequency bandwidth, i.e. increased bandwidth, and sinusoidal waveforms are just a special case of narrow band waveforms, having zero bandwidth, in which the amplitude and phase are constant.

For narrow band waveforms, which are common for sonar transmissions and radiations from targets, the variations in envelope and phase will usually be much slower than the carrier signal variations and the waveform A-scan will be sinusoidal in appearance but with variations in the amplitude and frequency. For more complicated waveforms, including explosive signals, this will not be true and the waveform A-scan can be very non-sinusoidal in appearance.

Equation (12) is a very useful form for representing and visualizing narrow band waveforms but it is inconvenient for analyzing waveforms because it does not have the property of linearity, i.e. the envelope and phase of the sum of two waveforms is not the sum of the envelopes and phases of the individual waveforms. This is rectified by the decomposition

$$p(t,x) = x(t)\cos(\omega(t \mp x/c)) - y(t)\sin(\omega(t \mp x/c)), \tag{13a}$$

$$x(t) = A(t)\cos(\theta(t)), \tag{13b}$$

$$y(t) = A(t)\sin(\theta(t)), \tag{13c}$$

where x(t) and y(t) are slowly varying functions if A and θ are slowly varying. They are called the quadrature components of the waveform because, as indicated in Figure 6, they can be represented as the "sides" of the waveform. The principal advantage of the quadrature component representation of the waveform is that it is linear, i.e. the quadrature components of the sum of two waveforms is the sum of the quadrature components of each waveform.

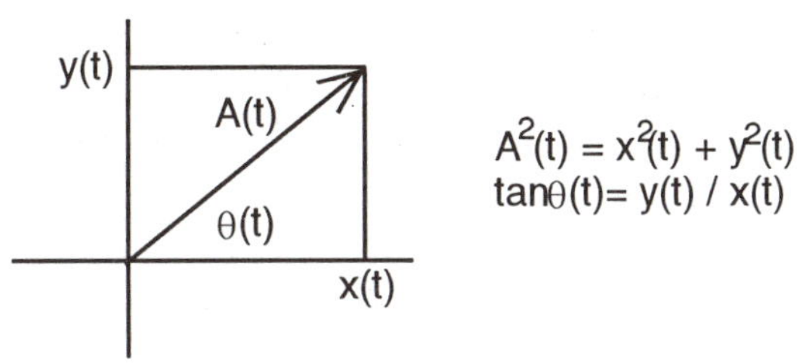

Figure 6 Quadrature Component Relations

The average intensity of narrow band waveforms in terms of their quadrature components is found, as in (11), to be

$$I(t) = \frac{x^2(t) + y^2(t)}{2\rho_0 c} = \frac{A^2(t)}{2\rho_0 c} = \frac{P(t)}{\rho_0 c}, \tag{13d}$$

where it is assumed that the quadrature components are approximately constant over one cycle of the carrier. The average intensity of narrow band waveforms, in (13d), can vary with time in contrast to sinusoidal waveforms.

The signal representations in (12) and (13) are referred to as narrow band waveform representations because it is usually assumed that the modulating terms vary slowly compared to the carrier signal. In fact, they are general signal representations, but for broad band signals, in which the modulation variation approaches or exceeds the variability of the carrier, the waveform shape loses its sinusoidal-like form and the concept of envelope and phase loses meaning. This is a matter of small concern to a signal processing algorithm and the quadrature component representation is used to represent both narrow and broad band signals.

The quadrature components of the received waveform carry the bulk of the information of the waveform. Consequently, signal and data processing operations are constructed to extract that information. Because of this, a more useful representation of the waveform is in terms of the analytic signal which, under a weak bandwidth restriction[4], is given by the complex expression

$$p(t, x) = m(t)\exp\left(j(\omega (t \mp x / c))\right), \tag{14a}$$

$$m(t) = x(t) + jy(t)\,. \tag{14b}$$

The quantity m(t) is the complex envelope of the signal. Analytic signals are useful because ASW system signal and data processing are most easily described in terms of and performed as operations on the complex envelope of the analytic signal. Because the analytic signal is complex, the representation of physical ocean borne waveforms is a little more complicated but the analytic signal can always be generated from the physical signal and physical quantities can always be extracted from analytic signals. More about this and a general discussion of analytic signals is given in Appendix 2-B. As an example, the real part of the pressure expression in (14) is just the real waveform representation in (13) and the intensity of the physical waveform, expressed in terms of the complex modulation, is seen from (13d) to be

$$I(t) = \frac{|m(t)|^2}{2\rho_0 c} = \frac{A^2(t)}{2\rho_0 c} = \frac{P(t)}{\rho_0 c}\,. \tag{14c}$$

Moreover, the wave equation is satisfied by complex waveforms if both the real and imaginary parts of the waveform satisfy the wave equation which is always the case for analytic signals. The power and elegance of complex variables simplifies much of the analysis to follow. These simplifications more than compensate for the need to relate our complex results to real world quantities.

The intensity of the analytic signal can be obtained by extending the intensity expression for real signals in (11a) to

$$I_a(t) = pu^*\,, \tag{15a}$$

where the superscript asterisk denotes complex conjugate. The intensity of the analytic narrow band signal in (14) is then given by

$$I_a(t) = \frac{A^2(t)}{\rho_0 c}\,, \tag{15b}$$

---

[4] See Appendix 2-B.

where (10) was used. Comparison with (14c) shows that the analytic signal intensity is double that of the real signal. This is because half of the intensity of the analytic signal is carried by the real signal and half by the imaginary signal.

Plane waves can be immediately generalized to cylindrical waves and spherical waves. These wave types can be represented in the forms

$$p(t,x) = \frac{m(t)}{\sqrt{\rho}} \exp\big(j(\omega(t \mp \rho / c))\big), \tag{16a}$$

$$p(t,x) = \frac{m(t)}{r} \exp\big(j(\omega(t \mp r / c))\big), \tag{16b}$$

where ρ and r denote, respectively, the cylindrical and spherical range coordinates. These expressions can be shown to satisfy the wave equation in (5).

Unlike plane waves, cylindrical and spherical waves are attenuated by geometric spreading. However, at long ranges from the acoustic source and for typical target and array sizes, the wave fronts incident upon the target or array are approximately plane and the incremental spreading loss over the target or array extent is negligible. This means the wave can be approximated as an attenuated plane wave within the target or array region and that all of the preceding results for plane waves can be directly applied. Spreading laws for ocean acoustic waves can frequently be approximated by cylindrical or spherical spreading but, for long range ocean propagation, spreading laws can be much more complicated which is discussed in Chapters 5 and 6.

General Waveforms and Spectra

Most waveforms encountered in ASW are more sophisticated than narrow band sinusoidal waveforms and exhibit complicated time and frequency structures. Depending on the range of associated frequencies, they are classified as broad band or narrow band with sinusoids being their limiting cases and also their basic building blocks. It is convenient to subdivide all waveforms into three general classes because of the nature of their mathematical representations. These classes and some of the ASW waveforms they include are

1. periodic infinite duration waveforms,
   - cw tones,
   - steady state machinery radiations,
2. finite duration waveforms,
   - modulated sonar transmissions,
   - explosive signals,
3. nonperiodic infinite duration waveforms,
   - ambient noise,
   - reverberation.

In these classifications the term "infinite duration" means, for practical purposes, long term compared to the duration of ongoing operations and need not be taken literally.

Periodic infinite duration waveforms are represented by Fourier series expansions. These expansions can be expressed as sums of complex sinusoids given by

$$p(t) = \sum_{n=-\infty}^{\infty} V_n \exp(j\omega_n t) , \tag{17a}$$

$$V_n = \frac{1}{T_p} \int_0^{T_p} p(t)\exp(-j\omega_n t)dt , \tag{17b}$$

$$\omega_n = \frac{2\pi n}{T_p} . \tag{17c}$$

The representation of the signal in (17) is seen to be periodic, with period T_p, by making the substitution $t \rightarrow t+T_p$. The Fourier coefficients V_n are in general complex and have the same unit of measure as the waveform p(t). They modulate the amplitude and phase of the component complex sinusoids and the nth Fourier coefficient is associated with the frequency ω_n and an angular frequency "bin" of width $2\pi/T_p$ as indicated in Figure 7. Note that the waveform period is different from a sinusoidal period which, for the nth Fourier component, is T_p/n. The exception being the n = 0 coefficient which is associated with a constant, i.e. a sinusoid of infinite period. This term is usually not present in representations of sonar signals since sonar signals cause no net displacement of water mass. The low order coefficients are associated with low frequency sinusoids and represent the gross form of the signal while higher order terms are associated with high frequency sinusoids and are required to give the sharp detail to the signal such as the leading and trailing edges of the signal. The frequency bin widths are uniform for all coefficients but shrink with increases in the waveform period which causes the density of the Fourier components in the frequency domain to increase. Both real and complex waveforms can be represented by the series in (17) but, for real waveforms, the Fourier coefficients are conjugate pairs, i.e. $V_{-n} = V_n^*$, as seen in (17b). Propagating waveforms are readily produced from (17) by the variable change, $t \rightarrow t - s/c$, where s is the appropriate range coordinate.

The intensity or power of the waveform can be expressed in terms of Parseval's theorem. The expression for the mean square pressure is given by

$$P = \frac{1}{T_p} \int_0^{T_p} |p(t)|^2 dt = \sum_{n=-\infty}^{\infty} |V_n|^2 . \tag{18a}$$

For p(t) real and equal to, say, acoustic pressure, division throughout by twice the acoustic impedance $2\rho c$ results in power per unit area, i.e. the intensity of the waveform. This result is comparable to (11b), (13d) and (14c). For p(t) equal to, say, voltage, the appropriate divisor is resistance, say 1 ohm, and the result is the power dissipated in a 1 ohm resistance. Following the general practice of assuming the appropriate divisor is implied, each term in the right member of (18a) is the intensity or power within the nth frequency bin and the sum is the intensity or power of the waveform. In the following, intensity will be the unit of choice unless contraindicated.

Computations of waveform power spectra are routinely performed during ASW operations but, since physical waveforms are real and their power spectra are symmetric, only the half of the power spectra supported by positive frequencies is displayed. The one sided power spectrum is given by

$$P = 2\sum_{n=1}^{\infty}|V_n|^2, \tag{18b}$$

where the V_0 term, the so-called DC term, is omitted since it is usually zero or out of the band of interest. The one-sided power spectrum is just the double of the positive frequency spectrum as indicated in Figure 7.

Example 1: For the case in which p(t) is the real sinusoid with angular carrier frequency ω_0 that is modulated to turn on and off periodically, the waveform duration T is less than the waveform period

$$p(t) = A\sin(\omega_0 t). \quad 0 \le t \le T < T_p \tag{18c}$$

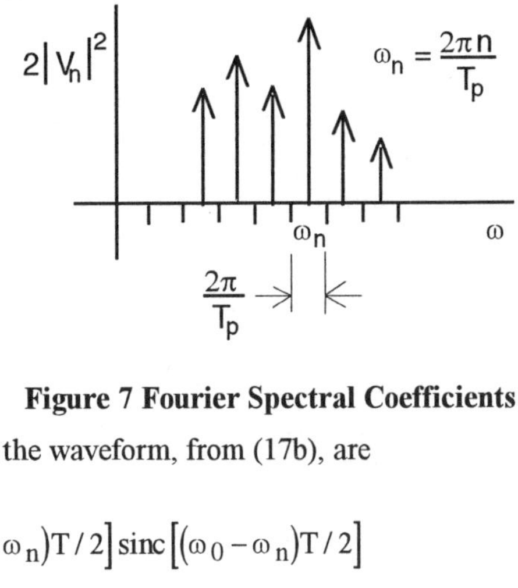

Figure 7 Fourier Spectral Coefficients

Such a signal might be an active sonar transmission or might be a signal emitted by machinery. The Fourier coefficients of the waveform, from (17b), are

$$V_n = \frac{AT}{2T_p}\left\{-j\exp[+j(\omega_0-\omega_n)T/2]\,\text{sinc}[(\omega_0-\omega_n)T/2]\right.$$
$$\left. + j\exp[-j(\omega_0+\omega_n)T/2]\,\text{sinc}[(\omega_0+\omega_n)T/2]\right\}, \tag{18d}$$

where the notation sinc(x) = sin(x)/x represents the sinc function. Since the waveform is represented by a real function, the Fourier coefficients are seen to satisfy the conjugate pair requirement, $V_{-n} = V_n^*$. The intensity spectral coefficients are found from (18b) to be

$$|V_n|^2 = \left(\frac{AT}{2T_p}\right)^2\left(\text{sinc}[(\omega_0-\omega_n)T/2] - \text{sinc}[(\omega_0+\omega_n)T/2]\right)^2, \tag{18e}$$

where we assume an integral number of cycles, i.e. $\omega_0 T = 2\pi K$ where K is an integer. By the nature of the sinc function, the spectral coefficients are seen to peak around the values $\pm\omega_0$. If the signal carrier frequency was, say, 1000 Hz with the duration of 1 second and the signal repeated every 10 seconds, the sinc function arguments would be

$$(\omega_0 \pm \omega_n)T/2 = \pi(1000 \pm n/10)\,. \tag{18f}$$

This shows that the main lobes of the spectrum have maximum values at the n = ±10,000 coefficients, corresponding to 1000 Hz, and that the zeros of the spectrum occur at intervals of $\Delta n = 10$, i.e. n = ⋯ ±9,980, ±9,990, ±10,010, ±10,020, ⋯ with minor lobes at intermediate positions. The magnitude of the spectrum away from the spectral peaks is bounded by $|10/\pi k|^2$, k = ±1,±2, ⋯. Therefore, at, say, k = ±10 spectral coefficients, i.e. ±1 Hz, from the spectral peaks, the spectrum is down by a factor of approximately 10^{-1}. Consequently, while the spectrum theoretically spans an infinite frequency range, for practical purposes the waveform energy is localized to a narrow spectral band and can be considered to be band limited with a one sided spectrum that spans the frequency range from, approximately, 9,999 Hz to 10,001 Hz and has an effective bandwidth of approximately 1 Hz.

Each Fourier coefficient in the one-sided spectrum is weighted by a common factor that can be expressed as $[(A^2/2)/(1/T)]\times(1/T_p)\times(T/T_p)$. The first factor is the intensity of the waveform, after division by ρc, divided by the signal bandwidth which gives the spectral density of the equivalent CW signal. The second factor is the bandwidth of each Fourier spectral bin which, when multiplied, gives the analysis band intensity of the equivalent CW signal. The last factor is the duty cycle of the signal which, when multiplied, gives the analysis band intensity of the signal. The mean intensity of the periodic signal is $(A^2/2)\times(T/T_p)$ which is the intensity of the CW waveform weighted by the waveform duty cycle. We note that, since the information in the signal is carried in the modulation, i.e. on and off in 1 second, and not the carrier signal, in practice the carrier would be stripped from the signal and only the, approximately, 200 spectral coefficients that span the band of interest would be computed instead of the 10,000 coefficients in this example.♦

Finite duration waveforms are represented by Fourier integral expansions which can be formally derived from the Fourier series expansions by letting the waveform period T_p go to infinity resulting in a representation for just one waveform. From (17)

$$p(t) = \lim_{T_p \to \infty} \frac{1}{2\pi} \sum_{n=-\infty}^{\infty} \frac{V_n}{\left(\frac{1}{T_p}\right)} \exp\left(j\frac{2\pi n t}{T_p}\right)\left(\frac{2\pi}{T_p}\right)$$

$$= \frac{1}{2\pi}\int_{-\infty}^{\infty} V(\omega)\exp(j\omega t)\,d\omega\,, \tag{19a}$$

where the exponent, $\omega_n = 2\pi n/T_p$, goes to the continuous variable ω as both the period T_p and the number of Fourier components tend to infinity. The function $V(\omega)$ is the Fourier transform of the waveform and is given by the limiting form

$$V(\omega) = \lim_{T_p \to \infty} \frac{V_n}{(1/T_p)} = \lim_{T_p \to \infty} \int_{-T_p/2}^{T_p/2} p(t)\exp\left(-j\frac{2\pi nt}{T_p}\right)dt$$

$$= \int_{-\infty}^{\infty} p(t)\exp(-j\omega t)dt \,. \tag{19b}$$

Since for finite duration waveforms the limit of the integral exists, the Fourier transform of the waveform is well defined. Also, since the quantity $1/T_p$ is the frequency resolution in the spectral domain, the Fourier transform is seen to be a spectral density function over the continuous frequency domain and its unit of measure is the waveform unit per Hertz.

The energy of the waveform is well defined, since it is of finite duration, and the energy flow per unit area E, i.e. energy flux density, where the divisor $2\rho c$ is implied, is given by

$$E = \int_{-\infty}^{\infty} |p(t)|^2 dt = \frac{1}{2\pi} \int_{-\infty}^{\infty} |V(\omega)|^2 d\omega \,. \tag{20a}$$

The integrand of the right member is seen to be the spectral density of the energy per unit area of the waveform and the unit of measure of the spectral density is energy flux density per Hertz. The average intensity I of the waveform is

$$I = \frac{1}{T} \int_{-\infty}^{\infty} |p(t)|^2 dt = \frac{1}{2\pi} \int_{-\infty}^{\infty} \frac{|V(\omega)|^2}{T} d\omega \,, \tag{20b}$$

where T is the finite time duration of the waveform. This time duration T for finite duration waveforms, is not to be confused with the waveform period T_p. The integrand of the right member is the intensity spectral density and its unit of measure is intensity per Hertz. For real waveforms, the one-sided spectral density is

$$P(\omega) = \frac{2|V(\omega)|^2}{T} \,, \tag{20c}$$

in analogy with (18b).

Example 2: For a finite duration sinusoid

$$p(t) = A\sin(\omega_0 t), \qquad 0 \le t \le T \tag{20d}$$

which is one period of the waveform in Example 1, the Fourier transform and spectrum are the same as in Example 1 with two important exceptions. From (19b) we obtain

$$V(\omega) = \frac{AT}{2}\{-j\exp[+j(\omega_0 - \omega)T/2]\,\mathrm{sinc}[(\omega_0 - \omega)T/2]$$

$$+j\exp[-j(\omega_0 + \omega)T/2]\,\mathrm{sinc}[(\omega_0 + \omega)T/2]\} \,, \tag{20e}$$

$$|V(\omega)|^2 = \left(\frac{AT}{2}\right)^2 (\mathrm{sinc}[(\omega_0 - \omega)T/2] - \mathrm{sinc}[(\omega_0 + \omega)T/2])^2 \,, \tag{20f}$$

where it is assumed that the waveform has a discrete number of cycles, i.e. $\omega_0 T = 2\pi K$ where K is some integer. It is seen that the discrete frequency variable ω_n is replaced with the continuous frequency variable ω, i.e. whereas periodic waveforms have discrete spectra, finite duration waveforms have spectral densities, and the leading coefficient indicates that the spectrum unit is energy per Hz. In fact, the one-sided energy spectrum is obtained by doubling the spectrum which gives a leading factor of $[(A^2T/2)/(1/T)]$. Dividing by the factor of ρc gives the energy flux density of the signal per Hz. Further dividing by the signal duration T gives the average intensity per Hz as shown in (20c).♦

Nonperiodic infinite duration waveforms present special problems. They cannot be represented in the time domain by either the Fourier series, because they are not periodic, or by the Fourier integral, because they are not finite duration. However, their power frequency spectra can be determined via Fourier analysis by means of the Wiener-Kinchine theorem. A brief derivation of this theorem follows.

For a finite duration segment of the waveform, the Fourier integral representation of the waveform and its Fourier transform are given by

$$p_T(t) = \frac{1}{2\pi} \int_{-\infty}^{\infty} V_T(\omega) \exp(j\omega t) d\omega \,, \tag{21a}$$

$$V_T(\omega) = \int_{-T/2}^{T/2} p_T(t) \exp(-j\omega t) dt \,, \tag{21b}$$

where the subscript T denotes the time interval of the finite duration segment. Letting $T \to \infty$ in these equations will in general not produce a limiting result and therefore there is no Fourier integral representation of the nonperiodic infinite duration waveform. However, the limit does exist for the power spectrum.

The waveforms of interest are viewed as random processes and averaging is performed to achieve the desired result. Forming the expected energy density function for the ensemble of finite duration time segments gives

$$\mathbf{E}|V_T(\omega)|^2 = \int_{-T/2}^{T/2} dt_1 \int_{-T/2}^{T/2} dt_2 C_T(t_1 - t_2) \exp\big(j\omega (t_1 - t_2)\big) \,, \tag{22a}$$

$$C_T(t_1 - t_2) = \mathbf{E}\left[p_T(t_1) p_T^*(t_2)\right] , \tag{22b}$$

where $\mathbf{E}$ denotes the expectation operator and $C_T(\cdot)$ is the autocovariance function of the ensemble of finite duration segments. With the variable change

$$\tau = \frac{t_1 - t_2}{\sqrt{2}} \,, \qquad \sigma = \frac{t_1 + t_2}{\sqrt{2}} \,,$$

where the domain of integration is shown in Figure 8, the expected energy density in (22) can be expressed, after some manipulation, as

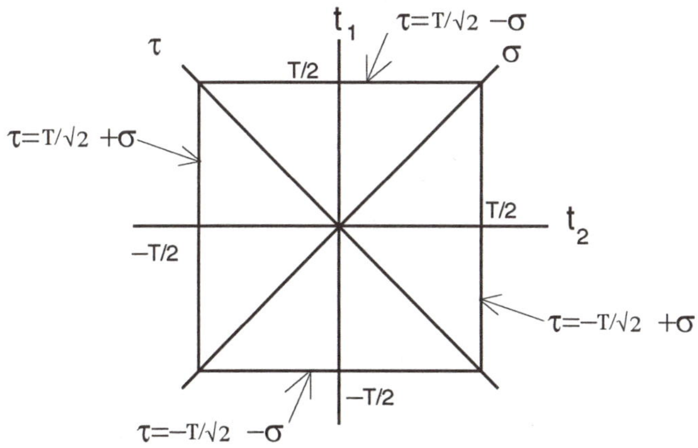

Figure 8 Coordinate Transformation

$$\mathbf{E}|V_T(\omega)|^2 = T\int_{-T}^{T} d\tau C_T(\tau)\exp(-j\omega t)\left(1-\frac{|\tau|}{T}\right). \tag{22c}$$

The autocovariance function of the finite segment exists and, in the limit $T \to \infty$, the autocovariance of the infinite duration process is given by

$$C(\tau) = \lim_{T\to\infty} C_T(\tau). \tag{22d}$$

Similarly, in the limit $T \to \infty$, the integral in (22c) exists and is the Fourier transform of the autocovariance. Consequently, the energy density in (22c) does not exist in the limit $T \to \infty$ but the power spectrum does exist and the one-sided power spectrum is given by

$$P(\omega) = \lim_{T\to\infty} \frac{2\mathbf{E}|V_T(\omega)|^2}{T}, \quad \omega \geq 0, \tag{22e}$$

$$= 2\int_{-\infty}^{\infty} d\tau C(\tau)\exp(-j\omega\tau), \quad \omega \geq 0. \tag{22f}$$

This is the Wiener-Kinchine theorem. In practice the power spectrum is estimated by one of two methods. Both require the ergodic hypothesis which asserts that the statistics of one time series are the same as the statistics of the ensemble of time series. In the first method, a sequence of magnitude squared Fourier transforms are averaged to approximate (22e). Since each transform is of windowed data, proper windowing is required to ensure accurate and stable estimates. In the second method, the autocorrelation function is substituted for the autocovariance function to approximate (22f). For this case the power spectral density is given by the Fourier transform of the autocorrelation function of a long time segment of the waveform.

Since the Wiener-Kinchine result is a Fourier transform, we can invert it to get

$$C(\tau) = \frac{1}{2\pi}\int_0^{\infty}\frac{P(\omega)}{2}\exp(j\omega\tau)d\omega = \frac{1}{2\pi}\int_{-\infty}^{\infty}N_0(\omega)\exp(j\omega\tau)d\omega , \quad (22g)$$

where $N_0(\omega)$ is the two-sided power spectrum. For white noise we have $N_0(\omega) = N_0$ and the correlation function becomes

$$C(\tau) = \frac{N_0}{2\pi}\int_{-\infty}^{\infty}\exp(j\omega\tau)d\omega = N_0\delta(\tau) = \frac{P_0}{2}\delta(\tau) . \quad (22h)$$

In (22h) P_0 is the one-sided power spectrum and $\delta(\tau)$ is the Dirac delta function which satisfies the Fourier transform relations

$$\delta(\tau) = \frac{1}{2\pi}\int_{-\infty}^{\infty}\exp(j\omega\tau)d\omega , \rightarrow \quad 1 = \int_{-\infty}^{\infty}\delta(\tau)\exp(-j\omega\tau)d\omega . \quad (22i,j)$$

Decibel Levels

Intensity and energy flux density determine whether or not a target can be detected which is the first step in all ASW operations. However, both quantities can be awkward to work with because they can be expressed in various units of measure and because, for many units of measure, ASW intensity and energy flux density values are very small numbers. Common practice has adopted the more convenient decibel measure in which intensity or energy flux density is referenced to a standard level. In equation form the intensity level is given by

$$I[B]L = 10\text{Log}\left(\frac{I}{I_0}\right) \quad \text{dB re } I_0[\text{per W at f}] , \quad (23a)$$

$$\begin{aligned} IL &= \text{intensity level} \\ IBL &= \text{intensity band level} \\ I &= \text{intensity} \\ I_0 &= \text{intensity reference} \end{aligned}$$

where Log denotes the base 10 logarithm. Decibel levels are always dimensionless quantities but are stated as decibel levels and are abbreviated: dB[5]. The qualifier, dB re I_0[per W at f] or dB//I_0[per W at f], indicates that the quantity is a decibel level and specifies the reference intensity which must have the same unit as the referenced intensity, i.e. we must compare "apples to apples". If the reference intensity is for a

[5] Intensity levels were originally measured in bels, defined to be $\text{Log}(I/I_0)$, after Alexander Graham Bell. The bel was found to be a large unit which gave low numerical values, so the decibel unit was defined such that, numerically, IL(dB) = 10x[IL(bels)]. This produces a scale on which one unit of IL(dB) corresponds to 0.1 units of IL(bels).

time domain signal the appropriate quantity is just intensity level IL with no reference to frequency. If the reference intensity is the intensity within a frequency band of W about the frequency f, i.e. the level is frequency dependent, the appropriate quantity is intensity band level, denoted IBL, and the qualifier "per W at f" is added. If the reference frequency band is one Hz, the appropriate quantity is intensity spectrum level, denoted ISL "at f". If the spectrum level is constant, the qualifier "at f" is omitted. For this case the relationship between the band level and spectrum level is

$$\mathrm{IBL} = \mathrm{ISL} + 10\mathrm{Log}(\mathrm{W}), \tag{23b}$$

since the intensity spectrum level is constant over the band W. If this is not the case then the total intensity within the band W is found by integrating the spectrum over the band W and the intensity band level is found from the expression 10Log((total intensity within W)/I_0). Note that the levels and the band of interest are for the one-sided spectrum and similarly for the reference intensity.

Energy flux density is the energy flow per unit area through a surface that is parallel to the acoustic wave fronts. It can be obtained from the time integral over intensity and, analogous to intensity level, the energy level is denoted

$$\mathrm{E[B]L} = 10\mathrm{Log}\left(\frac{\mathrm{E}}{\mathrm{E_0}}\right) \quad \mathrm{dB\ re\ E_0[per\ W\ at\ f]}. \tag{24a}$$

$$\begin{aligned} \mathrm{EL} &= \text{energy level} \\ \mathrm{EBL} &= \text{energy band level} \\ \mathrm{E} &= \text{energy} \\ \mathrm{E_0} &= \text{energy reference} \end{aligned}$$

The reference qualifiers are similar to those for intensity. If the reference is for a time domain signal, the appropriate quantity is just energy level EL with no reference to frequency. If the reference is for a frequency band of W, the appropriate quantity is the energy band level over W, denoted EBL, and, if the reference is for a frequency band of one Hz, the appropriate quantity is the energy spectrum level over a one Hz band, denoted ESL. For this case, the relationship between the band level and spectrum level is

$$\mathrm{EBL} = \mathrm{ESL} + 10\mathrm{Log}(\mathrm{W}), \tag{24b}$$

where, as for intensity, it is assumed that the energy spectrum level is constant over the band W. The energy levels are related to the intensity levels by the signal duration

$$\mathrm{EL} = \mathrm{IL} + 10\mathrm{Log}(\mathrm{T}), \tag{25a}$$

$$\mathrm{EBL} = \mathrm{IBL} + 10\mathrm{Log}(\mathrm{T}), \tag{25b}$$

$$\mathrm{ESL} = \mathrm{ISL} + 10\mathrm{Log}(\mathrm{T}), \tag{25c}$$

where the intensity is assumed constant over the signal duration. The energy band level is related to the intensity spectrum level by both the signal duration and the signal bandwidth

$$\text{EBL} = \text{ISL} + 10\text{Log}(\text{TW}). \tag{25d}$$

In each of (25) the levels are assumed to be constant over time and frequency. If this is not the case then in (25a,b,c) the total energy is found by integrating the signal intensity over time (within the appropriate band) and in (25d) the total energy is found by integrating the time dependent intensity spectrum over time and frequency. The corresponding energy level is 10Log((total energy within band W and duration T)/E_0). As for the case of spectral intensity levels, the spectral energy levels and the associated bands of interest are for one-sided spectra.

For plane waves the intensity level expression can be equivalently written in terms of pressures

$$\text{I}[\text{B}]\text{L} = 10\text{Log}\left(\frac{\langle |p|^2 \rangle}{\langle |p_0|^2 \rangle}\right) \text{ dB re } p_0 \left[\text{per W at f}\right], \tag{26}$$

p = sound pressure
p_0 = sound pressure reference

where the angular brackets <·> denote short term time average and the intensity level is sometimes referred to as the sound pressure level to indicate that the source of the measurement is acoustic pressure. The qualifier, dB re p_0 (per W at f), indicates that the quantity is a decibel level and states the reference pressure. However, this is misleading because the reference is, in fact, not pressure but, rather, the intensity that would be produced by a plane wave having rms pressure p_0. Moreover, if the quantity is a spectral density, the reference is the intensity in a one Hz band that would be produced by a plane wave, localized to that band, and having rms pressure p_0. Again, the reference is intensity per Hz, even though it is specified as pressure per Hz at f. If the pressure is constant over the band, the qualifier "at f" is omitted.

While levels given by (23a) and (24a) in terms of reference intensity and energy flux are always correct, levels given by (26) in terms of reference pressures are correct only for plane waves. Since the sensors used in most ASW work measure acoustic pressure and not intensity, the levels quoted are generally sound pressure level values and are the correct intensity level values only if the waveforms are planar. This is hardly ever a problem. For the case of received sonar signals, i.e. target echoes and target radiations, the received waveforms are always planar, or very nearly so, over the region of reception. Background noise and reverberation are more complicated because they are received from multiple directions simultaneously but, if the reception from each direction is planar and if the receptions are statistically independent, the measured sound pressure level is the sum of the received intensities. In this case the measured sound pressure level is the total background noise or reverberation intensity and is the correct IL. Unfortunately, noise received from nearby sources, e.g. biologics, may not be planar and other noise may not be independent, e.g. surface reflected versus direct path noise. Dependent noise components can be especially troublesome, e.g. whereas the intensity of two independent components is the sum of their intensities, the intensity of two identical components is the intensity of their sum which is twice as great as for the case of two independent components. Since the dependence typically causes constructive and destructive interference between multipath components, it is not likely that the dependence would maintain over a long period of time. However, a fluctuating dependence would cause the measure of the background noise level to have a larger than normal variance and could cause the average value to be biased. To qualify the sound pressure level measurement for these cases it is referred to as the equivalent plane-wave noise level. In sum, though, while small errors can arise using (26), generally speaking, they are almost always negligible compared to the larger errors that routinely arise from environmental uncertainties and other factors.

There are three decibel reference levels in common use in ASW sonar work which are listed in Table 2. Most modern ASW sonar work is conducted in terms of the micropascal (μPa) pressure reference. It was adopted by the U.S. Navy in 1970 by order of the Chief of Naval Operations to replace the larger microbar (μbar) pressure reference which was the older ASW reference pressure. The μbar was originally chosen, somewhat arbitrarily, for use in sonar work and its large size led to negative levels for many acoustic measurements. The adoption of the μPa reference has served to raise most common acoustic intensity levels to positive levels. Another commonly used reference is the auditory threshold which is the sound pressure in air at the threshold of hearing. This quantity is important in ASW sonar work because aural detection is still important in some systems and because the auditory level has been used as an underwater reference pressure in some of the older sonar work. The reference levels in Table 2 are given in both cgs and mks units. More correctly, following the common practice, the cgs reference intensities are reported in terms of watts which is an mks unit. Since the cgs unit of power, erg/second, has no single name assigned to it, the watt = joule/second is commonly used in its place but transformation to pure cgs units is easily performed by means of the conversion factor: watt = 10^7 erg/second.

Table 2 Reference Levels for Signals

Reference	Intensity (cgs/mks)	Intensity (mks)	Pressure (cgs)	Pressure (mks)
micropascal (μP)	$6.67 \times 10^{-23} \frac{watt}{cm^2}$	$6.67 \times 10^{-19} \frac{wat}{m^2}$	$10^{-5} \frac{dyne}{cm^2}$	$10^{-6} \frac{n}{m^2}$
microbar (μbar)	$6.67 \times 10^{-13} \frac{watt}{cm^2}$	$6.67 \times 10^{-9} \frac{watt}{m^2}$	$1 \frac{dyne}{cm^2}$	$10^{-1} \frac{n}{m^2}$
audibility thrshld(AT)				
air	$10^{-16} \frac{watt}{cm^2}$	$10^{-12} \frac{watt}{m^2}$	$2.04 \times 10^{-4} \frac{dyne}{cm^2}$	$2.04 \times 10^{-5} \frac{n}{m^2}$
water	$2.77 \times 10^{-20} \frac{watt}{cm^2}$	$2.77 \times 10^{-16} \frac{watt}{m^2}$	$2.04 \times 10^{-4} \frac{dyne}{cm^2}$	$2.04 \times 10^{-5} \frac{n}{m^2}$

The relationships between intensity levels having different references are

$$\mathrm{IL}(\mathrm{re}\ \mu\mathrm{P}) = \mathrm{IL}(\mathrm{re}\ \mu\mathrm{bar}) + 100\mathrm{dB}, \quad (27a)$$

$$\mathrm{IL}(\mathrm{re}\ \mu\mathrm{P}) = \mathrm{IL}(\mathrm{re\ AT}) + 26\mathrm{dB}, \quad (27b)$$

$$\mathrm{IL}(\mathrm{re\ AT}) = \mathrm{IL}(\mathrm{re}\ \mu\mathrm{bar}) + 74\mathrm{dB}, \quad (27c)$$

where all references are for water borne signals. Equation (27a) shows that the old μbar levels can be modernized to μP levels by the addition of 100 dB.

Example 3: For the sinusoidal waveform in Examples 1 and 2, assume the intensity level is 12 dB re μPa. The rms amplitude is found to be

$$\sqrt{A^2/2} = \sqrt{\rho c I_0 10^{1.2}} = 4 \times 10^{-5}\ \mathrm{dyne/cm^2}. \quad (28a)$$

This result can also be obtained more directly from the sound pressure level expression in (26). The energy level is found to be

$$\mathrm{EL} = 12\ \mathrm{dB}\ +\ 10\mathrm{Log}(10) = 22\ \mathrm{dB\ re}\ \mu\mathrm{Pa}. \quad (28b)$$

Note that the signal duration is necessarily measured in seconds within both the cgs and mks systems of units.♦

Example 4: For the sinusoidal waveform in Examples 1, 2 and 3, the bulk of the signal energy is localized to a ±1 Hz band with an effective bandwidth of 1 Hz around the carrier frequency and the energy band level and energy spectral level for this band is equal to the energy level, i.e. EBL = ESL = EL = 22 dB. Since the energy spectral density falls from a maximum at the center of the band to nearly zero at the band edges we assume that this energy band level is a mean level over the ±0.5 Hz band about the carrier, i.e. EBL = 22 dB re μPa per Hz at 1 kHz. Similarly, the intensity band level and intensity spectral level over this band are equal to the intensity level, i.e. IBL = ISL = IL = 12 dB re μPa per Hz at 1 kHz. From (25c), the intensity spectral level is ISL = ESL - 10Log(10) = 12 dB re μPa per 1 Hz at 1 kHz.♦

APPENDIX 2-A: WAVE EQUATION FOR OCEAN MEDIA

Simple examples of wave propagation are described in Chapter 2 and the basis for those descriptions is discussed in the following. During the passage of an acoustic wave through the ocean the motion of the medium is subject to physical laws, including Newton's laws of mechanics, the law of conservation of mass, and the thermodynamic laws, all of which govern the manner in which acoustic energy is transported by the medium. A brief derivation of the wave equation for fluids, which governs wave propagation in fluid media and which is based upon these physical laws, is presented in the following. In this derivation the medium is assumed to be subject to gravitational forces, and to be stationary. The ocean, of course, is in constant motion but the movements are slow compared to wave propagation and the effects of the motion, when important, can be included in an ad hoc manner.

2.1-A: Mathematical Preliminaries

Derivative of a Scalar Field Function - The Gradient

The properties of the fluid medium are defined by mathematical functions of time and space coordinates, such as F(t,x,y,z), where F is a scalar and might represent the pressure, density, etc. in the medium. The field described by F will have a unique value at each time and at each point of the medium and, for a continuous medium, there will be a surface through every point of the medium over which F is constant at each time, called a level surface. Displacements away from each point will encounter higher or lower values of F, depending upon which side of the level surface the displacements occur, but along displacement paths lying on the level surface the values will not change. Moreover, no two level surfaces touch or intersect or else F would be multi-valued at that point and, for displacements to successive level surfaces, the values of F are increasing or decreasing. For example, for the function $F = \exp(j\omega_0 t)/(x^2+ y^2+ z^2)^{1/2}$, the level surfaces are spheres that are concentric about the origin on which the level values are time dependent.

At each time, the directional derivative of F along some path in the medium denoted λ is given by

$$\frac{dF}{d\lambda} = \frac{\partial F}{\partial x}\frac{dx}{d\lambda} + \frac{\partial F}{\partial y}\frac{dy}{d\lambda} + \frac{\partial F}{\partial z}\frac{dz}{d\lambda} = \nabla F \cdot \mathbf{e}_\lambda \,, \tag{1a}$$

$$\nabla F = \mathbf{e}_x \frac{\partial F}{\partial x} + \mathbf{e}_y \frac{\partial F}{\partial y} + \mathbf{e}_z \frac{\partial F}{\partial z} \,, \tag{1b}$$

$$\mathbf{e}_\lambda = \mathbf{e}_x \frac{dx}{d\lambda} + \mathbf{e}_y \frac{dy}{d\lambda} + \mathbf{e}_z \frac{dz}{d\lambda} \,, \tag{1c}$$

where the symbol ∂ denotes partial differentiation and $\mathbf{e}_x$, $\mathbf{e}_y$ and $\mathbf{e}_z$ are the unit vectors along the x, y and z directions. The vector ∇F is the gradient of F, in which ∇ denotes the Del vector operator defined in (1b). The vector $\mathbf{e}_\lambda$ has components equal to the direction cosines in the (x,y,z) coordinate system and is the unit vector pointing along the path λ. Each term of ∇F gives the rate of change of F along one of the coordinate axes and the three terms give the vector rate of change of F at the point. To determine the direction of ∇F consider that the directional derivative is zero for a path lying on the level surface of F, i.e. when $\mathbf{e}_\lambda$ lies on the level surface, and the directional derivative is a maximum in the direction perpendicular to the level surface, i.e. when $\mathbf{e}_\lambda$ is perpendicular to the level surface. Since (1a) shows that the maximum directional derivative value is achieved when ∇F and $\mathbf{e}_\lambda$ are parallel, then ∇F is a vector perpendicular to the level surface and points in the direction of increasing F. Moreover, ∇F is the change in F per unit length in the direction perpendicular to the level surface.

The total derivative of a field function with respect to time is given by

$$\frac{dF}{dt} = \dot{F} + \left(\frac{\partial F}{\partial x}\frac{dx}{dt} + \frac{\partial F}{\partial y}\frac{dy}{dt} + \frac{\partial F}{\partial z}\frac{dz}{dt}\right) = \dot{F} + \nabla F \cdot \mathbf{v}\,, \qquad (2a)$$

$$\dot{F} = \frac{\partial F}{\partial t}, \qquad \mathbf{v} = \mathbf{e}_X \frac{dx}{dt} + \mathbf{e}_y \frac{dy}{dt} + \mathbf{e}_z \frac{dz}{dt}, \qquad (2b,c)$$

where **v** is the velocity of the particle of mass that passes through the point. During the passage, the F property of the particle changes in time. The first term in the right member of (2a) gives the time rate of change of F at the point, i.e. the F property of the mass changes at the point during the time interval whether the particle moves or not. The gradient term in the right member of (2a) is the spatial change in F at the point as viewed in a snapshot, i.e. the F property of the mass changes spatially over a sequence of particle positions that pass through the point and each component of the gradient gives the spatial change along one of the coordinates. The dot product between the gradient and the particle velocity gives the time rate of change of the F property as a particle moves through the point from one spatial position to another. The sum of the time rate of change of the particle, while at the point, and the time rate of change as the particle moves through the point is the total time rate of change in the F property at the point.

Derivative of a Vector Field Function - The Divergence

Vector field functions **F**(t,x,y,z) are used to represent displacements, forces and other vector properties of the medium. Operating on the vector field function with the vector Del operator in a vector dot product gives

$$\nabla \cdot \mathbf{F}(t,x,y,z) = \frac{\partial F_X(t,x,y,z)}{\partial x} + \frac{\partial F_y(t,x,y,z)}{\partial y} + \frac{\partial F_z(t,x,y,z)}{\partial z}, \qquad (3)$$

which is a scalar field that is termed the divergence of the field. The first term of the dot product is the spatial rate of change of the x-component of the field vector along the x-axis which is just a snapshot of the spatial rate of change of F_X over a sequence of particle positions that pass through the point along the x-axis. This represents the change in F_X as a particle moves through the point, i.e. it implies that the quantity F_X is being depleted, if the term is positive, or is being accumulated, if the term is negative, at the point. Similar interpretations apply to the y and z terms. The sum of the three terms represents the net change in the **F** field as the particle moves through the point, i.e. the net depletion or net accumulation of **F** at the point.

2.2-A: Physical Preliminaries

Newton's Second Law

For the important case in which the field described by the scalar F is the pressure p in the medium, the level surfaces are surfaces of constant pressure. The gradient ∇p of the pressure field is a vector field perpendicular to the level surfaces. The change in pressure along the gradient is the change in force per unit area, where the force acts on the incremental area that is perpendicular to the gradient, and, accordingly, the change in pressure per unit length along the gradient is the change in force per unit volume where the force acts on the incremental volume that is contained within the incremental area and the incremental length along the gradient. The change in force on the incremental volume is the net force on the volume and, by Newton's second law, is related to the acceleration of the volume element by the expression

$$-\nabla p + \mathbf{f} = \frac{d(\rho \mathbf{v})}{dt}. \tag{4a}$$

The negative sign on the gradient in (4a) arises because the force on the volume element is pointing in the direction contrary to that of increasing pressure. The vector quantity **f** represents the gravitational force per unit volume of the medium, ρ denotes the total density of the medium, including both the static density and density changes caused by the passage of the wave, and **v** denotes the velocity of the volume element due to the passage of the wave. The product $\rho\mathbf{v}$ is the mass flow per unit area in the direction of the particle velocity.

Equation (4a) can be immediately reduced to an equation for the acoustic wave only. The gravitational force acts only downward into the ocean, say the positive z direction, which causes a static pressure gradient in the amount

$$\mathbf{f} = \rho g z \mathbf{e}_Z, \tag{4b}$$

where the constant g is the gravitational acceleration, z is the depth into the medium and the unit vector is directed downward into the medium. Canceling the pressure gradient due to the gravitational field from the left member of (4a) gives

$$-\nabla p = \rho_0 \frac{d\mathbf{v}}{dt}, \tag{4c}$$

where the pressure p in (4c) is now for the acoustic wave only, as opposed to the total pressure in (4a), and the density has been expressed as the sum of the static density ρ_0 and the excess density due to the acoustic wave. The static density contains the effects of the gravitational field and, since the excess density and the particle velocity are both small terms, their product is a second order term and is ignored. The derivative of particle velocity can be simplified by an approximation. Applying (2a) to each component of the velocity and combining the three equations gives

$$\frac{d\mathbf{v}}{dt} = \dot{\mathbf{v}} + (\mathbf{v} \cdot \nabla)\mathbf{v} . \tag{5}$$

Since the last term in (5) is second order in the velocity it is ignored. Combining (4c) and (5) gives

$$-\nabla p = \rho_0 \dot{\mathbf{v}} , \tag{6}$$

which is Newton's second law for acoustic waves in a stationary medium under the force of gravity.

Conservation of Mass

The mass at a point in the medium changes because the density at the point changes; there is a flow through the point bringing in material of different density; and because mass is injected or absorbed at the point. The conservation of mass law is a balance among these changes. The change in mass due to a change in density at the point is $(\partial\rho/\partial t)\delta V$ where δV is the incremental volume element at the point. The mass flow through the point is $\rho v \delta A$ where δA is the incremental area perpendicular to the mass velocity through the point. Consequently, the change in mass at the point due to the mass flow in the x-direction is given by

$$\left(\frac{\delta(\rho v_x)}{\delta x}\right)\delta x \delta A_x , \tag{7a}$$

where v_x and δA_x are the velocity in the x-direction and the incremental area perpendicular to the x-direction. Similar expressions can be developed for the change in mass due to the flows in the y and z directions. The mass injected or absorbed at the point, which is effectively what a source does, is given by $q\rho_0\delta V$ where q is the source function with the units of fractional change per unit time. This term represents the injected or absorbed mass as a fraction of the static mass ρ_0 at the point.

Combining all mass change terms gives the mass change balance equation.

$$\dot{\rho}\,\delta V + \left(\frac{\delta(\rho v_x)}{\delta x}\delta x \delta A_x + \frac{\delta(\rho v_y)}{\delta y}\delta y \delta A_y + \frac{\delta(\rho v_z)}{\delta z}\delta z \delta A_z\right) = q\rho_0 \delta V \tag{7b}$$

in which each term contains the same incremental volume element. Canceling the volume element throughout, this equation can be expressed as the conservation of mass law

$$\dot{\rho} + \nabla \cdot (\rho \mathbf{v}) = q\rho_0 , \tag{7c}$$

where the Del operator, applied to the momentum per unit volume vector field ρ**v**, as a vector dot product is the divergence of the field as in (3). At each point in the medium, the divergence of the vector field represents a "snapshot" of the change in the mass at the point as the fluid moves through the point. If the divergence is positive, fluid is leaving the point faster than it arrives causing a depletion of fluid at the point. If the divergence is negative, fluid is leaving the point more slowly than it arrives causing an accumulation of fluid at the point. Note that for the case of an incompressible fluid, i.e. $\dot{\rho} = 0$, with no sources or sinks, i.e. q = 0, the divergence of the momentum vanishes $\nabla\cdot(\rho\mathbf{v}) = 0$, i.e. the mass entering and leaving the point are in balance.

For the case of acoustic propagation the conservation of mass law can be approximated

$$\dot{s} + \nabla\cdot\mathbf{v} = q\,, \tag{8a}$$

$$s = \frac{\rho - \rho_0}{\rho_0}\,, \tag{8b}$$

where the quantity s is the fractional change in density at the point and is called the condensation and, in the derivation of (8), the product between the excess density and particle velocity is second order and ignored.

Equation of State

The state of a simple medium can be described by the pressure, density and temperature of the medium because it is found experimentally that only two of these properties can be set independently with the remaining property set by the physics of the medium. Such a relationship, imposed by the physics of the medium on its properties, is the basis for the equation of state of the medium. While for many fluid media, the state variables: pressure, density and temperature are found to be sufficient to define the equation of state, for the case of the oceans the equation of state is found to be also dependent upon the time variation of density. This is because density changes in the ocean do not follow pressure changes in a lock-step manner, i.e. ocean water does not behave as an ideal elastic spring. A sudden increase or decrease in pressure in the ocean is not immediately translated into an increase or decrease in density. The density changes lag the pressure changes. This impacts the energy of the acoustic wave which is stored at each point in the medium alternately as kinetic and potential energy. The mass particles of the medium oscillate to-and-fro and reach states of maximum compression and rarefaction during the reversals. The to-and-fro motion of the fluid stores the energy of the wave as kinetic energy while the fluid elasticity during the compression and rarefaction phases stores the energy of the wave as potential energy. As the wave propagates, the energy of the wave at each point of the medium is continuously being exchanged between these states. For simple fluids the energy is exchanged without loss. However, for the ocean, since the density lags pressure, full compression and rarefaction is not achieved and the energy exchange occurs with loss. This is found to produce volume viscosity, which is more fully discussed in Chapter 5 and Appendix 5-A, and the lost energy is converted to heat.

The equation of state for the oceans is unknown in detail but, in a formal sense, it can be written

$$p = p(s, \dot{s}, \text{entropy}) . \tag{9}$$

In this form, the total pressure in the medium is expressed in terms of the total condensation and total entropy, i.e. density and temperature have been replaced, without loss of generality, with condensation and with entropy, which is a measure of heat growth in the medium. The advantage of this form is due to the fact that acoustic waves in the ocean propagate very nearly adiabatically, i.e. with constant entropy, even with volume viscosity. Consequently, acoustic pressures are essentially dependent only upon condensation and its time rate of change. Computing the differential of (9), while holding the entropy constant, we obtain

$$p = B_{ad} s + \sigma_{ad} \dot{s} , \tag{10a}$$

$$B_{ad} = \left(\frac{\partial p}{\partial s}\right)_{ad} , \quad \sigma_{ad} = \left(\frac{\partial p}{\partial \dot{s}}\right)_{ad} , \tag{10b,c}$$

where the differentials of total pressure, total condensation and its time rate of change are, respectively, identified in (10a) as the acoustic pressure, acoustic condensation and the time rate of change of the acoustic condensation. The coefficients, B_{ad} and σ_{ad}, are, respectively, the adiabatic bulk modulus and adiabatic viscosity of the medium in which the subscripts "ad" denote adiabatic changes. The adiabatic assumption asserts that, during the passage of the acoustic wave, no heat flows from the compressed regions into the rarefied regions. The validity of the assumption is borne out below where it is shown theoretically that the adiabatic bulk modulus is related to the sound speed by the expression $c = \sqrt{(B_{ad}/\rho_0)}$ and this has been verified experimentally.

2.3-A: The Wave Equation

Newton's second law, the conservation of mass and the equation of state of the medium are physical statements sufficient to determine the nature of wave propagation in the medium. Newton's second law and the conservation of mass law are four scalar equations in the five unknowns: p, v_x, v_y, v_z and s, where the static density ρ_0 is assumed known, and the equation of state of the medium provides the fifth relationship required to solve for all variables.

The wave equation derivation is given in the following in which the physical laws, which are the basis of the derivation, are repeated below and the steps for deriving the wave equation, which are relatively mechanical, are indicated by references to the relevant physical law, referenced as N2, MC and ES, and other equations. The references to adiabatic conditions in the Equation of State are omitted for convenience.

Newton's Second Law

$$-\nabla p = \rho_0 \dot{\mathbf{v}} \quad (6)$$

Mass Conservation

$$\dot{s} + \nabla \cdot \mathbf{v} = q \quad (8)$$

Equation of State

$$p = Bs + \sigma \dot{s} \quad (10)$$

$$\nabla^2 p = -\rho_0 \nabla \cdot \dot{\mathbf{v}} \quad (11\text{: N2})$$

$$\nabla^2 p = \rho_0 (\ddot{s} - \dot{q}) \quad (12\text{: 11 \& MC})$$

$$\dot{p} = B\dot{s} + \sigma \ddot{s} \quad (13\text{: ES})$$

$$\dot{p} = B\dot{s} - \sigma(\nabla \cdot \dot{\mathbf{v}} - \dot{q}) \quad (14\text{: 13 \& MC})$$

$$\dot{p} = B\dot{s} + \frac{\sigma}{\rho_0} \nabla^2 p + \sigma \dot{q} \quad (15\text{: 14 \& 11})$$

Wave Equation with Volume Viscosity

$$\nabla^2 p - \frac{1}{c^2}\left(\ddot{p} - \frac{\sigma}{\rho_0} \nabla^2 \dot{p}\right) = -\rho_0 \left(\dot{q} + \frac{\sigma}{B} \ddot{q}\right) \quad (16\text{: 12 \& 15})$$

$$c^2 = B / \rho_0$$

Basic Inviscid Wave Equation

$$\nabla^2 p - \frac{1}{c^2} \ddot{p} = -\rho_0 \dot{q} \quad (17\text{: 16})$$

The general solutions of the wave equations in (16) and (17) are uncountably infinite and, to select the solution applicable to the problem at hand, boundary conditions are imposed. More about that as we progress.

The wave equation in (16) applies to fluids subject to volume viscosity. The wave equation in (17) applies to inviscid media. Both apply to media subject to gravitational fields. For most ASW sonar applications, and for most discussions in this treatise, these wave equations are found to be sufficient. However, several phenomena were ignored to simplify their derivations. For example, shear viscosity, which is a small effect in the oceans, was omitted but this is rectified in Appendix 5-A. Similarly, the motions of the medium, sources and receivers were ignored but these effects can be included phenomenologically. Also, while variations in the background density have been included in the wave equation derivation, Chernov's term[1] $\nabla p \cdot \nabla \log(\rho_0)$ was omitted in (11) under the assumption that it is second order. Under conditions of sharp density gradients this may not be true and the omitted term can cause scattering and refraction of the signal. Other effects which have been ignored, such as heat conduction, are relatively unimportant in ASW work.

For the case of wave propagation within solid bodies, phenomena arise that, while not covered by the preceding theory, can be conceptually understood in terms of that theory and are addressed in some detail in Appendix 5-A. Wave propagation within a solid is, in many ways, more complicated than wave propagation in a fluid because, in addition to a pressure wave, a shear wave can also be generated. In fact,

---

[1] Chernov, Lev A., *Wave Propagation in a Random Medium*, Section 8, Translated from the Russian by R. A. Silverman, Dover Publications, New York, 1960.

as an aside regarding shear viscosity in fluids, the presence of a very weak shear wave is the source of the complexity in the derivation of the general wave equation for fluids in Appendix 5-A. However, while shear waves in fluids are not very important for ASW work, they can be very important for the case of solid media that can support substantial shearing stresses such as the ocean floor and submarine hulls. Whereas particle displacements in a longitudinal pressure wave are parallel to the direction of wave travel, and cause only pressure variations, particle displacements in a shear wave are perpendicular to the direction of wave travel and cause shearing stresses. Moreover, the speed of the shear wave is generally different than that of the pressure wave. One complexity caused by the presence of a shear wave concerns the coupling of a pressure wave and a shear wave at the boundary of a solid which can cause surface elastic waves to be generated. Such waves are infrequently encountered in operational ASW activities but one important case of surface waves for ASW applications is concerned with sonar transmissions that have penetrated ocean floors made of rock. It has been observed that a surface wave can be generated that propagates radially outward along the bottom, paralleling the in-water sonar transmission, that radiates energy back into the ocean medium as it propagates. Since the surface wave speed in rock is greater than the speed of the direct sonar transmission in water, the radiation from the surface wave can arrive at a down range receiver or target sooner than the direct signal which can produce an anomalous reception or detection.

For submarine targets, shear waves can similarly produce surface waves in the hull that can potentially reradiate into the medium with a period equal to the time of circumnavigation of the hull. Additionally, for wavelengths that are comparable in size to the target, acoustic energy can penetrate the target and cause structural vibrations that are, effectively, standing waves within the hull as opposed to propagating waves on the hull surface. In fact, modern high intensity low frequency sonar systems can, potentially, excite hull resonances in submarine targets that radiate energy that is spectrally modulated by the resonances of the submarine hull. The detection of such hull resonance information is a potential discriminant for submarine target classes.

While echo complexities caused by acoustic energy penetrating the target hull have the potential of being target discriminants, we would be remiss if we failed to note that, for most ASW scenarios, submarine targets are detected primarily on the basis of total energy reflected from the target surface and rarely on energy coupled into the submarine hull and reradiated. Consequently, while acoustic energy penetrating the target hull can, theoretically, become important if these waves reradiate detectable energy, for most operational systems and scenarios this is not the case. All target discriminants are not lost, though, because on occasion target "highlights" modulate the echo amplitude which can be discerned and used as discriminants. These amplitude modulations are the principal target effects on the signal and occur as the signal encounters the bow, sail and propeller screws of the target. Hence, we can generally infer that any spectral shaping on the returned echo is probably due to multi-path effects and not to target resonances. In sum, target size and shape are the primary determinants of detectability and discriminants in target echoes.

APPENDIX 2-B: THE ANALYTIC SIGNAL

Acoustic signals that propagate through the ocean and their electrical counterparts within the sonar signal and data processing systems are real signals, in a physical sense, and they can be expessed as real signals, in a mathematical sense. However, because modern signal processing algorithms are simplified by expressing them in terms of complex analytic signals, it is necessary that we develop the analytic signal representation for real physical signals.

Analytic signals are generalizations of the complex trigonometric function

$$\exp(j\omega t) = \cos(\omega t) + j\sin(\omega t)\,. \qquad (1)$$

In this expression, the complex exponential is the analytic signal while the cosine function is the real signal and the sine function is the Hilbert transform of the real signal as shown below. The Fourier transforms of these expressions are depicted in Figure 1.

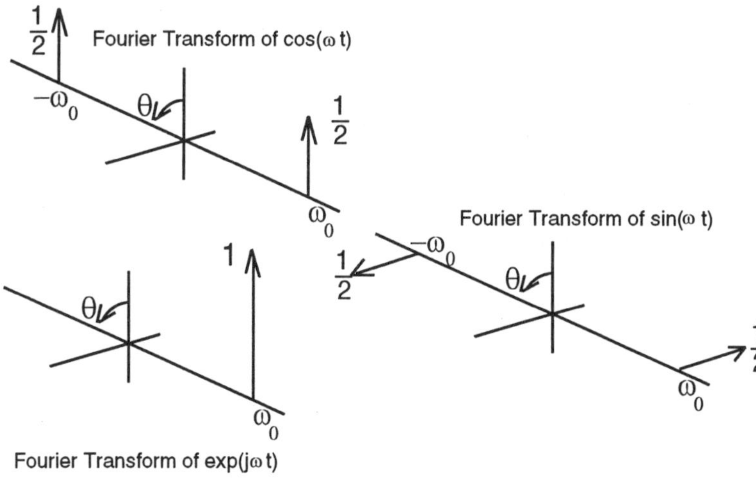

Figure 1 Fourier Transforms of Sinusoidal Components

The Fourier transform of the analytic signal is a single component of unit value at the carrier frequency ω_0 while the Fourier transforms of the cosine and sine signals are components at $\pm\omega_0$ having values 1/2. Both Fourier components for the cosine signal have zero phases while the components for the sine signal have phases of $-\pi/2$ and $+\pi/2$ for the components at, respectively, $+\omega_0$ and $-\omega_0$. Multiplying the Fourier transform of the sine signal by the imaginary quantity "j" adds $+\pi/2$ to the phase of each component increasing them to 0 and π respectively. With this rotation the sum of the Fourier transforms of the cosine and sine signals is seen to be the Fourier transform of the complex analytic signal as must be the case. The principal simplification gained from the analytic signal is to replace the two sided Fourier transform of the real signal with the one sided transform of the analytic signal.

2.1-B: General Representation of Analytic Signals

Fourier Transformations

The above elementary notions of the analytic signal can be extended to general sonar signals using Fourier transform techniques. Represent the real physical signal s(t) in terms of its Fourier transform

$$s(t) = \frac{1}{2\pi} \int_{-\infty}^{\infty} F(\omega) \exp(j\omega t) d\omega \ , \tag{2a}$$

$$F(\omega) = \int_{-\infty}^{\infty} s(t) \exp(-j\omega t) dt \ , \tag{2b}$$

where $F(\omega)$ is the Fourier transform of the signal, which is generally complex, and $\omega = 2\pi f$ is the angular frequency variable in the Fourier transform domain. For mathematically real signals, the Fourier transform is necessarily defined on both positive and negative domains of ω and (2b) shows that the transform satisfies the relationship

$$F(-\omega) = F^*(\omega) , \tag{2c}$$

where the superscript asterisk denotes complex conjugate.

Analytic Signal Representation by Fourier Transformations

A generalization of the expression in (1) can be written, using the Fourier transform of the real physical signal in (2), as

$$\frac{1}{2\pi} \int_{0}^{\infty} 2F(\omega) \exp(j\omega t) d\omega$$

$$= \frac{1}{2\pi} \int_{-\infty}^{\infty} F(\omega) \exp(j\omega t) d\omega + \frac{j}{2\pi} \int_{-\infty}^{\infty} -j\,\mathrm{sgn}(\omega) F(\omega) \exp(j\omega t) d\omega \ , \tag{3a}$$

$$\begin{aligned} \mathrm{sgn}(\omega) &= +1 \quad && \omega \geq 0 \\ &= -1 \quad && \omega < 0 \end{aligned} \tag{3b}$$

where $\mathrm{sgn}(\omega)$ is the signum function. Equation (3) is easily seen to be identically true for all cases. The left member is defined to be the analytic signal. Because it is an integral over only the positive domain of frequencies, the factor of 2 is required because $F(\omega)$ is the two sided Fourier transform. The leading term in the right member is just the real physical signal and the last term is shown below to be the Hilbert tranform of the real physical signal. Since the integrand of the last term satisfies the conjugate condition in (2c), this term also represents a real signal. Equation (3) can be abbreviated by the expression

$$s_a(t) = s(t) + j\hat{s}(t), \tag{4a}$$

where the left member is the analytic signal

$$s_a(t) = \frac{1}{2\pi}\int_0^{\infty} 2F(\omega)\exp(j\omega t)d\omega, \tag{4b}$$

and the terms in the right member are, respectively, the real signal and its Hilbert transform and are related to the analytic signal by the expressions

$$s(t) = \text{Real } s_a(t), \qquad \hat{s}(t) = \text{Imag } s_a(t). \tag{4c,d}$$

<u>Analytic Signal Representation by Hilbert Transformations</u>

The function $\hat{s}(t)$ is related to the real signal via its defining relation in (3)

$$\hat{s}(t) = \frac{1}{2\pi}\int_{-\infty}^{\infty} -j\text{sgn}(\omega)F(\omega)\exp(j\omega t)d\omega. \tag{5}$$

Since the integrand in (5) is a product, (5) can be shown to be equivalent to the convolution between the Fourier transform of –jsgn(ω) and the real signal. The transform of –jsgn(ω) is given by

$$\sigma(t) = \frac{1}{2\pi}\int_{-\infty}^{\infty} -j\text{sgn}(\omega)\exp(j\omega t)d\omega = \frac{1}{\pi}\int_0^{\infty}\sin(\omega t)d\omega. \tag{6a}$$

The latter integral is not an elementary integral but can be evaluated as the limit of the elementary integral[1]

$$\sigma(t) = \lim_{a\to 0}\frac{1}{\pi}\int_0^{\infty}\exp(-a\omega)\sin(\omega t)d\omega = \frac{1}{\pi t}. \tag{6b}$$

Therefore, the convolution between –jsgn(ω) and the real signal is given by

$$\hat{s}(t) = \frac{1}{\pi}\int_{-\infty}^{\infty}\frac{s(\tau)}{(t-\tau)}d\tau, \tag{7}$$

---

[1] Abramowitz and Stegun (4.3.136).

which is the Hilbert transform of the real signal.

Similarly, the inverse Fourier transform of (5) gives

$$F(\omega) = j\operatorname{sgn}(\omega) \int_{-\infty}^{\infty} \hat{s}(t)\exp(-j\omega t)dt\,, \tag{8a}$$

and the inverse Fourier transform or (8a) is

$$s(t) = \frac{1}{2\pi} \int_{-\infty}^{\infty} \hat{s}(\tau) \int_{0}^{\infty} 2\sin[\omega(\tau - t)]d\omega d\tau\,, \tag{8b}$$

where (6a) was used. Since the interior integral of (8b) was determined in (6b) we obtain the inverse Hilbert transform

$$s(t) = \frac{1}{\pi} \int_{-\infty}^{\infty} \frac{\hat{s}(\tau)}{(\tau - t)} d\tau\,. \tag{9}$$

Equations (7) and (9) show that the real signal and its Hilbert transform are a Hilbert transform pair[2].

A convenient and useful short hand notation for the kernel of the Hilbert transform is $h_{1,2} = [\pi(t_1 - t_2)]^{-1}$ from which we obtain the relations

Hilbert Transform:	$\int h_{1,2}s_2 = \hat{s}_1,$	$\int h_{1,2}\hat{s}_2 = -s_1,$	(10a,b)
Inverse Hilbert Transform:	$\int h_{1,2}\hat{s}_1 = s_2,$	$\int h_{1,2}s_1 = -\hat{s}_2.$	(10c,d)

Extending these relations to the analytic signal in (4a) we find the Hilbert transform of the analytic signal to be given by

$$\hat{s}_a = \hat{s} - js = -j(s + j\hat{s}) = -js_a\,. \tag{11}$$

2.2-B: Energy Considerations

The energy of the real signal is given by

$$E = \int_{-\infty}^{\infty} dt\, s^2(t)\,. \tag{12a}$$

Substituting the Fourier representation of the signal in (2a) gives

---

[2] Gradshteyn and Ryzhik, (3.722-6) and (3.722-8): to show that the cosine and sine functions are a Hilbert transform pair.

$$E = \frac{1}{2\pi}\int_{-\infty}^{\infty} d\omega_1 \int_{-\infty}^{\infty} d\omega_2 F(\omega_1) F^*(\omega_2)\delta(\omega_1 - \omega_2), \tag{12b}$$

with the Dirac delta function

$$\delta(\omega_1 - \omega_2) = \frac{1}{2\pi}\int_{-\infty}^{\infty} dt \exp(j(\omega_1 - \omega_2)t), \tag{12c}$$

and (12b) can be immediately evaluated to yield Parseval's theorem

$$E = \int_{-\infty}^{\infty} dt\, s^2(t) = \frac{1}{2\pi}\int_{-\infty}^{\infty} d\omega\, |F(\omega)|^2 . \tag{12d}$$

Similarly, the energy of the Hilbert transform of the real signal is given by

$$\hat{E} = \int_{-\infty}^{\infty} dt\, \hat{s}^2(t) = \frac{1}{2\pi}\int_{-\infty}^{\infty} d\omega |F(\omega)|^2 |-j\operatorname{sgn}(\omega)|^2 = \frac{1}{2\pi}\int_{-\infty}^{\infty} d\omega |F(\omega)|^2 , \tag{13}$$

which is the same as the energy of the real signal.

By similar arguments the cross energy between the real signal and its Hilbert transform are found to vanish

$$\int_{-\infty}^{\infty} dt\, s(t)\hat{s}(t) = \frac{-j}{2\pi}\int_{-\infty}^{\infty} d\omega |F(\omega)|^2 \operatorname{sgn}(\omega) = 0 , \tag{14}$$

because the integrand of the right member of (14) is odd.

The extension of these results to the analytic signal can be easily performed to obtain, respectively, the energy of the analytic signal and the energy of the Hilbert transform of the analytic signal

$$E_a = \int_{-\infty}^{\infty} dt\, s_a(t) s_a^*(t) = 2E , \qquad E_{\hat{a}} = \int_{-\infty}^{\infty} dt\, \hat{s}_a(t)\hat{s}_a^*(t) = 2E . \tag{15a,b}$$

From (12) and (13) the energies of the analytic signal and its Hilbert transform are seen to be equal. The cross energies of the analytic signal and its Hilbert transform are given by

$$\int_{-\infty}^{\infty} dt\, s_a(t)\hat{s}_a^*(t) = jE_a , \qquad \int_{-\infty}^{\infty} dt\, s_a(t)\hat{s}_a(t) = 0 . \tag{16a,b}$$

The preceding results are implicitly for finite energy signals. The case for infinite duration periodic signals and the case for infinite duration non-periodic signals is discussed in Chapter 2.

2.3-B: Analytic Signal Constuction

Real sonar signals can always be written in terms of their quadrature component representations

$$s(t) = x(t)\cos(\omega_0 t) - y(t)\sin(\omega_0 t), \qquad (17a)$$

where ω_0 is the angular carrier frequency and the quadrature components have the Fourier transforms

$$X(\omega) = \int_{-\infty}^{\infty} dt\, x(t)\exp(j\omega t), \qquad Y(\omega) = \int_{-\infty}^{\infty} dt\, y(t)\exp(j\omega t). \qquad (17b,c)$$

The Fourier transform of the real signal in (17a) can be expressed in terms of the quadrature component transforms as

$$F(\omega) = \frac{[X(\omega - \omega_0) + jY(\omega - \omega_0)]}{2} + \frac{[X(\omega + \omega_0) - jY(\omega + \omega_0)]}{2}. \qquad (17d)$$

The first term of the right member represents the concentration of energy around the frequency ω_0 and the second term represents the concentration of energy around the frequency $-\omega_0$. In general, the quadrature components are not strictly band limited and all terms are supported by the entire frequency domain.

To construct the corresponding analytic signal, it is always correct to Fourier transform the expression in (17d) over the positive band of frequencies but, for non-band limited signals, this can lead to complicated expressions. However, if the signal is band limited, as shown in Figure 2, simple results are obtained. As indicated in Figure 2 we constrain the quadrature component energy around ω_0 to the positive frequency band: $-\pi W < \omega < \pi W$ where $\pi W < \omega_0$.

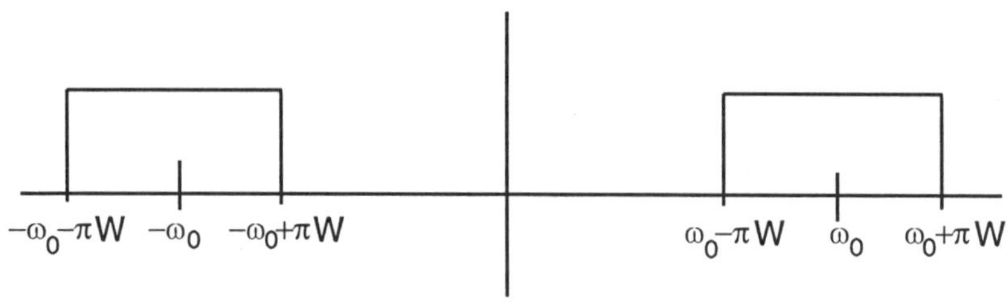

Figure 2 Real Signal Spectral Domain

The analytic signal, from (4b), is just twice the inverse Fourier transform of the real signal transform on the positive domain of frequencies and the support for the spectrum of the analytic signal is as indicated in Figure 3

$$s_a(t) = \frac{1}{2\pi}\int_{\omega_0-\pi W}^{\omega_0+\pi W} d\omega\left[X(\omega-\omega_0)+jY(\omega-\omega_0)\right]\exp(j\omega t),$$
$$= \left[x(t)+jy(t)\right]\exp(j\omega_0 t) = m(t)\exp(j\omega_0 t). \tag{18a}$$

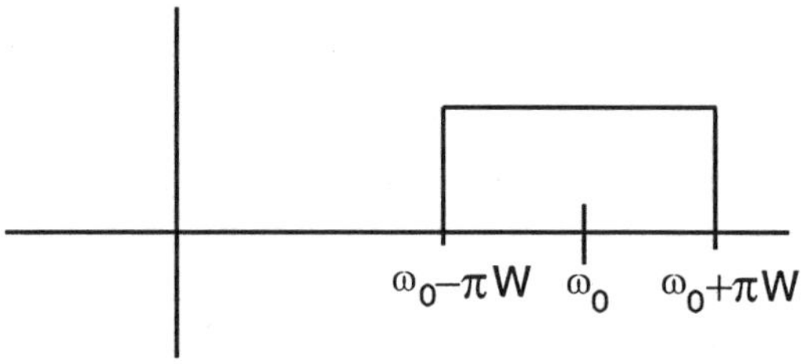

Figure 3 Analytic Signal Spectral Domain

The factor m(t) is the complex modulation on the carrier, where the carrier is represented by $\exp(j\omega_0 t)$. It's clear that the analytic signal representation can be immediately written from an inspection of the signal's quadrature component representation without the tedium of Fourier transformation. The complex modulation, in general, produces amplitude and phase variations on the carrier as indicated in (2.8) and (2.14).

The real signal and its Hilbert transform are, from (4c,d),

$$s(t) = \mathrm{Re}\, m(t)\exp(j\omega_0 t) = x(t)\cos(\omega_0 t) - y(t)\sin(\omega_0 t), \tag{18b}$$
$$\hat{s}(t) = \mathrm{Im}\, m(t)\exp(j\omega_0 t) = x(t)\sin(\omega_0 t) + y(t)\cos(\omega_0 t). \tag{18c}$$

Since the complex modulation and its quadrature components can be arbitrarily defined, within the constraints of the bandwidth limitation, a rich set of waveforms can be defined by the relations in (17) and (18).

The energy of the band limited analytic signal is given by

$$E_a = \int_0^T dt\left|m(t)\right|^2, \tag{19a}$$

and the real signal and its Hilbert transform each carry half of the energy

$$E = \frac{1}{2}\int_0^T dt\left|m(t)\right|^2, \quad \hat{E} = \frac{1}{2}\int_0^T dt\left|m(t)\right|^2. \tag{19b,c}$$

Example 1: For the real sinusoidal waveform discussed in Example 2.2

$$p(t) = A\sin(\omega_0 t), \qquad 0 \le t \le T \tag{20a}$$

with Fourier transform

$$V(\omega) = \frac{AT}{2}\left\{-j\exp\left[+j(\omega_0-\omega)T/2\right]\operatorname{sinc}\left[(\omega_0-\omega)T/2\right]\right.$$
$$\left. + j\exp\left[-j(\omega_0+\omega)T/2\right]\operatorname{sinc}\left[(\omega_0+\omega)T/2\right]\right\}. \tag{20b}$$

The first term of the Fourier transform represents the spectral energy centered around $+\omega_0$ and the second term represents the spectral energy centered around $-\omega_0$. For both terms the "tails" are down by about 50 dB at $\omega = 0$ and, to the degree that the energy in the tails can be ignored, the analytic signal is twice the inverse transform of the first term

$$s_a(t) = \frac{AT}{2\pi}\int_0^\infty d\omega\, \exp(j\omega t)\left\{-j\exp\left[j(\omega_0-\omega)T/2\right]\operatorname{sinc}\left[(\omega_0-\omega)T/2\right]\right\}. \tag{20c}$$

Straight forward manipulations yield the expression for the analytic signal

$$s_a(t) = -jA\exp(j\omega_0 t), \tag{20d}$$

which could have been written directly from the definition of the real signal. Moreover, the real part of the analytic signal is seen to be the real signal we started with as, of course, it must be. ♦

Example 2: The intensity of the analytic signal in Example 1 is, from (2.11),

$$I_a = \frac{|s_a(t)|^2}{\rho c} = \frac{A^2}{\rho c}. \tag{21a}$$

Assuming the 12 dB re μPa intensity level for the real signal, as in Example 2.3, the amplitude is found to be 5.6×10^{-5} dyne/cm^2 and the intensity is 21.2×10^{-22} watt/cm^2. Note that the intensity of the analytic signal is twice the intensity of the real signal. This is true even when the intensity of the real signal is expressed in terms of the the mean squared amplitude as in (2.11d) and (2.13d) which have a form similar to the analytic signal intensity expression. The energy of the analytic signal is the time integral over the intensity which, for the constant intensity in this example, yields

$$E_a = \frac{A^2 T}{\rho c}. \tag{21b}$$

For the 10 second duration signal in Examples 2.1, 2.2 and 2.3, the energy flux density is 21.2×10^{-21} joule/cm^2. ♦

CHAPTER 3
TRANSDUCERS

Mankind has long known of the potential for underwater acoustic systems because with simple acoustic sources, such as bells and the human voice, and the unaided ear as receiver, rudimentary active and passive underwater acoustic systems had been devised in antiquity. Until the early 1900s the realization of effective underwater acoustic systems was confounded by the problems of coupling signals into the medium and extracting signals out of the medium. But, within the backdrop of WWI and the early electronic revolution, the development of acoustic transducers accelerated and both problems were essentially solved at the outbreak of WWII.

In general terms, acoustic transducers perform transformations between in-water acoustic signals and shipboard electronic signals as indicated in Figure 1.

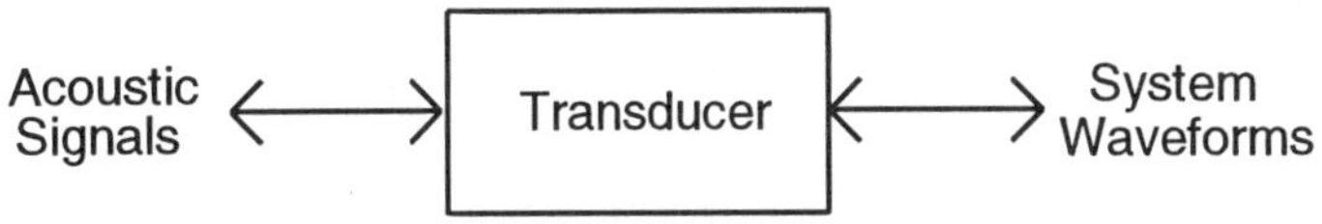

Figure 1 Transducer Functions

While transducers, in general, transform energy from one form to another, ASW sonar transducers are predominately electroacoustic transformers and have the properties of reversibility and linearity, i.e. they transform between acoustic signals and electronic signals in either direction and faithfully preserve the structure of the waveforms during their transformation. Such transducers are the focus of attention in the following. Other transducers, such as explosive charges and mechanical noise makers, are important sonar tools but are of secondary importance in ASW applications. Generally, source transducers are referred to as projectors or transmitters and receive transducers are referred to as hydrophones. In some systems both source and receive functions are performed by the same transducer while in other systems these functions are performed by separate specialized transducers. The transducer elements used in sonar systems are generally one of three types: piezoelectric, electrostrictive and magnetostrictive. While all can perform the source and receive functions, the details of their functioning differ.

Piezoelectric transducer elements are made from a crystal, such as quartz, and have the desirable transducer properties of strength and thermal stability. The atoms of the crystal are aligned in regular layers that are electrically neutral in the unstressed state. When stressed by an applied force, the atoms of the material are distorted causing a displacement of the positive and negative charges within each atom which results in regular layers of electric dipoles. The dipole charges of the adjacent internal layers cancel but, if the material is cut such that the external faces of the element are coincident with the polarized layers, these faces acquire positive and negative charges due to the dipole charges. A measure of the distortion, and hence the applied force, is obtained by measuring the electrical charge on the element faces. Similarly, if the element faces are coated with an electrical conductor and charged, the capacitive action of the faces causes the crystal to distort which, in turn, causes a force to be impressed on the surrounding medium. These concepts are indicated schematically in Figure 2a where either a varying force applied to the transducer element causes a varying current to flow in the associated circuitry or a varying current causes the element to exert a varying force on the surrounding medium.

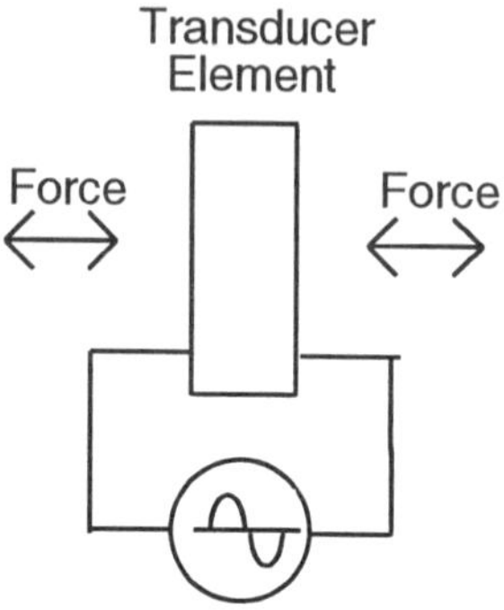

Figure 2a Transducer Functions - Piezoelectric/ Electrostrictive Elements

Electrostrictive transducer elements are made from a ferroelectric dielectric such as barium titanate or a mixture of lead zirconate and lead titanate. The elements are cast from the molten dielectric material, and therefore have the capability of being formed in a wide variety of shapes as opposed to piezoelectric transducer elements in which the shape is limited by the possibilities of the crystalline structure. During the fabrication process, microscopic clusters of atoms are spontaneously formed in which the atoms are aligned and each cluster exhibits an electric dipole. Since the cluster dipoles are randomly oriented, the raw material is electrically neutral and, in this state, the application of a force to its faces does not produce a net charge on those faces. However, the application of an electric potential to its faces causes a realignment of the clusters which results in a displacement of the faces. This displacement is independent of the polarity of the applied electric potential which means that the displacement obeys a magnitude law which is found to be a square law, e.g. if a sinusoidal potential of amplitude E is applied to the material, the displacements are proportional to $(E\sin\omega t)^2 = E^2(1-\cos 2\omega t)$ and the material acts as a nonlinear transducer with strength proportional to E^2. Hence, in contrast to crystals, the transduction capabilities of the raw ferroelectric material lacks reversibility and linearity. However, these properties can be imposed on the material if it is heated and then polarized by an intense constant electric field E_0 during an extended period of time and then cooled. The atomic clusters become somewhat aligned with the polarizing field causing a permanent distortion of the material when cooled. Such polarization is normally done after the material is shaped to the form desired for the transducer element and its faces coated with a conducting layer. As before, the capacitive action of a sinusoidal potential applied to the polarized element causes distortions in the material but, being permanently distorted, the displacements of the faces are now proportional to

$$\left(E_0 + E\sin\omega t\right)^2 = E_0^2\left(1 + 2\left(E/E_0\right)\sin\omega t + \left(E/E_0\right)^2\sin^2\omega t\right). \qquad (1)$$

If the polarizing field is very strong, i.e. $E/E_0 << 1$, the last term in (1) is negligible and the force exerted by the element on the medium is sinusoidal at frequency ω and proportional to E_0E. Conversely, the application of a sinusoidally varying force to the faces of the polarized element causes sinusoidal charge variations on the faces of the element, because of the partially aligned clusters, and the element functions similarly to the piezoelectric element as indicated in Figure 2a. Since the strength of an electrostrictive transducer depends upon the polarizing electric field, which weakens when the element is excessively heated, these transducers are not as thermally stable as piezoelectric transducers.

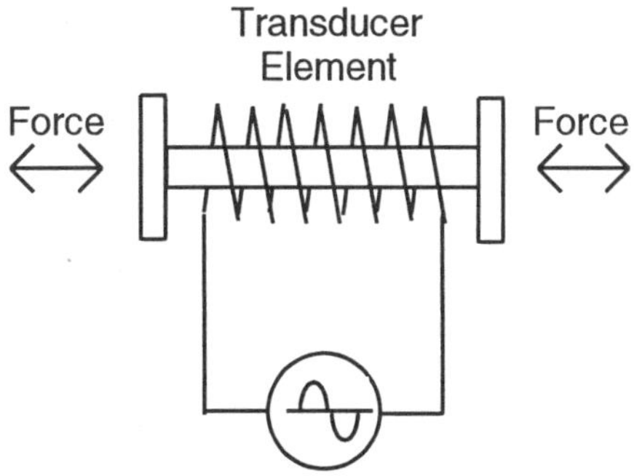

Figure 2b Transducer Functions for Magnetostrictive Transducer Element

Magnetostrictive transducer elements are made from a ferromagnetic material, usually nickel, and are analogous to electrostrictive devices. During the fabrication of the material, microscopic domains of atoms having magnetic moments spontaneously arise but, because the clusters are randomly oriented, the raw material does not exhibit a net magnetic field and the application of a force to its faces does not induce any net magnetic activity. If a magnetic field is applied to the raw material, it distorts as the clusters are realigned towards the direction of the applied field. As for the electrostrictive devices, the distortions of the raw material are independent of the polarity of the applied magnetic field and, for an applied sinusoidal field, the element responds at twice the frequency of the imposed field. Therefore the transduction capabilities of the raw ferromagnetic material, as for the electrostrictive material, lacks reversibility and linearity. However, these properties can be imposed on the element, using a methodology similar to that for electrostrictive materials, by applying an intense constant polarizing magnetic field to the material. The typical resulting magnetostrictive transducer element is indicated in Figure 2b in which the variable driving magnetic field is applied by electrical windings on a bar of magnetostrictive material. The polarizing field is generally provided either by additional windings that carry a constant current or by making the bar a permanent magnet.

3.1 Electroacoustic Transducer Analysis

The general functioning[1] of a transducer and the coupling between the electrical and mechanical sides can be represented by a four terminal network having two electrical and two mechanical terminals as indicated in Figure 3. Both sides of the transducer interact during operation such that when either side of a transducer is activated, the other side is also activated. Consequently, the transducer action must be analyzed by analyzing both sides jointly. This is done by forming a linear relationship between the response of the pure electrical side, characterized by the electrical impedance Z_e, and the response of the pure mechanical side of the transducer, characterized by the mechanical impedance Z_m.

The action of the pure electrical side of the transducer is determined when the transducer is clamped causing the velocity of the face to be zero, $v = 0$. Under

[1] Hunt, F. V., *Electroacoustics*, Harvard University Press, 1954, Chapter 2.

this condition the piezoelectric and electrostrictive transducers become capacitors and the magnetostrictive transducer becomes an inductor. The electrical impedance Z_e, referred to as the clamped electrical impedance, gives the current flow due to the impressed voltage. The action of the mechanical side of piezoelectric and electrostrictive transducers is determined when they are short circuited, $E = 0$. Under this condition no power flows in the electrical side and the transducer becomes a pure mechanical device for which the mechanical impedance gives the face velocity due to the impressed force. The mechanical impedance Z_m for this case is referred to as the short circuit mechanical impedance. Similarly, when a magnetostrictive transducer is open circuited, $I = 0$, it becomes a pure mechanical device, because no power flows in the electrical side, and, again, the mechanical impedance Z_m gives the face velocity due to the impressed force. The mechanical impedance for this case is referred to as the open circuit mechanical impedance. Prior to analyzing the total transducer dynamics, it is convenient to first consider the clamped electrical response of the transducer and the pure mechanical response of the transducer[2]. It is found that there are similarities and differences between piezoelectric and electrostrictive transducers, on the one hand, and magnetostrictive transducers, on the other hand. In the following sections these similarities and differences are emphasized.

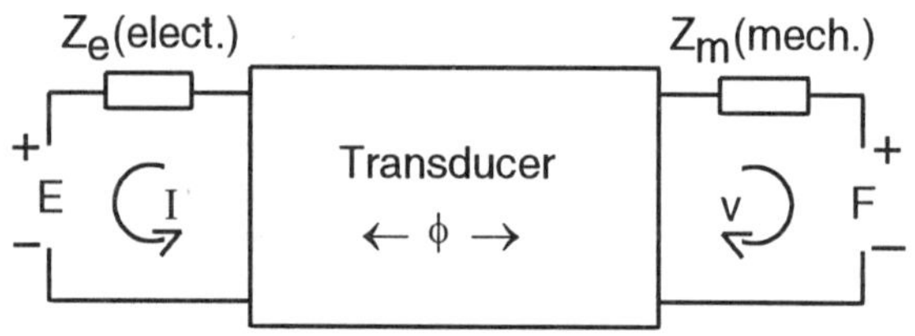

Figure 3 Schematic of a Sonar Transducer

<u>Clamped Transducers</u>

The clamped response of the piezoelectric and electrostrictive transducers is electrically equivalent to a capacitor and resistor in parallel as indicated in Figure 4a. The capacitance is formed by the charge on the conducting faces of the transducer and the resistance is due to charge leakage between the faces of the transducer. The clamped response of the magnetostrictive transducer is electrically equivalent to an inductor and resistor in series as indicated in Figure 4b. The inductance is due to the windings on the magnetostrictive bar and the resistance is due to the resistance of the windings. For the piezoelectric and electrostrictive transducers the current equation is found, from inspection of Figure 4a, to be

$$\frac{E}{Z_e} = \frac{E}{R_e} + C_e\dot{E}, \tag{2a}$$

---

[2] Tucker, D. G. and B. K. Gazey, *Applied Underwater Acoustics*, Pergamon Press, Oxford, 1966, Chapter 5.

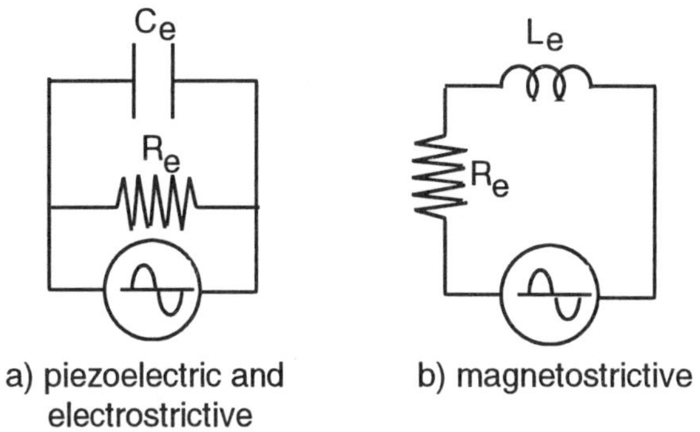

Figure 4 Equivalent Electric Circuits for Clamped Transducer Elements

and the steady state solution is easily verified to be

$$E = I_0 \exp(j\omega t) Z_e, \quad Z_e^{-1} = R_e^{-1} + j\omega C_e, \tag{2b}$$

where I_0 is the magnitude of the current. Similarly, for the magnetostrictive transducer the voltage equation is found, from inspection of Figure 4b, to be

$$IZ_e = IR_e + L_e \dot{I}, \tag{2c}$$

and the steady state solution is easily verified to be

$$I = \frac{E_0 \exp(j\omega t)}{Z_e}, \quad Z_e = R_e + j\omega L_e, \tag{2d}$$

where E_0 is the magnitude of the voltage. For both cases the electrical impedance is related to the properties of the electrical components and can be determined either analytically or by electrical measurements on the clamped transducer.

Short Circuit and Open Circuit Transducers

The action of the pure mechanical side of the transducer is equivalent to a vibrating mass with damping as indicated in Figure 5a. The mass M corresponds to the mass of the vibrating transducer element, the compliance C_m is a measure of the ease with which the transducer material is strained, i.e. the compliance is the reciprocal of the resistance to strain, and the mechanical resistance R_m is a measure of the resistance to the driving force that is due to dissipative losses. The resistive losses include viscous heating within the transducer element and its mountings and also the acoustic energy radiated by the transducer.

The equation of motion for the oscillator under a sinusoidal driving force is found, from inspection of Figure 5a, to be

$$M\dot{v} + R_m v + \frac{1}{C_m}\int v dt = F\exp(j\omega t), \tag{3a}$$

where $v = \dot{x}$. The steady state solution is readily verified to be

$$v = \frac{F\exp(j\omega t)}{Z_m}, \quad Z_m = R_m + j\left(\omega M - \frac{1}{\omega C_m}\right), \tag{3b}$$

where Z_m is the mechanical impedance. The electrical circuit equivalent to the vibrating mass is indicated in Figure 5b in which it is seen that mass corresponds to electrical inductance, i.e. it exerts a reaction to applied force, mechanical resistance corresponds to electrical resistance, i.e. both dissipate energy, and compliance corresponds to capacitance, i.e. it stores potential energy. The components of the mechanical impedance are related to the physical properties of the transducer element but it is difficult to specify these relationships exactly using analytical methods. An easier approach is to determine them using empirical methods as outlined below.

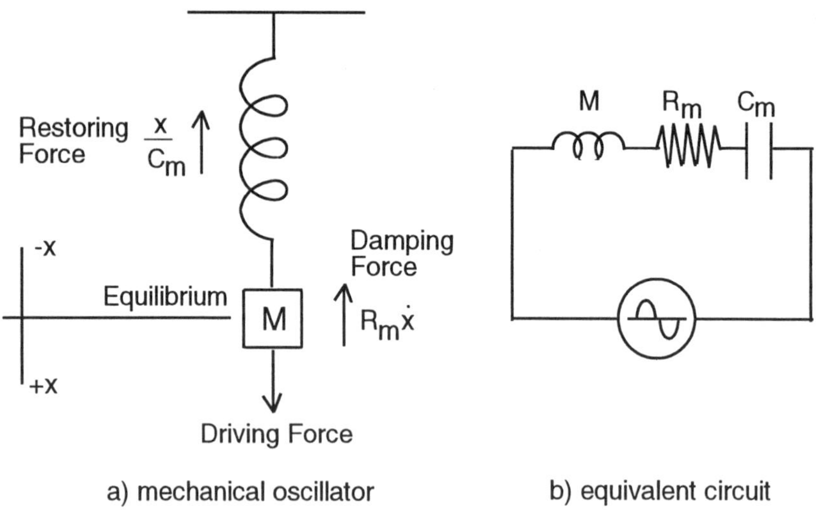

Figure 5 Mechanical Side of a Transducer Element

Operational Transducers

The dynamics of the full functioning operational transducer are obtained from linear relations between the clamped electrical response of the transducer and the pure mechanical response[3]. These relations assert that the total response of the transducer is provided by the superposition of the separate electrical and mechanical responses and are given by:

---

[3] Hunt, F. V., *Electroacoustics*, Harvard University Press, 1954, Chapter 2.

a) Piezoelectric and Electrostrictive Transducers;

$$\begin{bmatrix} I \\ F \end{bmatrix} = \begin{bmatrix} Y_e & \phi \\ -\phi & Z_m \end{bmatrix} \begin{bmatrix} E \\ v \end{bmatrix}. \tag{4a}$$

For the short circuit condition, E = 0, (4a) gives the rate of charge build up on the transducer faces due to the face velocity, i.e. $I = \phi v$, (and it also gives the pure mechanical response). For the clamped condition, v = 0, (4a) gives the force exerted by the clamped faces, i.e. $F = -\phi E$, (and it also gives the pure electrical response). Since for the short circuit condition the charge buildup and face velocity are in phase, and also for the clamped condition the force and electrical potential are in phase, the coupling coefficient ϕ for piezoelectric and electrostrictive transducers is real.

b) Magnetostrictive transducers;

$$\begin{bmatrix} E \\ F \end{bmatrix} = \begin{bmatrix} Z_e & \phi \\ -\phi & Z_m \end{bmatrix} \begin{bmatrix} I \\ v \end{bmatrix}. \tag{4b}$$

For the open circuit condition, I = 0, (4b) gives the electrical potential across the transducer windings due to the face velocity, $E = \phi v$, (and it also gives the pure mechanical response). For the clamped response, v = 0, (4b) gives the force exerted by the clamped faces, $F = -\phi I$, (and it also gives the pure electrical response). Because of hysteresis effects, for the open circuit condition, the potential across the transducer windings lags the face velocity slightly, and also, for the clamped condition, the force lags the electrical current slightly, and the coupling coefficient ϕ for magnetostrictive transducers is complex with a negative phase angle in contrast to piezoelectric and electrostrictive transducers.

The asymmetry in the signs of the coupling coefficients in both (4a) and (4b) is required for physical reasons. For example, under the sign convention in (4a) for the piezoelectric and electrostrictive transducers, the driving force under the open circuit condition, I = 0, is $F = (\phi^2 Z_e + Z_m)v$ which, as expected, shows that the driving force is increased, by the amount $\phi^2 Z_e$, due to the additional load of the capacitor leakage. Similarly, for the magnetostrictive transducer, the driving force under the short circuit condition, E = 0, is $F = (\phi^2/Z_e + Z_m)v$ which, as expected, shows that the driving force is increased by the amount ϕ^2/Z_e due to the windings current, i.e. the coupling coefficient phase angle in this relation is not found to be large enough to cause a sign reversal. Different sign conventions in (4a) and (4b) would lead to results that would not be consistent with physics. Also, for notational convenience, the same symbol for the electromechanical coupling coefficient was used in both (4a) and (4b). In fact, these coupling coefficients are different and unrelated. The units of the coupling coefficient for piezoelectric and electrostrictive transducers is electrical charge per unit length, whereas for magnetostrictive transducers it is impulse per unit electrical charge.

3.2 Projector Analysis

For projectors the force on the transducer faces is due to the medium opposing the motion of the transducer face and is related to the face velocity through the radiation impedance

$$F = -R_r v \,. \tag{5}$$

The radiation impedance is real and the minus sign in (5) asserts that the force of the medium on the transducer face always opposes the face velocity, i.e. they are 180° out of phase. Substituting into the general transducer equations for the piezoelectric and electrostrictive transducers in (4a) and for the magnetostrictive transducers in (4b), those matrix equations can each be reduced to a single equation:

a) Piezoelectric and Electrostrictive;

The electrical admittance at the in situ transducer input terminals is found from (4a) to be

$$\frac{I}{E} = Y_T = Y_e + \frac{\phi^2}{R_r + Z_m}, \tag{6a}$$

where Y_T is the total admittance. The leading term in the last member is just the admittance of the electrical circuit and would be the only contributing term if the electromagnetic coupling coefficient vanished, $\phi = 0$. For the general case in which the coupling coefficient does not vanish the remaining term is referred to as the motional admittance Y_{mot} and represents the additional admittance due to the vibrating transducer in the presence of the medium and includes the admittance due to radiation losses and internal damping losses. The motional admittance is the mechanical admittance multiplied by the squared electromechanical coupling coefficient ϕ^2 which results in the motional admittance having the electrical unit of mhos.

b) Magnetostrictive;

The electrical impedance at the in situ transducer input terminals is found from (4b) to be

$$\frac{E}{I} = Z_T = Z_e + \frac{\phi^2}{R_r + Z_m}, \tag{6b}$$

where Z_T is the total impedance. The leading term in the last member is just the impedance of the electrical circuit and would be the only contributing term if the electromagnetic coupling coefficient vanished, $\phi = 0$. For the general case in which the coupling coefficient does not vanish the remaining term is referred to as the motional impedance Z_{mot} and represents the additional impedance due to the vibrating transducer in the presence of the medium and includes the impedance due to radiation losses and internal damping losses. The motional impedance is the mechanical admittance multiplied by the squared electromechanical coupling coefficient ϕ^2 which results in the motional impedance having the electrical unit of ohms.

Motional Admittance/Impedance

Much insight into the operations of projectors can be determined by examining the projector motional admittance/impedance[4], i.e. the second terms in (6). Since the mechanically clamped electrical admittance/impedance is known, i.e. the first terms in (6), the motional admittance/impedance can be obtained by determining the electrical admittance/impedance at the projector input terminals. Measurements of the voltage and current over a range of driving frequencies at the input terminals give the electrical admittance/impedance and, subtracting out the clamped electrical admittance/impedance, gives the motional admittance/ impedance. It is found that, as the frequency is varied from low values to high values, the motional admittance/impedance traces out a circular locus of points in the complex plane.

To illustrate these concepts, consider first just the locus of points for the mechanical admittance with radiation loading that is common to both (6a) and (6b)

$$Y_{r,m} = \frac{1}{R_r + R_m + j\left(\omega M - \frac{1}{\omega C_m}\right)}, \tag{7a}$$

where $Y_{r,m}$ denotes the mechanical admittance with radiation loading. For frequency values that are vanishingly small or unbounded, the mechanical admittance vanishes but for intermediate frequency values we find

$$Y_{r,m} = |Y_{r,m}|\exp(j\theta), \tag{7b}$$

$$\cos\theta = \mathrm{Real}\left(Y_{r,m} / |Y_{r,m}|\right)$$
$$= |Y_{r,m}|(R_r + R_m). \tag{7c}$$

Consequently, the locus of points circumscribed by the mechanical admittance with radiation loading, expressed in cylindrical polar coordinates, is given by

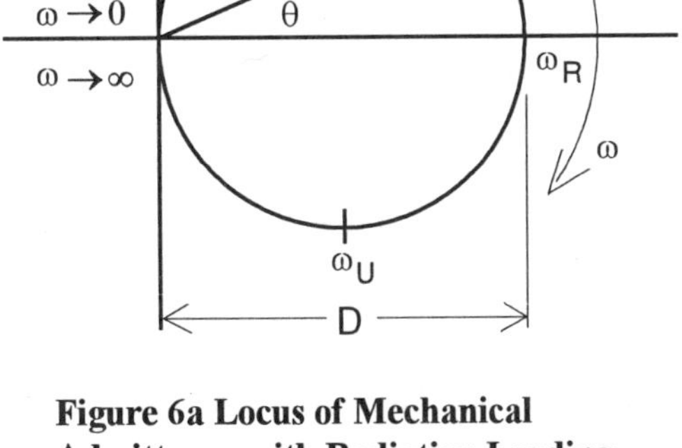

Figure 6a Locus of Mechanical Admittance with Radiation Loading

[4] Hunt, F. V., *Electroacoustics*, Harvard University Press, 1954, Chapter 4.

$$\left|Y_{r,m}\right| = \left(\frac{1}{R_r + R_m}\right)\cos\theta\,. \tag{7d}$$

Since the resistive impedances are real and constant, (7d) is the equation of a circle as indicated in Figure 6a. The locus starts and ends at the origin of coordinates and traces out a clockwise circular path as indicated.

The frequency that maximizes the magnitude of the mechanical admittance is the resonance frequency

$$\omega_R = \sqrt{\frac{1}{MC_m}}\,. \tag{8a}$$

At this frequency the mechanical admittance is real and, since $F = Y_{r,m}v$, the driving force on the projector face and the face velocity are in phase. This results in maximum face excursions and maximum energy coupled into the medium. As can be seen in Figure 6a, the mechanical admittance at resonance is a maximum and is equal to the diameter of the mechanical admittance locus, i.e. where D is the locus diameter, we have $|Y_{r,m}|_{max} = (R_r + R_m)^{-1} = D$.

The resonance frequency in (8a) relates both the effective mass and mechanical compliance of the projector and, with the help of the added mass method[5], explicit expressions can be developed for both quantities. An increment of mass m is added to the projector face which changes the resonant frequency to

$$\omega_{R,m} = \sqrt{\frac{1}{(M+m)C_m}}\,. \tag{8b}$$

Combining with (8a) gives the expressions for the effective mass and mechanical compliance of the projector

$$M = \left(\frac{\omega_{R,m}^2}{\omega_R^2 - \omega_{R,m}^2}\right)m\,, \qquad C_m = \frac{\omega_R^2 - \omega_{R,m}^2}{\omega_R^2\,\omega_{R,m}^2}\,. \tag{8c}$$

Since both resonance frequencies can be experimentally determined, these relations provide a means of estimating effective mass and compliance of the projector - quantities that can be calculated analytically only with great difficulty.

While the resonance frequency of a projector defines a "best" frequency for operations, it is sometimes desirable to operate at a different frequency or to operate over a band of frequencies. Consequently, it is important to know the frequency band over which the projector can transmit significant energy. The bandwidth of the transducer spans the frequency range between the half power points of the projector and this range is defined by the lower and upper frequencies ω_L and ω_U as indicated

---

[5] Hunt, F. V., *Electroacoustics*, Harvard University Press, 1954, pg. 119.

in Figure 6a. At these frequencies the real part of the mechanical admittance is equal to the imaginary part and we can write

$$R_r + R_m = -\left(\omega_L M - \frac{1}{\omega_L C_m}\right), \tag{9a}$$

$$R_r + R_m = +\left(\omega_U M - \frac{1}{\omega_U C_m}\right). \tag{9b}$$

Combining, we obtain the bandwidth expression

$$\omega_U - \omega_L = \frac{R_r + R_m}{M}. \tag{9c}$$

Since the effective mass of the projector can be estimated from (8c), (9c) provides a means of estimating the total resistive loading on the projector. By performing measurements on the projector in both water and air, i.e. when the radiation loading is effectively zero, $R_r = 0$, the radiation and internal resistive loadings can be separately determined from (9c).

The mechanical Q-factor (quality factor) of the projector provides a measure of the operational frequency range of the projector. The Q-factor is defined as the ratio of the mechanical resonance frequency to the projector bandwidth.

$$Q = \frac{\omega_R}{\omega_U - \omega_L} = \frac{\omega_R M}{R_r + R_m}. \tag{9d}$$

While narrow band projectors have Q-factors that are of the order of magnitude of 100 or more, possibly much more, broad band projectors are found to have Q-factors with magnitudes of only 10 or more. Of course, the Q-factor of a projector in air is substantially increased since $R_r = 0$ which means that the water loading broadens the bandwidth of the projector.

The motional admittance and impedance terms in (6a) and (6b) are obtained by multiplying the mechanical admittance by the squared electromechanical coupling coefficient which introduces a constant scaling factor and a phase factor.

$$\phi^2 = |\phi|^2 \exp(-j2\beta). \tag{10}$$

The scaling factor converts the mechanical admittance into electrical admittance/impedance and the phase factor rotates the locus of points in Figure 6a through the angle -2β to obtain the locus of points in Figure 6b. Of course, for piezoelectric and electrostrictive transducers, the phase is zero and only scaling of the mechanical admittance occurs. However, for magnetostrictive transducers the electromechanical coupling coefficient is complex and the mechanical impedance is both rescaled and rotated. The angle of rotation was seen to be negative because the magnetic hysteresis causes the response to lag. Plots of loci for two cases are indicated in Figure 6b, i.e. a projector in water; and a projector in air. The smaller circle, having diameter D_W, represents the case of the projector in water and the larger circle, having diameter D_A, represents the case of the projector in air. Since at mechanical resonance $Z_{mot} = |Y_{r,m}|_{max}\phi^2 = D\exp(-j2\beta)$, the diameter of each circular motional admittance or impedance locus, in Figure 6b, can be seen to correspond to the resonance cases

$$D_W = \frac{|\phi|^2}{R_r + R_m}, \tag{11a}$$

$$D_A = \frac{|\phi|^2}{R_m}. \tag{11b}$$

Since the resistive loading terms can be estimated, as discussed above following (9c), (11) provides a means for experimentally determining the magnitude of the electromechanical coupling coefficient. The phase of the coupling coefficient can be determined directly from the angle of inclination of the admittance/impedance locus that is indicated in Figure 6b.

Consequently, under the assumption that the clamped electrical admittance/impedance of the transducer is known, estimates of the motional admittance/impedance diagram can be determined experimentally and, by the preceding comments, some of the more important projector parameters can be determined directly from the motional admittance/impedance diagrams. These parameters can be used to determine many properties of projector operations. While a more complete discussion of transducer properties is beyond our intended scope, we show how these parameters can be used to determine projector efficiency. More extensive analysis would lead to a more complete description of projector operations[6].

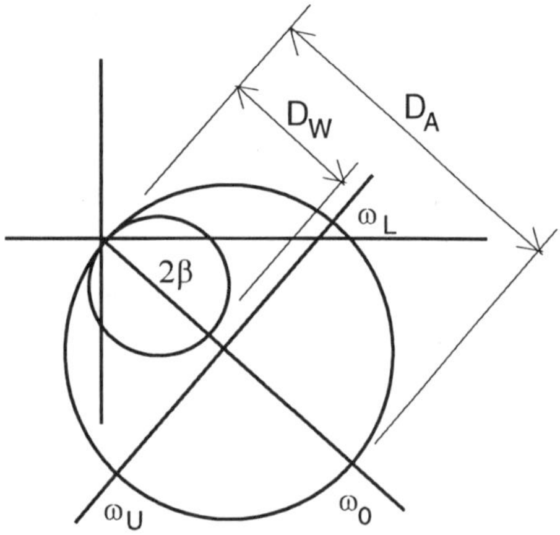

Figure 6b Motional Admittance/ Impedance for Transducers in Water and in Air

Projector Efficiency

Maximizing detection performance requires that targets be ensonified by maximum acoustic energy which means that, when power is a valuable commodity at sea, it is important to maximize the efficiency of the projector. The overall efficiency of the projector is defined to be the fraction of the total input electrical power that is radiated into the medium as acoustic power. It might be thought that

[6] Hunt, F. V., *Electroacoustics*, Harvard University Press, 1954.

maximum efficiency would be obtained at the mechanical resonance frequency since, as we found in the preceding, maximum energy is coupled into the medium at the mechanical resonance frequency. This is nearly correct but, because of the electrical side of the projector, increased input electrical power is required to drive the projector at mechanical resonance. Consequently, maximum efficiency is obtained by operating at a frequency slightly different (higher[7]) than the frequency of mechanical resonance. At the optimum driving frequency the radiated power is slightly reduced but the required electrical input power is reduced even more which has the net effect of raising the efficiency. In spite of these considerations it is generally sufficient to determine projector efficiency at the frequency of mechanical resonance because: it is simpler to compute the efficiency at mechanical resonance; the difference between the two estimates is small; and the efficiency measure of interest is an estimate for the entire frequency band of the projector. Of course, when power is plentiful operating at mechanical resonance is an optimal stategy.

a) Piezoelectric and Electrostrictive Projector Efficiency:

The radiated power is given by the product of the force on the projector face and the face velocity which defines the numerator of the efficiency expression. The total input power is the power converted to heat in the total resistive loading, i.e. the real part of the total impedance R_T = Real Z_T in (6a). For piezoelectric and electrostrictive projectors it is convenient to write the efficiency expression in terms of the electrical voltage power relation

$$\eta = \frac{R_r|v|^2}{R_T^{-1}|E|^2} = \frac{R_r}{R_T^{-1}}\left[\frac{|\phi|^2}{(R_r + Z_m)^2}\right] = \frac{R_r}{R_T^{-1}}\left[\frac{D_W}{(R_r + R_m)}\right], \tag{12a}$$

where (4a) and (5) were used to obtain the second equality; and (11a), with the assumption of mechanical resonance, was used to obtain the third equality. The operating frequency at maximum efficiency is found by minimizing the denominator of the second equality in (12a) but, for the efficiency at mechanical resonance, we omit this step and go directly to the third equality. The total resistive load at mechanical resonance is

$$R_T^{-1} = R_e^{-1} + \frac{\phi^2}{R_r + R_m} = R_e^{-1} + D_W, \tag{12b}$$

using (2b), (6a) and (11a). Assembling results, the expression for the overall efficiency of piezoelectric and electrostrictive transducers at mechanical resonance becomes

$$\eta = \frac{D_W(D_A - D_W)}{(R_e^{-1} + D_W)D_A}. \tag{12c}$$

---

[7] Hunt, F. V., *Electroacoustics*, Harvard University Press, 1954, pg. 126.

The power factor

$$\frac{R_r}{R_r + R_m} = \frac{D_A - D_W}{D_A}, \tag{12d}$$

was obtained from (11a) and (11b). The projector efficiency expression in (12c) is especially useful since it can be determined from measurements at the electrical input terminals to the transducer.

b) Magnetostrictive Projector Efficiency

The radiated power is given by the product of the force on the projector face and the face velocity which defines the numerator of the efficiency expression. The total input power is the power converted to heat in the total resistive loading, i.e. the real part of the total impedance R_T = Real Z_T in (6b). For magnetostrictive projectors it is convenient to write the efficiency expression in terms of the electrical current power relation

$$\eta = \frac{R_r|v|^2}{R_T|I|^2} = \frac{R_r}{R_T}\left[\frac{|\phi|^2}{(R_r + Z_m)^2}\right] = \frac{R_r}{R_T}\left[\frac{D_W}{(R_r + R_m)}\right], \tag{13a}$$

where (4b) and (5) were used to obtain the second equality; and (11a), with the assumption of mechanical resonance, was used to obtain the third equality. The operating frequency at maximum efficiency is found by minimizing the denominator of the second equality in (13a) but, for the efficiency at mechanical resonance, we omit this step and go directly to the third equality. The total resistive load at mechanical resonance is

$$R_T = R_e + \frac{|\phi|^2 \cos 2\beta}{R_r + R_m} = R_e + D_W \cos 2\beta, \tag{13b}$$

using (2d), (6b) and (11a). Assembling results, the expression for the overall efficiency of the magnetostrictive transducer at mechanical resonance becomes

$$\eta = \frac{D_W(D_A - D_W)}{(R_e + D_W \cos 2\beta) D_A}, \tag{13c}$$

where the power factor in (12d) was used. The projector efficiency expression in (13c) is especially useful since it can be determined from measurements at the electrical input terminals to the transducer.

Example 1: If we consider somewhat idealized piezoelectric and electrostrictive projectors for which the capacitive leakage is negligible, i.e. $(R_e)^{-1} \cong 0$, and magnetostrictive projectors for which the windings resistance is negligible, i.e. $R_e \cong 0$, both projector efficiency expressions reduce to the form $\eta = (D_A - D_W)/D_A \cos 2\beta$ where β is zero for the piezoelectric and electrostrictive projectors.

Consequently, for piezoelectric and electrostrictive projectors having an in-water motional admittance locus diameter of, say, 1/3 of the in-air locus diameter, the efficiency would be 67%. For magnetostrictive projectors having similar sized motional impedance locus diameters the efficiency would be increased by $\cos^{-1}2\beta$, i.e. the magnetic hysteresis has the effect of increasing efficiency.♦

3.3 Hydrophone Analysis

Hydrophones are electroacoustic transducers that transform acoustic pressure into an electrical signal that faithfully replicates the waveform of the acoustic field. Arrays of such hydrophones serve as the "ears" of ASW sonar systems and among these arrays might be included the acoustic projector. Since electroacoustic transducers can transduct in both directions, some active sonar system projectors serve as both projector and receiver. In such systems, the projector receptions are generally supplemented by other hydrophone receivers. For example, many submarine platforms have a spherical bow array that is employed as both projector and receiver and, in addition, they also have a towed array of passive only receive hydrophones, a conformal hull mounted array of receive hydrophones and other hydrophone receivers. In this way the reception capabilities of the platform can be considerably enhanced. Of course, in passive only systems all acoustic interceptions are acquired by arrays of hydrophone receivers.

To ensure that hydrophone elements sense the acoustic field without distorting it, they are generally designed to be omnidirectional with a flat frequency response. As such, hydrophones are generally small, compared to the wave lengths of the field being sensed to ensure that the response of the hydrophone is independent of its orientation in the field. They are also operated well below their mechanical resonance frequency, by low pass filtering of the electrical output, to ensure that the response of the hydrophone is independent of frequency. Such hydrophone receivers can be fabricated from piezoelectric, electrostrictive or magnetostrictive elements but, since piezoelectric and electrostrictive hydrophones are simple and can be readily configured to track acoustic pressure variations without phase shifting, we focus on those hydrophones.

Piezoelectric and electrostrictive hydrophones are taken to be configured as open circuit devices in which the voltage output represents the acoustic pressure field. The voltage output, as obtained from (4a), is given by

$$E = -\frac{F}{\phi}\left(\frac{Z_e}{Z_e + Z_{mot}}\right), \tag{14a}$$

where the force on the hydrophone is due to the acoustic pressure, i.e. $F = -pA$ where A is the area of the hydrophone face. The negative sign is required by our sign convention for transducer dynamics in (3). At frequencies well below mechanical resonance, the motional impedance is very nearly a pure capacitor, as can be discerned from (3b) and (6a), i.e. $Z_{mot} \cong -j/\omega C_{mot}$, where $C_{mot} = \phi^2 C_m$. Moreover, for a well designed piezoelectric or electrostrictive hydrophone, the leakage current is negligible and, as found from (2b), the clamped electrical impedance can also be approximated as a pure capacitor, i.e. $Z_e \cong -j/\omega C_e$. Substituting into (14a) we obtain

$$E = \frac{pA}{\phi}\left(\frac{C_{mot}/C_e}{C_{mot}/C_e + 1}\right). \tag{14b}$$

Equation (14b) gives the output voltage of the hydrophone due to the acoustic pressure on the hydrophone face. We note that the constant of proportionality is real and, therefore, the voltage output tracks the acoustic pressure input in phase and without delay which is a desirable property for precision sensing.

3.4 Transmitting and Receiving Response

Modern ASW sonar operations are based upon the sonar equations which require that the in-water transmitted and received signal levels be known in absolute terms. To readily determine these levels it is necessary to have simple conversion relations between the in-water acoustic levels and the system voltage and/or current levels. For both projectors and hydrophones the input-output response relationship is a proportionality and the constant of proportionality, referred to as the response factor, defines the conversion.

The transmitting response of a projector is the output acoustic pressure due to unit current input or unit voltage input to the projector. The acoustic pressure is taken to be the pressure on the projector beam axis at a reference range from the acoustic center of the projector which is usually taken to be one meter. The output acoustic pressure is proportional to the projector face velocity, a result that is demonstrated in Section 4.1, and the face velocity is proportional to the input voltage or current. For piezoelectric and electrostrictive projectors the relation between face velocity and input current and voltage is found from (4a) and (5) to be

$$v = \frac{I}{\phi}\left(\frac{Z_{mot}}{Y_e + Z_{mot}}\right) = \frac{E}{\phi} Z_{mot}. \tag{15a}$$

For magnetostrictive projectors the relation between face velocity and input voltage and current is found from (4b) and (5) to be

$$v = \frac{E}{\phi}\left(\frac{Z_{mot}}{Z_e + Z_{mot}}\right) = \frac{I}{\phi} Z_{mot}. \tag{15b}$$

Consequently, the transmitter response can be expressed in terms of either the electric current or voltage that is input to the projector. In the former case the output pressure is called the transmitting-current response and in the latter case it is called the transmitting-voltage response. The transmitting-current response is the more commonly cited response and has the form $p = AI$ where A is the constant of proportionality which is the transmitting response factor.

The transmitting response is usually expressed in decibels relative to the response factor of a standard projector. The standard projector response is defined to be an rms output of 1 μPa for an rms input of 1 amp. The ratio of the constants of proportionality for both projectors in decibels is

$$\alpha = 10\mathrm{Log}\left|\frac{A}{A_{std}}\right|^2 = 10\mathrm{Log}\left(\left|\frac{p}{I}\right|^2\left|\frac{\mathrm{amp}}{\mu\mathrm{Pa}}\right|^2\right), \tag{16}$$

where α is the transmitting response and A_{std} is the response factor for the standard projector which is equal to 1 μPa/amp.

Example 2: Consider the case of a projector having the transmitting response of 100 dB re μPa/amp and assume that it is driven by an rms current of 1 amp. Since the rms pressure corresponding to the input current, from (16), can be written

$$|p| = |I| \times 10^{\alpha/20} \frac{\mu\mathrm{Pa}}{\mathrm{amp}} \quad \text{at 1 meter}, \tag{17a}$$

the radiated rms pressure at one meter from the source would be 10^5 μPa.♦

The response of a hydrophone can be similarly determined in terms of output voltage or current but for most purposes the open circuit voltage response is used. In this case the receiving response is the hydrophone rms output voltage due to an input of unit rms acoustic pressure, as is suggested by (14b). It has the form $E = Bp$ where B is the constant of proportionality called the receiving response factor.

The receiving response is usually expressed in decibels relative to the response of a standard hydrophone where the response of the standard hydrophone is defined to be an rms output of 1 volt for an rms input of 1 μPa. The ratio of the constants of proportionality for both hydrophones in decibel form is

$$\beta = 10\mathrm{Log}\left|\frac{B}{B_{std}}\right|^2 = 10\mathrm{Log}\left(\left|\frac{E}{p}\right|^2\left|\frac{\mu\mathrm{Pa}}{\mathrm{volt}}\right|^2\right), \tag{17b}$$

where β is the receiving response and B_{std} is the receiving response factor for the standard projector which is equal to 1 volt/μPa.

Example 3: Consider the case of a hydrophone for which the receiving response is –80 dB re volt/μPa and assume it is driven by an rms pressure of 1 μPa. Since the voltage corresponding to the input pressure, from (16b), can be written

$$|E| = |p| \times 10^{\beta/20} \frac{\mathrm{volt}}{\mu\mathrm{Pa}}, \tag{17c}$$

the magnitude of the output voltage would be 10^{-4} volts.♦

CHAPTER 4
SOURCES AND RECEIVERS

Whales, dolphins and other marine animals have very effective biological sonar systems that they use to locate kith and kin, prey on their food chain inferiors, evade predators and generally navigate within the ocean. ASW sonar systems are designed to perform similar functions but, since their sources and receivers are based upon electroacoustic transducers rather than biological organs, their acoustic performance is not as constrained. ASW acoustic sources can be more powerful and they can transmit over broader frequency bandwidths than can any biological source. Their receivers, while generally not as sensitive as the biological ear, can have aperture sizes that are greater than that of any biological aperture because many ASW platforms are large compared to the largest whales and ASW sources and receivers can be distributed over apertures that are comparable to, or in excess of, the size of the platform. These hardware advantages over flesh, blood and bone can result in high performance systems that can have longer detection ranges and greater localization capabilities than any biological system.[1]

In the following we examine the concepts behind high performance ASW acoustic sources and receivers. We find that they both exhibit beam patterns and these beam patterns specify directional preferences for acoustic radiations and receptions. Further, we find that beam patterns largely define the performance of source and receive systems and we discuss these performance measures and the sonar equations. High performance ASW acoustic sources and receivers are generally configured as arrays of transducer elements but in some systems they consist of a single transducer element. Arrays introduce wiring complications but they also mitigate the requirement of fabricating large transducer elements and they provide the opportunity for sophisticated spatial processing. However, since single element sources and receivers are important for simpler systems, and to elaborate on acoustic principles, they are discussed in the following.

4.1 Radiation from Projectors

Target detection requires that sonar transmissions be concentrated in regions where targets are expected to be found. In order to achieve this goal, projectors are designed such that their radiations can be selectively directed. This is primarily accomplished through the configuration of the projector and the electrical waveforms driving the projector. Both affect the radiation patterns of projectors and

---

[1] ASW sonar systems can be superior to biological sonar systems in certain measures because of hardware versus wetware considerations. However, it would be a mistake judge ASW systems as generally superior because this ignores the brain based processing of biological organisms as compared to the software based processing of ASW systems. Considering that bottlenose dolphins have a brain to body mass ratio that is about 3/4 that of human beings and that much of the bottlenose brain is given over to acoustic signal processing, it appears that bottlenose processing systems may dwarf all ASW processing systems. Consequently, ASW processing systems may have, so far, only scratched the surface of what is possible and that the potential for improvements in ASW sonar processing may be substantial.

these patterns are called projector beam patterns. Beam patterns can enhance performance several ways. They enable the energy on the target to be maximized which raises the signal-to-noise ratio of the received echo. They focus transducer energy into only desired regions which conserves acoustic energy and increases the maximum range of the system. They can also minimize reverberation from unwanted directions which reduces the noise level at the receiver.

Simple Source

The simple source is the simplest of projectors and is important because all projectors can be viewed as collections of simple sources. The simple source is a small pulsating sphere which, from a theoretical viewpoint, is driven by the injection and/or extraction of mass at the sphere center causing the surrounding medium to expand and/or to contract. The result is an outgoing spherical acoustic wave. Practically speaking, such a source might be effected by a spherical shell that is hydraulically driven from within. However, acoustic sources of any configuration, if they are small compared to the wavelength of the acoustic radiation, behave as simple sources.

Following Morse[2], we consider the simple source indicated in Figure 1. When fluid is injected and extracted at the origin of coordinates, the mass of fluid crossing the surface of the sphere is equal to the mass of fluid injected at the sphere center, reduced by the mass buildup within the sphere due to the compression of the fluid. The flow of mass out of the sphere is then given by the balance equation (1a) where the fluid radial velocity is v(t). The mass flow through the sphere surface per unit area is $v(t)\rho_0$, the function M(t) is called the mass velocity of the fluid, which is the fluid mass per unit time that is injected/extracted at the origin, and the volume integral is the mass buildup/depletion within the sphere.

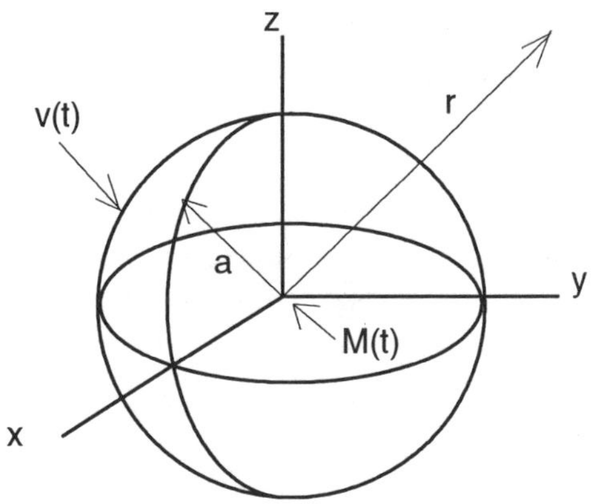

Figure 1 Simple Source

$$4\pi a^2 v(t)\rho_0 = M(t) - \int_V dv'\dot{\rho}(t,r') . \tag{1a}$$

The left member represents the mass rate crossing the sphere surface; the right member represents the mass rate injected/extracted at the sphere center reduced by the mass rate increases/decreases within the sphere volume due to compression.

The outgoing pressure wave in the region around the sphere satisfies the homogeneous wave equation in (2.17-A)

---

[2] Morse, Philip M., *Vibration and Sound*, McGraw-Hill, 1948, pg. 312.

$$\nabla^2 p(t,r) - \frac{1}{c^2}\ddot{p}(t,r) = 0. \quad r \neq 0 \tag{1b}$$

Since the radiated field is independent of direction, the solution has the form

$$p(t,r) = \frac{P(t - r/c)}{r}, \qquad r \neq 0 \tag{1c}$$

where r is the range between the field point and the center of the simple source. The function P(t) denotes the waveform of the outgoing wave and the function p(t,r) is the outgoing waveform attenuated by 1/r. The radiated pressure at each point on the sphere surface satisfies Newton's second law in (2.4c-A)

$$\lim_{r \to a}\left[-\frac{\partial p(t,r)}{\partial r}\right] = \rho_0 \dot{v}_r(t). \tag{1d}$$

This boundary condition gives, in the limit, the force per unit volume on each fluid particle at the surface of the pulsating sphere. Substituting into (1d) from (1a) and (1c) we obtain

$$\frac{P(t - a/c)}{a^2} + \frac{\dot{P}(t - a/c)}{ac} = \frac{\dot{M}(t) - \int dv' \ddot{\rho}(t, r')}{4\pi a^2}. \tag{1e}$$

For a vanishingly small sphere, the second term of the left member is negligible compared to the other terms and we obtain

$$P(t) = \frac{\dot{M}(t)}{4\pi} = \rho_0 \frac{\dot{Q}(t)}{4\pi}, \tag{1f}$$

where Q is the volume velocity of the fluid defined by $M = \rho_0 Q$. From (1c) the outgoing wave has the form

$$p(t,r) = \rho_0 \frac{\dot{Q}(t - r/c)}{4\pi r}. \tag{1g}$$

For the vanishing sphere (1g) asserts that the radiated pressure field at range is proportional to the acceleration at which mass is injected and extracted at the sphere center. For spheres, in general, the radiated field is different and, for completeness, the field radiated by a general sphere is determined in Appendix 4-A.

At range the radiation of the simple source is approximately planar over small regions and the instantaneous intensity of the radiated field is given by

$$I(t) = \frac{p^2(t,r)}{2\rho_0 c} = \frac{\rho_0 \dot{Q}^2(t - r/c)}{2(4\pi r)^2 c}, \tag{2a}$$

with the total radiated power

$$4\pi r^2 I(t) = \frac{\rho_0 \dot{Q}^2(t - r/c)}{8\pi c}. \tag{2b}$$

Example 1: Consider a simple sinusoidal source having source strength $Q = Q_0 \exp(j\omega t)$. We assume a peak mass velocity of $\rho_0 Q_0 = 1$ gm/sec and a driving frequency of 1 kHz. To determine the source level of the source we need the intensity at one meter from the acoustic center. The mean squared pressure at one meter is $P = 0.5(\rho_0 \omega Q_0/4\pi)^2 = 12.5$ dyne^2/cm^4 and the intensity at one meter is $I = P/\rho_0 c = 8.33 \times 10^{-12}$ watt/cm^2. The resulting source level is SL = 111 dB re μPa. Since the acoustic wavelength in water at 1 kHz is approximately 5 ft and the source is necessarily small compared to the wavelength, we assume a spherical source having diameter of, say, D = 30 cm and surface area of A = 2827 cm^2. The surface velocity is zero at the extremes of surface displacement and has the maximum value of $v_{r,max} = Q_0/\pi d^2 = 3.54 \times 10^{-4}$ cm/sec at zero surface displacement. The extreme surface displacement is $\delta r_{max} = \pm 5.63 \times 10^{-8}$ cm. The maximum volume displaced is $V_{max} = A\delta r_{max} = 1.59 \times 10^{-4}$ cm^3. The maximum volume velocity in terms of the maximum displacement volume is found by integrating the volume velocity over a quarter cycle to obtain $Q_0 = \omega V_{max} = 1$ cm^3/sec.♦

Operational Projectors

Operational projectors are typically configured as arrays of transducer elements that are held in a rigid frame, hung by flexible supports, etc. or can consist of a single transducer element. For all mounting configurations, the transducer elements can be of various sizes and shapes, e.g. planar, cylindrical, spherical, etc. but in all cases the transducer elements share the property of being spatially extended as opposed to being point sources as is the simple source. We can generalize the radiated field expression for simple sources in (1g) to extended sources by summing over a continuous distribution of simple sources. This sum can be expressed

$$p(t, \mathbf{r}) = \frac{\rho_0}{4\pi} \int_V dv' \frac{d\dot{Q}(t - r_V/c, \mathbf{r}')/dv'}{r_V}, \tag{3a}$$

where $r_V = |\mathbf{r} - \mathbf{r}'|$ is the range between the field point and the elemental volume point and the term $dQ(t - r_V/c, \mathbf{r}')/dv'$ is the volume velocity per unit volume. This is the source term that we denote $q(t, \mathbf{r}')$ to obtain

$$p(t, \mathbf{r}) = \frac{\rho_0}{4\pi} \int_V dv' \frac{\dot{q}(t - r_V/c, \mathbf{r}')}{r_V}. \tag{3b}$$

This result is derived from first principles in Appendix 4-B where the volume acceleration density $\partial q(t)/\partial t$ is found to be the source term in the wave equation in (2.17-A).

It is instructive to show directly, that the volume acceleration density in the integrand of (3b) is the source density in (2.17-A). To demonstrate this, we generalize the mass flow equation for the simple source in (1a) to the case of extended sources. The source is assumed to be extended in space and to be bound by a simple continuous surface. The sources within the volume are continuously distributed with the distributed mass velocity denoted m(t,**r**) and the mass flow equation is written

$$\int_S d\mathbf{s}' \cdot \rho_0 \mathbf{v}(t, \mathbf{r}') = \int_V dv' m(t, \mathbf{r}') - \int_V dv' \dot{\rho}(t, \mathbf{r}') . \tag{4a}$$

The left member is the rate of mass flow across the boundary of the extended source and the right member is the total mass injected into the volume reduced by the rate of mass buildup within the volume. The injected mass is defined by the mass velocity density m(t,**r**′) = dM(t,**r**′)/dv. Transforming the left member, using Gauss's divergence theorem from (4.3-B), and equating integrands we obtain

$$\dot{\rho}(t, \mathbf{r}') + \rho_0 \nabla \cdot \mathbf{v}(t, \mathbf{r}') = m(t, \mathbf{r}') . \tag{4b}$$

The wave equation source term in 2.17-A was introduced in the equation of continuity for fluids in (2.7c-A) by means of an equation similar to (4b) that can be expressed

$$\dot{\rho}(t, \mathbf{r}') + \rho_0 \nabla \cdot \mathbf{v}(t, \mathbf{r}') = \rho_0 q(t, \mathbf{r}') . \tag{4c}$$

A term-by-term comparison of (4b) and (4c) shows that the mass velocity density and the wave equation source term are related as are the mass velocity and volume velocity in (1f)

$$m(t, \mathbf{r}') = \rho_0 q(t, \mathbf{r}') . \tag{4d}$$

Consequently, the integrand of 3a can be written

$$\frac{d\dot{Q}(t, \mathbf{r}')}{dv'} = \frac{d\dot{M}(t, \mathbf{r}')}{\rho_0 dv'} = \frac{\dot{m}(t, \mathbf{r}')}{\rho_0} = \dot{q}(t, \mathbf{r}') , \tag{4e}$$

which confirms the transition from (3a) to (3b).

Operational projectors are extended sources but their distributions of simple sources are over surfaces and not within volumes. They can be generally viewed as shells over which acoustic source elements are mounted. For example, spherical bow arrays consist of numerous transducer elements distributed over the surface of a sphere but the shell can, in principle, take any shape. For such projectors all radiation originates at the projector surface and the radiated field expression in (3b) can be cast in terms of the surface velocity which is a more useful form. Since within the inner shell volume the source term vanishes and it is nonzero only within the shell itself we have for the integrand of (3b)

$$dv'\dot{q}(t,\mathbf{r}') = ds''\dot{v}(t,\mathbf{r}'')d\eta\delta(\mathbf{r}' - \mathbf{r}''), \tag{4f}$$

where the double primed coordinates are confined to the projector surface and $d\eta$ denotes a coordinate perpendicular to the surface of vanishingly small thickness. The left member is the rate at which volume is created at a point on the shell's bounding surface. The right member factor $ds''v(t,\mathbf{r}'')$ represents the rate at which volume is created by a surface element and the remaining factor represents a line integration with concentration at the surface. Substituting (4f) into (3b) and integrating over η we obtain

$$p(t,\mathbf{r}) = \frac{\rho_0}{4\pi}\int_S ds' \frac{\dot{v}(t - r_S/c, \mathbf{r}')}{r_S}. \tag{5}$$

where $r_S = |\mathbf{r} - \mathbf{r}'|$ is the range from a surface element to the field point. This is a particularly useful representation for the radiated field because the projector face velocity can be determined analytically or measured experimentally.

Transmit Beam Patterns - Narrow Band Signals

When a projector radiates a signal, the radiated field is the sum of spherical fields radiated by the projector's constituent simple sources and the projector beam pattern is the result of the constructive and destructive interference of these radiated fields. If the radiated signal is a pure tone, the total field transmitted in the various directions will have the same pure tone frequency but will, in general, be attenuated and phased shifted in azimuth which is the source of the beam pattern.

Representing the transmitted pressure field and source field by their Fourier transforms we obtain a representation of the transmitted signal in terms of sinusoidal components

$$p(t,\mathbf{r}) = \frac{1}{2\pi}\int_{-\infty}^{\infty} d\omega P(\omega,\mathbf{r})\exp(j\omega t), \tag{6a}$$

$$v(t,\mathbf{r}') = \frac{1}{2\pi}\int_{-\infty}^{\infty} d\omega V(\omega,\mathbf{r}')\exp(j\omega t), \tag{6b}$$

where $P(\omega,\mathbf{r})$ and $V(\omega,\mathbf{r}')$ are the respective Fourier transforms of the transmitted field and of the projector face velocity. Substituting into the radiated field expression in (5) we find for a single frequency component

$$P(\omega,\mathbf{r})\exp(j\omega t) = \frac{j\omega\rho_0}{4\pi}\int_S ds' \frac{V(\omega,\mathbf{r}')\exp(j\omega(t - r_S/c))}{r_S}. \tag{6c}$$

At long ranges from the transducer, as compared to the transducer size, the range vector between the source point and the field point is nearly parallel to the range vector between the acoustic center and the field point and the range can be expressed

$$r_S = |\mathbf{r} - \mathbf{r}'| = r - \mathbf{r}' \cdot \mathbf{e}_r \,, \qquad (7a)$$

as indicated in Figure 2. In this expression the unit vector $\mathbf{e}_r$ points at the field point. The approximation expresses the range between a point on the transducer face and the field point as the range from the acoustic center of the projector less the projection of the face point vector on the range vector $\mathbf{r}$. Substituting (7a) into (6c) gives the approximate expression for the radiated field

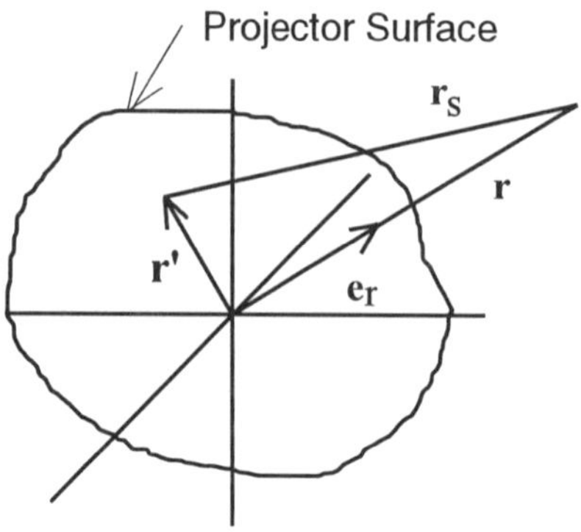

Figure 2 Range Relations

$$P(\omega, \mathbf{r})\exp(j\omega t) = \frac{\exp(j\omega(t - r/c))}{r}\left(\frac{j\omega\rho_0}{4\pi}\right)\int_S ds' V(\omega, \mathbf{r}')\exp\left(j\frac{\omega}{c}\mathbf{r}' \cdot \mathbf{e}_r\right). \qquad (7b)$$

For surface elements that are nearer to the field point than the acoustic center, the signal arrives at the field point at an earlier time and this is reflected in (7b) by the time shift, $\mathbf{r}' \cdot \mathbf{e}_r / c$, being positive and conversely.

For the case in which the projector consists of an array of transducer elements, we assume that the velocity over each element face is constant but, for purposes of beam forming, the velocity can vary from element-to-element. The transmission control system introduces the element velocities

$$V(\omega, \mathbf{r}') = w(\mathbf{r}')V(\omega)\exp\left(-j\frac{\omega}{c}\mathbf{r}' \cdot \mathbf{e}_{axis}\right). \qquad (7c)$$

The function $w(\mathbf{r}')V(\omega)$ is the weighted face velocity for the elements, where the first factor is a real weighting function that is applied to control the sidelobe levels and mainlobe width[3] and the second factor is the complex velocity of all elements. The element-to-element time delay in the exponential function is the means by which the radiation is directed and the unit vector $\mathbf{e}_{axis}$ defines the "look direction" of the projector. When the time delay $\mathbf{r}' \cdot \mathbf{e}_{axis}/c$ is positive, it counters a positive time shift in (7b), due to surface element position, and the intent of the introduced time delay is to effectively move each projector surface element onto a plane that passes through the acoustic center of the projector and is perpendicular to the axis direction. We note that for sinusoidal signals a time shift is the same as a phase shift and that the actual advancing and retarding of a sinusoidal transmission by each element is unnecessary because it is sufficient to simply phase shift each element transmission. Substituting (7c) into (7b) gives

---

[3] The projector beamformer weights are included in the following discussions because of their importance to the beamforming process but a discussion of their effects on the beam pattern is only given in Appendix 4C. The reader can simplify many of the remaining equations by setting the weights to unity.

$$P(\omega,\mathbf{r})\exp(j\omega t)=\frac{\exp(j\omega(t-r/c))}{r}R(\omega,\mathbf{e}_r-\mathbf{e}_{axis}), \tag{8a}$$

$$R(\omega,\mathbf{e}_r-\mathbf{e}_{axis})=\left(\frac{j\omega\rho_0 V(\omega)S}{4\pi}\right)\frac{1}{S}\int_S ds' w(\mathbf{r}')\exp\left(j\frac{\omega}{c}\mathbf{r}'\cdot(\mathbf{e}_r-\mathbf{e}_{axis})\right), \tag{8b}$$

where the function $R(\omega,\mathbf{e}_r-\mathbf{e}_{axis})$ is the directional response of the projector.

The transmitted signal described by (8) is a spherical sinusoidal wave that is time delayed by the travel time between the acoustic center of the source and the field point and is attenuated by 1/r. In addition, it is further attenuated and phase shifted by the directional response. The response along the look direction $\mathbf{e}_{axis}=\mathbf{e}_r$ is the direction of strongest response, which defines the beam axis direction, and is given by

$$R(\omega,\mathbf{0})=\frac{j\omega\rho_0 V(\omega)}{4\pi}\int_S ds' w(\mathbf{r}'). \tag{9a}$$

This result shows that the axial response of the projector is constant for all beam directions. The normalized response of the transducer is

$$\begin{aligned}R_N(\omega,\mathbf{e}_r-\mathbf{e}_{axis})&=R(\omega,\mathbf{e}_r-\mathbf{e}_{axis})/R(\omega,\mathbf{0})\\&=\int_S ds' w(\mathbf{r}')\exp\left(j\frac{\omega}{c}\mathbf{r}'\cdot(\mathbf{e}_r-\mathbf{e}_{axis})\right)/\int_S ds' w(\mathbf{r}'),\end{aligned} \tag{9b}$$

which has a maximum value of unity in the direction of the beam axis and decreases in magnitude for all look directions away from the beam axis. Note that, if the projector dimensions are small compared to the acoustic wavelength, i.e. $\omega r'/c \ll 1$, the normalized response is approximately unity for all beam directions and the projector appears to be a simple source regardless of the exact shape of the projector.

The intensity of the radiated field at long ranges is given by

$$I(\omega,\mathbf{r})=\frac{|P(\omega,\mathbf{r})|^2}{\rho_0 c}=\frac{|R(\omega,\mathbf{0})|^2}{\rho_0 c r^2}|R_N(\omega,\mathbf{e}_r-\mathbf{e}_{axis})|^2, \tag{10a}$$

where the magnitude squared of the normalized response is called the projector beam pattern

$$\begin{aligned}B(\omega,\mathbf{e}_r-\mathbf{e}_{axis})&=|R_N(\omega,\mathbf{e}_r-\mathbf{e}_{axis})|^2\\&=\left|\int_S ds' w(\mathbf{r}')\exp\left(j\frac{\omega}{c}\mathbf{r}'\cdot(\mathbf{e}_r-\mathbf{e}_{axis})\right)/\int_S ds' w(\mathbf{r}')\right|^2.\end{aligned} \tag{10b}$$

The maximum value of the beam pattern is on the beam axis and is equal to unity. At all other look directions the beam pattern has a value between unity and zero. In the event that the face velocities of the elements are all the same, i.e. there are no weightings, the beam pattern expression reduces to

$$B(\omega, \mathbf{e}_r - \mathbf{e}_{axis}) = \left| \frac{1}{S} \int_S ds' \exp\left(j\frac{\omega}{c} \mathbf{r}' \cdot (\mathbf{e}_r - \mathbf{e}_{axis}) \right) \right|^2, \tag{10c}$$

where S denotes the surface area of the array. See Appendix 4-C for discussions regarding the effect of the weightings on the beam pattern.

As an aside regarding notation, the directional vector argument in the response functions in (8) to (10) can be written

$$\begin{aligned} \mathbf{e}_r - \mathbf{e}_{axis} = {} & (\sin\theta\cos\varphi - \sin\theta_{axis}\cos\varphi_{axis})\mathbf{i} \\ & + (\sin\theta\sin\varphi - \sin\theta_{axis}\sin\varphi_{axis})\mathbf{j} \\ & + (\cos\theta - \cos\theta_{axis})\mathbf{k}. \end{aligned} \tag{11a}$$

While the vector form of the argument is more revealing of the physics of the beamforming process, (11a) shows that the vector argument is equivalent to the four coordinate angles that define the field point direction and beam axis direction and this angular form of the argument is more convenient from a computational viewpoint. For example, the beam pattern function in (10c) can be equivalently written

$$B(\omega, \theta, \varphi, \theta_{axis}, \varphi_{axis}) = \left| \frac{1}{S} \int_S ds' \exp\left(j\frac{\omega}{c} \mathbf{r}' \cdot (\mathbf{e}_r - \mathbf{e}_{axis}) \right) \right|^2, \tag{11b}$$

with (11a) substituted into the integrand. Similar expressions can be written for the other relations in (8) to (10).

Examples of Beam Patterns for Narrow Band Signals

Line Array:

Line arrays radiate with disc-like beam patterns perpendicular to the line. Consequently, vertical line arrays are well suited for performing horizontal searches. Small vertical line arrays are frequently used as short range navigation and collision avoidance aids and such line arrays can consist of a single element or an array of small elements which, being small, can be rigid and can be viewed as consisting of a continuous distribution of elements along the line. Large vertical line arrays that are designed for longer search ranges consist of discrete hanging elements that cannot be approximated as a continuous distribution of elements. We examine both cases.

Considering the small line array indicated in Figure 3, the source strength $V(\omega)$ is the maximum radial velocity of the projector elements along the line. The variation in time delay of the elements along the array is assumed to be continuous and the field point direction and beam axis directions are related to the time shifts by

$$\mathbf{r}' \cdot \mathbf{e}_r = z' \sin(\varphi)\,, \qquad (12a)$$

$$\mathbf{r}' \cdot \mathbf{e}_{axis} = z' \sin(\theta)\,, \qquad (12b)$$

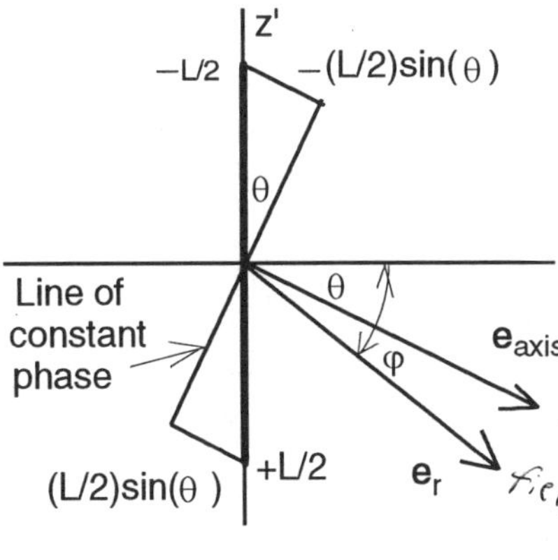

Figure 3 Line Array Projector

where φ and θ denote, respectively, the field point direction and beam direction. The effect of the beamformer time shifts at each transducer element, for the depressed beam direction as shown, is to advance the signal from the upper half of the array, i.e. transmit early, and to retard the signal from the lower half of the array, i.e. delay its transmission. The effect of all time shifts is to produce a transmission that is in phase at all points along the line of constant phase passing through the acoustic center that is indicated in Figure 3. Actually, since the line array is symmetrical about its axis the line of constant phase is more correctly a double cone of constant phase. From (8b) the directional response of the line transducer is

$$R(\omega,\varphi,\theta) = \left(\frac{j\omega\rho_0 V(\omega) S}{4\pi}\right)\frac{1}{L}\int_{-L/2}^{+L/2} dz' \exp\left(\left(\frac{j\omega z'}{c}\right)(\sin\varphi - \sin\theta)\right),$$

$$= \left(\frac{j\omega\rho_0 V(\omega) S}{4\pi}\right)\left[\frac{\sin\left(\frac{\omega L}{2c}(\sin\varphi - \sin\theta)\right)}{\left(\frac{\omega L}{2c}(\sin\varphi - \sin\theta)\right)}\right], \qquad (12c)$$

where S is the surface area of the line array, i.e. length×circumference. The radiation intensity is given by

$$I(\omega,\mathbf{r}) = \frac{\omega^2\rho_0 |V(\omega)|^2 S^2}{c(4\pi r)^2}\left[\frac{\sin\left(\frac{\omega L}{2c}(\sin\varphi - \sin\theta)\right)}{\left(\frac{\omega L}{2c}(\sin\varphi - \sin\theta)\right)}\right]^2, \qquad (12d)$$

and the beam pattern for the line array is given by

$$B(\omega,\varphi,\theta) = \frac{\sin^2\left(\frac{\omega L}{2c}(\sin\varphi - \sin\theta)\right)}{\left(\frac{\omega L}{2c}(\sin\varphi - \sin\theta)\right)^2}. \qquad (12e)$$

For the single element line array there are no time shifts and θ = 0. The beam width, defined to be the angle between the half power points, can be obtained from the solution of the equation: $\sin^2 x/x^2 = 1/2$ where x is the argument of (12e). An approximate solution for the argument is readily shown to be x = 1.4. Expressing the field point directions at the beam edges as $\varphi = \theta \pm \Omega/2$, where Ω is the beam width, we obtain the beam width from the solution for the argument

$$\Omega \cong \frac{2.8 \text{ radians}}{(kL/2)} , \tag{12f}$$

where it is assumed that kL/2 >> 1 and the beam elevation/depression angle is not too large. Sidelobe peaks and nulls, respectively, occur at the directions

peaks: $$\varphi = \arcsin\left[\left(\frac{2c}{\omega L}\right)\left(n+\frac{1}{2}\right)\pi + \sin\theta\right], \quad n = 0, \pm1, \pm2, \ldots \tag{12g}$$

nulls: $$\varphi = \arcsin\left[\left(\frac{2c}{\omega L}\right)n\pi + \sin\theta\right], \quad n = 0, \pm1, \pm2, \ldots \tag{12h}$$

where n does not exceed a value that would cause the axis direction to exceed $\pm\pi/2$.

A typical line array beam pattern can be observed in Figure 4. The displayed beam has a beam depression angle of 10° in which the bulk of the energy is seen to be confined to angles around the depression angle. A parameter value of kL/2 = 20 was arbitrarily chosen which corresponds to an array length of approximately 7 wave-lengths. The beam width is about 8° and the first sidelobe peak occurs at about ±14.6° from the main lobe peak. The radiation intensity from the first sidelobe is about 13 dB below the mainlobe radiation intensity, i.e. the sidelobe intensity is about 5% of the mainlobe intensity, where the mainlobe intensity of the array is given by

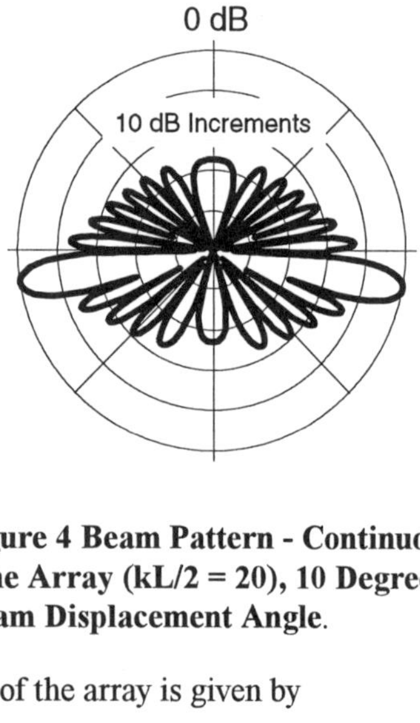

Figure 4 Beam Pattern - Continuous Line Array (kL/2 = 20), 10 Degree Beam Displacement Angle.

$$I(\omega, \mathbf{r}) = \frac{\omega^2 \rho_0 |V(\omega)|^2 S^2}{c(4\pi r)^2} . \tag{12i}$$

The source level of the array is just: SL = $10\text{Log}(I(\omega,1\text{m})/I_{ref})$ where I_{ref} is the reference intensity. This is usually the intensity of a plane wave having pressure of 1 μPa.

Considering a vertical line array with discrete hanging elements, the look direction and beam axis direction are as given in (12a) and (12b) where z′ is evaluated at each element position. Assuming an odd number N of uniformly spaced elements, the response function from (8b) becomes

$$R(\omega,\varphi,\theta)=\left(\frac{j\omega\rho_0 V(\omega)S}{4\pi}\right)\frac{1}{N}\sum_{n=-(N-1)/2}^{(N-1)/2}\exp\left(\left(\frac{j\omega n\Delta z}{c}\right)(\sin\varphi-\sin\theta)\right)$$

$$=\left(\frac{j\omega\rho_0 V(\omega)S}{4\pi}\right)\left[\frac{\sin\left(\frac{\omega N\Delta z}{2c}(\sin\varphi-\sin\theta)\right)}{N\sin\left(\frac{\omega\Delta z}{2c}(\sin\varphi-\sin\theta)\right)}\right], \tag{13a}$$

where Δz is the element spacing. The function in brackets is the normalized response function and its magnitude squared is the beam pattern for the array of hanging elements

$$B(\omega,\varphi,\theta)=\frac{\sin^2\left(\frac{\omega N\Delta z}{2c}(\sin\varphi-\sin\theta)\right)}{N^2\sin\left(\frac{\omega\Delta z}{2c}(\sin\varphi-\sin\theta)\right)^2}. \tag{13b}$$

A plot of (13b) showing a beam pattern graph for a discrete line array is shown in Figure 5. For purposes of comparing to the continuous line array beam pattern graph in Figure 4, the beam depression angle was taken to be 10° and 13 elements at half wavelength spacing was assumed giving the parameter $kL/2 \cong 20$. As can be seen the beam patterns for both arrays are nearly the same but the discrete element array has vertical sidelobes levels that are about 3 dB higher than for the continuous line array beam pattern.

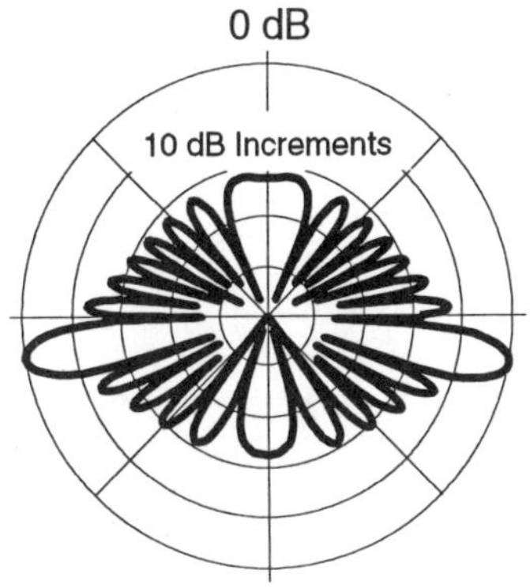

Figure 5 Beam Pattern - Discrete Line Array (13 Elements @ λ/2), 10 Degree Beam Depression Angle.

The similarity between the continuous and discrete beam patterns is found to be strongly dependent upon the choice of half wavelength element spacing for the discrete array. Comparing the continuous line array beam pattern expression in (12e) to the beam pattern expression for a discrete line array in (13b) shows a significant difference in the denominators which, since for both cases the envelope of the beam pattern is controlled by the denominator, can result in substantial beam pattern differences. Considering the case of a horizontal beam direction $\theta = 0$, we see that for the continuous array the denominator is always a maximum and the beam pattern envelope is always a minimum for the vertical look directions $\varphi = \pm\pi/2$. However, for the array of discrete elements the beam envelope is minimized in the vertical look directions only if $\omega\Delta z/2c = \pi/2$, i.e. if the element spacing is half a wavelength. For other element spacings the level of the vertical sidelobes increase. For example, if the element spacing is a full wavelength, the vertical sidelobes are found to have a level of 0 dB and to exceed the mainlobe in width. For the more general case of nonzero beam elevation/depression angles these conclusions are in need of slight modifications but the general conclusion remains valid that the half wavelength element spacing minimizes the radiation in the vertical look directions. This is important since the vertical lobes are the source of bathymetric echoes that are generated by repeated signal bounces between the surface and bottom. These echoes are generally not automatically filtered by detection systems and can appear as false nearby targets or can obscure actual nearby targets. Consequently, for this reason half wavelength spacing is standard design practice for vertical line source arrays of discrete elements.

Rectangular Projector:

Consider a projector consisting of a single rectangular element as indicated in Figure 6a and the single circular element in Figure 6b. In both cases the projectors are assumed to be set in an infinite baffle such that they radiate to only one side of the baffle. Since for single elements no spatially distributed time delays are introduced, the beam directions are perpendicular to the plate. For the rectangular element in Figure 6a, the look direction and beam direction vectors and the surface coordinate vector can be written in terms of rectangular coordinates and substituting into (8b) the directional response is found to be

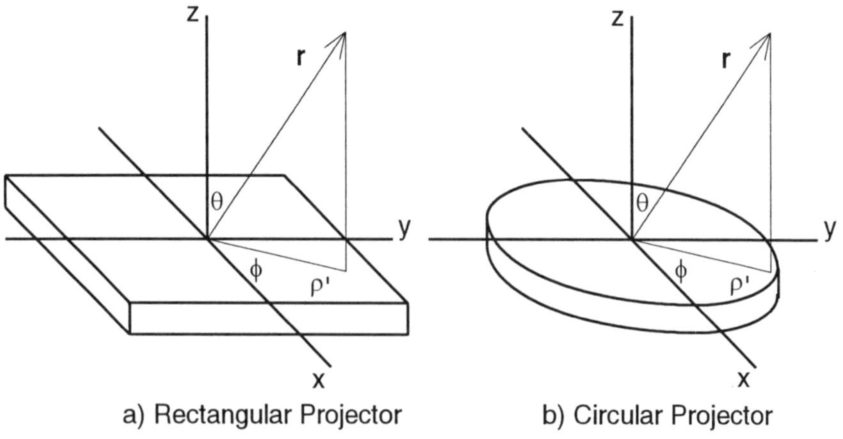

Figure 6 Coordinates for a Rectangular and Circular Projector

$$\mathbf{e}_r = \sin\theta\cos\phi\,\mathbf{i} + \sin\theta\sin\phi\,\mathbf{j} + \cos\theta\,\mathbf{k}\,, \qquad \mathbf{e}_{axis} = \mathbf{k}\,, \qquad \mathbf{r}' = x'\mathbf{i} + y'\mathbf{j}\,, \quad (14a)$$

$$R(\omega,\theta,\varphi) = \frac{j\omega\rho_0 V(\omega)}{4\pi}\int_{-W/2}^{W/2} dx \int_{-H/2}^{H/2} dy\, \exp\big(jk(x\sin\theta\cos\phi + y\sin\theta\sin\phi)\big)$$

$$= \left(\frac{j\omega\rho_0 V(\omega) HW}{4\pi}\right)\left[\frac{\sin(kW\sin\theta\cos\phi/2)}{(kW\sin\theta\cos\phi/2)}\,\frac{\sin(kH\sin\theta\sin\phi/2)}{(kH\sin\theta\sin\phi/2)}\right], \quad (14b)$$

where H and W are the height and width of the element and $k = \omega/c$. The leading factor is the axial response of the array and the beam pattern is immediately found to be given by

$$B(\omega,\theta,\varphi) = \frac{\sin^2(kW\sin\theta\cos\phi/2)}{(kW\sin\theta\cos\phi/2)^2}\,\frac{\sin^2(kH\sin\theta\sin\phi/2)}{(kH\sin\theta\sin\phi/2)^2}\,. \qquad (14c)$$

The beam pattern cross sections at $\phi = 0$ and $\phi = \pi/2$ are given, respectively, by the first and second factors. A graph of a such a cut through the beam pattern for a rectangular projector is shown in Figure 7a which shows the first sidelobes to be down approximately 13 dB. The intensity of the beam on the axis is given by

$$I_{Axis}(\omega,r) = \frac{\omega^2\rho_0|V(\omega)|^2(HW)^2}{c(4\pi r)^2}\,. \qquad (14d)$$

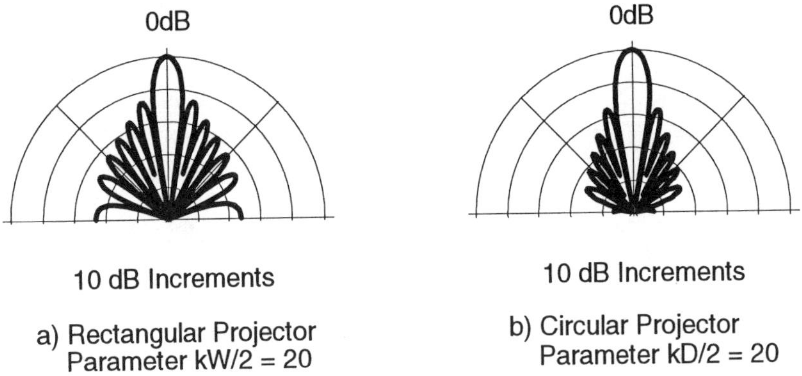

Figure 7 Beam Pattern Cuts for Rectangular and Circular Projectors

Circular Projector:

For the circular projector in Figure 6b, the look direction and beam direction vectors and surface coordinate vector can be written in terms of cylindrical coordinates and substituting into (8b) the directional response is found to be

$$\mathbf{e}_r = \sin\theta\mathbf{e}_\rho + \cos\theta\mathbf{k}, \qquad \mathbf{e}_{axis} = \mathbf{k}, \qquad \mathbf{r}' = r'\mathbf{e}_\rho, \qquad (15a)$$

$$R(\omega,\theta) = \frac{j\omega\rho_0 V(\omega)}{4\pi}\int_0^{2\pi} d\varphi \int_0^{D/2} dr' r' \exp(jkr'\sin\theta)$$

$$= j\omega\rho_0 V(\omega)\left(\frac{\pi D^2}{4}\right)\frac{2J_1(kD\sin\theta/2)}{(kD\sin\theta/2)}\,, \qquad (15b)$$

where D is the disk diameter and $k = \omega/c$. The integral in the response function is immediately recognized as a standard form[4] and the beam pattern is immediately found to be

$$B(\omega,\theta) = \left[\frac{2J_1(kD\sin\theta/2)}{(kD\sin\theta/2)}\right]^2 . \tag{15c}$$

A graph of the beam pattern for the circular projector is shown in Figure 7b which shows the first sidelobes to be down approximately 18 dB. The intensity of the beam on the axis is given by

$$I_{axis}(\omega,r) = \frac{\omega^2 \rho_0 |V(\omega)|^2 (\pi D^2/4)^2}{c(4\pi r)^2} . \tag{15d}$$

A comparison of the beam pattern crosscuts in Figure 7 for the rectangular and circular projectors suggests that the sidelobe problem for the rectangular projector is more severe than for the circular projector. However, viewed in a 3-dimensional format as in Figure 8, we see that the sidelobe problems for the two projectors are different but of comparable severity. Also we find that for rectangular and circular projectors having equal widths and diameters the main lobe width and sidelobe displacements of the rectangular projector are less than for the circular projector. This occurs because of the greater area of the rectangular array. The x-y coordinates in Figure 8a are the sinc function arguments in (14c) and the radial polar coordinate in Figure 8b is the Bessel function argument in (15c). The beams of single element transducers are steered by mechanically steering the element.

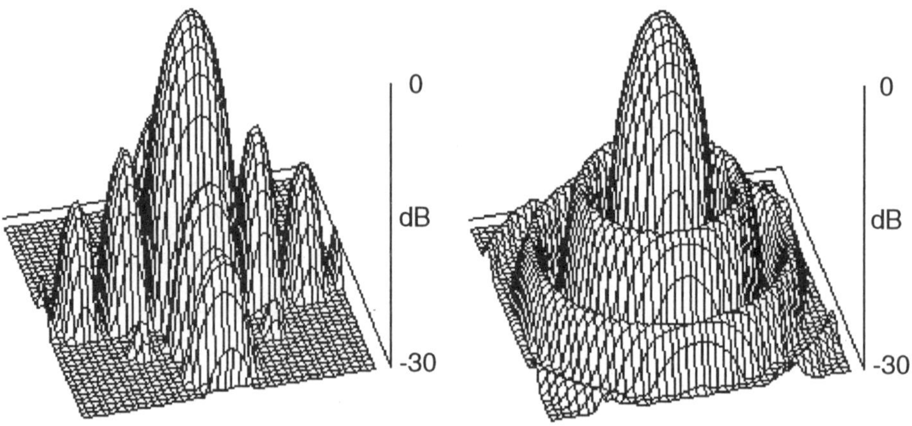

a) Rectangular Projector b) Circular Projector

Figure 8 Three Dimensional Beam Patterns for Rectangular and Circular Projectors

[4] Gradshteyn and Ryzhik, (5.52-1).

<u>Transmitting Directivity Index</u>

The transmitting directivity index[5] is a measure of the directionality of a projector. It is the difference between the intensity level on the projector beam axis and the intensity level that would be obtained by an equivalent power projector for which the total power is transmitted omnidirectionally. In mathematical terms

$$DI_T = 10\text{Log}\frac{I_a}{I_o}, \tag{16a}$$

where I_a is the axial projector intensity at range and I_o is the equivalent omnidirectional intensity at range. By its definition the transmitting directivity index is always greater than or equal to 0 dB.

Both intensities can be obtained from the general expression for radiated intensity in (10a). The axial intensity is immediately found to be

$$I_a = \frac{|R(\omega,0)|^2}{\rho_0 c r^2}. \tag{16b}$$

The omnidirectional intensity is defined by the relationship for total radiated power

$$4\pi r^2 I_o = \frac{|R(\omega,0)|^2}{\rho_0 c r^2}\int_{4\pi} d\Omega R^2 B(\omega, \mathbf{e}_r - \mathbf{e}_{axis}). \tag{16c}$$

where the integration in (16c) is over a sphere of radius r and the differential solid angle $d\Omega$ subtends a differential area at range r, i.e. $dS = d\Omega\, r^2$.

Collecting results and substituting into (16a) gives an expression for the transmitting directivity index in terms of the projector beam pattern

$$\begin{aligned} DI_T &= 10\text{Log}\left(4\pi \Big/ \int_{4\pi} d\Omega B(\omega, \mathbf{e}_r - \mathbf{e}_{axis})\right) \\ &= 10\text{Log}\left(4\pi \Big/ \int_0^{2\pi} d\varphi \int_{-\pi/2}^{\pi/2} d\theta \cos\theta B(\omega, \varphi, \theta)\right). \end{aligned} \tag{16d}$$

The second equality in (16d) provides a computational procedure for determining the transmitting directivity index which, for simple beam patterns, can be determined analytically.

Example 2: The transmitting directivity index for a continuous line array can be readily computed from (16d). The beam pattern is given in (12e) and the

---

[5] Urick, Robert J., *Principles of Underwater Sound*, McGraw-Hill, 1975, pg. 66.

integral over the beam pattern in the denominator of (16d) can be written in the form[6]

$$\frac{2}{kL}\int_{-(1-\sin\varphi)kL/2}^{(1+\sin\varphi)kL/2} du \frac{\sin^2 u}{u^2} = \frac{\lambda}{L}. \qquad (16e)$$

Substituting into the receiving directivity index expression in (16d) we obtain

$$DI_T = 10\text{Log}\left(\frac{2L}{\lambda}\right). \qquad (16f)$$

If the line array consists of elements at half wavelength spacing, this expression reduces to

$$DI_T = 10\log N, \qquad (16g)$$

and a 10 element line array would have a directivity index of 10 dB.♦

The transmitting directivity index is related to the projector source level which provides a convenient method for determining source level[7]. The axial intensity at range can be written

$$I_a(r) = \alpha I_o(r) = \frac{\alpha P}{4\pi r^2}, \qquad (17a)$$

where the range r is made explicit and where the first equality defines α, which is just the linear transmitting directivity index, and the second equality expresses the omnidirectionally radiated intensity at range in terms of the total radiated power P. Since the total power is the same at all ranges, the axial intensity of the directional projector at one meter is found by letting r = 1 meter in the last member. Consequently, we can write

$$\frac{I_a(1m)}{I_{ref}} = \frac{\alpha P}{4\pi I_{ref}}, \qquad (17b)$$

and taking 10 times the logarithm of both sides gives

$$SL = DI_T + \left(10\text{LogP} - 10\text{Log}4\pi I_{ref}\right). \qquad (17c)$$

The total power in (17c) is just the fraction of total input electrical power that is radiated as acoustic power which is given by the projector efficiency as determined in Section 3.1. When evaluating (17c) it is required that the power and the reference power at one meter, i.e. $4\pi(1m)^2 I_{ref}$, be determined in the same units of measure. For a reference having an rms pressure of 1 μPa at one meter we obtain

---

[6] Gradshteyn and Ryzhik, (3.821-9).

[7] Urick, Robert J., *Principles of Underwater Sound*, McGraw-Hill, 1975, pg. 67.

$$SL = DI_T + \left(10Log(P) + 170.7\right), \tag{17d}$$

where P is measured in watts. For a system having a transmitting directivity index of 10 dB and radiated power of 10 kwatts, the source level is 220.7 dB re μPa at one meter which is a typical source level for an operational system.

Transmit Beam Patterns - Broad Band Signals

In the preceding, the discussions of beam patterns were restricted to the case of narrow band transmissions and their beam patterns were found to be dependent upon signal frequency. Because of this dependence, we can infer that for broad band transmissions beam patterns can be expected to be a mixture of narrow band beam patterns.

The broad band radiated signal can be represented by the Fourier transform of (8a) which, with the aid of (9b), gives the broad band radiated field expression

$$p(t,\mathbf{r}) = \frac{1}{2\pi r}\int_{-\infty}^{\infty} d\omega R(\omega,\mathbf{0}) R_N(\omega, \mathbf{e}_r - \mathbf{e}_{axis}) \exp\left(j\omega(t - r/c)\right). \tag{18a}$$

This is the Fourier transform of a product which can be expressed as the convolution of the inverse transforms. The inverse transform of the first factor in the integrand is obtained from (9a) and is found to be just the undistorted signal that drives the projector

$$P(t) = \frac{1}{2\pi}\int_{-\infty}^{\infty} d\omega R(\omega,\mathbf{0}) \exp(j\omega t) = \frac{\rho_0 \dot{v}(t)}{4\pi} S_W, \tag{18b}$$

where v(t) is the broad band projector face velocity and S_W is the weighted projector face area. This is just the signal that drives the projector as can be inferred from (1f). The inverse transform of the second factor is a little more complicated but, inverse transforming (9b) via straight forward techniques, we find

$$\begin{aligned} r(t, \mathbf{e}_r - \mathbf{e}_{axis}) &= \frac{1}{2\pi}\int_{-\infty}^{\infty} d\omega R_N(\omega, \mathbf{e}_r - \mathbf{e}_{axis}) \exp(j\omega t) \\ &= \frac{1}{2\pi S_W}\int_S ds' w(\mathbf{r}') \int_{-\infty}^{\infty} d\omega \exp\left(j\omega\left(t + \mathbf{r}'\cdot(\mathbf{e}_r - \mathbf{e}_{axis})/c\right)\right) \\ &= \frac{1}{S_W}\int_S ds' w(\mathbf{r}') \delta\left(t + \mathbf{r}'\cdot(\mathbf{e}_r - \mathbf{e}_{axis})/c\right). \end{aligned} \tag{18c}$$

Combining these results, the transmitted signal in (18a) can be expressed as the convolution between the functions in (18b) and (18c)

$$p(t,\mathbf{r}) = \int_{-\infty}^{\infty} d\tau P(t - r/c - \tau) r(\tau, \mathbf{e}_r - \mathbf{e}_{axis})$$

$$= \frac{\rho_0}{4\pi r} \int_S ds' w(\mathbf{r}') \dot{v}\left(t - R/c + \mathbf{r}' \cdot (\mathbf{e}_r - \mathbf{e}_{axis})/c\right). \tag{18d}$$

This result shows that, while the axial signal expression reduces to the undistorted transmitted signal, analogous to (5), the off-axis signal is a filtered version of the transmitted signal. The integral performs a smoothing of the signal over a time window of duration that is comparable to the time of signal transit across the projector. Since most transmissions have a spatial extent that is large compared to the projector dimensions, we can expect this effect to be small. In fact, for broad band signals we can write

$$\dot{v}(t) = m(t) \exp(j\omega_0 t), \tag{18e}$$

in which the complex modulation is slowly varying compared to the carrier. If the time of modulation variation is greater than the time of signal transit over the array, the transmitted signal in (18d) can be approximated as

$$p(t,\mathbf{r}) = \left[\frac{\rho_0 \dot{v}(t - r/c) S_W}{4\pi r}\right] B(\omega_0, \mathbf{e}_r - \mathbf{e}_{axis}), \tag{18f}$$

where the leading factor gives the axial response of the array and the beam pattern is evaluated at the carrier frequency. Equation (18f) shows that the beam pattern concept can be approximately extended to broad band signals.

For most broadband ASW transmissions the bandwidth to carrier frequency ratio is small enough that the distortion of the off-axis transmission is negligible. However, for transmitted signals having durations that are comparable to the signal transit time over the array, the array spatial effects become more significant. In any event, it is important to remember that the broad band axial transmissions are always undistorted.

4.2 Reception by Hydrophone Arrays

Hydrophone arrays perform two functions. They enhance the detection of signals by increasing the received signal-to-noise ratio and they determine the direction of propagation of the received signals. Both functions are performed by time delaying and summing the signals from the hydrophones of the array. The received signal-to-noise ratio is increased because the signal components of the receptions from all hydrophones, when the outputs are properly time delayed, add constructively, while the noise components of the receptions add destructively. The signal directionality is determined by converting the time delays that maximize signal-to-noise ratio into the direction of signal propagation. These functions are performed by the beamformer which also weights the hydrophone outputs for the purpose of controlling beam sidelobe levels. In modern ASW systems the beamformer is frequently implemented by means of Fourier transform techniques.

Beam Forming Hydrophone Array Receivers

The acoustic field at a hydrophone array is generated by acoustic sources that are located at different azimuths and ranges from the array and the field in the vicinity of the array consists of waves received from various directions having various frequencies, bandwidths and strengths. These signals include distant shipping, biologics and sea surface noise and also include reflections from targets, reflections from surface and bottom and reverberation. Collectively, they make up the background noise within which the receive array operates and they are statistically independent from azimuth-to-azimuth. By means of time delay and summing techniques, arrays have the capability of repressing signals received from all directions away from a preferred direction and can raise the signal-to-noise ratio of the signal that is received from the preferred direction. Based upon these concepts, the acoustic field at the array can be modeled as a signal component coming from the preferred direction plus a noise component that consists of the sum of signals from all other directions. The signal plus noise within the vicinity of the array can be written

$$p(t,\mathbf{r}) = m(t+\mathbf{r}\cdot\mathbf{e}_r/c)\exp(j\omega_0(t+\mathbf{r}\cdot\mathbf{e}_r/c)) + n(t,\mathbf{r}), \tag{19}$$

where $\mathbf{r}$ is an arbitrary point in the vicinity of the array, the unit vector $\mathbf{e}_r$ points toward the direction from which the signal is received, which is the preferred direction, and the last term is additive noise. The incoming signal is depicted in Figure 9 as an acoustic wave sweeping over the array.

The hydrophone array indicated in Figure 9 consists of a collection of single hydrophone elements which, as we saw in Section 3.1, are designed to be small compared to the wavelength of the acoustic field, to ensure that the element response is constant in all directions, and their electrical output is filtered to a band of frequencies away from the resonance frequency of the elements, to ensure that their response is constant over the receive frequency band. The resulting hydrophone elements introduce minimal distortion to the received acoustic signals and their electrical outputs were shown in (3.14b) to be proportional to pressure, i.e. $E_n(t)=Cp(t,\mathbf{r}_n)$ where C is the constant of proportionality which is real.

The beamformer attempts to combine all hydrophone outputs coherently by time delaying the output of each hydrophone element and then summing over all outputs. Additionally, each hydrophone output is weighted to control beam sidelobes and each hydrophone output has the form

$$E_n(t-\tau_n) = Cw_n p(t-\tau_n,\mathbf{r}_n). \tag{20a}$$

In this expression, the quantities w_n are the beamformer weights, which are real, and the time delays introduced by the beamformer shift the output along the direction $\mathbf{e}_{axis}$ and are given by

$$\tau_n = \frac{\mathbf{r}_n \cdot \mathbf{e}_{axis}}{c}. \quad (20b)$$

These time delays effectively shift all hydrophone elements from the planes indicated by dotted lines in Figure 9 to the plane of constant phase passing through the acoustic center of the array. The total array output is the sum of the weighted and time delayed hydrophone outputs. Considering just the signal component, which is the expected value of the array output, the signal output is given by

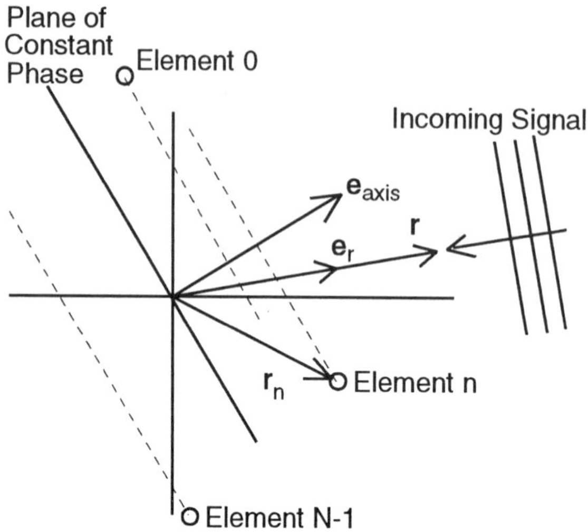

Figure 9 Hydrophone Array Geometry

$$E(t, \mathbf{e}_r - \mathbf{e}_{axis}) = \frac{C}{N}\sum_{n=0}^{N-1} m_n\left(t + \frac{\mathbf{r}_n}{c}\cdot(\mathbf{e}_r - \mathbf{e}_{axis})\right)\exp\left(j\omega_0\left(t + \frac{\mathbf{r}_n}{c}\cdot(\mathbf{e}_r - \mathbf{e}_{axis})\right)\right), \quad (21)$$

where $m_n = w_n m$ are the weighted hydrophone outputs[8]. Examination of this time domain beamformer implementation shows that when the array axis is aligned with the signal source direction, $\mathbf{e}_{axis} = \mathbf{e}_r$, the signal components combine coherently and the array output is just the incoming signal, i.e. $E(t,\mathbf{0})=C'm(t)\exp(j\omega_0 t)$. The constant in this expression C' is the weighted receiving response factor, i.e. $C'=C(\Sigma w_n)/N$. For other axis directions the signal components sum incoherently and the array output is reduced in magnitude.

Alternatively, beamforming can be performed in the frequency domain. The Fourier transforms of the weighted hydrophone signal outputs from (19) is

$$\begin{aligned} E_n(\omega) &= C\int dt\, m_n(t + \mathbf{r}'_n \cdot \mathbf{e}_r / c)\exp(-j(\omega - \omega_0)t)\exp(j\omega_0 \mathbf{r}'_n \cdot \mathbf{e}_r / c) \\ &= CM_n(\omega - \omega_0)\exp\left(j\frac{\omega \mathbf{r}'_n \cdot \mathbf{e}_R}{c}\right), \quad \omega_0 - \pi W \le \omega \le \omega_0 + \pi W \qquad (22a) \end{aligned}$$

where $M_n = w_n M$ denotes the Fourier transform of the weighted signal modulation. Phase shifting each frequency component of the hydrophone outputs by the amount $\tau_n = \omega \mathbf{r}_n' \cdot \mathbf{e}_{axis}/c$, as required by (20), and summing over hydrophones gives

---

[8] The receiver beamformer weights are included in the following discussions because of their importance to the beamforming process but a discussion of their effects on the beam pattern is given only in Appendix 4b. The reader can simplify many of the remaining equations by assuming the weights are unity, i.e. there is no beamformer weightings.

$$E(\omega, \mathbf{e}_r - \mathbf{e}_{axis}) = \frac{CM(\omega - \omega_0)}{N} \sum_{n=0}^{N-1} w_n \exp\left(j\frac{\omega}{c} \mathbf{r}_n \cdot (\mathbf{e}_r - \mathbf{e}_{axis}) \right). \qquad (22b)$$

$$\omega_0 - \pi W \le \omega \le \omega_0 + \pi W$$

The inverse Fourier transformation of the frequency domain beamformer implementation in (22b) shows it to be equivalent to the time domain beamformer implementation in (21). The advantage of the frequency domain implementation is that it is more easily performed in modern sampled data systems. Whereas time shifting in the time domain requires elaborate interpolation techniques, time shifting in the frequency domain requires simple complex multiplications. The problem with this implementation is the high sampling rates needed to satisfy the Nyquist theorem.

Alternatively, heterodyning the hydrophone outputs to baseband, prior to frequency domain beamforming, reduces the required sampling rates. For the analytic signal, heterodyning is simply performed by multiplying the hydrophone output by the conjugate of the carrier signal, i.e. the quantity $\exp(-j\omega_0 t)$. Then, instead of (22a), the Fourier transform of the hydrophone output is given by

$$\begin{aligned} E_n(\omega) &= C\int dt\, m_n(t + \mathbf{r}_n' \cdot \mathbf{e}_r / c) \exp(-j\omega t) \exp(j\omega_0 \mathbf{r}_n' \cdot \mathbf{e}_r / c) \\ &= CM_n(\omega) \exp\left(j \frac{(\omega + \omega_0)\mathbf{r}_n' \cdot \mathbf{e}_r}{c} \right). \quad -\pi W \le \omega \le \pi W \end{aligned} \qquad (23a)$$

Performing the time delay operation in the frequency domain by phase shifting by the amount $(\omega+\omega_0)\rho'_n \cdot \mathbf{e}_{axis}/c$, the beamformed output is

$$E(\omega, \mathbf{e}_r - \mathbf{e}_{axis}) = \frac{CM(\omega)}{N} \sum_{n=0}^{N-1} w_n \exp\left(j\frac{(\omega + \omega_0)}{c} \mathbf{r}_n \cdot (\mathbf{e}_r - \mathbf{e}_{axis}) \right). \qquad (23b)$$

$$-\pi W \le \omega \le \pi W$$

A comparison of the beamformed outputs in (22b) and (23b) shows them to be equivalent but the latter output is at baseband while the former is at the band around ω_0. In fact, for both expressions, the output Fourier transform at the beam steering angle $\mathbf{e}_{axis} = \mathbf{e}_r$, i.e. the preferred direction, is C′M and the inverse transform produces, respectively, the analytic signal and its complex modulation which are the desired ouputs. Since the sampling rate required to compute the baseband Fourier transforms for the heterodyned signal is equal to the signal bandwidth, i.e. $f_s = W$, whereas, the sampling rate required to compute the Fourier transform for the high frequency signal is $f_s = W + 2f_0$, it follows that implementation of the baseband expression in (23b) is computationally less intensive.

The beam forming sums in (22b) and (23b) can both be performed by means of the Discrete Fourier Transform algorithm for the linear array shown in Figure 3. For the beam forming sum in (22b), the weighted and Fourier transformed data out of each hydrophone can be written $w_n'(\omega) = w_n \exp(j\omega \mathbf{r}_n \cdot \mathbf{e}_r/c)$ excluding the factors that are constant over the array. Defining the hydrophone displacements $\mathbf{r}_n \cdot \mathbf{e}_{axis} = n\Delta z \sin(\theta)$, (22b) can be written

$$E(\omega, \mathbf{e}_r - \mathbf{e}_{axis}) = \frac{CM(\omega - \omega_0)}{N} \sum_{n=0}^{N-1} w_n'(\omega) \exp\left(-j \frac{\omega \mathbf{r}_n' \cdot \mathbf{e}_{axis}}{c}\right)$$

$$= \frac{CM(\omega - \omega_0)}{N} \sum_{n=0}^{N-1} w_n'(\omega) \exp\left(\frac{-j\omega n \Delta z \sin\theta}{c}\right), \qquad (24a)$$

$$\omega_0 - \pi W \le \omega \le \omega_0 + \pi W$$

where θ is the steering angle of the beam, in Figure 3, and $w_n' = w_n \exp(j\omega \mathbf{r}_n \cdot \mathbf{e}_r/c)$ is the weighted phase of the received signal. Defining the kth beam steering angle

$$\frac{\omega \Delta z \sin\theta_k}{c} = \frac{2\pi k}{N} \rightarrow \theta_k = \sin^{-1}\left(\frac{2k}{N(\omega/\omega_0)}\right), \qquad (24b)$$

where the element spacing is assumed to be half wave length spacing, i.e. $\Delta z = \lambda_0/2$ where λ_0 is the wavelength at the carrier frequency. We see that the beam steering angles are given by the integers

$$0 \le k \le K(N, \omega) = \left[\frac{N(\omega/\omega_0)}{2}\right], \qquad (24c)$$

where K is the number of beam pattern points to be computed which increases with both the number of hydrophone elements and the component frequency and where the square brackets in the right member denote the greatest integer less than or equal to the argument. Collecting results into (24a) yields

$$E(\omega, k) = \frac{CM(\omega - \omega_0)}{N} \sum_{n=0}^{N-1} w_n'(\omega) \exp\left(\frac{-j2\pi nk}{N}\right), \qquad (24d)$$

$$k = 1, 2, \cdots, K \qquad \omega_0 - \pi W \le \omega \le \omega_0 + \pi W$$

which is the discrete Fourier transform of the weighted hydrophone outputs at each frequency. For a signal from the direction $\mathbf{e}_R$, this transform produces K beam pattern points for each frequency. If the number of hydrophone elements is zero-filled to a power of 2, i.e. if N is chosen to be larger than the actual number of hydrophones and equal to a power of 2, the discrete Fourier transform can be computed using a Fast Fourier Transform algorithm. The result gives N beam pattern points for each frequency.

The entire beamforming process can be performed by Fourier transform techniques as indicated in Figure 10. The weighted time sample sets from N hydrophones are Fourier transformed to produce N frequency domain transforms. Each output Fourier frequency component is then Fourier transformed where the integration is performed across N hydrophones. In this computation N can be zero-filled and the computation performed via the FFT. If the computation is performed by the FFT algorithm, N beam pattern points are produced. Since K is dependent on N and frequency, the number of beam steering angles obtained at each frequency can be different and the angles θ_k will be different for different frequencies. However, since the beam pattern varies slowly in azimuth, a standard set of K_S steering angles is obtained by interpolation. This facilitates the inverse Fourier transformation of the frequency samples at each beam steering angle to arrive at the signal modulation that is received at each angle.

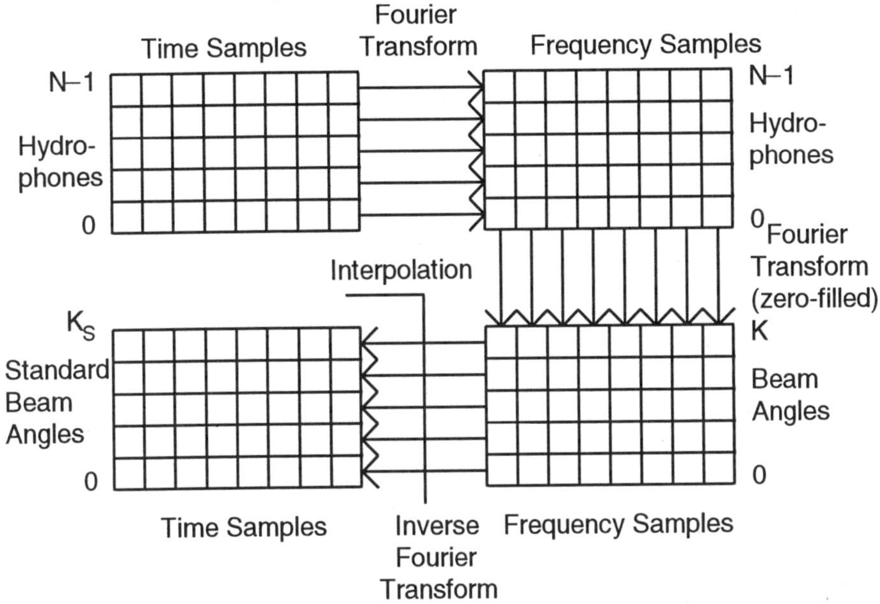

Figure 10 Beam Forming via the Fourier Transform

Receiver Beam Patterns

Hydrophone array receivers have beam patterns analogous to transmitters. In fact, rewriting (22b) in the form

$$E(\omega, \mathbf{e}_r - \mathbf{e}_{axis}) = E(\omega, \mathbf{0}) R_N(\omega, \mathbf{e}_r - \mathbf{e}_{axis}), \tag{25a}$$

the first factor is the axial response along the look direction, $\mathbf{e}_{axis} = \mathbf{e}_r$, with a value given by

$$E(\omega, \mathbf{0}) = C' M(\omega - \omega_0), \tag{25b}$$

which is analogous to (9b) for transmitters. The factor C′ is the weighted response factor, defined after (21), and the second factor is the normalized response of the receive array with a value given by

$$R_N(\omega, \mathbf{e}_r - \mathbf{e}_{axis}) = \sum_{n=0}^{N-1} w_n \exp\left(j\frac{\omega}{c}\mathbf{r}'_n \cdot (\mathbf{e}_r - \mathbf{e}_{axis})\right) / \left(\sum_{n=0}^{N-1} w_n\right), \quad (25c)$$

which is analogous to (9b) for transmitters.

The power of the output electrical signal is obtained from (25) to be

$$P(\omega, \mathbf{e}_r - \mathbf{e}_{axis}) = |E(\omega, \mathbf{e}_r - \mathbf{e}_{axis})|^2 = |E(\omega, \mathbf{0})|^2 |R_N(\omega, \mathbf{e}_r - \mathbf{e}_{axis})|^2, \quad (26a)$$

where the first factor in the right member is the weighted power of the signal and the magnitude squared of the normalized response is called the array receive beam pattern

$$B(\omega, \mathbf{e}_r - \mathbf{e}_{axis}) = |R_N(\omega, \mathbf{e}_r - \mathbf{e}_{axis})|^2$$

$$= \left| \left(\sum_{n=1}^{N} w_n\right)^{-1} \sum_{n=1}^{N} w_n \exp\left(j\frac{\omega}{c}\mathbf{r}'_n \cdot (\mathbf{e}_r - \mathbf{e}_{axis})\right) \right|^2 . \quad (26b)$$

The maximum value of the beam pattern is in the direction of the incoming signal and is equal to unity. At all other look directions the beam pattern has a value between unity and zero.

In the event that the beamformer weights are all unity, the beam pattern expression reduces to

$$B(\omega, \mathbf{e}_r - \mathbf{e}_{axis}) = \left| \frac{1}{N} \sum_{n=1}^{N} \exp\left(j\frac{\omega}{c}\mathbf{r}'_n \cdot (\mathbf{e}_r - \mathbf{e}_{axis})\right) \right|^2, \quad (26c)$$

which is analogous to (10c) for transmitters. Because the beam pattern expression for receivers has the same form as for transmitters, the beam pattern graphs discussed in the proceeding for transmitters can also be viewed as receiving beam patterns.

The receive beam pattern concept does not extend well to broad band signals because, as shown in (21), the beam pattern is a function of the received signal. However, for narrow band signals the modulation is relatively constant over time periods comparable to the time of signal transit across the array and (21) can be approximated

$$E(t, \mathbf{e}_r - \mathbf{e}_{axis}) = C' m(t) \exp(j\omega_0 t) R_N(\omega_0, \mathbf{e}_r - \mathbf{e}_{axis}) . \quad (27a)$$

The power is immediately found to be

$$P(t, \mathbf{e}_r - \mathbf{e}_{axis}) = |E(t, \mathbf{e}_r - \mathbf{e}_{axis})|^2 = C'^2 |m(t)|^2 B(\omega_0, \mathbf{e}_r - \mathbf{e}_{axis}), \quad (27b)$$

where the beam pattern is evaluated at the carrier frequency. When the array look direction is the source direction the output power is the scaled acoustic power.

The validity of the approximation in (27) hinges upon the relative magnitudes of the time over which the modulation is constant and the time of signal transit over the array. Considering the modulation to be constant over the time period 1/W, where W is the signal bandwidth, and the time of signal transit over the array to be D/c, where D is the array dimension along the direction of propagation, the narrow band approximation requires the ratio WD/c to be less than unity. For a typical towed array surveillance scenario in which the bandwidth is, say, less than 10 Hz and an appropriate array dimension is, say, less than 150m, the ratio would be less than unity and the approximation in (27) would be valid. However, if the approximation were not valid, beamforming would proceed as normal and the actual beam pattern would be found to be signal dependent and somewhat more complicated than that implied by (27). In any event, the broad band signal is always received undistorted along the beam axis.

Beamformed Noise Power

In addition to the signal component out of the array there is a noise component also. Since the two are statistically independent their powers add. The noise power out of the array is given by

$$\mathbf{E}|E_n|^2 = \left(\frac{C}{N}\right)^2 \sum_{n=1}^{N} \sum_{m=1}^{N} w_n w_m \mathbf{E}\left(n_n(t-\tau_n) n_m^*(t-\tau_m)\right), \tag{28a}$$

where the operator $\mathbf{E}$ is the expectation operator and E_n is the electrical output due to noise. The term in the double sum, which is the expectation of the product of the noise at two hydrophone elements, is defined to be

$$\mathbf{E}\left(n_n(t-\tau_n) n_m^*(t-\tau_m)\right) = \sigma_a^2 \rho_{n,m}, \tag{28b}$$

where σ_a^2 is the power of the analytic noise at each hydrophone element and $\rho_{n,m}$ is the cross correlation coefficient between the time delayed analytic noise at hydrophone elements n and m. The cross correlation coefficient is unity at n = m, i.e. $\rho_{n,n} = 1$, and is complex with a magnitude between ±1 for $n \neq m$. It also satisfies the conjugate relation $\rho_{n,m} = (\rho_{m,n})^*$. Collecting results, (28a) gives

$$\begin{aligned}\mathbf{E}|E_N|^2 &= \sigma_a^2 \left(\frac{C}{N}\right)^2 \sum_{n=1}^{N} \sum_{m=1}^{N} w_n w_m \rho_{n,m} \\ &= \frac{\sigma_a^2 C^2}{N} \left(\frac{1}{N} \sum_{n=0}^{N-1} w_n^2\right) \left(1 + 2\left(\sum_{n=0}^{N-1} w_n^2\right)^{-1} \sum_{n=2}^{N} \sum_{m=1}^{n-1} \mathrm{Re}\, w_n w_m \rho_{n,m}\right).\end{aligned} \tag{28c}$$

We see that, if the element-to-element correlation were zero, the output noise power would be the transducted acoustic noise reduced by, nominally, 1/N which means the array beamformer rejects all but about 1/N of the in-water acoustic noise. This suggests that increasing the number of hydrophone elements would further reduce the noise at the beamformer output. This is only partly correct since increasing the number of hydrophone elements within the array aperture could also increase the element-to-element correlation of the noise which would tend to nullify the benefit. For example, if all beamformer weights and correlation coefficients were unity, the output noise would be $C^2\sigma_a^2$ which asserts that the beamformer would provide no noise rejection. To minimize the output noise power, the array aperture must contain a maximum number of hydrophone elements but they must be spaced sufficiently far apart to decorrelate the element-to-element noise. This is found to be at least half an acoustic wave length which is discussed in Section 8.2.

Combining with the signal power, i.e. the magnitude squared of the inverse transform of (25b), the broad band signal-to-noise ratio out of the array is

$$\mathrm{SNR}_A = \frac{\left(|m|^2/2\sigma^2\right)\left(\sum_{n=0}^{N-1} w_n\right)^2 \Big/ \left(\sum_{n=0}^{N-1} w_n^2\right)}{\left(1 + 2\sum_{n=2}^{N}\sum_{m=1}^{n-1} \mathrm{Re}\, w_n w_m^* \rho_{n,m} \Big/ \sum_{n=0}^{N-1} |w_n|^2\right)}, \tag{29a}$$

where σ^2 is the power of the real noise and the leading factor in the numerator is the signal-to-noise ratio out of one hydrophone element. The trailing factor is the processing gain factor and the processing gain is

$$\mathrm{PG} = 10\mathrm{Log}\left[\left(\sum_{n=0}^{N-1} w_n\right)^2 \Big/ \left(\sum_{n=0}^{N-1} w_n^2\right)\right]. \tag{29b}$$

For the case of unity hydrophone weights the output signal-to-noise reduces to

$$\mathrm{SNR}_A = \frac{N\left(|m|^2/2\sigma^2\right)}{\left(1 + \frac{2}{N}\sum_{n=2}^{N}\sum_{m=1}^{n-1} \mathrm{Re}\,\rho_{n,m}\right)}. \tag{29c}$$

In the event of uncorrelated noise at all hydrophones, i.e. half wavelength spacing of the array elements, the output signal-to-noise ratio reduces to

$$\mathrm{SNR}_A = N\left(\frac{|m|^2}{2\sigma^2}\right). \tag{29d}$$

The output signal-to-noise ratio of the array is increased by the number of hydrophone elements because the array suppresses the noise by a factor of 1/N.

However, if aperture weighting is applied, since the processing gain is bounded by the number of hydrophone elements, $(\Sigma w_n)^2/(\Sigma w_n^2) \le N$ by Cauchy's inequality[9], the effect of the aperture weighting is to suppress less noise and reduce the output signal-to-noise ratio.

Array Gain and Receiving Directivity Index

Array gain is a measure of array performance. It is the difference between the signal-to-noise level out of the array and the average signal-to-noise level out of the individual hydrophones

$$AG = 10\text{Log}\left(\frac{SNR_A}{SNR_H}\right). \tag{30a}$$

From (29a)

$$AG = 10\text{Log}\left(\frac{\left(\sum_{n=0}^{N-1} w_n\right)^2 / \left(\sum_{n=0}^{N-1} w_n^2\right)}{\left(1 + 2\sum_{n=2}^{N}\sum_{m=1}^{n-1} \text{Re}\, w_n w_m^* \rho_{n,m} / \sum_{n=0}^{N-1} |w_n|^2\right)}\right), \tag{30b}$$

which, for half wave length hydrophone element spacing reduces to AG = PG. If the aperture weights are unity

$$AG = 10\text{LogN}. \tag{30c}$$

In general the array gain is improved by increasing the number of hydrophone elements as long as the element separation is great enough that the element-to-element noise is decorrelated. For a typical towed array the number of hydrophone elements is on the order of 100 for an array gain of 20 dB.

Alternative expressions for the signal and noise power out of the array beamformer can be developed in terms of the azimuthal and frequency distributions of signal and noise in the vicinity of the array. Since these quantities have been tabulated for many ocean basins by various naval research programs and, in principle, can be measured at sea on an operational basis, these expressions are particularly useful. The spectral densities of the signal and noise output are given by

$$S_A(\omega) = \int_{4\pi} S(\omega, \mathbf{e}_r) B(\omega, \mathbf{e}_r - \mathbf{e}_{axis}) d\Omega, \tag{31a}$$

$$N_A(\omega) = \int_{4\pi} N(\omega, \mathbf{e}_r) B(\omega, \mathbf{e}_r - \mathbf{e}_{axis}) d\Omega, \tag{31b}$$

where $S(\omega,\mathbf{e}_r)$ and $N(\omega,\mathbf{e}_r)$ represent the azimuthal-frequency distributions of signal and noise power in the region of the array. More explicitly, these functions give,

[9] Abramowitz and Stegun (3.2.9).

respectively, the signal and noise power, per unit solid angle and per unit Hertz, of the acoustic energy that is propagating from the direction $\mathbf{e}_R$ at the frequency ω.

For an incoming plane wave signal that is propagating along the beam axis the azimuthal-frequency spectrum is

$$S(\omega, \mathbf{e}_r) = \frac{|A|^2}{2} \delta(\mathbf{e}_r - \mathbf{e}_{axis}) \,, \tag{32a}$$

where $|A|^2/2$ is the power of the real signal and the signal power out of the array from (31a) is

$$S_A(\omega) = \frac{|A|^2}{2} \int_{4\pi} \delta(\mathbf{e}_r - \mathbf{e}_{axis}) B(\omega, \mathbf{e}_r - \mathbf{e}_{axis}) d\Omega = \frac{|A|^2}{2} \,. \tag{32b}$$

Similarly, for the case of a single hydrophone element the beam pattern is unity in all directions

$$S_H(\omega) = \frac{|A|^2}{2} \int_{4\pi} \delta(\mathbf{e}_r - \mathbf{e}_{axis}) d\Omega = \frac{|A|^2}{2} \,. \tag{32c}$$

which shows that the beamformer has no effect on the received signal. The noise power out of the array is given by (31b) and which also shows that the noise out of a single hydrophone is just the spectral density of the background acoustic noise

$$N_H(\omega) = \int_{4\pi} N(\omega, \mathbf{e}_r) d\Omega = N(\omega) \,, \tag{32d}$$

where $N(\omega)$ is obtained from a standard spectral analysis of a single hydrophone output. Collecting results the array gain can be expressed

$$AG = 10 \mathrm{Log}\left(\frac{S_A / N_A}{S_H / N_H} \right) = 10 \mathrm{Log}\left(N(\omega) \left(\int_{4\pi} N(\omega, \mathbf{e}_r) B(\omega, \mathbf{e}_r - \mathbf{e}_{axis}) d\Omega \right)^{-1} \right) . \tag{32e}$$

This result shows that for an omnidirectional receiver the array gain is zero and that for a receiver with a beam pattern the array gain is always greater than zero.

For the case of uniform noise within the region of the array the azimuthal-frequency spectrum is

$$N(\omega, \mathbf{e}_r) = \frac{N(\omega)}{4\pi} \,, \tag{33a}$$

and the array gain expression in (32e) reduces to the receiving directivity index

$$AG = DI_R = 10\text{Log}\left(4\pi / \left(\int_{4\pi} B(\omega, \mathbf{e}_r - \mathbf{e}_{axis})d\Omega\right)\right)$$

$$= 10\text{Log}\left(4\pi / \left(\int_0^{2\pi} d\varphi \int_{-\pi/2}^{\pi/2} d\theta \cos\theta B(\omega, \theta, \varphi)\right)\right), \qquad (33b)$$

which is identical in form to the transmitting directivity index in (16d). The receiving directivity index is the difference between the array output signal-to-noise level for the beamformer and the output signal-to-noise level for a hydrophone within an omnidirectional noise field. The receiving directivity index provides an approximate measure of array performance but it is the array gain that is the real measure of array performance within the ocean environment and background noise field. The transmitting directivity index for a line array projector, found in Example 2, applies also to a hydrophone line array. From that example we obtain the expression DI_R = 10logN which is the same as (30c) that was derived for a general array.

APPENDIX 4A: RADIATION FROM A PULSATING SPHERE

In Chapter 4 we determined the radiation from a vanishingly small pulsating sphere. That solution was then used as an elemental source to develop an expression for the radiation from an arbitrary distribution of sources. Because of the general importance of the spherical source, in the following we determine the radiation from a pulsating sphere of arbitrary size.

The field radiated by the sphere satisfies the homogeneous wave equation in (2.17-A)

$$\nabla^2 p(t,r) - \frac{1}{c^2}\ddot{p}(t,r) = 0. \tag{1a}$$

For a sphere pulsating at the angular frequency ω, the surface velocity can be written

$$v(t) = v_0 \exp(j\omega t), \tag{1b}$$

where v_0 is a constant complex amplitude.

Let $p(t,r) = p(r)\exp(j\omega t)$ and substitute into (1) to get the Helmholtz equation

$$\frac{1}{r^2}\frac{\partial}{\partial r}\left(r^2 \frac{\partial p(r)}{\partial r}\right) + k^2 p(r) = 0, \tag{2a}$$

where $k = \omega/c$ is the wave number. Making the variable change $z = kr$ in (2a), we get the differential equation

$$z^2 p''(r) + 2zp'(r) + z^2 p(r) = 0. \tag{2b}$$

The solution of this equation for outgoing waves is the spherical Hankel function of the second kind[1]

$$p(r) = Ah_0^{(2)}(kr), \tag{2c}$$

where A is a constant to be determined.

Since the radiated pressure matches the pressure exerted by the pulsating sphere at each point on the surface, we obtain from Newton's second law in (2.4c-A) the boundary condition

$$\lim_{r \to a}\left[-\frac{\partial p(t,r)}{\partial r}\right] = \rho_0 \dot{v}_r(t). \tag{2d}$$

This boundary condition gives, on the face of a sphere of radius a, the force per unit volume on each fluid particle. Substituting from (1b) and (2c) this boundary condition becomes

---

[1] Abramowitz and Stegun (10.1.1).

$$Akh_1^{(2)}(kr) = j\omega\rho_0 v_0, \tag{2e}$$

where[2]

$$\frac{\partial h_0^{(2)}(kr)}{\partial r} = -kh_1^{(2)}(kr). \tag{2f}$$

Solving for A, we obtain the expression for the radiated field

$$p(t,r) = \frac{j\omega\rho_0 v_0 h_0^{(2)}(kr)}{kh_1^{(2)}(ka)} \exp(j\omega t). \tag{3}$$

This is the general solution for the radiated field of a sphere pulsating uniformly at the angular frequency ω.

At a range from the sphere for which kr >> 1 the asymptotic form of the Hankel function[3] gives the result

$$p(t,r) = \frac{j\omega\rho_0 v_0}{kh_1^{(2)}(ka)r} \exp(j\omega(t - r/c)). \tag{4}$$

The Hankel function in the denominator can be expanded to obtain two important limiting cases. For very large spheres such that ka >> 1 replace the Hankel function by its asymptotic expansion to obtain

$$p(t,r) = \frac{\rho_0 c v_0 a}{r} \exp(j\omega(t - (r-a)/c)). \tag{5a}$$

For very small spheres such that ka << 1 replace the Hankel function by its small argument expansion[4] to obtain

$$p(t,r) = \frac{j\omega\rho_0 v_0 a^2}{r} \exp(j\omega(t - r/c)). \tag{5b}$$

Comparing the two limits, we see that the pressure amplitude radiated from the large sphere is independent of frequency but for the small sphere it is proportional to frequency. Comparing (5b) to the radiated pressure for a vanishingly small sphere in (4.1g) shows them to be equivalent where $Q(t) = 4\pi a^2 v_0 \exp(j\omega t)$ in (4.1g).

---

[2] Abramowitz and Stegun, (10.1.24).

[3] Abramowitz and Stegun, (10.1.1) and (9.2.4).

[4] Abramowitz and Stegun, (10.1.1) and (9.1.9).

APPENDIX 4-B: INVERSION OF THE HELMHOLTZ EQUATION

We obtained an expression for the acoustic field radiated by an extended source using heuristic methods in (4.3b). In the following we rederive that result from first principles and also generalize it to include the acoustic fields in the presence of scattering surfaces. We accomplish this by inverting the Helmholtz differential equation into an integral equation. The differential equation form of the wave equation is especially useful when the acoustic source configuration and/or scattering surfaces are coincident with one of the orthogonal surfaces defined by a curvilinear coordinate system, e.g. rectangular, cylindrical, spherical, etc. However, while some acoustic transducer configurations and scattering bodies can be approximated by such surfaces, many cannot be. Consequently, it is useful to integrate the differential form of the wave equation to produce an integral form in which the radiated field is represented by a volume integral over sources and the scattered field is represented by a surface integral over scattering surfaces. The integration of the differential form, performed in the following, leads to the Helmholtz integral formula with free-field Green's function kernels. These concepts are developed in the following and the concept of the Green's function is further developed in Appendix 7-A.

The wave equation for ocean media was found in (2.17-A), to be

$$\nabla^2 p(t,\mathbf{r}) - \frac{1}{c^2}\ddot{p}(t,\mathbf{r}) = -\rho_0 \dot{q}(t,\mathbf{r}). \tag{1a}$$

The solutions of this equation are matched to the problem of interest by means of the boundary conditions

$$a p(t,\mathbf{r}_S) + b\nabla p(t,\mathbf{r}_S)\cdot\mathbf{e}_S = f(t,\mathbf{r}_S), \tag{1b}$$

$$p(t_0,\mathbf{r}) = u_0(\mathbf{r}), \qquad \dot{p}(t_0,\mathbf{r}) = v_0(\mathbf{r}). \tag{1c}$$

The wave equation in (1a) is inhomogeneous because of the source term in the right member. This means that, when the field coordinates t and $\mathbf{r}$ are within the domain of the active source, the source term is nonzero but, when the field coordinates are exterior to the domain of the active source, the right member is zero and the equation is homogeneous. An infinity of solutions are found to satisfy the wave equation and it is necessary to select the solution that is appropriate to the problem at hand. This is done by choosing that solution that satisfies the boundary conditions. The boundary condition in (1b) specifies the field and its gradient on the scattering surfaces at positions $\mathbf{r}_S$ and for times t. Further, if these surfaces should happen to be at infinity, i.e. there are no scattering surfaces within the region of interest, the acoustic pressure and its gradient are assumed to vanish on the boundary and, it follows, that this boundary condition also vanishes. The boundary condition in (1c) gives conditions for the pressure and its rate of change at one point in time. These boundary conditions play a role in the development of the integral formulation of the wave equation. Prior to developing this representation, we derive Gauss' Divergence Theorem and Green's Theorem for bounded volumes.

4.1-B: Gauss' Divergence Theorem

Consider an arbitrary vector field within a volume enclosed by a surface which, for the development of Gauss' divergence theorem, we assume to be of finite extent. Later, we let the bounding surface extend to infinity. The volume can be taken to be the ocean for which the principle bounding surfaces would be the surface, which is generally rough and in motion, and the bottom, which is defined by its topography and materials. Other surfaces within the volume would be the exterior surfaces of submarines, biologics, bubbles, particles, etc.

From (2.3-B), the divergence of the vector field at each point of the medium is given by

$$\nabla \cdot \mathbf{F}(t,x,y,z) = \frac{\partial F_x(t,x,y,z)}{\partial x} + \frac{\partial F_y(t,x,y,z)}{\partial y} + \frac{\partial F_z(t,x,y,z)}{\partial z} . \tag{2a}$$

This expression represents the net depletion or net accumulation of the vector quantity **F** at the point, e.g. if the vector **F** represented the fluid velocity at the point, the divergence would give the spatial change in velocity as a fluid particle passed through the point. Particles are arriving from various directions and leaving in various directions and, if the divergence were positive, the particles would, on the average, be leaving the point faster than they were arriving at the point.

Multiplying (2a) by the size of a volume element gives

$$\nabla \cdot \mathbf{F}\, dxdydz = \partial_x F_x\, dydz + \partial_y F_y\, dxdz + \partial_z F_z\, dxdy , \tag{2b}$$

where $\partial_x F_x$ denotes the partial differential with respect to x of the x-component of the vector and similarly for the other terms. Expanding the x-component of (2b) gives

$$\partial_x F_x\, dydz = F_x(x+\delta x, y, z)dydz - F_x(x,y,z)dydz . \tag{2c}$$

Each term in the right member of (2c) can be viewed as the inner product between the x-component of the vector field and a surface vector pointing outward from the volume as is depicted in Figure 1. In equation form this can be written

$$\partial_x F_x\, dydz = F_x(x+\delta x, y, z)\mathbf{i} \cdot \mathbf{e}_s ds_x + F_x(x,y,z)\mathbf{i} \cdot \mathbf{e}_s ds_x , \tag{2d}$$

where **i** is the unit vector pointing in the positive x-direction, $\mathbf{e}_S$ is a unit vector that points outward from the volume element on each face and ds_x = dydz is the magnitude of the face area. Note that in the second term of the right member of (2d) the vectors **i** and $\mathbf{e}_S$ are opposed and their inner product is −1. Collecting all terms in (2b) gives

$$\nabla \cdot \mathbf{F}\, dv(x,y,z) = \sum_{n=1}^{6} \mathbf{F} \cdot \mathbf{s}_n , \tag{2e}$$

where dv(x,y,z) = dxdydz is the volume of the volume element at the point x, y, z and where $\mathbf{s}_n$, n = 1,2,···,6 denotes the nth surface vector that is defined to be pointing outward from the nth face of the volume element and which has magnitude equal to the area of the face.

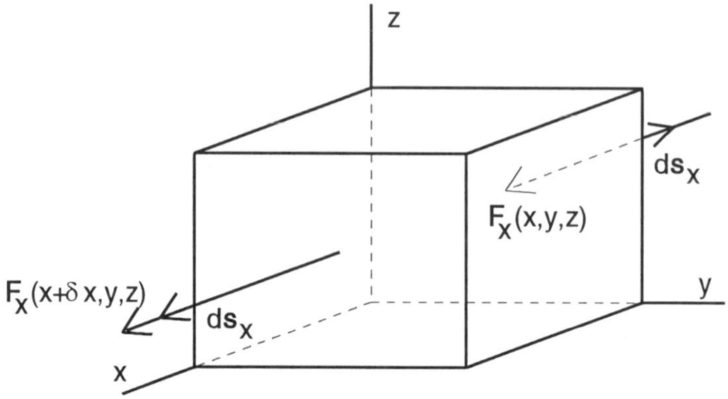

Figure 1 Vector Component Change at a Point within Medium

The result in (2e) can be readily extended to arbitrary volumes. Consider such a volume that is bounded by a surface exterior to the volume and that may be bounded by one or more surfaces interior to the volume. Summing both members of (2e) over all volume elements within the volume, it is seen that the terms in the left member simply accumulate over all volume elements. However, for the terms in the right member, the sum of terms for element faces that face each other cancel because the area vectors of the these faces are opposed. This leaves only the terms for the element faces that lie in the exterior and interior bounding surfaces to accumulate. The result obtained is Gauss' Divergence Theorem

$$\int_V dv \nabla \cdot \mathbf{F} = \int_S d\mathbf{s} \cdot \mathbf{F}, \tag{3}$$

where V and S denote, respectively, that the integrations are to be carried out over the entire volume and over all bounding surfaces both exterior and interior to the volume and **ds** points outward from the volume.

4.2-B: Green's Theorem

Green's Theorem is obtained directly from Gauss' Divergence Theorem. The vector in Gauss' Divergence Theorem in (3) can be constructed from two scalar functions, that are functions of the space coordinates within the volume and on its bounding surfaces. The construction is achieved by multiplying one of the functions by the gradient of the other function. For the two functions u and w, the resulting vector will be different depending upon which function that the gradient is applied to and we choose both possibilities

$$\mathbf{F}_1 = u\nabla w, \qquad \mathbf{F}_2 = w\nabla u. \tag{4a}$$

The divergences of these vectors are

$$\nabla\cdot\mathbf{F}_1 = u\nabla^2 w + \nabla u\cdot\nabla w, \qquad \nabla\cdot\mathbf{F}_2 = w\nabla^2 u + \nabla w\cdot\nabla u, \tag{4b}$$

where the divergence of the gradient is the Laplacian, $\nabla\cdot\nabla = \nabla^2$. Substituting (4b) alternately into Gauss' Divergence Theorem in (3) gives

$$\int_V dv\left(u\nabla^2 w + \nabla u\cdot\nabla w\right) = \int_S d\mathbf{s}\cdot u\nabla w\,, \tag{5a}$$

$$\int_V dv\left(w\nabla^2 u + \nabla w\cdot\nabla u\right) = \int_S d\mathbf{s}\cdot w\nabla u\,. \tag{5b}$$

Subtracting gives Green's Theorem

$$\int_V dv\left(u\nabla^2 w - w\nabla^2 u\right) = \int_S d\mathbf{s}\cdot\left(u\nabla w - w\nabla u\right). \tag{6}$$

Because of continuity requirements on the functions u and w and their derivatives, it would be a stretch to say that (6) is a mathematical identity that is correct for arbitrary volumes enclosed by arbitrary surfaces in which arbitrary functions u and w are defined. However, for the acoustic fields and media encountered in ASW ocean problems, Green's theorem can always be taken as valid.

4.3-B: Integral Formulation for the Inhomogeneous Helmholtz Equation

We now turn to the development of an integral solution of the Helmholtz equation. The process we follow is to first Fourier transform the wave equation into the time independent inhomogeneous Helmholtz equation. We develop and solve the Helmholtz equation for the case of a point source which leads to the free-field Green's function. We then use the free-field Green's function to invert the general Helmholtz equation into integral form. Note that in this process we are implicitly assuming that the source function and boundary conditions are known for all times or, at least, for all times during a sonar transmission cycle. Practically speaking, this is a matter of small consequence but, from a logical standpoint, it implies that own-ship and target movements are known during the next transmission interval - an assumption of little validity in an ASW scenario. For the present, we accept the assumption and rectify this theoretical problem in Appendix 7-A.

Helmholtz Equation

The acoustic fields and source functions in the original wave propagation problem in (1) can be expressed by their Fourier transforms

$$p(t,\mathbf{r}) = \frac{1}{2\pi} \int_{-\infty}^{\infty} P(\omega,\mathbf{r}) \exp(j\omega t) d\omega \,, \tag{7a}$$

$$q(t,\mathbf{r}) = \frac{1}{2\pi} \int_{-\infty}^{\infty} Q(\omega,\mathbf{r}) \exp(j\omega t) d\omega \,. \tag{7b}$$

Substituting these expressions into the wave equation in (1a) gives the Helmholtz equation

$$\nabla^2 P(\omega,\mathbf{r}) + k^2 P(\omega,\mathbf{r}) = -j\omega\rho_0 Q(\omega,\mathbf{r}), \tag{8}$$

where $k = \omega/c$ is the wave number. The solution $P(\omega,\mathbf{r})$ represents a snapshot of the acoustic field at $t = 0$ that would be obtained if the acoustic sources were radiating continuously at the angular frequency ω. Summing this solution over all frequencies would give a snapshot of the total field at $t = 0$ and inverse Fourier transforming gives the total field at all times. In general, this field varies in range and bearing depending upon the source scattering surface configuration. Prior to solving for this general field we consider the related free-field solution for the case of a point source.

Free-field Solution of the Helmholtz Equation for a Point Source

The free-field solution for a point source is necessarily a function of range-only that is independent of direction from the source. For this case the Helmholtz equation in (8) becomes

$$\nabla^2 P(\omega,r) + k^2 P(\omega,r) = -j\omega\rho_0 Q(\omega,r), \tag{9a}$$

where r is the spherical range coordinate which is independent of direction and the source term is zero everywhere except at $r = 0$. Using the direction independent Laplacian in spherical coordinates, the homogeneous form of (9a) can be written

$$\frac{1}{r^2}\frac{d}{dr}\left(r^2 \frac{dP(\omega,r)}{dr}\right) + k^2 P(\omega,r) = 0. \quad r \neq 0 \tag{9b}$$

Expanding and manipulating (9b) by simple methods, it can be expressed as the elementary second order differential equation

$$\frac{d^2(rP(\omega,r))}{dr^2} + k^2(rP(\omega,r)) = 0. \quad r \neq 0 \tag{9c}$$

The solution is

$$P(\omega,r) = \frac{\exp(\pm jkr)}{r}, \qquad r \neq 0\,, \tag{9d}$$

where either sign in the exponent is valid. However, in our formalism the negative sign represents an outgoing wave, which we choose, while the positive sign represents a wave collapsing on the point r = 0. Since the latter field has little relevance for ASW scenarios we ignore it.

We need to determine the source term in the inhomogeneous equation in (9a) that corresponds to the free-field solution in (9d). Since both the left member and source term of (9a) vanish everywhere except at the point at r = 0, the volume integral of (9a) over all space reduces to an integral of both members over the vanishingly small sphere in Figure 2. Performing this integration we show that the source term is a delta function.

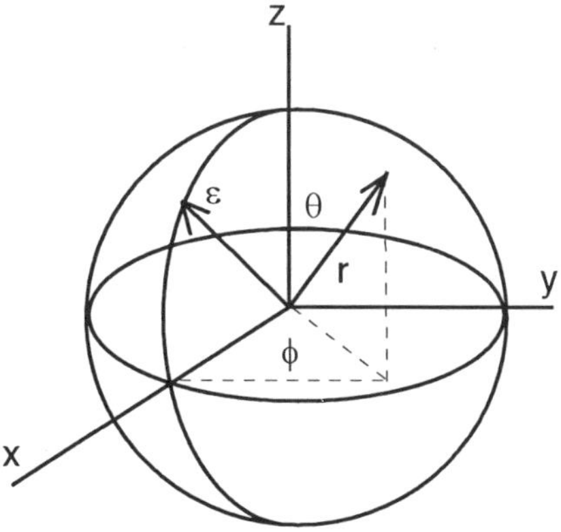

Figure 2 Excluded Point at r = 0.

Substituting (9d) into (9a) and integrating over the small sphere of radius ε we obtain

$$\int_{\text{sphere}} dv\left(\nabla^2\left(\frac{1}{r}\right)+\frac{k^2}{r}\right) = -j\omega\rho_0 \int_{\text{sphere}} dvQ(\omega, r)\,, \tag{10a}$$

where the exponential function in the left member becomes unity within the vanishingly small sphere. The first integral of the left member is computed using Gauss' theorem in (3) and the second integral is found to vanish. These integrals are, respectively,

$$\varepsilon^2 \int_0^{2\pi} d\varphi \int_0^{\pi} d\theta \sin(\theta)\left[\nabla\left(\frac{1}{r}\right)\right]_{r=\varepsilon} = -4\pi\,, \tag{10b}$$

$$\int_0^{\varepsilon} dr \int_0^{2\pi} d\varphi \int_0^{\pi} d\theta \sin(\theta)\left[r^2\left(\frac{k^2}{r}\right)\right] = 0\,. \tag{10c}$$

Collecting terms, (10a) becomes

$$-4\pi = -j\omega\rho_0 \int_{\text{sphere}} dvQ(\omega, r)\,. \tag{10d}$$

Since $Q(\omega,r)$ vanishes everywhere except at r = 0 and its integral over all space is nonzero we infer that the source term is given by

$$-j\omega\rho_0 Q(\omega, r) = -4\pi\delta(r)\,. \tag{11}$$

This shows that the free-field Helmholtz equation for a point source satisfies the equation

$$\nabla^2\left(\frac{\exp(-jkr)}{r}\right)+k^2\left(\frac{\exp(-jkr)}{r}\right)=-4\pi\delta(r)\,. \tag{12}$$

The usefulness of this result is that we can obtain the solution to the Helmholtz equation for distributed sources and scattering surfaces. The general method is known as the Green's function method.

Helmholtz Integral Formulation

The Helmholtz integral formula for distributed sources is found by the following manipulations. We first define $\mathbf{r}'$ to be an arbitrary point within the medium and express the Helmholtz equation in (8) in terms of this point

$$\nabla'^2 P(\omega,\mathbf{r}')+k^2P(\omega,\mathbf{r}')=-j\omega\rho_0Q(\omega,\mathbf{r}')\,, \tag{13a}$$

where ∇'^2 is the Laplacian operator acting on the primed coordinates. We next modify the point source Helmholtz equation in (12) by substituting $r=|\mathbf{r}-\mathbf{r}'|$

$$\nabla'^2\left(\frac{\exp(-jk|\mathbf{r}-\mathbf{r}'|)}{|\mathbf{r}-\mathbf{r}'|}\right)+k^2\left(\frac{\exp(-jk|\mathbf{r}-\mathbf{r}'|)}{|\mathbf{r}-\mathbf{r}'|}\right)=-4\pi\delta(|\mathbf{r}-\mathbf{r}'|)\,. \tag{13b}$$

Note that, by symmetry, (13b) represents the field due to a point source at either $\mathbf{r}'$ or $\mathbf{r}$. We choose the former interpretation in which the wave spreading from the source point at $\mathbf{r}'$ to the field point at $\mathbf{r}$ carries information regarding the processes at the source point. This point is more fully developed in the following. Equations (13a) and (13b) are combined by multiplying (13a) by the free-field solution in (13b); multiplying (13b) by the field solution in (13a); and subtracting the latter from the former. The difference is then integrated with respect to the $\mathbf{r}'$ coordinates over all space to obtain

$$\int_{\text{All Space}} dv'\left[\left(\frac{\exp(-jk|\mathbf{r}-\mathbf{r}'|)}{|\mathbf{r}-\mathbf{r}'|}\right)\nabla'^2P(\omega,\mathbf{r}')-P(\omega,\mathbf{r}')\nabla'^2\left(\frac{\exp(-jk|\mathbf{r}-\mathbf{r}'|)}{|\mathbf{r}-\mathbf{r}'|}\right)\right]=$$

$$\int_{\text{All Space}} dv'\left[-j\omega\rho_0Q(\omega,\mathbf{r}')\left(\frac{\exp(-jk|\mathbf{r}-\mathbf{r}'|)}{|\mathbf{r}-\mathbf{r}'|}\right)+4\pi\delta(|\mathbf{r}-\mathbf{r}'|)P(\omega,\mathbf{r}')\right]. \tag{13c}$$

The left member can be transformed to a surface integral by means of Green's theorem in (6). In the right member the term with the delta function can be immediately integrated to give

$$P(\omega,\mathbf{r}) = \frac{j\omega\rho_0}{4\pi}\int_V dv' Q(\omega,\mathbf{r}')\left(\frac{\exp(-jk|\mathbf{r}-\mathbf{r}'|)}{|\mathbf{r}-\mathbf{r}'|}\right)$$
$$+\frac{1}{4\pi}\int_S d\mathbf{s}'\left[\left(\frac{\exp(-jk|\mathbf{r}-\mathbf{r}'|)}{|\mathbf{r}-\mathbf{r}'|}\right)\nabla' P(\omega,\mathbf{r}') - P(\omega,\mathbf{r}')\nabla'\left(\frac{\exp(-jk|\mathbf{r}-\mathbf{r}'|)}{|\mathbf{r}-\mathbf{r}'|}\right)\right]. \quad (14)$$

Equation (14) is the Helmholtz integral formula. It is a solution of the Helmholtz equation once the source term and boundary conditions are specified. The integrals in (14) are weighted sums of Green's free-field solutions. The weightings are the amplitude and phase of the source elements and the pressure gradient and pressure on the scattering surfaces. Each free-field solution carries the weighting information to the observation point and the sums give the total field at the observation point. The surface element vectors point outward from the volume. If the scattering surfaces are infinitely distant from the source, (14) represents only the radiated field and inverse Fourier transforming that field produces (4.3b).

Equation (14) is particularly useful for solving steady state problems involving a single frequency component or, at most, a narrow band signal. We perform such calculations for target scattering in Chapter 7. Of course, the broad band solution of the inhomogeneous wave equation can be determined by inverse Fourier transforming (14). We defer this calculation until Appendix 7-A in which the Kirchoff broad band integral solution is developed directly.

APPENDIX 4-C: APERTURE WEIGHTING

Projector and receive array beam patterns were discussed in Chapter 4 in which we included weights for the purpose, stated there without proof, of controlling the sidelobes of the beam pattern. In the following we pursue this topic in more detail. Specifically, we discuss Harris'[1] and Nuttal's[2] criteria for "good" windows and Nuttal's methods for determining transducer weights. We then examine Hann and Hamming weightings, and their effects on the beam pattern.

The normalized response function for projectors was found, in (4.9b), to be given by

$$R_N(\omega, \mathbf{e}_r - \mathbf{e}_{axis}) = \int_S ds' w(\rho') \exp\left(j\frac{\omega}{c}\rho' \cdot (\mathbf{e}_r - \mathbf{e}_{axis})\right) / \int_S ds' w(\rho'), \qquad (1a)$$

where S is the surface of the projector face. The normalized response function for receive arrays was found, in (4.25c), to be given by

$$R_N(\omega, \mathbf{e}_r - \mathbf{e}_{axis}) = \sum_{n=0}^{N-1} w_n \exp\left(j\frac{\omega}{c}\rho_n \cdot (\mathbf{e}_r - \mathbf{e}_{axis})\right) / \left(\sum_{n=0}^{N-1} w_n\right), \qquad (1b)$$

where the sum is over hydrophone elements. In both cases the function w denotes the aperture weightings over the array. These expressions were derived, respectively, for projectors and receivers with arbitrary configuration but it's clear that either expression can apply to either projectors or receivers and that the distinction between the two is that (1a) applies to continuous arrays with continuous phase shifts while (1b) applies to discrete arrays with discrete phase shifts. To examine the effect of the aperture weights on the response function we will consider just the continuous linear array for analytical convenience.

The normalized response for the continuous linear array is readily obtained from (4.9b) and (4.12a,b) as

$$R_N(\omega, \varphi, \theta) = \int_{-L/2}^{+L/2} dz' w(z') \exp\left(\left(\frac{j\omega z'}{c}\right)(\sin\varphi - \sin\theta)\right) / \int_{-L/2}^{+L/2} dz' w(z'). \qquad (2a)$$

For the case of a rectangular weighted array, i.e. an unweighted array, this was found in (4.12c) to reduce to

$$R_N(\omega, \varphi, \theta) = \frac{\sin\left(\frac{\omega L}{2c}(\sin\varphi - \sin\theta)\right)}{\left(\frac{\omega L}{2c}(\sin\varphi - \sin\theta)\right)}. \qquad (2b)$$

---

[1] Harris, Frederic J. "On the Use of Windows for Harmonic Analysis with the Discrete Fourier Transform," Proceedings of the IEEE, Vol. 66, No. 1, January 1978.

[2] Nuttal, Albert H., "Some Windows with Very Good Sidelobe Behavior," IEEE Transactions on Acoustics, Speech and Signal Processing, Vol. ASSP-29, No. 1, February 1981.

This expression establishes a performance base line for the continuous line array against which we can compare other proposed weightings. The beam pattern is given by $|R_N(\omega,\varphi,\theta)|^2$ and a plot of the beam pattern for the unweighted line array is presented in Figure 1. To produce this plot the argument of (2b) was expressed as $\pi Lf(\sin\varphi - \sin\theta)/c$ where L is the array length and f is the frequency. The nulls in the plot are seen to occur at arguments of $\pm n\pi$, n = 1,2,⋯ and the sidelobe peaks occur at the mid-points, i.e. at the angles φ at which the numerator of (2b) is unity. The first sidelobe is found to be down about 13 dB from the mainlobe and the remaining sidelobes fall off at the rate 6 dB per octave where the fall off is determined by the denominator of (2b).

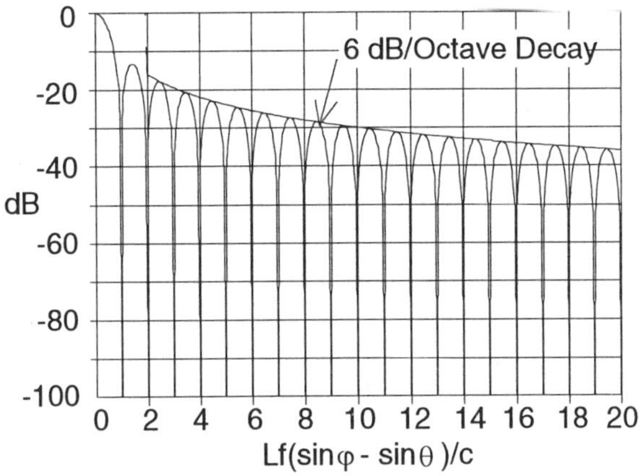

Figure 1 Beam Sidelobes - Rectangular Weighting

The precepts underlying the design of aperture weightings are based partly in science and partly in art and several sets of aperture weights have been devised using various methods - notably by Harris and Nuttal. These weightings produce the desirable properties of low and/or rapidly decaying sidelobes as compared to rectangular weighting shown in Figure 1. To examine the effects of aperture weighting we consider Nuttal's class of aperture weights represented by

$$w(z) = \frac{1}{L}\sum_{k=0}^{K} a_k \cos\left(\frac{2\pi kz}{L}\right). \tag{3a}$$

These weights are real and symmetric about the array center and are capable of representing a number of commonly used aperture weight sets as well as more sophisticated weighting sets. These weights are continuous with continuous derivatives over the array aperture but are zero outside the aperture. It is easy to show that the values of the aperture weighting and its derivatives at the array ends are given by

$$w^{(2n)}(L/2) = \frac{1}{L}\left(\frac{2\pi}{L}\right)^{2n}\sum_{k=0}^{K}(-1)^k k^{2n} a_k, \tag{3b}$$

where n = 1,2,3 ⋯ and all the odd derivatives are zero. In this calculation, z is allowed to approach the array ends from the interior of the aperture. Since at points outside the array aperture the weighting and all of its derivatives are zero, discontinuities occur at the array ends and it will be shown that these discontinuities control the rate of sidelobe decay in the beam pattern.

Substituting (3a) into (2a) we obtain the normalized response function

$$R_N(\omega,\varphi,\theta)=\frac{\sin(\alpha L/2)}{a_0(\alpha L/2)}\sum_{k=0}^{K}\frac{(-1)^k\alpha^2 a_k}{\left(\alpha^2-(2\pi k/L)^2\right)}, \tag{4a}$$

where

$$\alpha=\omega(\sin\varphi-\sin\theta)/c\,.$$

We see that, as is required, the response is unity on the beam axis, $\sin\varphi = \sin\theta$, since in the limit $\alpha = 0$ only the zeroth term of (4a) is nonzero.

Expanding (4a) in powers of $(2\pi k/\alpha L)$ we obtain

$$\begin{aligned}R_N(\omega,\varphi,\theta)&=\frac{\sin(\alpha L/2)}{a_0(\alpha L/2)}\sum_{n=0}^{\infty}\left(\frac{2\pi}{\alpha L}\right)^{2n}\sum_{k=0}^{K}(-1)^k k^{2n}a_k\\&=\frac{\sin(\alpha L/2)}{a_0(\alpha L/2)}\sum_{n=0}^{\infty}\frac{w^{(2n)}(L/2)}{\alpha^{2n}},\end{aligned} \tag{4b}$$

where the derivative expression in (3b) was used. To ensure convergence of the infinite series, the number of coefficients of the aperture weighting must be bounded, i.e. $K < \alpha L/2\pi$. The first nonzero term in (4b), i.e. the first discontinuity in the weighting or its array-end derivatives, controls the rate at which the sidelobes falloff and the falloff rate is given by $1/\alpha^{2n+1}$.

The simplest set of aperture weights is obtained when just the first coefficient in (3a) is nonzero. For this case only a_0 is nonzero and the weights are constant and equal to $w(z) = 1/L$ where, as shown below, we choose $a_0 = 1$ in this instance. These weights correspond to the unweighted case which is the rectangular weighting case. The response function for the unweighted continuous linear array is shown in (2b) and the sidelobe falloff is seen to be $1/\alpha$. A plot of (2b), on which the rate of sidelobe fall-off is indicated, is given in Figure 1.

Aperture weights that yield substantially improved sidelobe characteristics can be developed using two Fourier coefficients in (3a). To determine these coefficients we observe that, in principle, the amplitude of the aperture weighting is arbitrary and that only the shape of the weighting matters. In fact, the normalized array response is unaffected by the amplitude of the aperture weighting and, consequently, the amplitude can be chosen arbitrarily. We choose the weight at the array center such that

$$w(0)=\frac{1}{L}\sum_{k=0}^{K}a_k=\frac{1}{L},\quad\rightarrow\quad\sum_{k=0}^{K}a_k=1\,. \tag{5}$$

For the case of two coefficients this gives the relationship for the coefficients

$$a_0+a_1=1, \tag{6a}$$

and, from (3b), the weight at the array edge gives the second relationship

$$a_0 - a_1 = w(L/2)/w(0)\,, \tag{6b}$$

where the edge weight must be chosen.

For the case of Hann aperture weighting, the edge weight is in (6b) taken to be zero and the weighting is continuous at the edge boundary but its first derivative is not. Consequently, the sidelobe fall-off, from (4b), is given by $1/\alpha^3$. The Fourier coefficients are found to be $a_0 = a_1 = 0.5$ and, from (3a), the normalized weights are given by the cosine

$$\frac{w(z)}{w(0)} = 0.5 + 0.5\cos(2\pi z/L)\,. \tag{6c}$$

A plot of the array beam pattern is given in Figure 2 and is seen to be substantially better than rectangular weighting. The first sidelobe is down approximately 31 dB and the sidelobe decay is 18 dB per octave as indicated. The beam pattern nulls occur at $\alpha L/2 = \pm n\pi$, $n = 1,2,\cdots$ and the sidelobe peaks occur at intermediate points.

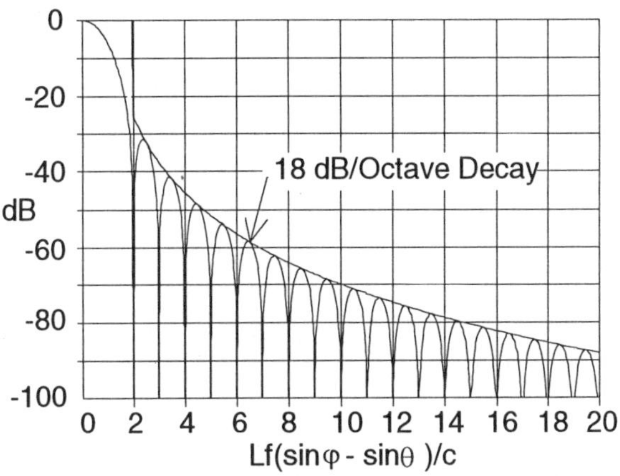

Figure 2 Beam Sidelobes - Hann Weighting

The case for a nonzero edge weight in (6b) is a little more complicated because an optimal edge weight must be determined. However, the reward is a window with greater sidelobe reduction. Expanding the response function in (4a) for two coefficients we obtain

$$R_N(\omega,\varphi,\theta) = \frac{\sin(\alpha L/2)}{\alpha L/2} + \frac{a_1}{2a_0}\left[\frac{\sin(\alpha L/2-\pi)}{(\alpha L/2-\pi)} + \frac{\sin(\alpha L/2+\pi)}{(\alpha L/2+\pi)}\right], \tag{7a}$$

where we must determine the coefficients a_0 and a_1. Examination of (7a) shows that all the sinc functions are null at $\alpha L/2 = \pm n\pi$, $n = 2,3,\cdots$ which indicates that the first opportunity for a sidelobe of the response function occurs at approximately $\alpha L/2 = 2.5\pi$. Substituting into (7a) and setting the response function to zero to suppress this first sidelobe we obtain $a_1/a_0 = 21/25$. Solving this equation simultaneously with (6a) we obtain $a_0 = 25/46 \cong 0.543478$ and $a_1 = 21/46 \cong 0.456522$. These weights are found to generate lower first sidelobes than the Hanning weights but it is also found that further reductions in the sidelobes can be obtained by "tweaking" these weights to the values $a_0 = .54$ and $a_1 = 0.46$. For these values of the coefficients the weighting is called Hamming weighting. The normalized weights are then given by

$$\frac{w(z)}{w(0)} = 0.54 + 0.46\cos(2\pi z / L)\,. \tag{7b}$$

The normalized edge weight in (7b) is found to be $w(L/2)/w(0) = 0.08$ which shows that the cosine is a raised cosine sitting on a pedestal that is about 12% of the maximum weight. Hamming weights reduce the first sidelobes of the beam pattern but, because the weights are discontinuous at the array edge, the sidelobe fall-off is only at the rate $1/\alpha$. The beam pattern with Hamming weighting is shown in Figure 3. The first sidelobe is seen to be down approximately 43 dB and the sidelobe fall-off is 6 dB per octave. The beam pattern nulls occur at $\alpha L/2 = \pm n\pi$, $n = 2,3,\cdots$, and the sidelobe peaks occur at the intermediate points.

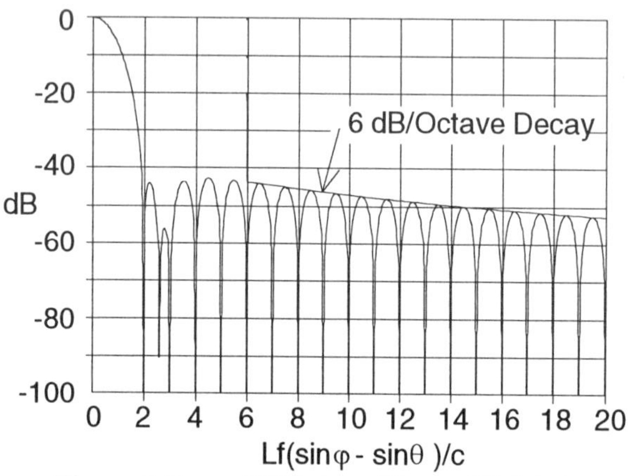

Figure 3 Beam Sidelobes - Hamm Weighting

We have derived two of the more basic aperture weightings using a technique that is not only simple but is one that is sufficiently capable of developing very effective weightings. However, in terms of methodologies for determining weightings and the available range of good weightings, we have only scratched the surface. The reader is encouraged to pursue the cited references and their references for further information on weightings.

CHAPTER 5
ACOUSTICS IN THE OCEAN

The ocean depths throughout much of the world have been mapped during the twentieth century, using mostly acoustic means, and especially equipped deep diving vessels, e.g. Trieste I and II, Alvin and Deep Quest, have actually explored the bottoms of some of the deeper regions. In contrast, virtually all of the ocean depths are unreachable by military submarines. The deep diving titanium hulled Alpha and Mike submarines can dive to approximately 1000 m, but for most submarines, including the advanced nuclear boats, dives to depths greater than about 400 m are beyond their operational capability and, for most diesel submarines, dives are limited to depths of less than about 300 m. With typical ocean depths in the range of 4000 m to 6000 m, the shallowness of the operational depths of submarines dramatizes just how limited humanities forays into the ocean have been.

The depth limitations of operational submarines might suggest that the ocean can be viewed, from the standpoint of ASW sonar operations, as a vast medium that is bounded only above by the air-sea surface. Such a model was valid in the early days of sonar when active sources could not penetrate into the ocean very far and passive receivers lacked the sensitivity required to detect targets at long ranges. However, modern sonar systems are capable of achieving detections over ranges of many 10s of nautical miles and even 100s of nautical miles both of which are distances that are large compared to the ocean depths. For these systems the ocean must be viewed as a water mass that is bounded, not only above by the air-sea interface, but also below by the ocean floor. Moreover, since one way transmissions of active SURTASS-like systems have been shown to propagate half way around the world, the ocean must sometimes be viewed as a layer of fluid on a globe that is inlaid with continental and island masses. This complicates the problem considerably because, since the depths of the oceans are approximately 1/1000 of the earth radius, the ocean depths on a global scale are comparable to the skin on Newton's apple.

Acoustic propagation in the ocean is complicated by the geometry of its bounding surfaces but its dependence upon the physics and chemistry of the ocean medium brings in even greater complications. In fact, it is found that acoustic propagation from source to target and/or target to receiver does not generally obey the "line-of-sight" rule that usually prevails in radar detection scenarios but, rather, it is found that numerous refracted propagation paths, possibly involving reflections from both surface and bottom, can exist between any two points. The refraction of the acoustic waves is due to spatial and temporal sound speed variations which cause the direction of propagation to deviate from straight lines. The net result is that acoustic signals can be concentrated within certain regions of the medium and can be absent in other regions which produces "hot spots" and "blind spots" within the medium. Since search strategies that place targets within hot spots would be rewarded by more frequent and reliable detections, while evasion strategies that place targets within blind spots would be rewarded by reducing the frequency and reliability of detections, an understanding of the acoustics of the undersea environment is important to effective operations. In addition to the problems introduced by boundaries and refraction, once an acoustic wave is set in motion it also undergoes acoustic absorption by the medium and leakages out of the path of propagation which serve to weaken the signal intensity beyond that due to geometrical spreading. In sum, ASW sonar searches for targets in the ocean are somewhat like looking through an imperfect opaque glass pane - from on edge - and the propagation conditions that are commonly found in most ocean environments are the primary reason that ASW searches for submarine targets are so difficult.

Acoustic propagation models have been developed for the dual purposes of forecasting detection performance for any given ocean basin and for optimizing target searches. These models are based upon well understood principles of ocean acoustics and have attained very high levels of sophistication but they frequently produce propagation predictions that differ disquietingly from the measured propagation values. While the basic physical principles are, mostly, not in question, the models do make errors because: their propagation equations are usually approximated to make them tractable; iterative computer based numerical algorithms entail errors that accumulate with range; and, mostly, because many of the environmental parameters that are required for accurate forecasting are unknown and are approximated. As a result the models and data collection procedures are in an on-going state of development in attempts to improve performance.

In the following sections we discuss the physics and chemistry of ocean acoustics and examine the principles of acoustic propagation in the ocean. However, rather than attempt a discussion of the latest acoustic propagation models, which is beyond the scope of our intent, we examine propagation from the standpoint of ray theory which is the basis for many propagation models. In the context of ray theory, acoustic propagation is dependent primarily upon the sound variations in the ocean, reflections at the surface and bottom boundaries and acoustic absorption within the medium. The results obtained are relatively simple but they are capable of representing the principal characteristics of ocean acoustic propagation. Chapter 6 is devoted to discussions of applications of these concepts.

5.1 Sound Speed Variations

Sound speed in the ocean can vary both spatially and temporally and the mechanisms causing the variations are due primarily to variations within the water column of:

- temperature,
- salinity,
- pressure.

Due largely to gravitational forces, these quantities are structured, roughly speaking, as level surfaces layered in horizontal strata and, consequently, it is found that the variations in sound speed are generally more strongly depth dependent than range dependent. Accordingly, the sound speed can be characterized by a set of spatially and temporally distributed sound speed depth profiles in which the sound speed variations within any profile are generally more pronounced than the sound speed variations between profiles.

Empirical measures of sound speed depth profiles throughout the world's oceans show that they are typically structured as shown in Figure 1 in which the following four distinct regions can be identified:

- surface layer,
- seasonal thermocline,
- main thermocline,
- deep isothermal layer.

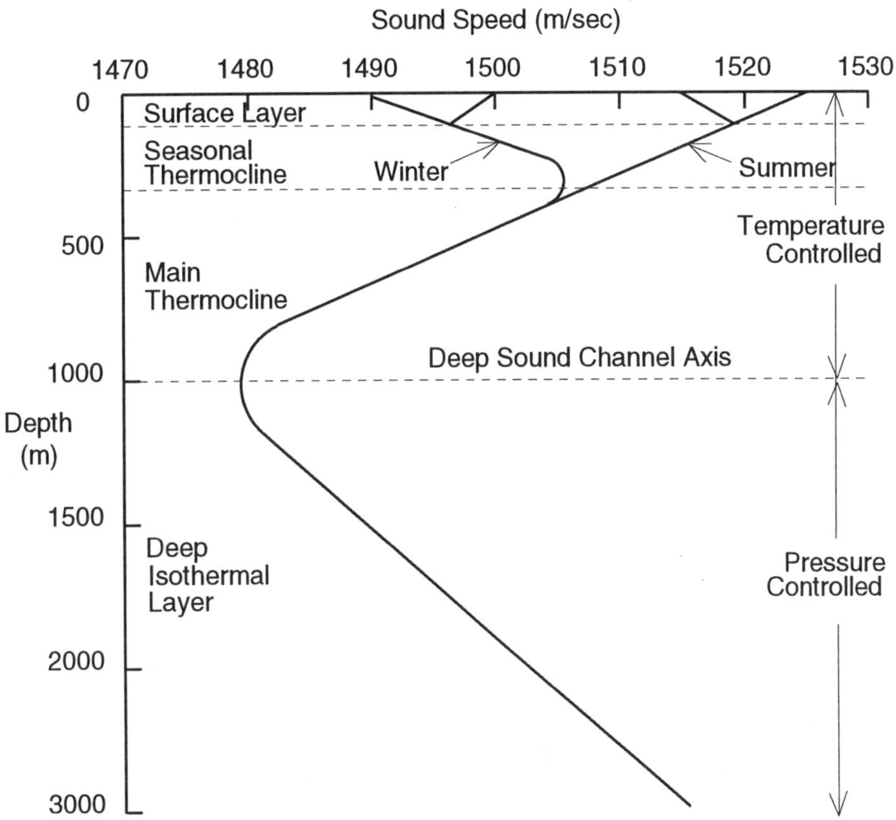

Figure 1 Typical Sound Speed Depth Profiles - Mid Latitudes

In the upper three layers the sound speed is controlled predominately by temperature while pressure is the controlling factor in the lowest layer. Salinity is generally found to be constant throughout the water column and its effects on sound speed variations in the open ocean are generally small. However, salinity effects can become dominant in the vicinity of fresh water outflows from rivers, ice packs, etc. and in these regions the flows may even cause the medium to lose its horizontal stratification.

The temperature structure above the deep isothermal layer is due to surface solar heating and mixing due to surface dynamics. The heating effect is most pronounced at equatorial latitudes, where it extends to depths of approximately 1300 m, and decreases to negligible strength and depth at polar latitudes which results in the deep sound channel axis approaching the surface as latitude increases. Below the deep sound channel axis within the deep isothermal layer, which is below the depths to which surface heating effects reach, the temperature is approximately 4 °C which is the temperature of maximum water density. Any variations in temperature at these depths, due to heat flows from geothermal sources, inflows of ice water, etc., causes the heated/cooled water mass to rise to less dense layers which maintains the temperature equilibrium at depth.

Pressure within the ocean increases linearly with depth and is due to the gravitational compression caused by the weight of the overlying water mass. This effect is operant throughout the water column and produces a linear increase in sound speed with depth. In water of constant temperature and salinity, such as in the deep isothermal layer, it is the dominant cause of sound speed variations. Other such instances include a turbulently mixed surface layer in which a positive linear sound speed gradient caused by pressure extends to the bottom of a layer and, also, in ice covered polar regions in which a positive linear sound speed gradient can extend from surface to bottom.

The surface layer is the upper most layer in the ocean and extends down to approximately 30 m to 150 m, depending upon latitude, and is caused by daily variations in weather as suggested by Figure 1. Diurnal solar heating causes an increase in surface temperature, and sound speed, in the upper parts of the layer which can result in a negative sound speed gradient in the surface layer. Similarly, nocturnal cooling causes a decrease in the surface temperature, and sound speed, in the upper parts of the layer which can produce a positive sound speed gradient in the layer. The effects of nocturnal cooling are most pronounced under clear night skies or during a snow storm. During stormy periods, which usually occur more frequently in winter, strong surface dynamics cause mixing of the surface layer which can produce constant temperature and salinity throughout the surface water column in which case the sound speed will increase with depth because of the pressure effect. The depth of the surface layer can vary on the order of 10 m over a period of hours due to internal waves.

The seasonal thermocline extends down to depths of approximately 300 m and is due to seasonal solar heating which produces temperature and sound speed structures similar to the structures in the surface layer. During summer periods of strong surface heating and quiescent surface dynamics, strong negative thermal and sound speed gradients can be established and during winter periods of severe surface cooling and quiescent surface dynamics strong positive thermal and sound speed gradients can be established as indicated in Figure 1. During seasonal transitions and strong surface dynamics, the sound speed profile can become more varied in the upper 300 m and, in general, the upper 300 m of the ocean exhibits the most variability in sound speed structure and a wide range of sound speed profiles are found for this region thoughout the world's oceans.

The main thermocline depends strongly upon geographic location but only weakly upon the season and is a relatively permanent feature in the ocean. It's lower edge ranges from approximately 1300 m at the equator to approximately 1200 m to 900 m in the middle latitudes and rises to the surface in the polar regions. At the lower edge of the main thermocline the sound speed ceases to decrease and begins to increase as the sound speed changes from a temperature controlled variable to a pressure controlled variable and the minimum sound speed defines the deep sound channel axis.

The deep isothermal layer lies below the main thermocline, beginning at the deep sound channel axis, and extending to the ocean bottom. Within the deep isothermal layer the sound speed increases with depth due to gravitational pressure which is an essentially invariant condition throughout the world's oceans. Because the sound speed increases with depth, the sound speed, within the deeper ocean basins, can be greater at the bottom than at the surface. For long range acoustic propagation this phenomena has important consequences leading to convergence zone propagation and enhanced target detectability at the surface.

The dependence of the sound speed on temperature, salinity and pressure has been empirically determined by several investigators and Leroy's equation[1] for sound speed is one such sound speed formula that is simple enough to be amenable to hand calculation. It is given by

$$\begin{aligned} c = 1492.9 + 3(T-10) - 6\times 10^{-3}(T-10)^2 \\ -4\times 10^{-2}(T-18)^2 + 12(S-35) - 10^{-2}(T-18)(S-35) + Z/61 . \end{aligned} \quad (1)$$

c = sound speed (m/s)
T = temperature (°C)
S = salinity (parts per thousand, ppt)
Z = depth (m)

Sample calculations from Leroy's equation give the surface sound speed values

c = 1534.6 m/s (5034.7 ft/s) for T = 25 °C, S = 35 ppt,
c = 1471.0 m/s (4826.1 ft/s) for T = 5 °C, S = 35 ppt.

These values of sound speed bracket the range of sound speeds that are found in the near surface regions over much of the world's oceans with 1500 m/s (4920 ft/s) being a nominal value. Typical sound speed gradients are given by[2]

$$\frac{\Delta c}{\Delta T} = 2.74 \text{ m/s per } {}^{\circ}\text{C } (5 \text{ ft/s per } {}^{\circ}\text{F}), \quad (2a)$$

---

[1] Leroy C. C., "Development of Simple Equations for Accurate and More Realistic Calculation of the Speed of Sound in Sea Water," J. Acoust. Soc. Am., Vol 46, pg. 216, 1969.
[2] Urick, Robert J., *Principles of Underwater Sound for Engineers*, McGraw-Hill Book Co. New York, NY, 1975, pg. 108.

$$\frac{\Delta c}{\Delta S} = 1.22 \text{ m/s per ppt (4 ft/s per ppt)}, \tag{2b}$$

$$\frac{\Delta c}{\Delta Z} = 0.017 \text{ m/s per m (0.017 ft/s per ft)}. \tag{2c}$$

The positive signs of these gradients shows that sound speed increases with increases in temperature, salinity and depth where, it is important to note, the depth is increasing downward into the water column.

The foregoing descriptions of sound speed variations in the ocean are generally applicable to all oceans but each ocean basin will be found to exhibit these features in a unique way. Furthermore, it may be found that the exceptions to the above descriptions, which can be caused by eddies, currents, upwelling, landmass outflows, etc., can have significant effects on acoustic propagation.

5.2 Ray Theory

Acoustic propagation in the ocean is governed by the wave equation but exact solutions of the wave equation for arbitrary ocean environments have eluded all investigators and are yet to be developed. The ray solution is an approximate solution to the wave equation that has increasing accuracy with increased acoustic signal frequency and is the basis of many propagation models. It is not capable of explaining all the nuances of acoustic propagation but it does capture the dominant propagation features for most scenarios. It is expressed in terms of acoustic wave fronts, which are defined by surfaces of constant number of wavelengths from the source, and rays, which are normals to the acoustic wave fronts and which are parallel to the medium particle displacements.

Wave Fronts and Rays - A Simple Example

Consider initially the simple free field solution having the form of a spherical wave radiating from the origin

$$p(t, r) = \frac{A}{r} \exp\left(j(\omega_0 t - k_0 r)\right). \tag{3a}$$

The amplitude A, angular frequency ω_0 and wave number $k_0 = 2\pi/\lambda_0$ are all constant. The acoustic wave fronts are defined by the surfaces of constant phase

$$\omega_0 t - k_0 r = 2\pi\,(f_0 t - r/\lambda_0) = \text{constant}. \tag{3b}$$

The simplicity of (3b) leads to simple physical interpretations that describe important properties of wave kinematics.

Equation (3b) asserts that a snap shot of a wave at a given time t, taken by a global observer, would show that the range to a wave front in any direction is constant. The value of the constant is left arbitrary in the above expression since the value chosen defines a particular wave front among the infinity of possibilities. For example, the phase of zero value applies to the wave front that radiated away from the source at r = 0 at the time t = 0. For this case, positive values of the constant apply to wave fronts radiated after t = 0 and represent the total number of cycles that were radiated after that time. Similarly, negative values of the constant apply to wave fronts radiated before t = 0 and represent the total number of cycles to be radiated by the source prior to that time. Equation (3b) also asserts that sensor measurements at a given range r, obtained by a local observer, would show that the travel time of each wave front is constant for all directions from the source. In both cases, time and range are related by the chosen constant, i.e. the wave front under consideration. The relationship is such that when both time and range are fixed they jointly define a wave front over all azimuths which, for this example, is a spherical surface nested within an infinity of concentric spherical surface wave fronts. Furthermore, the distance to each wave front with respect to the origin, in terms of acoustic wave lengths, is r/λ_0 and the total number of cycles radiated with respect to t = 0, in terms of cycles, is f_0t. The perpendiculars to the wave fronts are acoustic rays which, for a spherical wave, are the radials. The sound pressure magnitude A/r over each wave front is also constant but it decreases as the wave fronts propagate away from the source. Similar discussions apply for cylindrical and planar waves.

Wave Fronts and Rays

The simplicity of the wave and ray expressions in the preceding example was due to the assumption that the sound speed within the medium was constant. For the case in which the sound speed varies within the medium, different portions of each wave front will travel at different speeds which causes the wave front to deviate from simple spherical, cylindrical or planar shapes. For the case in which the sound speed varies spatially but is constant in time the wave fronts and rays can be represented by the generalization of the spherical wave in (3)

$$p(t, \mathbf{r}) = A(\mathbf{r})\exp\big(j(\omega_0 t - k_0 W(\mathbf{r}))\big) . \qquad (4a)$$

In this expression, the amplitude A(**r**) and the eikonal[3] function W(**r**) are allowed to vary spatially. The physical meaning of the amplitude function is analogous to the spherical wave amplitude with the exception that it can represent azimuthal variations as well as range variations. The eikonal function represents the wave fronts and, as for the spherical wave discussed above, the wave fronts are defined by the surfaces of constant phase

$$\omega_0 t - k_0 W(\mathbf{r}) = 2\pi\left(f_0 t - W(\mathbf{r}) / \lambda_0\right) = \text{constant} . \qquad (4b)$$

The number of acoustic wave lengths along a ray to the wave front is given by $W(r)/\lambda_0$ which is constant over the entire wave front. Note, though, that in a medium having a spatially varying sound speed, the wave lengths vary along a ray,

---

[3] The term 'eikon' is the German equivalent of the word icon and, in the context of ray theory, means 'an image of' or 'representation of' the wave front.

depending on the local sound speed, and the designated wavelength, denoted by λ_0, has no physical counterpart. This wavelength is sometimes defined to be the "average" wavelength found throughout the medium. However, since the average wave length is generally unknown, this is not a satisfactory definition. In fact, the designated wavelength is an arbitrary quantity which can be conveniently chosen to be, say, an approximation of the average wavelength found throughout the medium or the wave length at the sound channel axis, etc. However chosen, the designated wavelength scales the eikonal function in the sense that the magnitude of the eikonal function divided by λ_0 gives the number of physical wavelengths between the wave front and the source. Consequently, while the eikonal function represents the wave fronts of the field functionally, its value, as for the value of the designated wavelength, is a constructed quantity and does not represent a physical quantity. Having raised this issue for the purpose of gaining insight into the physics and mathematics of ray theory, we can now set it aside because, while the eikonal function leads to the ray equation, the eikonal function is not computed. However, it is important to note that the designated wavelength is an arbitrary quantity.

Ray Equations

The eikonal function represents a continuous sequence of nested wave fronts. Each wave front is defined by the locus of spatial coordinates over which the eikonal function is constant. One such wave front is defined by

$$W(\mathbf{r}) = c_0 t + \text{constant}, \tag{5a}$$

where $c_0 = f_0\lambda_0$ is the sound speed corresponding to the designated wave length. Since the designated wavelength is an arbitrary quantity, we see that the c_0 is also an arbitrary sound speed. After a short time interval dt, the locus of spatial coordinates corresponding to this wave front will change by small amounts denoted $d\mathbf{r}$. The change is dependent upon the local sound speed and is given by

$$d\mathbf{r} = \mathbf{e}_s c(\mathbf{r})dt, \tag{5b}$$

where $\mathbf{e}_s$ is the unit normal vector to the wave front which lies along a ray and $c(\mathbf{r})$ is the local sound speed in the medium. This expression asserts that the wave fronts expand normal to their surfaces along rays as is required by the definition of a wave front. Evaluating the eikonal function in (5a) on the adjacent wavefront gives

$$W(\mathbf{r} + d\mathbf{r}) = W(\mathbf{r} + \mathbf{e}_s c(\mathbf{r})dt) = c_0(t + dt) + \text{constant}, \tag{5c}$$

and a linear expansion of this expression gives

$$W(\mathbf{r}) + \nabla W(\mathbf{r}) \cdot \mathbf{e}_s c(\mathbf{r})dt = c_0(t + dt) + \text{constant}. \tag{5d}$$

Subtracting (5a) from (5d) gives

$$\nabla W(\mathbf{r})\cdot \mathbf{e}_S = \frac{c_0}{c(\mathbf{r})} = n(\mathbf{r})\,, \tag{5e}$$

where n(**r**) is called the index of refraction of the medium. This result implies $\nabla W(\mathbf{r}) = n(\mathbf{r})\mathbf{e}_S$ and the magnitude squared of this equation, $|\nabla W(\mathbf{r})|^2 = n^2(\mathbf{r})$, is the historically important eikonal equation. To determine the ray equations, we find it convenient to reexpress (5e) in terms of the definition of the directional derivative

$$\frac{d}{ds}W(\mathbf{r}) = n(\mathbf{r})\,, \tag{5f}$$

where $d/ds = \mathbf{e}_S\cdot\nabla$ is the derivative in the direction of $\mathbf{e}_S$. Computing the gradient of (5f) gives

$$\frac{d}{ds}\left(n(\mathbf{r})\mathbf{e}_S\right) = \nabla n(\mathbf{r})\,, \tag{5g}$$

which is the vector form of the ray equation. Since the arbitrary sound speed c_0 appears as a constant on both sides of this equation, we see that the particular choice of designated wavelength has no bearing on the ray solutions. We can expand (5g) into its vector components by expanding the unit normal vector to the wave fronts in terms of its direction cosines

$$\mathbf{e}_S = \alpha\mathbf{i} + \beta\mathbf{j} + \chi\mathbf{k}\,. \tag{5h}$$

In this expression, α, β and χ are the cosines of the angles between the unit vector and, respectively, the x, y and z axes. Separating (5g) into its vector components gives the scalar ray equations

$$\frac{d}{ds}\left(n(\mathbf{r})\alpha\right) = \frac{\partial n(\mathbf{r})}{\partial x}\,, \tag{6a}$$

$$\frac{d}{ds}\left(n(\mathbf{r})\beta\right) = \frac{\partial n(\mathbf{r})}{\partial y}\,, \tag{6b}$$

$$\frac{d}{ds}\left(n(\mathbf{r})\chi\right) = \frac{\partial n(\mathbf{r})}{\partial z}\,. \tag{6c}$$

The solutions of these equations give the direction cosines which define the locus of points along a ray in terms of an arbitrary sound speed profile that can vary in three dimensions. They are the basis of most computer implemented ray tracing algorithms that, effectively, solve these equations iteratively. General closed formed solutions of these equations are not available, but for special cases such solutions do exit.

For example, consider a linear sound speed profile in which the sound speed increases linearly downward into the medium, as indicated in Figure 2. The sound speed and index of refraction in the medium are given by

$$c = c_0 + gz, \qquad n = \frac{c_0}{c_0 + gz}, \tag{7a,b}$$

where c_0 is taken to be the surface sound speed. From the ray equations in (6a) and (6b) we obtain

$$\frac{d}{ds}(n\alpha) = 0, \tag{8a}$$

$$\frac{d}{ds}(n\beta) = 0. \tag{8b}$$

Figure 2 Circular Ray Path for a Linear Sound Speed Profile

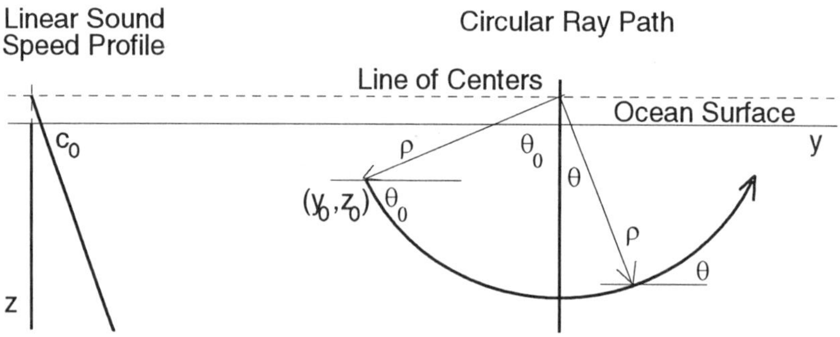

Equations (8a) and (8b) integrate to constants and, dividing the solution of (8a) by the solution of (8b), gives α/β = constant which shows that the direction of propagation of a ray in the x-y plane is constant. Consequently, without loss of generality we can consider just the rays propagating in a plane perpendicular to the horizontal x-y plane along the y axis. For these rays (8b) yields

$$n\cos\theta = \cos\theta_0, \tag{8c}$$

where θ is the angle the ray makes with the horizontal, as indicated in Figure 2, and (8c) is Snell's law. The direction cosine with respect to the z axis is $\chi = \sin\theta$ and (6c) becomes

$$\frac{d}{ds}(n\sin\theta) = \frac{dn}{dz}. \tag{9a}$$

Expanding (9a) gives the differential equation for θ

$$\frac{d\theta}{ds} = \frac{1}{n}\frac{dn}{dz}\cos\theta, \tag{9b}$$

where we note that $dz/ds = \sin\theta$. The left member of (9b) is the reciprocal of the local radius of curvature of the ray, i.e. $d\theta/ds = 1/\rho$. Substituting the linear sound speed profile from (7) into (9b) we find that the radius of curvature of the rays are constant and given by

$$\frac{1}{\rho} = -\frac{g}{c_0}\cos\theta_0 \,. \tag{9c}$$

Since the right member is constant, the ray paths in a linear sound speed profile environment are circular.

We note that for a positive sound speed gradient $g > 0$, i.e. sound speed increasing downward, the radius of curvature is negative which means the horizontal angle θ decreases along the ray. Since the definition of the direction cosines implies that θ is positive downward into the medium, these rays are refracted upward toward slower sound speed regions. Similarly, for a negative sound speed profile $g < 0$, i.e. sound speed decreasing downward, the radius of curvature is positive which means the horizontal angle θ increases along the ray and the rays are refracted downward toward slower sound speed regions. A typical circular ray for an upward refracting medium is depicted in Figure 2. We see that $\rho\cos\theta_0 = -c_0/g$ is constant for all θ_0 and, therefore, all rays emanating from the point (y_0,z_0) are defined by circular arcs having radii lying on the line of centers. Since $c = c_0 + gz$, the line of centers is positioned at the z coordinate at which the sound speed would be zero, i.e. $z(c = 0) = -c_0/g$. For small angles in the range of, say, ±10 degrees, the radii of curvatures are approximately $\rho \cong -c_0/g$. For nominal sound speed and gradient values from (1) and (2), we obtain for the radii of curvatures $\rho \cong -(1500\ \text{m/sec})/(0.017\ /\text{sec}) \cong 88$ km. Since the depth of the ocean is generally in the range 4 km to 6 km, we see that the rays follow very nearly straight line paths between encounters with the surface and bottom. Very nearly but not quite and the deviations from straight line propagation can produce dramatic effects on acoustic intensity and propagation loss.

Intensity and Transmission Loss

The intensity along a ray path is conceptually easy to understand. For a source at position x_0, y_0 and z_0, power is radiated into a bundle of rays at the launch angle θ_0 as indicated in Figure 3. The total power in the ray bundle is $P(\theta_0)d\Omega$ where $P(\theta_0)$ denotes the power per unit solid angle radiated in the launch angle direction and $d\Omega$ is the solid angle subtended by the ray bundle at launch. The intensity in the ray bundle at range is the power in the ray bundle divided by the cross sectional area of the bundle at that range where the cross section is coincident with the wave front. Accordingly, we have the total power in the ray bundle at launch equaling the total power at range

$$P(\theta_0)d\Omega = I(x - x_0, y - y_0, z - z_0)dA \,. \tag{10}$$

The coordinates x, y and z define positions along the ray path and they are functions of the initial launch angle and the time of propagation. Intensity calculations for the case of arbitrary sound speed profiles have been developed and are generally implemented as iterative computer based ray tracing algorithms because closed form solutions don't exist for the general case. However, for the purpose of gaining insight into ocean acoustic propagation, we restrict the discussions to the case of sound speed profiles that are only depth dependent for which closed form solutions do exist. Following Ugincius,[4,5] the ray structure for the depth dependent only case has axial symmetry about the z axis and the solid angle at unit distance along the ray is

$$d\Omega = 2\pi \cos\theta_0 d\theta_0 . \tag{11a}$$

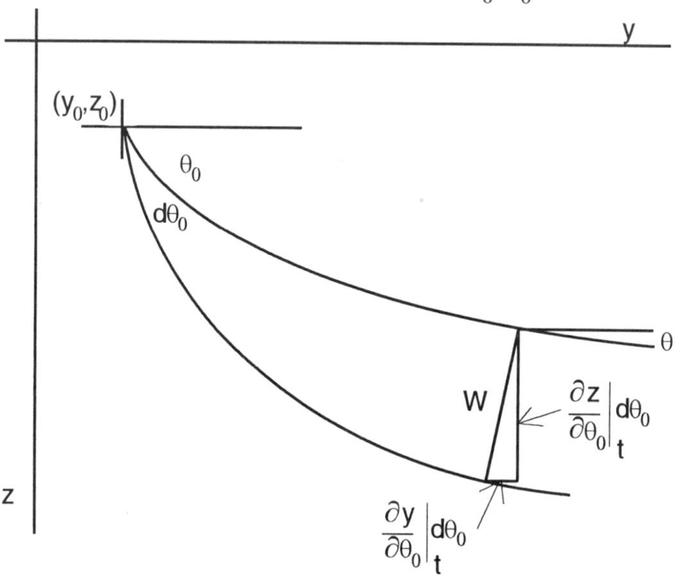

Figure 3 Ray Bundle Intensity Dimensions

Since for axial symmetry we can restrict ourselves, without loss of generality, to the y, z plane, the cross sectional area of the ray bundle at the down range point is written

$$dA = 2\pi(y - y_0)W, \tag{11b}$$

where y is the horizontal coordinate of a point on the ray path and W is the width of the ray bundle that is coincident with the wave front. The travel time to the cross section along all rays within the bundle is the same. Expressing the z coordinate along the ray path as $z(t,\theta_0)$, we see in Figure 3 that the ray bundle width is given by

$$W = \left| \left(\frac{\partial z}{\partial \theta_0} \right)_t \right| \frac{d\theta_0}{\cos\theta}, \tag{11c}$$

[4] Ugincius, Peter, "Intensity Equations in Ray Acoustics. I," J. Acoust. Soc. Am., Vol. 45, No. 1, pg. 193-205, 1969.

[5] Ugincius, Peter, "Intensity Equations in Ray Acoustics. II," J. Acoust. Soc. Am., Vol. 45, No. 1, pg. 206-209, 1969.

where, notwithstanding Figure 3, the absolute value in required for the general case. Substituting (11) into (10), the intensity along a ray path is found to be

$$I = \frac{P(\theta_0)\cos\theta_0\cos\theta}{(y-y_0)\left|\left(\frac{\partial z}{\partial\theta_0}\right)_t\right|}. \tag{12a}$$

By similar arguments the intensity at unit range from the source is found to be

$$I_0 = P(\theta_0). \tag{12b}$$

The transmission loss along a ray path is then given by

$$TL = -10\log\left(\frac{I}{I_0}\right) = 10\log\left[\frac{(y-y_0)}{\cos\theta_0\cos\theta}\left|\left(\frac{\partial z}{\partial\theta_0}\right)_t\right|\right]. \tag{13}$$

To evaluate the intensity and transmission loss expressions in (12) and (13), it is convenient to eliminate the time dependency from those relations.

The coordinate variables along a ray path can be expressed in terms of the propagation time along the ray and the launch angle of the ray or, alternatively, in terms of the distance along the ray path and the launch angle of the ray. The relationship between these representations for the z coordinate can be expressed as $z(t,\theta_0) = \zeta(s,\theta_0)$. The propagation time along the ray path and distance along the ray path are linked to make this relationship valid, i.e. $s = s(t,\theta_0)$. The functions z and ζ are numerically equal but functionally different. Differentiating this relation gives the relationship

$$\left(\frac{\partial z}{\partial t}\right)_{\theta_0} dt + \left(\frac{\partial z}{\partial\theta_0}\right)_t d\theta_0 = \left(\frac{\partial\zeta}{\partial s}\right)_{\theta_0} ds + \left(\frac{\partial\zeta}{\partial\theta_0}\right)_s d\theta_0, \tag{14a}$$

where the subscript denotes the variable held constant. Since on the wavefront z is constant over time, the first term vanishes and we can abbreviate the relationship to the form

$$Z_t d\theta_0 = \zeta'_{\theta_0} ds + \mathcal{Z}_s d\theta_0, \tag{14b}$$

where capitals denote partial differentiation with respect to θ_0; prime denotes partial differentiation with respect to s; and the subscripts denote the variables held constant in the differentiations. Manipulating we obtain

$$Z_t - \mathcal{Z}_s = \zeta'_{\theta_0}\left(\frac{\partial s}{\partial\theta_0}\right)_t. \tag{14c}$$

Since the differentials in (14a) were performed at constant time, their ratio in (14c) is expressed as a partial derivative at constant time. A similar equation can be written for the y coordinate along the ray path

$$Y_t - \Psi_s = \psi'_{\theta_0}\left(\frac{\partial s}{\partial \theta_0}\right)_t , \tag{14d}$$

and the ratio of (14c) and (14d) eliminates the partial derivative of s

$$\frac{Z_t - \mathcal{Z}_s}{Y_t - \Psi_s} = \frac{\zeta'_{\theta_0}}{\psi'_{\theta_0}} . \tag{14e}$$

Equation (14e) would give an expression for Z_t in terms of the ray path length and launch angle if Y_t could be eliminated. This can be done using the fact that at constant time the wave fronts are perpendicular to the rays. The position vector along the ray path, as indicated in Figure 4, can be expressed alternatively as

$$\mathbf{r} = y(t,\theta_0)\mathbf{i} + z(t,\theta_0)\mathbf{k} = \psi(s,\theta_0)\mathbf{i} + \zeta(s,\theta_0)\mathbf{k} . \tag{15a,b}$$

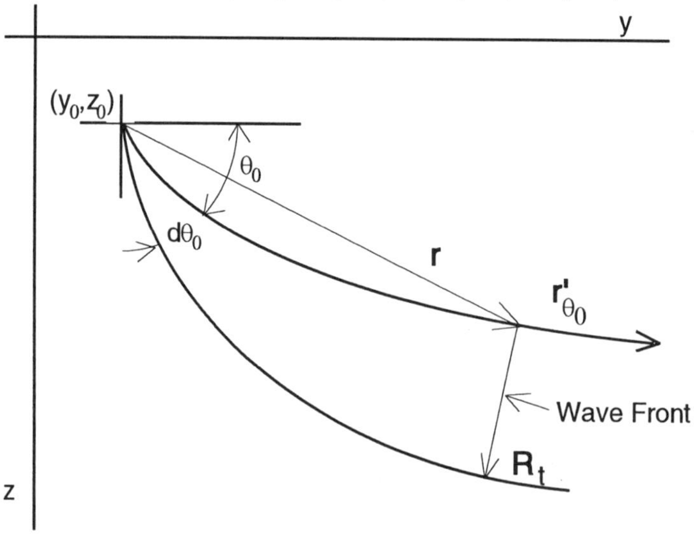

Figure 4 Position Vector Along Ray Path

Form the partial derivatives

$$\mathbf{R}_t = Y_t\mathbf{j} + Z_t\mathbf{k} , \tag{15c}$$

$$\mathbf{r}'_{\theta_0} = \psi'_{\theta_0}\mathbf{j} + \zeta'_{\theta_0}\mathbf{k} . \tag{15d}$$

Equation (15c) is a vector that is perpendicular to the ray path and (15d) is a vector that is parallel to the ray path. Consequently, forming the vector inner product, i.e. the vector dot product, we obtain the relation

$$Y_t \psi'_{\theta_0} + Z_t \zeta'_{\theta_0} = 0 . \qquad (15d)$$

This gives the required relation for Y_t. Substituting this relation into (14e) gives desired expression for Z_t

$$Z_t = \psi'_{\theta_0}\left(\psi'_{\theta_0} \mathcal{Z}_s - \zeta'_{\theta_0} \Psi_s\right). \qquad (15e)$$

Recognizing that the prime derivatives in (15e) are the direction cosines, i.e. $\psi'_{\theta_0} = \cos\theta$ and $\zeta'_{\theta_0} = \sin\theta$, and substituting (15e) into the transmission loss expression in (13) gives

$$TL = -10\log\left[\frac{(y-y_0)}{\cos\theta_0}\left|\cos\theta Z_s - \sin\theta Y_s\right|\right]. \qquad (16)$$

In (16), all functions are dependent only on the distance along a ray path and the launch angle of the ray. Also, for notational convenience, we revert to the Latin correspondents of the Greek characters, i.e. $\mathcal{Z}_s \rightarrow Z_s$ and $\Psi_s \rightarrow Y_s$.

For the case of the linear sound speed profile in (7) the ray coordinates can be determined from an inspection of Figure 2

$$y - y_0 = +\rho(\sin\theta_0 - \sin\theta), \qquad (17a)$$

$$z - z_0 = -\rho(\cos\theta_0 - \cos\theta), \qquad (17b)$$

$$\theta = \theta_0 - s/\rho . \qquad (17c)$$

We note that for the positive sound speed gradient in this example, the ray paths are upgoing and θ is decreasing. The partial derivatives of (17) are readily found to be

$$Y_s = \rho(\cos\theta_0 - \cos\theta\Theta_s) + P_s(\sin\theta_0 - \sin\theta), \qquad (18a)$$

$$Z_s = \rho(\sin\theta_0 - \sin\theta\Theta_s) - P_s(\cos\theta_0 - \cos\theta), \qquad (18b)$$

$$\Theta_s = 1 + \frac{s}{\rho^2} P_s, \qquad (18c)$$

where $P_s = (\partial\rho/\partial\theta_0)_s$. Equations (18) give all the quantities required to evaluate the transmission loss in (16) except for the partial derivative of the ray path radii. This quantity can be obtained from (9b) from which we obtain

$$\rho = -\frac{c_0}{g\cos\theta_0}, \qquad P_s = r\tan\theta_0 . \qquad (18d,e)$$

Substituting (18) into the transmission loss expression in (16) gives

$$TL = 10\log\left[\frac{(y-y_0)^2}{\cos^2\theta_0}\right]. \qquad (19)$$

This result can be shown to be valid for linear sound speed profiles having both positive and negative gradients. It shows that in a linear sound speed environment that the upgoing rays and downgoing rays incur different transmission losses. For the positive sound speed gradient in our example, the upgoing rays are found to incur a smaller loss than the downgoing rays as indicated in Figure 5. This figure shows that, for equal displacements along the ray path from the source, the spread of the upgoing rays is less than that of the downgoing rays, i.e. $(y-y_0)_U < (y-y_0)_D$. However, for these short ray paths the difference in losses over the upgoing and downgoing paths is usually negligible. In fact, to the degree that the radial distance from the source can be approximated $r \cong (y-y_0)$, we see that the short range spreading is nearly spherical. For longer ranges involving surface and bottom interactions the situation changes substantially. More about this in Chapter 6.

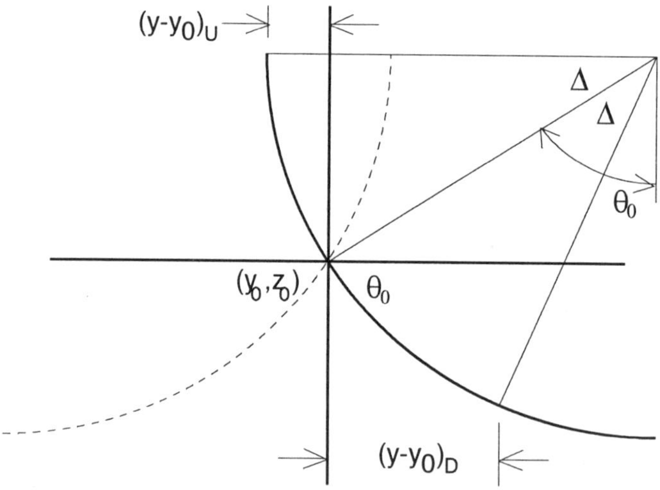

Figure 5 Spread of Upgoing and Down go -ing Rays for a Linear Sound Speed Profile

5.3 Surface Reflections

When acoustic waves from an undersea source interact with the sea surface, the energy is reflected back into the medium. Actually, while the energy is mostly reflected back into the medium, a portion of the energy is transmitted through the air-sea interface into the atmosphere above. However, we find that the transmitted portion is negligible for practical purposes, i.e. we don't search for submarines using airborne acoustic sensors. Conversely, though, when the radiated noise field from an aircraft strikes the sea surface, that small portion of the energy that is transmitted through the air-sea interface to the water below is detectable by submarine sensors. Since these detections can be used by the submarine to evade detection by the aircraft or to attack the aircraft, it is important to examine them closely and we discuss this problem in Appendix 5-B. In the following, we assume the sea surface to be impenetrable to acoustic radiations and examine the field reflected by the surface and the surface interference problem.

An acoustic source at depth radiates a spherical sound field that can propagate directly to the target and that can also reflect from the sea surface. The reflected field and the direct path field can combine to produce interference effects that can significantly affect detection performance. This analysis of the surface interference problem requires that the sea surface be calm, i.e. flat in the sense that the sea surface irregularities are small compared to an acoustic wavelength. In the more usual case that the sea surface irregularities are large or comparable to the acoustic wavelength, the phenomena of reflection from the surface (and transmission through the surface) still occurs but it does so in a random manner and the analysis becomes more difficult. Random scattering from the sea surface and sea surface reverberation are discussed in Chapter 8.

For a source at depth d, as indicated in Figure 6, the radiated field propagates to the field point over a direct path and a reflected path. For short ranges the propagation can be assumed to be line-of-sight. For a flat surface the radiated field reflects specularly and the reflected field appears to come from an image source located at elevation d above the surface. For this reason, the surface interference effect is sometimes referred to as the image interference effect. The radiated field can be represented by

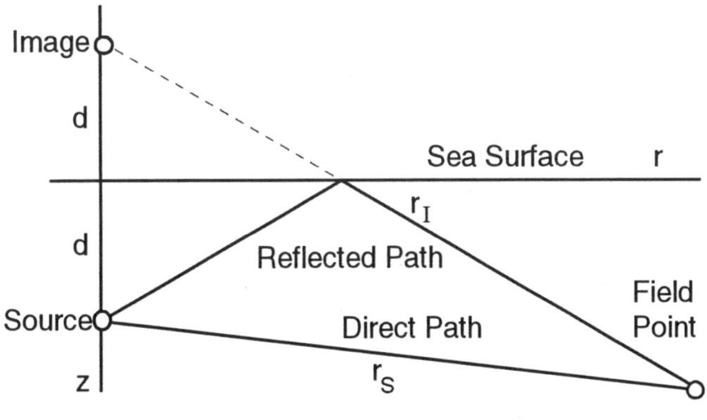

Figure 6 Reflection from Surface - Image Method

$$p_S = \frac{A_S B_S}{r_S} \exp\left(j\omega\left(t - r_S / c\right)\right), \tag{20a}$$

where A_S is the source amplitude, B_S is the source beam pattern amplitude in the direction of radiation and r_S is the range. The image field can be similarly represented by

$$p_I = \frac{A_I B_I}{r_I} \exp\left(j\omega\left(t - r_I / c\right)\right). \tag{20b}$$

Since the sea surface is a pressure release surface, sometimes called a "soft" surface, the total pressure vanishes at the surface. Then, since at the surface $B_S = B_I$, $r_S = r_I$ and $p_S + p_I = 0$, we find the image field coefficient to be given by $A_I = -A_S$.

At the field point in the medium we can approximate the ranges in terms of the cylindrical polar coordinates

$$r_S = \sqrt{r^2 + (z-d)^2} \cong r + \frac{(z-d)^2}{2r}, \tag{21a}$$

$$r_I = \sqrt{r^2 + (z+d)^2} \cong r + \frac{(z+d)^2}{2r}, \tag{21b}$$

where r is the horizontal range and z is positive downward into the medium.

Substituting the reflected field coefficient and the approximations in (21) into (20), the total field in the medium can be expressed

$$p = \frac{A}{r}\exp\left(j\omega(t - r/c)\right)\exp\left(-j\frac{k\left(z^2+d^2\right)}{2r}\right)\left(B_S\exp\left(j\frac{kzd}{r}\right) - B_I\exp\left(-j\frac{kzd}{r}\right)\right), \tag{22}$$

where $k = \omega/c$ and $A = A_S = -A_I$. The intensity in the medium is readily found from the field expression in (22) and is given by

$$I = \frac{A^2}{2\rho_0 c r^2}\left(B_S^2 + B_I^2\right)\left[1 - \frac{2B_S B_I}{B_S^2 + B_I^2}\cos\left(\frac{2kzd}{r}\right)\right]. \tag{23}$$

There are three important cases to consider in the analysis of the field intensity. The simple case occurs when neither the source beam nor image beam is directed at the field point, i.e. $B_S = B_I = 0$, and the field intensity vanishes. In this instance the target would escape detection.

Slightly more complicated is the case in which either the source beam or the image beam is directed at the field point but not both. For this case the field intensity reduces to the intensity of just one component, e.g. if $B_S = 0$, we obtain the intensity of the image component

$$I = \frac{A^2 B_I^2}{2\rho_0 c r^2}, \tag{24a}$$

where the transmission loss between source and field point is spherical. This can occur, for example, if the source beam is directed at the surface and the field point is out of the beam of the direct path energy but in the path of the reflected energy.

In the event that both the source beam and image beam are directed at the field point, the intensity in (23) fluctuates with range causing the transmission loss to vary from purely spherical spreading. For example, if $B_S = B_I = 1$, the difference between the transmission loss due to spherical spreading and the actual transmission loss can be written

$$TL - TL_{Spherical} = 10\left|\mathrm{Log}\left[\sin^2\left(\frac{kzd}{r}\right)\right]\right| - 6\ \mathrm{dB}. \tag{24b}$$

This is called the transmission anomaly due to surface interference and was historically called the Lloyd mirror effect. It is represented by a fluctuating term and a 6 dB bias term. The bias term arises because the reflection of energy back into the medium doubles the acoustic power at the field point and reduces the transmission loss that would be found due to spherical spreading. A graph of the transmission anomaly in (24b) is given in Figure 7 which shows that it fluctuates between –6 dB and infinity at short ranges. At longer ranges, where the argument of the sine function in (24b) approaches 0, i.e. $r > 4zd/\lambda$, the anomaly increases monotonically to infinity. This happens because the phase difference between the direct path signal and the reflected signal asymptotically approaches π radians causing total destructive interference. In the long range monotonic fall-off region we can rewrite (24b) by means of the small angle approximation to obtain

$$TL - TL_{Spherical} = 10\text{Log}\left(\frac{r}{kzd}\right)^2 - 6\text{ dB}. \tag{24c}$$

Inspection of (24c) shows that the transmission anomaly produces a spreading loss of 20Log(r). Added to the 20Log(r) loss due to spherical spreading, we see that the surface interference effect can produce a 40Log(r) fall-off in signal intensity at long ranges. Inspection of Figure 7 shows that the range at which the 40Log(r) fall-off begins is at the graph argument: $10[\text{Log}(r/kzd) + 1] \cong 8$.

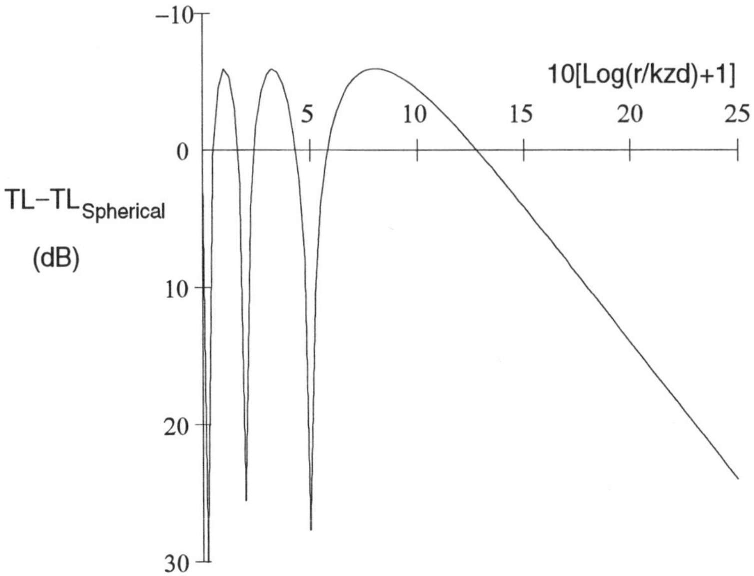

Figure 7 Transmission Anomaly due to Surface Interference Effect

Example: 1. Consider the detection of a target in an environment that is free of surface effects. We assume that the target is radiating a 100 Hz tonal at 100 dB re μPa and we assume the receiver array gain is 20 dB. For a detection threshold of 0 dB[6] and a background noise level of 70 dB at 100 Hz, the maximum allowable transmission loss would be 50 dB which corresponds to a maximum detection range of 100 km. If the surface interference effect were operative, we find for target and receiver depths of, say, 100 m, that 10[Log(r/kzd) + 1] ≅ 23.8. Scaling Figure 7 shows that an additional loss of about 22 dB would drop the received intensity level well below the detection threshold and the target would be undetectable. However, since the acoustic wavelength at 100 Hz is about 15 m, wave heights exceeding this height would disrupt the image interference effect and sporadic target detections at ranges approaching 100 km would be expected.♦

5.4 Bottom Reflections

The interaction between acoustic waves from an undersea source and the ocean bottom is, in general, complicated because the material structure of the ocean bottom is highly variable. It is mostly stratified in sedimentary layers of mud, sand, clay, rock and other materials but the layers are typically uneven in thickness and depth and have limited extent in range. Additionally, portions of the ocean floor are found to be broken by rock outcroppings, sea mounts, chasms, pockets of disparate materials and other irregularities. To deal with these complexities, surveys have subdivided the ocean floor into domains based upon the characteristics of the upper materials. The acoustic properties of each domain differ but they all share the common property of having a complicated acoustic environment in which acoustic energy incident upon the ocean floor is partly reflected or scattered back into the medium and partly transmitted into the ocean floor. In addition, the transmitted energy can be partly reflected or scattered back into the medium from subsurface layers and, if the bottom material supports shear waves, it can generate surface waves on the ocean floor. In the following, we discuss the simplest problem of a fluid-like bottom that is flat. Such bottoms are composed of sand, silt, mud, clay and various soft materials. The problem of random scattering and reverberation from irregular bottoms is touched upon in Chapter 8.

Acoustic energy from a source in the ocean that is incident on the ocean bottom, as indicated in Figure 8, is partly reflected from the bottom and partly transmitted into the bottom. For a flat bottom the reflection is specular. Since the bottom is generally a large distance from the source, it is convenient to approximate the incident field on the bottom as a locally plane wave. For this case the incident field can be represented by the expression

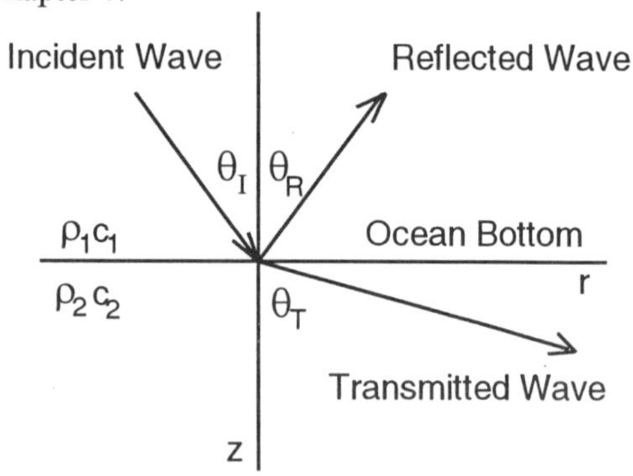

Figure 8 Reflection and Transmission from Fluid-Like Ocean Bottoms

---

[6] See Chapters 9, 10, and 11 for discussions of detection criteria.

$$p_I = A \exp j(\omega t - \mathbf{k}_I \cdot \mathbf{r}_I)\,, \tag{25a}$$

$$\mathbf{k}_I = k_I(\sin\theta_I \mathbf{j} + \cos\theta_I \mathbf{k})\,, \qquad \mathbf{r}_I = r\mathbf{j} + z\mathbf{k}\,, \tag{25b,c}$$

where A is the source amplitude. The reflected field and transmitted field can be similarly represented by the expressions

$$p_R = B \exp j(\omega t - \mathbf{k}_R \cdot \mathbf{r}_R)\,, \tag{26a}$$

$$\mathbf{k}_R = k_R(\sin\theta_R \mathbf{j} - \cos\theta_R \mathbf{k})\,, \qquad \mathbf{r}_R = r\mathbf{j} + z\mathbf{k}\,, \tag{26b,c}$$

$$p_T = C \exp j(\omega t - \mathbf{k}_T \cdot \mathbf{r}_T)\,, \tag{27a}$$

$$\mathbf{k}_T = k_T(\sin\theta_T \mathbf{j} + \cos\theta_T \mathbf{k})\,, \qquad \mathbf{r}_T = r\mathbf{j} + z\mathbf{k}\,. \tag{27b,c}$$

At the ocean floor, the pressure and the particle velocity normal to the bottom, in the water mass above, and, in the bottom below, must balance. These two conditions imply that the three fields must satisfy the boundary conditions

$$p_I + p_R = p_T\,, \tag{28a}$$

$$v_I + v_R = v_T\,. \tag{28b}$$

Substituting the fields in (25), (26) and (27) into the boundary conditions gives

$$A + B = C\,, \qquad \frac{A\cos\theta_I}{\rho_1 c_1} - \frac{B\cos\theta_R}{\rho_1 c_1} = \frac{C\cos\theta_T}{\rho_2 c_2}\,, \tag{29a,b}$$

where (2.10) was used and we found $k_I\cos\theta_I = k_R\cos\theta_R = k_T\cos\theta_T$ because the factors involving the r coordinate must cancel if the boundary conditions are to be valid for all points on the surface. Moreover, since $k_I = k_R$, this gives the law of specular reflection $\theta_I = \theta_R$ from the first equality. Snell's law of refraction follows from the second equality. Solving (29) for the reflection and transmission coefficients gives

$$R_R = \frac{B}{A} = \frac{\rho_2 c_2 \cos\theta_I - \rho_1 c_1 \cos\theta_T}{\rho_2 c_2 \cos\theta_I + \rho_1 c_1 \cos\theta_T}\,, \quad R_T = \frac{C}{A} = \frac{2\rho_2 c_2 \cos\theta_I}{\rho_2 c_2 \cos\theta_I + \rho_1 c_1 \cos\theta_T}\,, \tag{30a,b}$$

where R_R and R_T denote the Rayleigh reflection and transmission coefficients for plane waves incident on a fluid-fluid interface[7].

Conservation of energy requires that the sum of the power reflected from the surface and the power transmitted through the surface be equal to the power incident on the sea surface at each point. This can be expressed as

$$I_R \cos\theta_R + I_T \cos\theta_T = I_I \cos\theta_I\,. \tag{31}$$

[7] Rayleigh, J. W. S., *The Theory of Sound*, **II**, pg. 79-81, Dover Publications, 1945.

Dividing both members by the incident power, (31) can be expressed as

$$\alpha_R + \alpha_T = 1, \tag{32a}$$

$$\alpha_R = \frac{(\rho_2 c_2 \cos\theta_I - \rho_1 c_1 \cos\theta_T)^2}{(\rho_2 c_2 \cos\theta_I + \rho_1 c_1 \cos\theta_T)^2}, \quad \alpha_T = \frac{4\rho_2 c_2 \rho_1 c_1 \cos\theta_I \cos\theta_T}{(\rho_2 c_2 \cos\theta_I + \rho_1 c_1 \cos\theta_T)^2}, \tag{32b,c}$$

where α_R and α_T are the sound power reflection and transmission coefficients.

Graphs of the sound power reflection and transmission coefficients are shown in Figures 9a,b,c,d. Figures 9a,b are for the case of the sound speed in the ocean bottom exceeding the water sound speed, i.e. $c_1/c_2 < 1$. This is the more usual case since fluid-like bottoms usually consist of highly elastic materials such as sand and gravel. Figures 9c,d are for the case of the sound speed in water exceeding the sound speed in the ocean bottom, i.e. $1 < c_1/c_2$. This occurs when the bottom consists of soft plastic materials such as clay and biological materials. Each of these cases is subdivided into two cases depending on the relative magnitudes of the density of the bottom material and the water density. All cases of relative densities are shown in Figures 9 but the cases in which the bottom material is denser than water, i.e. $1 < \rho_2/\rho_1$, are of primary interest. The contrary case is highly improbable in the ocean because materials that are lighter than water float to the surface while materials that are heavier than water end up on the bottom. Consequently, the case of the density ratio less than unity is viewed as being an unlikely occurrence that could exist only as a temporary condition.

The sound power reflection and transmission coefficients shown in Figures 9a,b, for $c_1/c_2 < 1$, show that a critical angle exists beyond which all sound is reflected from the bottom and no sound is transmitted into the bottom. In this figure, case A is probably the most typical situation encountered in the ocean while case B represents our unlikely density case that is included for completeness. The graphs in Figure 9a,b were computed for the sound speed ratio $c_1/c_2 = 0.75$ and we see that, for this sound speed ratio, the critical angle of total reflection, from Snell's law, is $\theta_I = \sin^{-1}(c_1/c_2) \cong 48.6°$. For case A the density ratio was chosen to be $\rho_2/\rho_1 = 3$. For this case, maximum energy is seen to be transmitted into the bottom at normal incidence and the reflected energy is seen to increase with increasing angle of incidence. For case B the density ratio was chosen to be $\rho_2/\rho_1 = 0.5$.

The sound power reflection and transmission coefficients shown in Figures 9c,d, for $c_1/c_2 > 1$, show that sound is transmitted into the ocean floor at all incident angles. These plots were produced for the sound speed ratio $c_1/c_2 = 1.5$. For case A in Figures 9c,d an angle of intromission exists at which all sound is absorbed and none reflected. This case can be viewed as an extension of case A in Figures 9a,b in which the sound speed ratio approaches unity and the critical angle moves to 90°. When the sound speed ratio exceeds unity, an angle of intromission is introduced at 90° that approaches zero degrees as the sound speed ratio is increased toward the density ratio. For case A the density ratio was chosen to be $\rho_2/\rho_1 = 3$ and the angle of intromission is found to be

$$\theta_I = \sin^{-1}\sqrt{\frac{1-(\rho_1 c_1/\rho_2 c_2)^2}{1-(\rho_1/\rho_2)^2}} \cong 66.7° . \tag{33}$$

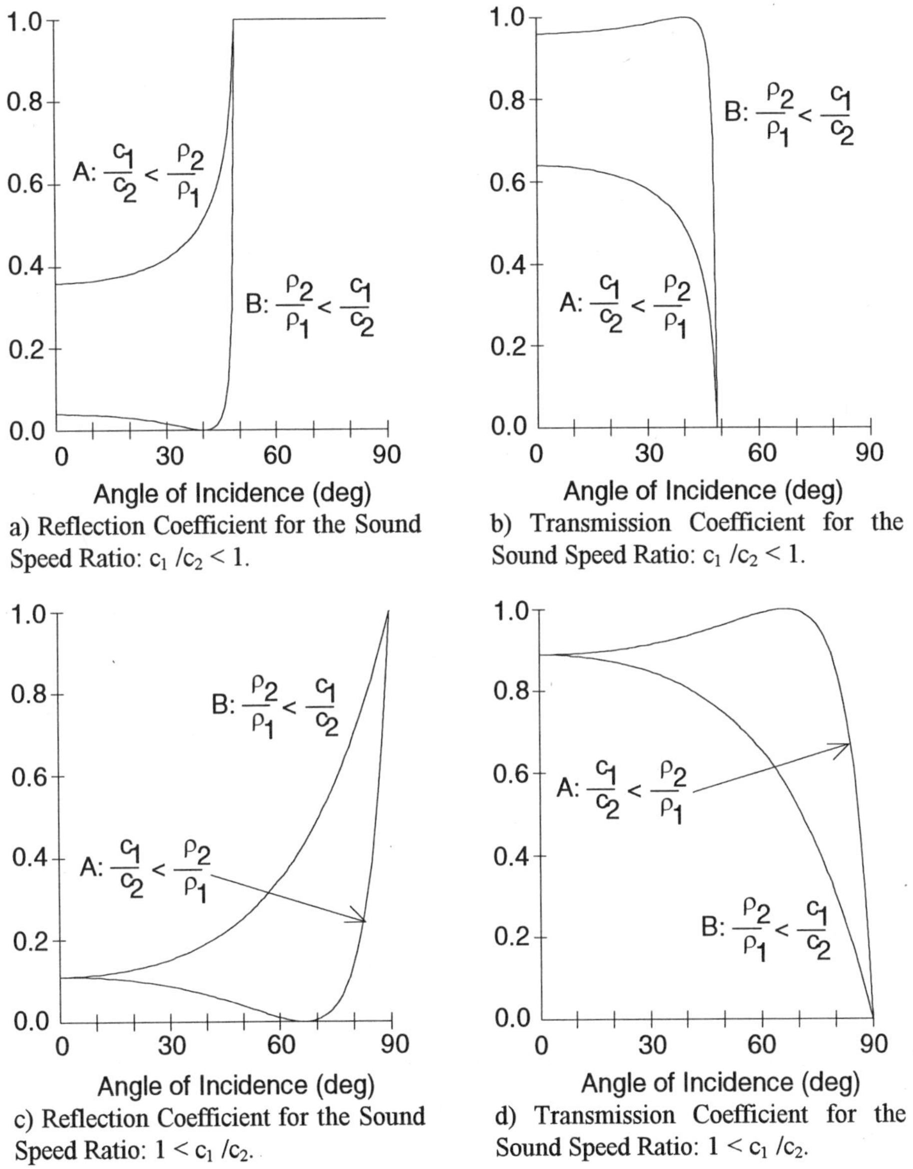

a) Reflection Coefficient for the Sound Speed Ratio: $c_1/c_2 < 1$.

b) Transmission Coefficient for the Sound Speed Ratio: $c_1/c_2 < 1$.

c) Reflection Coefficient for the Sound Speed Ratio: $1 < c_1/c_2$.

d) Transmission Coefficient for the Sound Speed Ratio: $1 < c_1/c_2$.

Figure 9 Reflection & Transmission Coefficients for a Fluid-Like Ocean Bottom

Case B can be viewed as an extension of case A as the sound speed ratio exceeds the density ratio. For this case, maximum energy is seen to be transmitted into the bottom at normal incidence and the reflected energy is seen to increase with increasing angle of incidence. For case B the plots were produced with the sound speed and density ratios reversed being, respectively, 3 and 1.5.

Real fluid-like ocean bottoms are, of course, more complicated than our fluid-fluid model would suggest. Fluid-like bottoms of sand, silt and clays exist but they usually exhibit acoustic absorption. For such bottoms, the general structure of the reflection and transmission coefficients, as shown in Figure 9, are still valid but the graphs tend to be smoothed. The critical angle and the angle of intromission are less well defined and reflection and transmission at these angles may not be total. If the bottom and its layers are irregular in shape and are not smooth planes, acoustic energy is scattered randomly and the notions of angles of reflection and transmission lose their meaning. Additionally, the sound speed within the ocean bottom may increase continuously with depth, due to the compaction of materials, causing the acoustic rays to be refracted upward. Any number of these phenomena can be operative simultaneously and even fluid-like bottoms can produce an exceedingly complicated acoustic environment in the real ocean.

5.5 Acoustic Absorption

Acoustic waves that propagate in the ocean lose intensity for many reasons. Spreading losses arise because the radiated energy is spread over an increasingly broader wave front as the wave propagates further from the source. Also, leakage out of the medium and scattering are also contributors to reductions in intensity as the wave propagates. Absorption losses occur because acoustic energy is transformed into heat energy which is absorbed by the medium. In this section we examine the effect of absorption on acoustic propagation and discuss the principal absorption mechanisms that are found in the ocean. Acoustic attenuation due to geometric spreading losses in the ocean and leakage as an attenuation mechanism are discussed in Chapter 6.

Absorption Losses

In the early days of sonar between the world wars and during WW II[8], acoustic waves that propagated through the ocean were found to attenuate faster than expected for the known geometric spreading laws. Upon investigation, the additional losses were found to be proportional to the propagation range increment. By that we mean the additional fractional loss in intensity over, say, a 10 nm range increment was found to be the same whether the 10 nm was the first range increment or a range increment at extended range. This suggested that absorption was the mechanism causing the propagation anomaly and also suggested the mathematical model

$$\frac{1}{I}\left(\frac{dI}{dr}\right) = -\alpha_f \,, \tag{34a}$$

[8] *Physics of Sound in the Sea*, NDRC Div. 6 Summary Tech. Rpt., Vol. 8, Chap. 5, Dept of Navy, Naval Material Command, 1969.

where α_f is the fractional intensity loss. In many long range surveillance scenarios this loss is on the order of 1/4% per kyd. This mathematical model implies an exponential intensity fall off given by

$$I = I_0 \exp\left[-\alpha_f\left(r - r_0\right)\right], \tag{34b}$$

where I_0 is the intensity at a reference range r_0. From this expression, the transmission loss due to absorption is given by

$$\begin{aligned} TL_A &= -10\log\left(\frac{I}{I_0}\right), \\ &= \alpha_f 10\mathrm{Log}_{10}(e)\left(r - r_0\right), \\ &= \alpha\left(r - r_0\right), \end{aligned} \tag{34c}$$

where $\alpha[=\alpha_f 10\mathrm{Log}_{10}(e) \cong 4.34\alpha_f]$ is the standard sonar equation absorption coefficient and is normally given as dB per km or kyd. For the above case of an intensity loss of 1/4% per kyd, the absorption coefficient is found to be approximately 0.01 dB per kyd. The reference range in (34c) is normally taken to be 1 m or 1 yd. This choice makes the term αr_0 negligible compared to the transmission loss and (34c) is usually expressed as $TL_A = \alpha r$.

Absorption losses have been studied extensively and have been found to be due to several causes and to be dependent upon frequency and ocean characteristics in a relatively complicated manner. The principal mechanisms that cause absorption in the ocean are:

1. viscosity(bulk and shear),
2. chemical dissociation($MgSO_4$ and $B(OH)_3$),

A graph of absorption losses versus frequency is given in Figure 10 in which the dominant absorption mechanisms are indicated over the frequency range. Many ASW scenarios fall in the frequency range of approximately 100 Hz to 5000 Hz for which all absorption mechanisms contribute to the attenuation of the signal. Over this band of frequencies, the graph in Figure 10 gives a corresponding range of absorption coefficients of approximately 0.002 dB per kyd to 0.6 dB per kyd. For the above example of an intensity loss of 1/4% per kyd and absorption coefficient of 0.01 dB per kyd, the corresponding frequency is seen to be approximately 300 Hz which is in the low end of this range. This frequency range is important for passive surveillance of target radiations.

The absorption coefficient values for operationally important ASW scenarios are numerically small but, because absorption loss is proportional to range, as compared to spreading losses which are proportional to the logarithm of range, these losses can be severe. Consequently, spreading losses can be dominated by absorption losses at sufficiently long ranges. This is shown in Figure 11 for the absorption coefficients spanning the frequency range 100 Hz to 5 kHz. From Figure

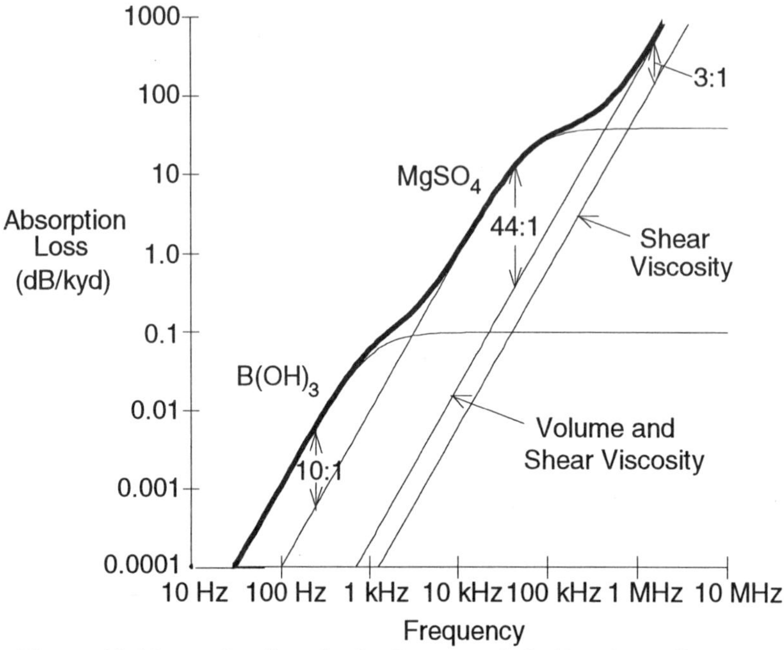

Figure 10 Absorption Loss in the Ocean and the Dominant Processes

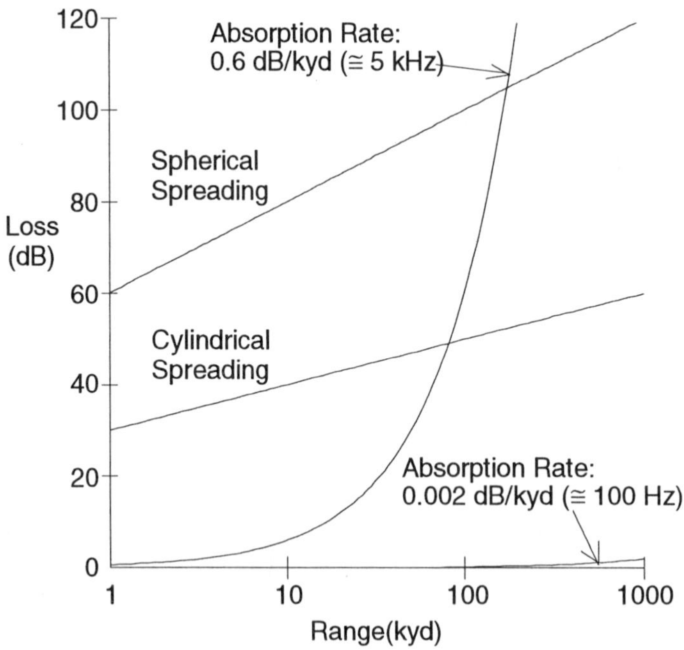

Figure 11 Absorption and Spreading Losses Compared

11 we see that long range operations require low frequencies. To convert the absorption coefficients in Figures 10 and 11 from dB/kyd to dB/km, multiply the coefficients by 1.1.

Absorption Mechanisms

Absorption of acoustic energy by the medium occurs when the acoustic wave performs net work on the medium. In this process, the acoustic energy, stored by the medium alternately as kinetic and potential energy, is not wholly propagated but is partially converted to heat. The conversion results because the response of the medium to the "upstream" mass motions is slightly delayed in time which, in turn, causes the "downstream" mass motions to be reduced in amplitude. Since at low frequencies signal wave periods are long compared to the medium response time, the medium follows the acoustic pressure wave closely resulting in very low absorption. As the signal frequency increases, the signal wave periods approach the medium response time and eventually become shorter than the medium response time. This phase mismatch causes the absorption to increase and to eventually saturate. At frequencies for which signal wave periods are less than the medium time delay, the absorption reaches a maximum and remains constant as the signal frequency increases.

The general functional dependence of the absorption coefficient for all mechanisms is found to be of the form

$$\alpha = AP\left(\frac{f_m}{f_m^2 + f^2}\right)f^2, \tag{35}$$

where f is the signal frequency and the parameter f_m is the reciprocal of the medium response time, i.e. the upper edge of the band of signal frequencies that will propagate with less than maximum absorption. The parameters A, P and f_m are primarily dependent upon the absorption mechanisms, which are discussed below, but A and f_m are also dependent upon the temperature, salinity and pH of the medium and P is dependent upon the static pressure. The factor P is unity at the surface and decreases with depth.

For the case of low frequencies the absorption coefficient has the form

$$\alpha = \left(\frac{AP}{f_m}\right)f^2, \quad f \ll f_m \tag{36a}$$

which shows that the absorption coefficient increases as the frequency squared. For the case of high frequencies the absorption coefficient has the form

$$\alpha = APf_m, \quad f \gg f_m \tag{36b}$$

which is frequency independent. Together these limits result in the frequency dependence shown in Figure 10.

Bulk Viscosity and Shear Viscosity:

Bulk viscosity and shear viscosity act in subtle and historically controversial ways to create acoustic absorption in the ocean. If we consider an ideal fluid in which no viscosity exists, we would find that only compressional and rarefactional stresses would be supported by the medium. Moreover, these stresses would obey Hook's law. Stresses would be proportional to strains and, during the application of stress, the medium would immediately respond by deforming and there would be no lag time between stress and strain. Since shearing stresses would not be supported by this ideal medium, it would behave as the perfect lubricant. For example, during the passage of a body through the fluid, force would be required to displace fluid around the body but no force would be required to overcome frictional drag on the body. Unfortunately, real media, including the ocean, are non-ideal media in which both bulk and shear viscosities occur and both mechanisms are sources of acoustic absorption. Bulk and shear viscosity are caused by two different mechanisms but they share the property of being caused by mechanical "frictions" within the material of the fluid medium. In the following we discuss how these mechanisms cause acoustic absorption in the ocean in terms of longitudinal waves, in which the directions of propagation and particle motion are collinear, and shear waves, in which the directions of propagation and particle motion are perpendicular. By way of preview, we note that both bulk and shear viscosity are found to occur in longitudinal waves but only shear viscosity occurs in shear waves.

Bulk viscosity occurs in a nonideal fluid when the fluid is compressed and decompressed. It is theorized that fluids have multiple states such that during compression the fluid relaxes allowing the molecules to crowd closer together which results in a reduction in compressional stress with no reduction in strain. Similarly, during decompression the fluid relaxes allowing the molecules to separate which results in a reduction in decompressional stress with no reduction in strain. Such fluids behave analogously to leaky pumps in contrast to ideal Hook's law media that behave as springs. Since bulk viscosity does not depend upon shearing stresses and strains, it is sometimes referred to as longitudinal viscosity or volume viscosity.

The relaxation that causes bulk viscosity is the primary source of the lag in the medium response to the acoustic pressure. This lag causes the reduction in the pressure wave amplitude which is identified as an acoustic absorption loss due to bulk viscosity. The multiple state explanation of the bulk viscosity mechanism is somewhat speculative because it has not been directly verified and it is not predicted by classical acoustic theory. In fact, one of the historical mysteries in the acoustic absorption story was that classical acoustic theory predicted viscous absorption losses, based upon a shear viscosity loss mechanism for longitudinal waves which we next discuss, but the predicted losses were only about 1/4 to 1/2 as great as the measured viscous absorption losses. The concept of bulk viscosity was proposed to account for the anomalously high absorption losses found and the fact that the absorption observed in the ocean is explained by the multiple state theory has been a strong factor in its acceptance as fact.

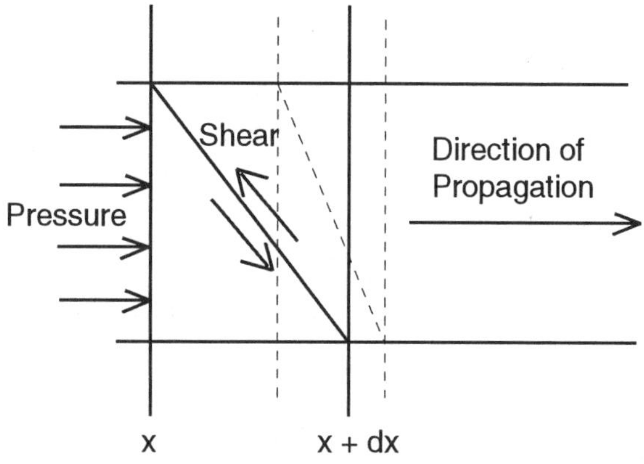

Figure 12 Shear in a Longitudinal Wave

Shear viscosity resists slippage along planes. Longitudinal waves, at first thought, might appear to cause only longitudinal stresses and strains and to endure no slippage along planes. For example, for the case of the longitudinal plane wave, as indicated in Figure 12, all of the stresses and stains on planes perpendicular to the direction of propagation are in the direction of propagation. However, if we consider a volume element, at rest between x and x + dx, we see that the element diagonal is foreshortened as it is compressed and displaced. The upstream fluid along the diagonal, pushing on the downstream fluid, establishes shearing stresses along the diagonal. In a medium that supports shearing stresses, the shear stress along the diagonal balances a portion of the upstream pressure and the downstream pressure is reduced. This reduction in downstream pressure is the cause of the acoustic absorption due to shear viscosity. In the event that the fluid is inviscid, the stresses along the diagonal would be immediately released and the downstream pressure would be greater resulting in no absorption. An alternative way of viewing this process is that the upstream pressure is communicated downstream more slowly if shear stresses are supported. In the event that the fluid is inviscid, the communication of the upstream pressure would be instantaneous. Note that during this process, no lateral straining occurs for either a viscid or inviscid fluid because the adjacent fluid is undergoing the same processes and the lateral pressures balance. Therefore no lateral flows occur but, rather, the longitudinal flows are delayed.

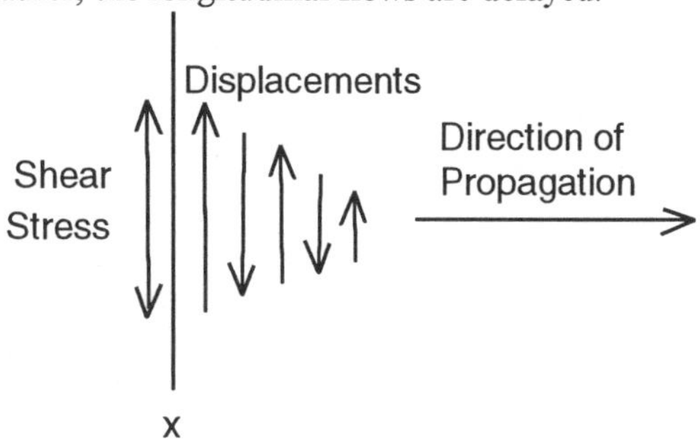

Figure 13 Shear Wave Propagation

The case of pure shear waves exists in the ocean but they are weak to the point of being insignificant and they propagate very slowly. Shear waves are set in motion by lateral motions, as indicated in Figure 13, in which shearing stresses are applied to the x plane. The material displaces laterally in response to the shearing stresses and the direction of propagation of the shear wave is perpendicular to the particle motion. The lateral motions are transmitted to adjacent layers of the medium by shear viscosity which causes the adjacent layers of the fluid to be "dragged along". For the case of an inviscid fluid the lateral motion is not communicated to the adjacent layer since the medium is a perfect lubricant with no drag at its boundaries. In weakly viscous fluids, such as the ocean, the dragging stresses are weak causing these waves to be strongly attenuated.

Bulk viscosity and shear viscosity are always irreversible processes that convert acoustic energy to heat but it is important to distinguish these mechanisms from the mechanism of absorption caused by heat conduction that can occur during acoustic compression. All fluids, whether viscous or nonviscous, undergo a temperature increase when compressed and a temperature decrease when rarefied. In principle, during the passage of an acoustic wave, the flow of heat from regions of higher temperature to regions of lower temperature can cause the pressure variations to be smoothed resulting in an irreversible loss of mechanical energy to heat energy. However, for water, this loss mechanism is negligible because the compressions and rarefactions caused by an acoustic wave occur too rapidly for a significant heat flow to occur. For all practical purposes, during acoustic propagation in water the acoustic compressions and rarefactions are adiabatic and reversible.

The wave equation for an acoustic wave propagating in a viscid fluid was found in (2.16-A) to be

$$\nabla^2 p - \frac{1}{c^2}\left(\ddot{p} - \frac{\sigma}{\rho_0}\nabla^2 \dot{p}\right) = 0, \quad c^2 = B/\rho_0 \tag{37}$$

where σ was defined to be an effective viscosity. If we consider sinusoidal solutions with time dependence $\exp(j\omega t)$, (37) can be written in the form

$$\ddot{p} - \left(c^2 + \frac{j\sigma\omega}{\rho_0}\right)\nabla^2 p = 0 . \tag{38}$$

This is a standard wave equation where the quantity in parentheses is a complex sound speed. Denoting this complex sound speed as c' and substituting the trial solution $p = A\exp[j\omega(t \pm x/c')]$ into (38) we find c' to be given by

$$\frac{1}{c'} = \frac{\pm 1}{c\sqrt{1 + j\sigma\omega / c^2\rho_0}} \cong \frac{\pm 1}{c}\left(1 - \frac{j\sigma\omega}{2c^2\rho_0}\right). \tag{39a}$$

The approximation in the last member is possible for ocean propagation because, as shown below, the imaginary term is approximately $f \times 10^{-12}$, where f is the frequency in Hertz, which is small for all frequencies of any practical interest. Substituting this complex sound speed into the trial solution gives the expression for the propagating wave

$$p = A\exp\left(\pm\frac{x\sigma\omega^2}{2c^3\rho_0}\right)\exp\left[j\omega\left(t \pm \frac{x}{c}\right)\right]. \tag{39b}$$

This solution represents a plane wave traveling in either the negative or positive x direction that is attenuated by viscosity[9]. The intensity of this wave is attenuated with the fractional absorption coefficient

$$\alpha_f = \frac{\sigma\omega^2}{c^3\rho_0}. \tag{39c}$$

In Appendix 5-A we discuss bulk viscosity and shear viscosity and find in (5.42d-A) that they are related to the effective viscosity by

$$\sigma = \hat{\eta} + \frac{4\eta}{3}, \tag{40}$$

where $\hat{\eta}$ and η represent, respectively, the bulk viscosity and shear viscosity of the medium. By Liebermann,[10] the bulk viscosity is approximately two to four times the shear viscosity. For the median value $\hat{\eta} \cong 3\eta$, the sonar equation absorption coefficients for bulk and shear viscosity are

Bulk and Shear Viscosity: $$\alpha = 4.34\left(\frac{52\pi^2\eta}{3\rho_0 c^3}\right)f^2, \tag{41a}$$

Shear Viscosity Only: $$\alpha = 4.34\left(\frac{16\pi^2\eta}{3\rho_0 c^3}\right)f^2. \tag{41b}$$

These expressions have the form of low frequency approximations for absorption given in (36a). To determine the frequency of the medium response for water, consider the condensation expression in (2.10-A). If the pressure were to suddenly drop to zero at time $t = 0$, the condensation for $t > 0$ is obtained from the solution of the homogeneous form of (2.10-A) and is found to decay as $s = p(0)\exp(-\rho_0 c^2 t/\sigma)$. The factor multiplying the time t in the exponent of this solution is a measure of the saturation frequency $f_m = \rho_0 c^2/\sigma$ which is the reciprocal of the relaxation time. Substituting appropriate values into this expression we find the saturation frequency to be $f_m \cong 1.7\times10^{12}$ Hz. Consequently, the response of the medium to shear stress is a very short time period and the absorption coefficient expressions in (41) are valid over all practical ranges of signal frequencies, i.e. for all practical purposes the absorption due to bulk and shear viscosity never saturates.

The bulk and shear viscosity absorption coefficient expressions in (41) will vary, depending on ocean conditions, because of variations, primarily, in the bulk viscosity. In fact, at great depths, since water is more nearly in a saturated compressed state, absorption by bulk viscosity is expected to be reduced. It is estimated that the reduction will be as low as 70% or less of the surface value. This

[9] Rayleigh, J. W. S., *The Theory of Sound*, **II**, pg. 316, Dover Publications, 1945.
[10] Liebermann, Phys. Rev., **75**, pg 1415, 1949.

can provide an increase in detection performance for situations that support propagation at depth such as convergence zone propagation discussed in Chapter 6.

Chemical Dissociation ($MgSO_4$ and $B(OH)_3$):

The ocean is a complicated mixture of chemicals but only two appear to have a pronounced effect on the propagation of sound. These chemicals are magnesium sulfate ($MgSO_4$) and boric acid ($B(OH)_3$). Both exist in the ocean in concentrations much weaker than many other chemicals, e.g. common salt (NaCl), but both can cause significant absorption of acoustic energy.

$MgSO_4$ and $B(OH)_3$, in a water solution, are decomposed into their ionic components Mg, SO_4, B and OH and form bonds with the water ions H and OH. During the passage of an acoustic wave these ions are driven alternately to reassociate and disassociate. The physical phenomenon behind the absorption of acoustic energy due to this mechanism is analogous to the case of absorption by bulk viscosity. Energy is required to overcome the ionic chemical forces and the medium response lags the acoustic driving pressure. The important differences between acoustic absorption due to chemical dissociation and that due to bulk viscosity lies in the relative magnitudes of the absorption and the response times of chemical dissociation. The magnitude of the absorption due to $MgSO_4$ dissociation, measured in dB, is approximately 44 times greater than the absorption due to viscosity as indicated in Figure 10. Moreover, the magnitude of absorption due to $B(OH)_3$ dissociation, measured in dB, is approximately 10 times greater than the absorption due to $MgSO_4$ dissociation as indicated in Figure 10. The medium response times associated with chemical dissociation are much slower than the more nearly instantaneous response times due to viscosity. This reduces the saturation frequencies substantially but, at approximately 64 kHz for $MgSO_4$ and 1 kHz for $B(HO)_3$, the saturation frequencies are well above the frequency bands important for many, but not all, ASW operations. Consequently, most ASW operations can be expected to be affected by acoustic absorption dominated by chemical dissociation. This effect can only be mitigated through reductions in operating frequencies.

Data fitting studies show that, analogous to the case of volume viscosity, the absorption due to chemical dissociation is found to be reduced at depth because the pressure at depth impedes the dissociation process. The absorption at depth can be reduced to approximately 70% of the surface values which is significant for long range propagation at depth. These data studies also show a dependence upon the pH of the medium. The greater the pH of the medium, the greater the absorption. These results indicate that the Atlantic Ocean absorbs energy at approximately twice the rate as the Pacific Ocean, i.e. the Atlantic is more basic than the Pacific. This effect has found support from a limited amount of experimental data.

The Total Absorption Coefficient

Various mathematical expressions for absorption due to chemical dissociation have been developed from "best fits" to absorption data collected both at-sea and in the laboratory for wide ranges of temperature, salinity, pressure and pH levels. All of these expressions have the general form of (35). They differ only in the detailed mathematical structures describing how their parameters depend upon temperature, salinity, pressure and pH level[11].

An approximate expression for the absorption coefficient that includes terms for viscosity and chemical dissociation is given by[12]

$$\alpha = 0.00022f^2 + \frac{40f^2}{4100+f^2} + \frac{0.1f^2}{1+f^2}, \tag{42}$$

where f is in kHz. The terms in this expression are for, respectively, absorption caused by bulk and shear viscosities, $MgSO_4$ and $B(OH)_3$. We have modified the coefficient of the leading term from Urick's value of 0.000275 to correspond to the median bulk and shear viscosity given in (41a). The temperature, salinity, pressure and pH dependence have been represented by nominal values to produce this expression and, therefore, (42) is not adequate for parametric studies but it is simple and useful for obtaining approximate absorption coefficient values. Approximate values can also be scaled off the plot of this function in Figure 10.

---

[11] Urick, Robert J., *Sound Propagation in the Sea*, Chap. 4, Def. Adv. Res. Proj. Agency, 1979.

[12] Urick, Robert J., *Principles of Underwater Sound for Engineers*, pg. 102, McGraw-Hill, 1975.

APPENDIX 5-A: WAVE EQUATIONS FOR SOLIDS AND VISCOUS FLUIDS

In Appendix 2-A we discussed the wave equation for a fluid in terms of the relaxation time of the fluid. In the following we discuss the wave equation for solid elastic media, in terms of stresses and strains within the medium, and generalize the results to viscid fluid media. For both elastic solids and viscid fluids the media are found to support both longitudinal and shear waves, in contrast to the discussions in Appendix 2-A, and we determine some of the elementary properties of both. For viscid fluid media it is found that the longitudinal wave equation is equivalent to the wave equation for inviscid fluid media that was developed in Appendix 2-A and that shear waves in the ocean are inconsequential.

5.1-A: Stress and Strain Relations in Elastic Solids

When an elastic solid is deformed by an acoustic wave, stresses and strains are setup within the material and at all points the material will be displaced from its undistorted position. Note, though, that the stresses and strains are dependent upon the relative displacements between each point and its neighboring points and not just by the displacement of the point itself. For example, if two points in the deformed body are simply translated or rotated, while keeping a constant separation, there would be no elastic deformation in the region of the two points. In the following we are only concerned with deformations and not with simple translations and rotations. The displacements of the distorted material throughout the body are represented by the vector field

$$\mathbf{d}(\mathbf{r}) = \mathrm{u}(\mathbf{r})\mathbf{i} + \mathrm{v}(\mathbf{r})\mathbf{j} + \mathrm{w}(\mathbf{r})\mathbf{k}\,, \tag{1a}$$

where u, v and w represent, respectively, the displacements in the x, y and z directions due to the deformation of the material. The position of the undistorted point is given by the usual position vector

$$\mathbf{r} = \mathrm{x}\mathbf{i} + \mathrm{y}\mathbf{j} + \mathrm{z}\mathbf{k}\,. \tag{1b}$$

Equation (1a) represents the displacement of a point that is initially located at the position $\mathbf{r}$ when the material is in the nondeformed state to a position $\mathbf{r} + \mathbf{d}$ when the material is deformed. For all cases of interest, the deformation is small compared to the dimensions of the material.

We want to determine the stresses and strains at a particular point. Consider the point O in Figure 1, which is located at position $\mathbf{r}_O$, and a neighboring point at $\mathbf{r}$ which is removed from O by $\mathbf{dr}$. We can express the displacement of the latter in terms of the former by the expression[1]

$$\mathbf{d} = \mathbf{d}_O + \mathbf{dd}\big|_O\,, \tag{2a}$$

[1] Lass, Harry, *Vector and Tensor Analysis*, McGraw-Hill, Chapter 7, 1956.

where the differential is evaluated at the point O. We obtain for the differential

$$\mathbf{dd}\big|_O = \left(\frac{\partial u}{\partial x}\bigg|_O dx + \frac{\partial u}{\partial y}\bigg|_O dy + \frac{\partial u}{\partial z}\bigg|_O dz\right)\mathbf{i} + \left(\frac{\partial v}{\partial x}\bigg|_O dx + \frac{\partial v}{\partial y}\bigg|_O dy + \frac{\partial v}{\partial z}\bigg|_O dz\right)\mathbf{j} + \left(\frac{\partial w}{\partial x}\bigg|_O dx + \frac{\partial w}{\partial y}\bigg|_O dy + \frac{\partial w}{\partial z}\bigg|_O dz\right)\mathbf{k}. \tag{2b}$$

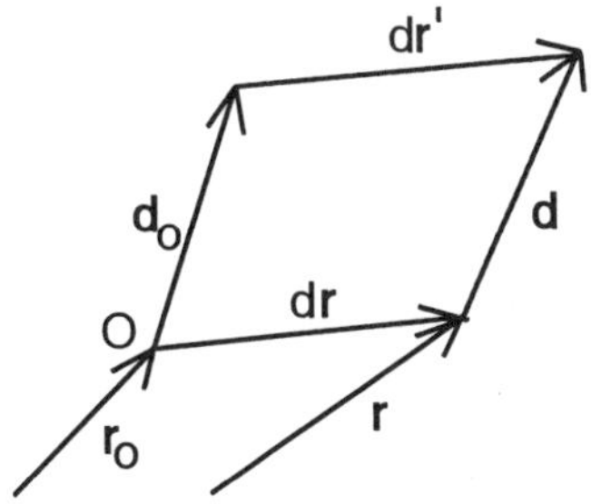

Figure 1 Displacements

More simply, (2) becomes

$$\mathbf{d} = \mathbf{d}_O + (\mathbf{dr} \cdot \nabla)\mathbf{d}\big|_O. \tag{3}$$

After the displacement, the separation between the points changes from **dr** to **dr′**. As a result of the deformation, we see that the relative separation between the points has changed and the stress-strain relations can be determined from the relative displacement vector

$$\mathbf{dr}' - \mathbf{dr} = \mathbf{d} - \mathbf{d}_O = (\mathbf{dr} \cdot \nabla)\mathbf{d}\big|_O. \tag{4}$$

This can be transformed into a more useful form by the following operations.

Given the vector identity[2]

$$\nabla(\mathbf{f} \cdot \mathbf{g}) = (\mathbf{f} \cdot \nabla)\mathbf{g} + (\mathbf{g} \cdot \nabla)\mathbf{f} + \mathbf{f} \times (\nabla \times \mathbf{g}) + \mathbf{g} \times (\nabla \times \mathbf{f}), \tag{5a}$$

let

$$\mathbf{f} = \mathbf{dr}, \quad \mathbf{g} = \mathbf{d} - \mathbf{d}_O = (\mathbf{dr} \cdot \nabla)\mathbf{d}\big|_O. \tag{5b}$$

Substituting and expanding, the first two terms of (5a) are

$$(\mathbf{dr} \cdot \nabla)\mathbf{g} = \left(dx\frac{\partial}{\partial x} + dy\frac{\partial}{\partial y} + dz\frac{\partial}{\partial z}\right)\left(dx\frac{\partial \mathbf{d}}{\partial x}\bigg|_O + dy\frac{\partial \mathbf{d}}{\partial y}\bigg|_O + dz\frac{\partial \mathbf{d}}{\partial z}\bigg|_O\right) = \mathbf{g}, \tag{5c}$$

where the partial derivatives in the second factor are constants. Also

$$(\mathbf{g} \cdot \nabla)\mathbf{dr} = \left(g_x\frac{\partial}{\partial x} + g_y\frac{\partial}{\partial y} + g_z\frac{\partial}{\partial z}\right)(\mathbf{dxi} + \mathbf{dyj} + \mathbf{dzk}) = \mathbf{g}. \tag{5d}$$

Since $\nabla \times \mathbf{dr} = 0$, these expressions reduce the vector identity in (5a) to

[2] Gradshteyn and Ryzhik, (10.31-3).

$$\mathbf{g} = \frac{1}{2}\nabla(\mathbf{dr}\cdot\mathbf{g}) + \frac{1}{2}(\nabla\times\mathbf{g})\times\mathbf{dr}\,. \tag{5e}$$

Expanding the last term we obtain for the x component

$$(\nabla\times\mathbf{g})_x = \frac{\partial}{\partial y}\left(dx\frac{\partial d_z}{\partial x}\bigg|_O + dy\frac{\partial d_z}{\partial y}\bigg|_O + dz\frac{\partial d_z}{\partial z}\bigg|_O\right)$$
$$-\frac{\partial}{\partial z}\left(dx\frac{\partial d_y}{\partial x}\bigg|_O + dy\frac{\partial d_y}{\partial y}\bigg|_O + dz\frac{\partial d_y}{\partial z}\bigg|_O\right) = (\nabla\times\mathbf{d})_x\Big|_O\,, \tag{6a}$$

where the partial derivatives in each term are constants. This can be generalized to

$$(\nabla\times\mathbf{g}) = (\nabla\times\mathbf{d})\big|_O\,. \tag{6b}$$

Substituting results, the relative displacement expression in (4) becomes

$$\mathbf{dr}' - \mathbf{dr} = \frac{1}{2}\nabla(\mathbf{dr}\cdot\mathbf{g}) + \frac{1}{2}(\nabla\times\mathbf{d})\big|_O\times\mathbf{dr}\,, \tag{7a}$$

where

$$\mathbf{g} = dx\frac{\partial\mathbf{d}}{\partial x}\bigg|_O + dy\frac{\partial\mathbf{d}}{\partial y}\bigg|_O + dz\frac{\partial\mathbf{d}}{\partial z}\bigg|_O\,. \tag{7b}$$

The second term in (7a) represents a rigid body rotation as shown in Section 7.2-B. In this instance the rotation is due to the deformation of the material, and not the rotation of the whole body, but it is not part of the elastic deformation so we ignore this term in determining the stress-strain relations.

To better examine the strains within the material we expand the first term of (7a) to obtain the relative displacement expression

$$\mathbf{dr}' - \mathbf{dr} = \frac{1}{2}\nabla\left[(dx)^2\frac{\partial u}{\partial x}\bigg|_O + dxdy\frac{\partial u}{\partial y}\bigg|_O + dxdz\frac{\partial u}{\partial z}\bigg|_O\right]$$
$$+\frac{1}{2}\nabla\left[dxdy\frac{\partial v}{\partial x}\bigg|_O + (dy)^2\frac{\partial v}{\partial y}\bigg|_O + dydz\frac{\partial v}{\partial z}\bigg|_O\right]$$
$$+\frac{1}{2}\nabla\left[dxdz\frac{\partial w}{\partial x}\bigg|_O + dydz\frac{\partial w}{\partial y}\bigg|_O + (dz)^2\frac{\partial w}{\partial z}\bigg|_O\right]$$

$$= \mathbf{i}\left[dx\frac{\partial u}{\partial x}\bigg|_O + \frac{dy}{2}\left(\frac{\partial v}{\partial x} + \frac{\partial u}{\partial y}\right)\bigg|_O + \frac{dz}{2}\left(\frac{\partial w}{\partial x} + \frac{\partial u}{\partial z}\right)\bigg|_O\right]$$

$$+\mathbf{j}\left[\frac{dx}{2}\left(\frac{\partial u}{\partial y}+\frac{\partial v}{\partial x}\right)\bigg|_O+dy\frac{\partial v}{\partial y}\bigg|_O+\frac{dz}{2}\left(\frac{\partial w}{\partial y}+\frac{\partial v}{\partial z}\right)\bigg|_O\right]$$

$$+\mathbf{k}\left[\frac{dx}{2}\left(\frac{\partial u}{\partial z}+\frac{\partial w}{\partial x}\right)\bigg|_O+\frac{dy}{2}\left(\frac{\partial v}{\partial z}+\frac{\partial w}{\partial y}\right)\bigg|_O+dz\frac{\partial w}{\partial z}\bigg|_O\right]. \tag{8}$$

In matrix form the relative displacements in (8) can be written

$$\mathbf{dr}'-\mathbf{dr}=\mathbf{S}\mathbf{dr}\,, \tag{9a}$$

where the strain matrix S is given by

$$S=\begin{bmatrix} e_{x,x} & e_{x,y} & e_{x,z}\\ e_{y,x} & e_{y,y} & e_{y,z}\\ e_{z,x} & e_{z,y} & e_{z,z}\end{bmatrix}=\begin{bmatrix} \dfrac{\partial u}{\partial x} & \dfrac{1}{2}\left(\dfrac{\partial v}{\partial x}+\dfrac{\partial u}{\partial y}\right) & \dfrac{1}{2}\left(\dfrac{\partial w}{\partial x}+\dfrac{\partial u}{\partial z}\right)\\ \dfrac{1}{2}\left(\dfrac{\partial u}{\partial y}+\dfrac{\partial v}{\partial x}\right) & \dfrac{\partial v}{\partial y} & \dfrac{1}{2}\left(\dfrac{\partial w}{\partial y}+\dfrac{\partial v}{\partial z}\right)\\ \dfrac{1}{2}\left(\dfrac{\partial u}{\partial z}+\dfrac{\partial w}{\partial x}\right) & \dfrac{1}{2}\left(\dfrac{\partial v}{\partial z}+\dfrac{\partial w}{\partial y}\right) & \dfrac{\partial w}{\partial z}\end{bmatrix}_O . \tag{9b}$$

The subscript O in (9b) indicates that the matrix is evaluated at the point O and each term in the matrix represents a strain at point O. As we discuss below, the diagonal elements represent the longitudinal strains and the off-diagonal terms represent the shearing strains.

Longitudinal strain for a volume element under tensile stress is given by the change in length of the volume element divided by the length of the volume element. For example, for the volume element under tension, as indicated in Figure 2a, the stress is applied in the x direction to the face perpendicular to the x axis. Denoting the direction by the first subscript and the face by the second subscript, the strain is given by

$$e_{x,x}=\frac{u(x+dx,y,z)-u(x,y,z)}{dx}=\frac{\partial u(x,y,z)}{\partial x}, \tag{10a}$$

which, of course, is the first element of the strain matrix. Similar results are obtained for the other diagonal elements.

Shear strain for a volume element under shear stress is given by the angle through which the volume element is distorted. For example, for the volume element under shear, as indicated in Figure 2b, stresses are applied in the x and y directions to the faces perpendicular, respectively, to the y and x axes. Both stresses are necessarily present and equal if the volume element is to be in equilibrium. The shear strain is seen to be the angle γ and we also see that

$$\tan(\gamma)=\frac{u(x,y+dy,z)-u(x,y,z)}{dy}=\frac{v(x+dx,y,z)-v(x,y,z)}{dx}. \tag{10b}$$

Consequently, for small shear strain angles we obtain

$$\gamma = \frac{\partial u(x,y,z)}{\partial y} = \frac{\partial v(x,y,z)}{\partial x}. \tag{10c}$$

Since both terms necessarily contribute equally to the shear strain we obtain

$$e_{x,y} = \frac{1}{2}\left(\frac{\partial v}{\partial x} + \frac{\partial u}{\partial y}\right) = e_{y,x}, \tag{10d}$$

which corresponds to the off-diagonal terms in the strain matrix. Again, the first subscript denotes the direction of the force and the second subscript denotes the face on which it acts. Actually, while the form of both the longitudinal and shear strain terms are of crucial importance, their multiplicative coefficients are open to choice since stress is proportional to strain and the coefficients in the stress matrix simply adjust to the choice.

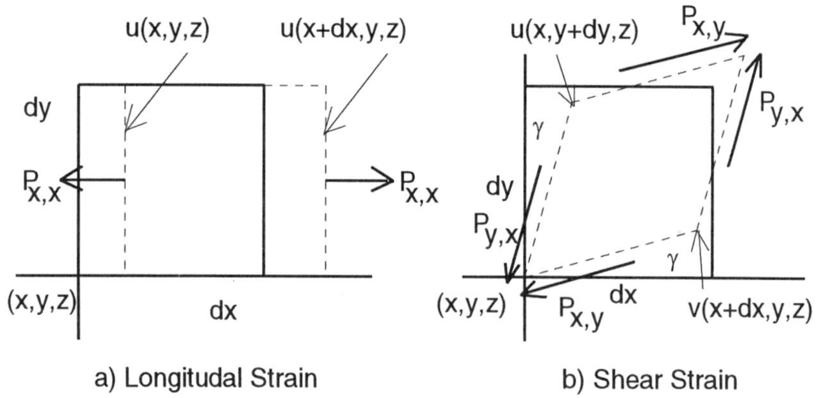

Figure 2 Longitudinal and Shear Strains in an Elastic Solid

For each element of strain there is a corresponding element of stress, as is indicated in Figure 2, and the stress matrix has the form

$$P = \begin{bmatrix} P_{x,x} & P_{x,y} & P_{x,z} \\ P_{y,x} & P_{y,y} & P_{y,z} \\ P_{z,x} & P_{z,y} & P_{z,z} \end{bmatrix}. \tag{11}$$

The stress is defined to be the force acting on a volume element face divided by the face area. The subscript convention for stresses is the same as for strains with the first subscript denoting the direction of force and the second subscript denoting the face on which the force acts. The diagonal elements represent tensile stresses and the off diagonal elements represent shear stresses. The task we need to complete is to relate the stresses to the strains.

Stresses and Strains along the Principal Axes

The relationship between the stress and strain matrices is easiest to develop when both are diagonal, i.e. there are no shearing stresses and strains. For this case the stress and strain matrices are expressed in terms of the principal axes and have the forms

$$P = \begin{bmatrix} P_{x,x} & 0 & 0 \\ 0 & P_{y,y} & 0 \\ 0 & 0 & P_{z,z} \end{bmatrix}, \tag{12a}$$

$$S = \begin{bmatrix} e_{x,x} & 0 & 0 \\ 0 & e_{y,y} & 0 \\ 0 & 0 & e_{z,z} \end{bmatrix} = \begin{bmatrix} \frac{\partial u}{\partial x} & 0 & 0 \\ 0 & \frac{\partial v}{\partial y} & 0 \\ 0 & 0 & \frac{\partial w}{\partial z} \end{bmatrix}_O . \tag{12b}$$

In the event that the stress and strain matrices are not diagonal at point O, we can always diagonalize them, as we show later. The stress-strain relations for the case of no shearing stresses are given by Hook's law, which states that stress is proportional to strain,

$$\frac{\partial u}{\partial x} = \frac{P_{x,x} - \sigma\left(P_{y,y} + P_{z,z}\right)}{Y} = \frac{(1+\sigma)}{Y} P_{x,x} - \frac{\sigma}{Y}\left(P_{x,x} + P_{y,y} + P_{z,z}\right), \tag{13a}$$

$$\frac{\partial v}{\partial y} = \frac{P_{y,y} - \sigma\left(P_{x,x} + P_{z,z}\right)}{Y} = \frac{(1+\sigma)}{Y} P_{y,y} - \frac{\sigma}{Y}\left(P_{x,x} + P_{y,y} + P_{z,z}\right), \tag{13b}$$

$$\frac{\partial w}{\partial z} = \frac{P_{z,z} - \sigma\left(P_{y,y} + P_{x,x}\right)}{Y} = \frac{(1+\sigma)}{Y} P_{z,z} - \frac{\sigma}{Y}\left(P_{x,x} + P_{y,y} + P_{z,z}\right), \tag{13c}$$

where Y is Young's modulus and σ is Poisson's ratio. The form of these stress-strain relations arises because of the way elastic materials deform under tensile stress. For the case of simple longitudinal stress-strain, a longitudinal stress is applied to a material, say the stress $P_{x,x}$ in Figure 2a, and the material stretches without any change in cross section as indicated in Figure 2a. The strain is proportional to the stress and the constant of proportionality is Young's modulus. For real materials, though, a lateral strain occurs during the stretching which results in a "necking down" as the cross section is decreased. The ratio of lateral strain to longitudinal strain is given by Poisson's ratio. If lateral stresses $P_{y,y}$ and $P_{z,z}$ are also applied, the necking down is decreased which serves to reduce the longitudinal stretching. Consequently, the effective longitudinal strains are the simple strains reduced by a fraction of the lateral strains as indicated in (13).

We want to solve (13) to get the stresses in terms of the strains. This is easily done by first summing over (13) to obtain

$$\left(\frac{\partial u}{\partial x}+\frac{\partial v}{\partial y}+\frac{\partial w}{\partial z}\right)=\frac{(1-2\sigma)}{Y}\left(P_{x,x}+P_{y,y}+P_{z,z}\right). \tag{14}$$

Substituting (14) back into (13) we can easily solve for the stress terms to obtain

$$P_{x,x}=\frac{Y}{(1+\sigma)}\left[\frac{\partial u}{\partial x}+\frac{\sigma}{(1-2\sigma)}\left(\frac{\partial u}{\partial x}+\frac{\partial v}{\partial y}+\frac{\partial w}{\partial z}\right)\right], \tag{15a}$$

$$P_{y,y}=\frac{Y}{(1+\sigma)}\left[\frac{\partial v}{\partial y}+\frac{\sigma}{(1-2\sigma)}\left(\frac{\partial u}{\partial x}+\frac{\partial v}{\partial y}+\frac{\partial w}{\partial z}\right)\right], \tag{15b}$$

$$P_{z,z}=\frac{Y}{(1+\sigma)}\left[\frac{\partial w}{\partial z}+\frac{\sigma}{(1-2\sigma)}\left(\frac{\partial u}{\partial x}+\frac{\partial v}{\partial y}+\frac{\partial w}{\partial z}\right)\right]. \tag{15c}$$

It is useful to put the stress and strain relations in (13) and (15) into matrix form.This is readily done by inspection and leads, respectively, to the matrix expressions

$$S=\frac{(1+\sigma)}{Y}P-\frac{\sigma}{Y}\mathrm{Tr}(P)I, \tag{16a}$$

$$P=\frac{Y}{(1+\sigma)}\left[S+\frac{\sigma}{1-2\sigma}\mathrm{Tr}(S)I\right]. \tag{16b}$$

In these relations the notation Tr( ) denotes the trace of a matrix and the quantity I denotes the identity matrix.

In the preceding analyses leading to (16), we have achieved the desired stress-strain relations at the point O - for the case of no shear stresses - which leaves us with two limitations to overcome. On the one hand, we will want to know what the shear stresses and strains are at point O for axis orientations other than the principal axes orientation and, on the other hand, the principal axes coordinate system at point O will not, in general, be principal axes coordinate systems at other points in the medium. To rectify both limitations we need to be able to transform from one coordinate system to another at a point which is another way of saying rotate the coordinate system.

Coordinate System Rotational Transformations

Coordinate system rotational transformations can be represented in the matrix form

$$A=\begin{bmatrix} a_{1,1} & a_{1,2} & a_{1,3} \\ a_{2,1} & a_{2,2} & a_{2,3} \\ a_{3,1} & a_{3,2} & a_{3,3} \end{bmatrix}. \tag{17}$$

For example, a vector **x** can be transformed to a primed coordinate system from its representation in an unprimed coordinate system by means of the operation

$$\mathbf{x}' = \mathrm{A}\mathbf{x}\,. \tag{18a}$$

In this transformation the ith component of the transformed vector is given by

$$x_i' = \sum_{n=1}^{3} a_{i,n} x_n\,. \tag{18b}$$

For the transformation to be a pure rotation that does not change the vector length we must have

$$|\mathbf{x}'|^2 = \mathbf{x}^{\mathrm{T}}\mathrm{A}^{\mathrm{T}}\mathrm{A}\mathbf{x} = |\mathbf{x}|^2, \tag{18c}$$

where superscript T denotes transpose and we used the matrix operation $(\mathrm{A}\mathbf{x})^{\mathrm{T}} = \mathbf{x}^{\mathrm{T}}\mathrm{A}^{\mathrm{T}}$. This result implies

$$\mathrm{A}^{\mathrm{T}}\mathrm{A} = \mathrm{I}, \quad \rightarrow \quad \mathrm{A}^{\mathrm{T}} = \mathrm{A}^{-1}. \tag{18d}$$

From (18d) we obtain

$$\mathrm{A}^{\mathrm{T}}\mathrm{A} = \left[\sum_{i=1}^{3} a_{i,n} a_{m,i}\right]_{3\times 3} = \left[\delta_{n,m}\right]_{3\times 3}, \tag{18e}$$

which shows that the columns of the transformation matrix are orthogonal.

Consider the transformation of a matrix from one coordinate system to another. More specifically, consider the transformation of the strain matrix for the relative displacements in (9) which has the form $\mathbf{r} = \mathrm{S}\mathbf{t}$. Transforming the vectors to the primed coordinate system gives

$$\mathbf{r}' = \mathrm{A}\mathbf{r} = \mathrm{AS}\mathbf{t} = \mathrm{ASA}^{-1}\mathrm{A}\mathbf{t} = \mathrm{S}'\mathbf{t}'\,. \tag{19a}$$

The matrix $\mathrm{S}' = \mathrm{ASA}^{-1}$ is the transformed matrix and its components are given by

$$\mathrm{S}' = \mathrm{ASA}^{-1} = \mathrm{ASA}^{\mathrm{T}} = \left[\sum_{n=1}^{3}\sum_{m=1}^{3} a_{i,n} e_{n,m} a_{m,j}\right]_{3\times 3}, \tag{19b}$$

where $e_{n,m}$ is an element of S and where $i = 1,2,3$ and $j = 1,2,3$ give the elements of the transformed matrix. Each of these elements has the form

$$e_{i,j}' = \sum_{n=1}^{3}\sum_{m=1}^{3} a_{i,n} \frac{1}{2}\left(\frac{\partial u_m}{\partial x_n} + \frac{\partial u_n}{\partial x_m}\right) a_{m,j}\,. \tag{20a}$$

To evaluate this sum, let u_j denote the jth vector component in the unprimed coordinates and u_j' denote the jth vector component in the primed coordinates. Partial differentiation gives

$$\frac{\partial u'_j}{\partial x'_i} = \sum_{n=1}^{3} \frac{\partial u'_j}{\partial x_n} \frac{\partial x_n}{\partial x'_i} = \sum_{n=1}^{3} \frac{\partial u'_j}{\partial x_n} a_{i,n} , \tag{20b}$$

where the vector transformation

$$x_n = \sum_{i=1}^{3} a_{i,n} x'_i , \tag{20c}$$

(compare to (18b)) was differentiated to obtain the coefficients. Expressing the u'_j vector component in (20a) as a series of u_m components, similar to (18b), gives

$$\frac{\partial u'_j}{\partial x'_i} = \sum_{n=1}^{3} a_{i,n} \frac{\partial u'_j}{\partial x_n} = \sum_{n=1}^{3} \sum_{m=1}^{3} a_{i,n} \frac{\partial u_m}{\partial x_n} a_{j,m} . \tag{20d}$$

Substituting this result into (20a) gives

$$e'_{i,j} = \frac{1}{2}\left(\frac{\partial u'_j}{\partial x'_i} + \frac{\partial u'_i}{\partial x'_j} \right) , \tag{20e}$$

which demonstrates that the transformed strain matrix has the form of a strain matrix.

Diagonalization of the Stress and Strain Matrices

In the preceding we asserted that the stress and strain matrices could always be diagonalized. At this point we demonstrate the validity of that assertion and find in the process that the solution of the diagonalization problem leads to the eigenvalue problem. Since the diagonalization procedures are the same for both the stress and strain matrices we consider just the strain matrix in the following. To diagonalize a given strain matrix S, we need to find, as for (19b), a transformation matrix A such that

$$A^{-1} S A = D = \begin{bmatrix} d_1 & 0 & 0 \\ 0 & d_2 & 0 \\ 0 & 0 & d_3 \end{bmatrix} , \tag{21a}$$

where D is the transformed strain matrix which by definition is a diagonal matrix. For this demonstration it is convenient to define A in terms of its column vectors

$$A = \begin{bmatrix} \mathbf{c}_1 & \mathbf{c}_2 & \mathbf{c}_3 \end{bmatrix} \quad \mathbf{c}_j = \begin{bmatrix} a_{1,j} & a_{2,j} & a_{3,j} \end{bmatrix}^T . \tag{21b}$$

Multiplying (21a) from the left by A gives

$$\begin{bmatrix} S\mathbf{c}_1 & S\mathbf{c}_2 & S\mathbf{c}_3 \end{bmatrix} = \begin{bmatrix} d_1\mathbf{c}_1 & d_2\mathbf{c}_2 & d_3\mathbf{c}_3 \end{bmatrix} , \tag{21c}$$

and equating the respective column vectors in (21c) gives eigenvalue[3] equations with the characteristic values d_j. Expanding the jth eigenvalue equation gives the system of equations

$$\begin{aligned}(e_{1,1}-d_j)a_{1,j}+e_{1,2}a_{2,j}+e_{1,3}a_{3,j}&=0\\ e_{2,1}a_{1,j}+(e_{2,2}-d_j)a_{2,j}+e_{1,3}a_{3,j}&=0\\ e_{3,1}a_{1,j}+e_{3,2}a_{2,j}+(e_{3,3}-d_j)a_{3,j}&=0\end{aligned} \tag{21d}$$

where d_j and the $a_{i,j}$ are to be determined. For there to be any solutions to this system of equations, the determinant of coefficients must vanish

$$\det\begin{bmatrix}(e_{1,1}-d_j) & e_{1,2} & e_{1,3}\\ e_{2,1} & (e_{2,2}-d_j) & e_{2,3}\\ e_{3,1} & e_{3,2} & (e_{3,3}-d_j)\end{bmatrix}=0, \tag{21e}$$

which gives a cubic equation for d_j and the 3 solutions correspond to the three column vectors $\mathbf{c}_j$ where $j = 1,2,3$. Substituting the d_j into (21d) and solving for the $a_{i,j}$, we obtain the three column vectors - up to an arbitrary constant. The arbitrary constant arises because (21d) can be multiplied by an arbitrary constant without changing the meaning of the system of equations. This reflects a property of eigenvectors, i.e. only the direction of eigenvectors can be determined and not the magnitude without an additional requirement. For us, that requirement is that the desired transformation matrix be a rotation without stretching, as in (18c), and we obtain

$$A^TA=\begin{bmatrix}\mathbf{c}_1^T\\ \mathbf{c}_2^T\\ \mathbf{c}_3^T\end{bmatrix}\begin{bmatrix}\mathbf{c}_1 & \mathbf{c}_2 & \mathbf{c}_3\end{bmatrix}=\begin{bmatrix}\mathbf{c}_1^T\cdot\mathbf{c}_1 & 0 & 0\\ 0 & \mathbf{c}_2^T\cdot\mathbf{c}_2 & 0\\ 0 & 0 & \mathbf{c}_3^T\cdot\mathbf{c}_3\end{bmatrix}=\begin{bmatrix}1&0&0\\0&1&0\\0&0&1\end{bmatrix}. \tag{21f}$$

The second equality is obtained because of the orthogonality of the eigenvectors, which implies $A^T = A^{-1}$, as can be shown by taking the transpose of (21a). The third equality ensures that the transformation is a pure rotation and the conditions $\mathbf{c}_j^T\cdot\mathbf{c}_j = 1$, $j = 1,2,3$; establish the lengths of the eigenvectors.

Trace of the Strain Matrix

From (19b) the trace of the strain matrix in the primed coordinate system is given by

$$\mathrm{Tr}(S')=\sum_{i=1}^{3}\sum_{n=1}^{3}\sum_{m=1}^{3}a_{i,n}e_{n,m}a_{m,i}=\sum_{n=1}^{3}\sum_{m=1}^{3}e_{n,m}\sum_{i=1}^{3}a_{i,n}a_{m,i}\,. \tag{22a}$$

From (18e)

[3] Eigen is the German word for characteristic.

$$\mathrm{Tr}(S') = \sum_{n=1}^{3} \sum_{m=1}^{3} e_{n,m}\delta_{n,m} = \sum_{n=1}^{3} e_{n,n} = \mathrm{Tr}(S), \tag{22b}$$

which shows that the trace of the strain matrix is an invariant that has the same value in all coordinate systems at any given point. Of course, the trace will, in general, vary from point to point. Similarly, the trace of the stress matrix is also invariant at a point.

Transformation of the Stress-Strain Matrices

Having examined rotational transformations and their properties, we can now return to the stress-strain relations in (16), that were developed for the principal axes, and determine their form for arbitrary systems of coordinates. If we let A denote an arbitrary rotational transformation of coordinates, then multiplying the strain relation in (16a) from left and right by, respectively, A^T and A gives

$$A^T S A = \frac{(1+\sigma)}{Y} A^T P A - \frac{\sigma}{Y} \mathrm{Tr}(P) A^T I A, \tag{23a}$$

which reduces to

$$S' = \frac{(1+\sigma)}{Y} P' - \frac{\sigma}{Y} \mathrm{Tr}(P') I. \tag{23b}$$

Manipulating in a similar manner with the stress relation in (16b) we find

$$A^T P A = \frac{Y}{(1+\sigma)} \left[A^T S A + \frac{\sigma}{1-2\sigma} \mathrm{Tr}(S) A^T I A \right], \tag{24a}$$

which reduces to

$$P' = \frac{Y}{(1+\sigma)} \left[S' + \frac{\sigma}{1-2\sigma} \mathrm{Tr}(S') I \right]. \tag{24b}$$

Equations (23b) and (24b) give the stress and strain matrices in terms of arbitrary coordinate systems in which both longitudinal and shear components are generally found. We see that the forms of the stress and strain matrices are invariant in all coordinate systems but, of course, the value of the elements change with coordinate system and, of course, from point-to-point within the medium. Young's modulus and Poison's ratio in the stress and strain matrices in (23) and (24) are generally constant throughout the medium but need not be. These constants are related to the bulk modulus and shear modulus and it is useful to determine these relations.

Bulk Modulus

When a volume element is subjected to a uniform pressure on all sides, the volume of the element changes. The ratio of the stress to strain is the bulk modulus

$$B = \frac{\Delta p}{\Delta V / V}, \tag{25}$$

where the change in pressure at a point is the mean longitudinal stress at the point

$$\Delta p = \left(\frac{P_{x,x} + P_{y,y} + P_{z,z}}{3} \right) = \frac{1}{3} \mathrm{Tr}(P) . \tag{26a}$$

The fractional change in volume at a point can be determined from the relative displacement vector in (4) and Figure 1. Each of the vectors **dr**′and **dr**, considered as diagonals of volume elements, define the respective volumes by the product of their vector components. To first order we obtain

$$dV' = \left(1 + \frac{\partial u}{\partial x} + \frac{\partial v}{\partial y} + \frac{\partial w}{\partial z} \right)\bigg|_O dxdydz, \tag{26b}$$

$$dV = dxdydz, \tag{26c}$$

and the fractional change in volume is found to be the trace of the strain matrix

$$\frac{dV' - dV}{dV} = \left(\frac{\partial u}{\partial x} + \frac{\partial v}{\partial y} + \frac{\partial w}{\partial z} \right)\bigg|_O = \mathrm{Tr}(S) . \tag{26d}$$

Similarly, the condensation at point O, defined in (2.8-A), can be expressed as the negative of the trace

$$s = \frac{\rho - \rho_0}{\rho_0} = -\left(\frac{\partial u}{\partial x} + \frac{\partial v}{\partial y} + \frac{\partial w}{\partial z} \right)\bigg|_O = -\mathrm{Tr}(S) . \tag{26e}$$

Since the traces of the stress and strain matrices are related by (14), which can be written in the matrix form

$$\mathrm{Tr}(S) = \frac{(1 - 2\sigma)}{Y} \mathrm{Tr}(P), \tag{26f}$$

the bulk modulus can be written

$$B = \frac{1}{3} \frac{\mathrm{Tr}(P)}{\mathrm{Tr}(S)} = \frac{Y}{3(1 - 2\sigma)} . \tag{27}$$

<u>Shear Modulus</u>

The shear terms in the stress matrix in (24b) have the form

$$P_{x,y} = \frac{1}{2} \left(\frac{Y}{1 + \sigma} \right) \left(\frac{\partial v}{\partial x} + \frac{\partial u}{\partial y} \right), \tag{28}$$

where the multiplicative coefficient is related to the shear modulus. From elementary considerations the stresses in Figure 2b can be written in terms of the shear modulus as

$$P_{x,y} = 2\mu \frac{\partial u}{\partial y}, \qquad P_{y,x} = 2\mu \frac{\partial v}{\partial x}, \tag{29a,b}$$

where μ denotes the shear modulus. The factor of 2 arises because the total shear strain is 2γ. To see this, rotate the distorted volume element in Figure 2b, alternatively, to the x and y axes and note that for both cases the total shear strain is 2γ. To put (29) into the form required by the stress matrix we note that, since the equilibrium of the volume element requires that $P_{x,y} = P_{y,x}$, we can add (29a,b) to obtain

$$P_{x,y} = \mu\left(\frac{\partial v}{\partial x} + \frac{\partial u}{\partial y}\right). \tag{30}$$

Comparing (28) and (30) gives the relationship

$$Y = 2\mu(1+\sigma). \tag{31}$$

Example 1: For steel we have the nominal values: $Y = 20 \times 10^{11}$ dyne/cm^2, $\sigma = 0.25$, $B = 13.3 \times 10^{11}$ dyne/cm^2 and $\mu = 8 \times 10^{11}$ dyne/cm^2. These values can be shown to satisfy (27) and (31). Also, for longitudinal and shear stresses of, say, one dyne/cm^2, we have the strains $\Delta V/V = 7.5 \times 10^{-13}$ and $e_{i,j} = 6.3 \times 10^{-13}$ which indicate the size of the deformations due to an acoustic wave.♦

General Stress and Strain Matrices

The stress and strain matrices in (23) and (24) can be written, alternatively, in terms of the bulk and shear moduli

$$S' = 2\mu P' - \frac{1}{6\mu B}\left(B - \frac{2\mu}{3}\right)\mathrm{Tr}(P')I, \tag{32a}$$

$$P' = 2\mu S' + \left(B - \frac{2\mu}{3}\right)\mathrm{Tr}(S')I, \tag{32b}$$

where the quantity $\Lambda = (B - 2\mu/3)$ is called the Lamé constant. Expanding (32) gives

$$S = \begin{bmatrix} 2\mu P_{x,x} - \frac{\Lambda}{6\mu B}\mathrm{Tr}(P') & 2\mu P_{x,y} & 2\mu P_{x,z} \\ 2\mu P_{y,x} & 2\mu P_{y,y} - \frac{\Lambda}{6\mu B}\mathrm{Tr}(P') & 2\mu P_{y,z} \\ 2\mu P_{z,x} & 2\mu P_{z,y} & 2\mu P_{z,z} - \frac{\Lambda}{6\mu B}\mathrm{Tr}(P') \end{bmatrix}, \tag{33a}$$

$$P = \begin{bmatrix} 2\mu e_{x,x} + \Lambda Tr(S) & 2\mu e_{x,y} & 2\mu e_{x,z} \\ 2\mu e_{y,x} & 2\mu e_{y,y} + \Lambda Tr(S) & 2\mu e_{y,z} \\ 2\mu e_{z,x} & 2\mu e_{z,y} & 2\mu e_{z,z} + \Lambda Tr(S) \end{bmatrix}. \tag{33b}$$

In the preceding we have determined the stress and strain matrices for an elastic solid and many of their properties. We have also established Hook's law which asserts that stress is proportional to strain. These relations are essentially static relations and to extend them to the description of waves in an elastic medium we turn next to Newton's second law and determine the wave equation for elastic solids.

5.2-A: Wave Equation for Elastic Solids

From Newton's second law we can write

$$\rho_0 \frac{\partial^2 \mathbf{d}}{\partial t^2} = \nabla P^T, \tag{34a}$$

where $\mathbf{d}$ is the displacement vector in (1); P is the stress matrix; and , in contrast to the case of Newton's law for acoustic pressure in (2.4c-A), the stresses are positive when directed away from the volume element. Expanding the right member of this equation gives

$$\rho_0 \frac{\partial^2 \mathbf{d}}{\partial t^2} = \mu \nabla^2 \mathbf{d} + \left(B + \frac{\mu}{3} \right) \nabla (\nabla \cdot \mathbf{d}), \tag{34b}$$

where the trace of the strain matrix was expressed in terms of the divergence of the displacement field, i.e. $Tr(S) = \nabla \cdot \mathbf{d}$. Equation (34b) can be put into a more useful form by the vector identity[4]

$$\nabla \times \nabla \times \mathbf{d} = \nabla (\nabla \cdot \mathbf{d}) - \nabla^2 \mathbf{d}, \tag{34c}$$

to obtain

$$\rho_0 \frac{\partial^2 \mathbf{d}}{\partial t^2} = \left(B + \frac{4\mu}{3} \right) \nabla (\nabla \cdot \mathbf{d}) - \mu \nabla \times \nabla \times \mathbf{d}. \tag{34d}$$

Two general solutions of (34d) are of importance.

For the case of irrotational motion we have $\nabla \times \mathbf{d} = 0$, as is shown in Section 7.2-B, and (34d) reduces to

$$\rho_0 \frac{\partial^2 \mathbf{d}}{\partial t^2} = \left(B + \frac{4\mu}{3} \right) \nabla^2 \mathbf{d}. \tag{35a}$$

This is recognized as a vector wave equation. Since the differential operators commute with the components of the vector it is also a wave equation for each

[4] Gradshteyn and Ryzhik, (10.31-8).

component of displacement. The solutions of (35a) are also seen to be waves that travel at the sound speed

$$c = \sqrt{\frac{B + 4\mu/3}{\rho_0}}\,. \tag{35b}$$

Since by assumption $\nabla\times\mathbf{d} = 0$, then for the case of displacements only in the x direction, i.e. $\mathbf{d} = u\mathbf{i}$, we must have $\partial u/\partial y = \partial u/\partial z = 0$. This asserts that u is dependent only upon x. Consequently, the solutions of (35a) have the plane wave form $u = f(t \pm x/c)$ which represent plane waves propagating in the x direction. Therefore, the direction of wave propagation is the same as the direction of particle motion and the waves are longitudinal waves.

For the case of solenoidal motion $\nabla\cdot\mathbf{d} = 0$ (34d) reduces to

$$\rho_0 \frac{\partial^2 \mathbf{d}}{\partial t^2} = \mu\nabla^2\mathbf{d}, \tag{36a}$$

which is seen to be a vector wave equation. Again, since the differential operators commute with the components of the vector, it is also a wave equation for each component of displacement. The solutions of (36a) are also seen to be waves that travel at the sound speed

$$c = \sqrt{\frac{\mu}{\rho_0}}\,. \tag{36b}$$

Since by assumption $\nabla\cdot\mathbf{d} = 0$, then for the case of displacements only in the x direction, i.e. $\mathbf{d} = u\mathbf{i}$, , we must have $\partial u/\partial x = 0$. This asserts that u is dependent only upon y and z. Consequently, the solutions of (36a) have the form $u = f(t \pm y/c \pm z/c)$ which represent waves having a direction of propagation in the y-z plane. Therefore, the direction of wave propagation is perpendicular to the direction of particle motion and the waves are shear waves.

Irrotational and solenoidal waves do not exhaust the range of possible solutions of (34d). Other solutions are possible, e.g. surface waves that result when irrotational and solenoidal wave couple at the surface of an elastic solid.

5.3-A: Wave Equation for Fluids

The stress-strain relations for solids can be generalized to liquids with certain changes. Since the principal difference between liquids and solids is that, under shearing stresses, liquids deform continuously, the displacement terms in the strain matrix for solids must be changed to velocity terms. Also the coefficients in the matrices must be changed to coefficients that are suitable for fluids. For example, in the stress matrix for solids in (33b) the shearing stress in the x direction on an x-z plane is given by

$$P_{x,y} = \mu\left(\frac{\partial u}{\partial y} + \frac{\partial v}{\partial x}\right). \tag{37a}$$

For a fluid, the shearing stress is given by

$$P_{x,y} = \eta\left(\frac{\partial \dot{u}}{\partial y} + \frac{\partial \dot{v}}{\partial x}\right), \tag{37b}$$

where the shear modulus is replaced by the coefficient of viscosity, i.e. $\mu \rightarrow \eta$. In the place of the Lamé constant in (33b) we make the substitution $\Lambda \rightarrow \lambda$ where the new constant must be determined. In addition, we include a term that represents the total pressure at a point. With these changes, the stress matrix for solids in (32b) becomes

$$P = 2\eta\dot{S} + \left(\lambda \operatorname{Tr}\left(\dot{S}\right) - p\right)I, \tag{38}$$

where the total pressure p is positive when directed in toward the volume element. The total pressure term is required so that, for the case of an inviscid medium in which both η and λ are zero, or, for the case of a stationary medium in which the time derivative of S is zero, the normal stresses reduce to the negative of total pressure. Under dynamic conditions, we see that the normal stresses generally differ from total pressure. How this occurs has been historically controversial and we now turn to an examination of that controversy.

Bulk Viscosity

The trace of the stress matrix in (38) yields the expression

$$\operatorname{Tr}(P + pI) = (2\eta + 3\lambda)\operatorname{Tr}\left(\dot{S}\right), \tag{39a}$$

where the value of the left member is the net normal stress at a point and the value of the right member is the net rate of expansion at the point. Consequently, the coefficient in the right member represents a volume viscosity coefficient. If the average stress at a point is assumed to be equal to the total pressure

$$\frac{P_{x,x} + P_{y,y} + P_{z,z}}{3} = -p, \tag{39b}$$

then the left member of (39a) vanishes which implies that the volume viscosity coefficient vanishes giving the constant

$$\lambda = -\frac{2\eta}{3}. \tag{39c}$$

This is the classical interpretation[5] of how the total pressure relates to the stresses but more recent investigators have challenged this interpretation. This later work

---

[5] Attributed by many commentators to Stokes.

suggests that for many fluids, including water, (39b) does not hold instantaneously and that, because of a nonzero volume viscosity, there is a time lag between the applied stresses and the response of the fluid. Consequently, the total pressure at a point is the average of the normal stresses that are applied to the fluid at the point - but with a time lag. If we define the constant in (38) as

$$\lambda = \hat{\eta} - \frac{2\eta}{3}, \tag{39d}$$

where $\hat{\eta}$ is the nonzero volume viscosity to be determined, then from (39a) we obtain

$$\hat{\eta} = \frac{\mathrm{Tr}(P + pI)}{3\mathrm{Tr}(\dot{S})}. \tag{39e}$$

In analogy with the bulk modulus for solids in (27) this is also known as the bulk viscosity

While (39b) is not valid in the event of a nonzero bulk viscosity, it is reasonable to expect that the time averaged net normal stress at a point will vanish and similarly for the time rate of condensation

$$\frac{\langle P_{x,x}\rangle + \langle P_{y,y}\rangle + \langle P_{z,z}\rangle}{3} + \langle p\rangle = 0\,, \quad \langle \mathrm{Tr}(\dot{S})\rangle = 0\,, \tag{39f,g}$$

where the brackets <·> denote time average. For this case, the classical interpretation of (39a) is preserved, if it is time averaged, but with a volume viscosity coefficient that is nonzero. The frequency dependence of bulk viscosity values are not well understood but experimental data suggests that for water and sea water they are relatively constant over frequency as indicated in Figure 5.10.

Bulk Modulus for Fluids

The bulk modulus for fluids is defined the same as for solids in (27)

$$B = \frac{1}{3}\frac{\mathrm{Tr}(P)}{\mathrm{Tr}(S)}. \tag{39h}$$

It is valid for all fluids having a vanishing bulk viscosity regardless of the frequency of pressure variation since the condensation follows stress instantaneously. However, for fluids exhibiting a nonvanishing bulk viscosity, it might be expected that static bulk modulus values would be of declining accuracy with increasing frequency of pressure variations. This is primarily a theoretical issue since for operational sonar signals in water the errors in the static bulk modulus values appear to be negligible.

Stress Matrix for Fluids with Bulk Viscosity

The stress matrix in (38) can be written for the case of fluids exhibiting a nonzero bulk viscosity

$$P = 2\eta\dot{S} + \left(\left(\hat{\eta} - \frac{2\eta}{3}\right)Tr\left(\dot{S}\right) - p\right)I. \tag{40a}$$

This can be solved for the strain, for the case of a fluid exhibiting a bulk viscosity, by means of the bulk viscosity definition in (39e) giving

$$\dot{S} = \frac{1}{2\eta}\left[(P + pI) - \left(1 - \frac{2\eta}{3\hat{\eta}}\right)Tr(P + pI)I\right]. \tag{40b}$$

For fluids with vanishing bulk viscosity (40a) can be solved for the stress using the time derivative of the bulk modulus expression in (27).

Navier-Stokes Equation and Wave Equation

From Newton's second law we can write

$$\rho_0 \frac{\partial^2 \mathbf{d}}{\partial t^2} = \nabla P^T, \tag{41a}$$

where P is the stress matrix and, in comparison to the case of Newton's law for acoustic pressure in (2.4c-A), the stresses are positive when directed away from the volume element. Expanding the stress matrix in (40a) we obtain

$$P = \begin{bmatrix} 2\eta\dot{e}_{x,x} + \left(\hat{\eta} - \frac{2\eta}{3}\right)Tr\left(\dot{S}\right) - p & 2\eta\dot{e}_{x,y} & 2\eta\dot{e}_{x,z} \\ 2\eta\dot{e}_{y,x} & 2\eta\dot{e}_{y,y} + \left(\hat{\eta} - \frac{2\eta}{3}\right)Tr\left(\dot{S}\right) - p & 2\eta\dot{e}_{y,z} \\ 2\eta\dot{e}_{z,x} & 2\eta\dot{e}_{z,y} & 2\eta\dot{e}_{z,z} + \left(\hat{\eta} - \frac{2\eta}{3}\right)Tr\left(\dot{S}\right) - p \end{bmatrix}, \tag{41b}$$

from which the right member of (41a) is immediately obtained to give the Navier-Stokes equation

$$\rho_0 \frac{\partial^2 \mathbf{d}}{\partial t^2} = -\nabla p + \eta\nabla^2\dot{\mathbf{d}} + \left(\hat{\eta} + \frac{\eta}{3}\right)\nabla\left(\nabla \cdot \dot{\mathbf{d}}\right). \tag{41c}$$

In the derivation of (41c) the trace of the strain matrix was expressed in terms of the divergence of the displacement field, i.e. $Tr(S) = \nabla\cdot\mathbf{d}$. Equation (41c) can be put into a more useful form by means of the vector identity in (34c) to yield

$$\rho_0 \frac{\partial^2 \mathbf{d}}{\partial t^2} = -\nabla p + \left(\hat{\eta} + \frac{4\eta}{3}\right)\nabla\left(\nabla \cdot \dot{\mathbf{d}}\right) - \eta \nabla \times \nabla \times \dot{\mathbf{d}}. \tag{41d}$$

Two general solutions of (41d) are of importance.

For the case of irrotational motion $\nabla\times\mathbf{d} = 0$, (41d) reduces to

$$\rho_0 \frac{\partial^2 \mathbf{d}}{\partial t^2} = -\nabla p + \left(\hat{\eta} + \frac{4\eta}{3}\right)\nabla^2 \dot{\mathbf{d}}. \tag{42a}$$

Assuming that the total pressure is the sum of an excess acoustic pressure and a hydrostatic pressure, we have $p \rightarrow p + \rho_0 gz$ where p, now, represents the acoustic pressure, g is the acceleration of gravity and z denotes the depth into the ocean. Applying the divergence operator to both sides of (42a) we obtain terms involving the divergence of the displacements

$$\nabla \cdot \mathbf{d} = \frac{\mathrm{Tr}(P)}{3B} = -\left(\frac{p + \rho_0 gz}{B}\right). \tag{42b}$$

In this expression, the bulk modulus in (39h) was used and we made the first order approximation that the diagonal elements of the stress matrix are equal to the negative of the total pressure, i.e. that bulk viscosity is second order. Substituting into (42a) gives the wave equation

$$\nabla^2 p - \frac{1}{c^2}\left(\frac{\partial^2 p}{\partial t^2} - \frac{\left(\hat{\eta} + 4\eta/3\right)}{\rho_0}\nabla^2 \dot{p}\right) = 0, \tag{42c}$$

where $c^2 = B/\rho_0$ is the sound speed. This is the same as the wave equation in (2.16-A) where the effective viscosity in that equation is seen to be related to the bulk and shear viscosity terms by

$$\sigma = \hat{\eta} + \frac{4\eta}{3}. \tag{42d}$$

Solutions to (42c) and the effect of viscosity on acoustic absorption are discussed in Section 5.4.

For the case of solenoidal motion $\nabla\cdot\mathbf{d} = 0$, Equation (41d) reduces to

$$\rho_0 \frac{\partial^2 \mathbf{d}}{\partial t^2} = -\nabla p + \eta \nabla^2 \dot{\mathbf{d}}. \tag{43a}$$

Taking the divergence of both sides of (43a) gives $\nabla^2 p = 0$. Since no solutions of Laplace's equation represent propagating waves we infer that the pressure is constant everywhere and obtain

$$\rho_0 \frac{\partial^2 \mathbf{d}}{\partial t^2} = \eta \nabla^2 \dot{\mathbf{d}}, \tag{43b}$$

which is seen to be a vector wave equation. Since the differential operators commute with the components of the vector it is also a wave equation for each component of displacement. Since by assumption $\nabla\cdot\mathbf{d} = 0$, then for the case of displacements only in the x direction, i.e. $\mathbf{d} = u\mathbf{i}$, we must have $\partial u/\partial x = 0$. This asserts that u is dependent only upon y and z. Consequently, the solution of (43b) has the plane wave form $u = f(t\pm y/c\pm z/c)$ which represents a wave having a direction of propagation in the y-z plane. Therefore, the direction of wave propagation is perpendicular to the direction of particle motion and the wave is a shear wave. Assuming a sinusoidal solution, $u \to u\exp(j\omega t)$ the wave equation reduces to

$$\frac{\partial^2 u}{\partial t^2} = \frac{j\omega\eta}{\rho_0}\frac{\partial^2 u}{\partial z^2}. \tag{43c}$$

Substituting the solution $u = A\exp[j\omega(t \pm z/c)]$, where $\pm$ corresponds to a wave traveling in the negative and positive z directions, we obtain

$$\frac{1}{c} = \pm\sqrt{\frac{\rho_0}{2\omega\eta}}(1-j), \tag{43d}$$

and the solution is

$$u = A\exp\left(\pm\sqrt{\frac{\omega\rho_0}{2\eta}}z\right)\exp\left[j\omega\left(t\pm\sqrt{\frac{\rho_0}{2\omega\eta}}z\right)\right]. \tag{43e}$$

For a wave moving in either the positive or negative z direction the wave is attenuated at the rate $\alpha = \sqrt{(\omega\rho_0/2\eta)}$ and has a phase velocity $c = \sqrt{(\omega\eta/\rho_0)}$.

Example 2: For a wave propagating in water[6] with a frequency of, say, 1000 Hz, the attenuation rate is 793/cm, which is usually referred to as 793 nepers/cm, and the phase velocity is 11.2 cm/sec. These values correspond to a very strongly attenuated and slowly propagating wave. In fact, the amplitude of this wave would be down approximately 6888 dB in one cm and it would propagate at less than 1% of the speed of the longitudinal wave. Consequently, shear waves in the ocean are of small consequence in ASW sonar applications.♦

[6] Nominal viscosity of water is $\eta = 10^{-2}$ poise, where one poise equals one dyne-second/cm^2.

APPENDIX 5-B: AIRCRAFT DETECTION BY SUBMARINES

Sound does not easily pass through the sea surface. Nevertheless, a small portion of the incident energy does get passed and this transmitted energy can be detected at depth. For example, during the search for submarines by aircraft, a small fraction of the engine and propeller noise radiated by the aircraft will penetrate the ocean surface and can reveal the presence of the airborne platform to the submarine crew. Since they can use this information to either evade detection or to attack the aircraft, the transmitted energy is a small effect that can have enormous consequences. In the following, we examine the reflection and transmission of an airborne sound field at the surface of the sea and discuss the source detectability.

If we consider an airborne source at d units above the surface, as indicated in Figure 1, the sound field radiated by the source will be partly reflected from the sea surface and partly transmitted by the sea surface. We show that the reflected field above the surface appears to come from a mirror image source located at d units below the sea surface. The transmitted field below the surface is refracted by the surface and we show that it appears to come from an image that is located above the surface at cylindrical coordinates (ρ′,d′), as shown in Figure 1, where positive z is downward into the medium. We also show that, while the position of the image source of the reflected field is fixed for all observation points, the position of the image source for the transmitted field varies depending upon the observation point.

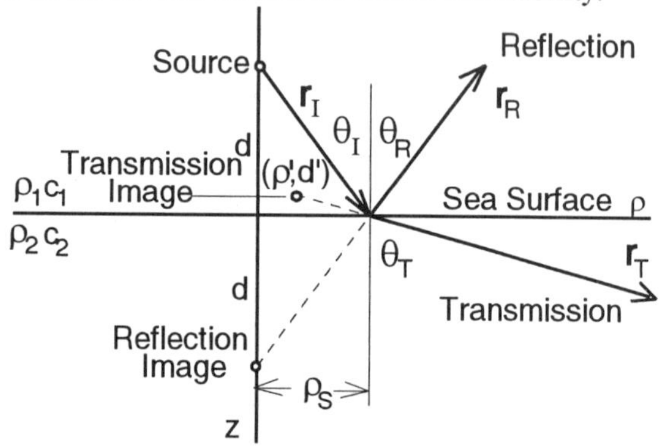

Figure 1 Reflected and Transmitted Waves at the Sea Surface from an Airborne Source

The incident field radiated by the source is given by

$$p_I = \frac{A}{r_I} \exp j(\omega t - \mathbf{k}_I \cdot \mathbf{r}_I), \tag{1a}$$

$$\mathbf{k}_I = k_I(\sin\theta_I \mathbf{e}_\rho + \cos\theta_I \mathbf{k}), \quad \mathbf{r}_I = \rho\mathbf{e}_\rho + (z+d)\mathbf{k}, \tag{1b,c}$$

where k_I and r_I are the magnitudes of the corresponding vectors, and $\mathbf{e}_\rho$ and $\mathbf{k}$ are unit vectors in the cylindrical coordinate system. Similarly, the fields radiated by the images of the reflected and refracted fields are given, respectively, by

$$p_R = \frac{B}{r_R} \exp j(\omega t - \mathbf{k}_R \cdot \mathbf{r}_R), \tag{2a}$$

$$\mathbf{k}_R = k_R(\sin\theta_R \mathbf{e}_\rho - \cos\theta_R \mathbf{k}), \quad \mathbf{r}_R = \rho\mathbf{e}_\rho + (z-d)\mathbf{k}, \tag{2b,c}$$

$$p_T = \frac{C}{r_T} \exp j(\omega t - \mathbf{k}_T \cdot \mathbf{r}_T), \tag{3a}$$

$$\mathbf{k}_T = k_T(\sin\theta_T \mathbf{e}_\rho + \cos\theta_T \mathbf{k}), \quad \mathbf{r}_T = (\rho - \rho')\mathbf{e}_\rho + (z - d')\mathbf{k}. \tag{3b,c}$$

At the sea surface the atmospheric pressure must balance the pressure in the sea water. Also, the atmospheric particle velocity normal to the surface must equal the sea water particle velocity normal to the surface. These two conditions imply that, in mathematical terms, the three fields must satisfy the boundary conditions

$$p_I + p_R = p_T, \tag{4a}$$

$$v_I + v_R = v_T. \tag{4b}$$

For distances from the source and image points that are large compared to a wavelength, the acoustic fields can be approximated as locally plane and the particle velocity is given by $v = p/\rho_0 c$, by (2.10), where ρ_0 is the density and c is the sound speed. Substituting the field expressions from (1), (2) and (3) into the boundary conditions in (4) gives

$$\frac{A}{r_I}\exp(-jk_I d\cos\theta_I) + \frac{B}{r_R}\exp(-jk_R d\cos\theta_R) = \frac{C}{r_T}\exp\bigl(jk_T(\rho'\sin\theta_T + d'\cos\theta_T)\bigr), \tag{5a}$$

$$\frac{A\cos\theta_I}{\rho_1 c_1 r_I}\exp(-jk_I d\cos\theta_I) - \frac{B\cos\theta_R}{\rho_1 c_1 r_R}\exp(-jk_R d\cos\theta_R) = \frac{C\cos\theta_T}{\rho_2 c_2 r_T}\exp\bigl(jk_T(\rho'\sin\theta_T + d'\cos\theta_T)\bigr), \tag{5b}$$

where $k_I\sin\theta_I = k_R\sin\theta_R = k_T\sin\theta_T$ since the exponential factors involving the ρ coordinate must cancel if (5) are to be valid for all points on the surface. Moreover, since $k_I = k_R$, this gives the law of specular reflection $\theta_I = \theta_R$ from the first equality and Snell's law of refraction follows from the second equality. Solving (5) for the reflection and transmission coefficients gives

$$\frac{B}{A} = R_R, \quad \frac{C}{A} = R_T\left(\frac{r_T}{r_I}\right)\exp\bigl[-j\bigl(k_I d\cos\theta_I + k_T(\rho'\sin\theta_T + d'\cos\theta_T)\bigr)\bigr], \tag{6a,b}$$

where R_R and R_T denote the Rayleigh reflection and transmission coefficients for plane waves incident on a fluid-fluid interface,[1] as in (5.30),

---

[1] Rayleigh, J. W. S., *The Theory of Sound*, **II**, pg. 79-81, Dover Publications, 1945.

$$R_R = \frac{\rho_2 c_2 \cos\theta_I - \rho_1 c_1 \cos\theta_T}{\rho_2 c_2 \cos\theta_I + \rho_1 c_1 \cos\theta_T}, \qquad R_T = \frac{2\rho_2 c_2 \cos\theta_I}{\rho_2 c_2 \cos\theta_I + \rho_1 c_1 \cos\theta_T}. \tag{7a,b}$$

Here, though, the angles of incidence and transmission are not constant but vary over the surface of the sea.

We can gain additional insight into the reflected and transmitted fields by substituting the coefficients in (6) back into the field expressions in (2) and (3) obtaining, with some manipulation,

$$p_R = \frac{AR_R}{r_I(\rho_S,0)}\left[\frac{r_R(\rho_S,0)}{r_R(\rho,z)}\right]\exp j\left[\omega t - \mathbf{k}_I\cdot\mathbf{r}_I(\rho_S,0) - \mathbf{k}_R\cdot\left(\mathbf{r}_R(\rho,z) - \mathbf{r}_R(\rho_S,0)\right)\right], \tag{8a}$$

$$p_T = \frac{AR_T}{r_I(\rho_S,0)}\left[\frac{r_T(\rho_S,0)}{r_T(\rho,z)}\right]\exp j\left[\omega t - \mathbf{k}_I\cdot\mathbf{r}_I(\rho_S,0) - \mathbf{k}_T\cdot\left(\mathbf{r}_T(\rho,z) - \mathbf{r}_T(\rho_S,0)\right)\right]. \tag{8b}$$

From these field representations the underlying physics can be easily discerned.

For the reflected field the amplitude of the source is attenuated by the range from the source to the surface reflection point, denoted by $r_I(\rho_S,0)$; by the plane wave reflection coefficient; and by the fractional range from the image source of the reflected field to the observation point. The fractional range first amplifies the field strength, by projecting back from the surface reflection point to the image source, and then attenuates the field strength by projecting forward from the image source to the observation point. Of course, $r_I(\rho_S,0) = r_R(\rho_S,0)$ in this expression. The phase of the reflected field at the observation point is the sum of the phase change along the path from the source to the surface reflection point and the phase change along the reflected path from the surface reflection point to the observation point.

For the transmitted field the amplitude of the source is attenuated by the range from the source to the surface transmission point, which is the same as the surface reflection point denoted by $r_I(\rho_S,0)$; by the plane wave transmission coefficient; and by the fractional range from the image source of the transmitted field to the observation point. As for the reflected wave, the fractional range first amplifies the field strength, by projecting back from the surface transmission point to the image source, and then attenuates the field strength by projecting forward from the image source to the observation point. The phase of the transmitted field at the observation point is the sum of the phase changes along the path from the source to the surface transmission point and along the refracted path from the surface transmission point to the observation point.

Conservation of energy requires that the sum of the power reflected from the surface and the power transmitted through the surface be equal to the power incident on the sea surface at each point. This can be expressed as

$$I_R(\rho_S,0)r_R(\rho_S,0)d\theta_R + I_T(\rho_S,0)r_T(\rho_S,0)d\theta_T = I_I(\rho_S,0)r_I(\rho_S,0)d\theta_I. \tag{9}$$

Dividing both members by the incident power this can be expressed as

$$\alpha_R + \alpha_T = 1, \tag{10a}$$

$$\alpha_R = R_R^2 \frac{r_R(\rho_S,0)d\theta_R}{r_I(\rho_S,0)d\theta_I}, \qquad \alpha_T = R_T^2 \frac{\rho_1 c_1}{\rho_2 c_2} \frac{r_T(\rho_S,0)d\theta_T}{r_I(\rho_S,0)d\theta_I}, \tag{10b,c}$$

where α_R and α_T are, respectively, the sound power reflection coefficient and sound power transmission coefficient. At the surface point we have

$$r_I(\rho_S,0)d\theta_I = \cos\theta_I d\rho_S, \tag{11a}$$

$$r_R(\rho_S,0)d\theta_R = \cos\theta_R d\rho_S, \tag{11b}$$

$$r_T(\rho_S,0)d\theta_T = \cos\theta_T d\rho_S. \tag{11c}$$

Substituting the appropriate ratios from (11) into (10) gives the sound power reflection and transmission coefficients, as in (5.32),

$$\alpha_R = \frac{(\rho_2 c_2 \cos\theta_I - \rho_1 c_1 \cos\theta_T)^2}{(\rho_2 c_2 \cos\theta_I + \rho_1 c_1 \cos\theta_T)^2}, \quad \alpha_T = \frac{4\rho_2 c_2 \rho_1 c_1 \cos\theta_I \cos\theta_T}{(\rho_2 c_2 \cos\theta_I + \rho_1 c_1 \cos\theta_T)^2}. \tag{12a,b}$$

We see that the sound power reflection and transmission coefficients for spherical waves incident upon a fluid-fluid interface are the same as for plane waves, as found in Section 5.4, but for the spherical case the angles of incidence, reflection and transmission vary over the surface. Again, we see that the loss through the surface is the same whether the source is above or below the surface.

The reflected and transmitted fields can be usefully represented in terms of the sound power reflection and transmission coefficients. As a preliminary we find from (11), Snell's law and geometrical considerations the expression

$$\frac{r_T(\rho_S,0)}{r_I(\rho_S,0)} = \sqrt{\frac{c_1|d'|\cos\theta_T}{c_2 d\cos\theta_I}}. \tag{11d}$$

Substituting from (11) and (12) into (8) gives the desired results

$$p_R = \frac{A\sqrt{\alpha_R}}{r_R(\rho,z)} \exp j\left[\omega t - \mathbf{k}_I \cdot \mathbf{r}_I(\rho_S,0) - \mathbf{k}_R \cdot \left(\mathbf{r}_R(\rho,z) - \mathbf{r}_R(\rho_S,0)\right)\right], \tag{13a}$$

$$p_T = \frac{A\sqrt{\alpha_T}}{r_T(\rho,z)} \sqrt{\frac{\rho_2 c_2}{\rho_1 c_1}} \sqrt{\frac{c_1|d'|}{c_2 d}} \exp j\left[\omega t - \mathbf{k}_I \cdot \mathbf{r}_I(\rho_S,0) - \mathbf{k}_T \cdot \left(\mathbf{r}_T(\rho,z) - \mathbf{r}_T(\rho_S,0)\right)\right]. \tag{13b}$$

The intensity levels at the observation points of the reflected and refracted waves can be determined from (13) and are given, respectively, by

$$\mathrm{IL_R}(\mathrm{AT}) = \mathrm{SL}(\mathrm{AT}) + 10\mathrm{Log}(\alpha_R) - 20\mathrm{Log}(r_R(\rho, z)), \tag{14a}$$

$$\mathrm{IL_T}(\mu\mathrm{Pa}) = \mathrm{SL}(\mathrm{AT}) + 10\mathrm{Log}(\alpha_T) - 20\mathrm{Log}(r_T(\rho, z)) + 10\mathrm{Log}\left(\frac{c_1|d'|}{c_2 d}\right) + 10\mathrm{Log}\left(\frac{I_0(\mathrm{AT})}{I_0(\mu\mathrm{Pa})}\right), \tag{14b}$$

where $I_0(\mathrm{AT})$ denotes the intensity of the audibility threshold in air and $I_0(\mu\mathrm{Pa})$ denotes the intensity of a plane wave having pressure of 1 μPa. Equations (14) represent the received intensities in terms of the fields radiated by the image sources. The last two terms of the transmitted intensity expression represent corrections due, respectively, to the apparent displacement of the transmitted field image source and increased ray spreading at depth and to the difference in reference intensities in air and water.

To calculate the intensities in (14) we need to determine the positions of the image sources. Since the position of the reflected field image source is fixed and known for all observation points, the calculation of the reflected intensity is relatively straight forward. For the transmitted field, however, the image source position depends upon the observation point position and its determination will require some effort. Preliminary to these calculations we examine the important case of an overhead source for which all calculations are easily done.

Example 1: For an overhead source, the range to the surface reflection and refraction point is zero as are the angles of incidence, reflection and refraction. From (11), the position of the image of reflected field is at d, as is expected, and the position of the image of the transmitted field is given by $|d'|/d = d\theta_I/d\theta_T$. From Snell's law we obtain $|d'|/d = c_1/c_2$. Substituting these results into the intensity relations in (14), the reflected and transmitted intensity levels are given, respectively, by

$$\mathrm{IL_R}(\mathrm{AT}) = \mathrm{SL}(\mathrm{AT}) - 5\times10^{-3}\ \mathrm{dB} - 20\mathrm{Log}(d - z), \tag{15a}$$

$$\mathrm{IL_T}(\mu\mathrm{Pa}) = \mathrm{SL}(\mathrm{AT}) - 29.6\ \mathrm{dB} - 20\mathrm{Log}\left(z + \frac{c_1}{c_2}d\right) - 12.8\ \mathrm{dB} + 61.8\ \mathrm{dB}. \tag{15b}$$

We see that the loss upon reflection from the surface is 5×10^{-3} dB, which is negligible, while the loss upon transmission through the surface is −29.6 dB, which is substantial. The correction for the image position of the transmitted field and the increased spreading of the rays is −12.8 dB and the correction between the auditory threshold and μPa reference levels for the transmitted field is 61.8 dB. If, say, we assume a helicopter is hovering at 100 m altitude with source level of 120 dB re AT per Hz over the band of frequencies up to, say, 5 kHz, the reflected intensity back at the helicopter would be 74 dB which is negligible compared to the source field at the helicopter. If we further assume that a submarine is directly beneath the helicopter at 100 m below the surface, the intensity level at the submarine would be approximately 98 dB re μPa per Hz. Since the ambient noise level in the ocean is approximately 60 dB to 80 dB re μPa per Hz over the band of assumed frequencies, the helicopter could be easily detected by the submarine detection system. However, since there is a loss of approximately 30 dB through the submarine hull, the received level of approximately 6.2 dB re AT per Hz within the submarine interior would not likely be detected aurally by the crew in the submarine. However, for a helicopter that is hovering just above the surface, $d \cong 0$, and a submarine at periscope depth of, say, 20 m the received level of approximately 22 dB re AT per Hz within the interior of the submarine could be detected by an alert crew within a quiet submarine.♦

Example 2: As an aside, we ask what the intensity level at the helicopter would be due to acoustic radiation from the submarine. For this case (14b) becomes

$$IL_T(AT) = SL(\mu Pa) - 29.6\text{ dB} - 20\text{Log}\left(d + \frac{c_2}{c_1} z\right) + 12.8\text{ dB} - 61.8\text{ dB}. \qquad (16)$$

For a submarine source level of 130 dB re μPa per Hz over the band of frequencies up to, say, 5 kHz, the received intensity level at the helicopter would be approximately −3 dB which, for all practical purposes, would be wholly undetectable even without the aircraft noise. The primary difference between the two situations is that the 120 dB source level in air of the helicopter represents a much greater source of sound power than the 130 dB source level in water of the submarine, i.e. submarines are just much quieter than helicopters.♦

Returning to the problem of determining a general solution for the intensity at depth we need to determine the range to the point at which the ray penetrates the surface and the image source position for the transmitted field. As indicated in Figure 2, when a cone of radiation from an airborne source penetrates the sea surface, the rays of the cone are refracted in accordance with Snell's law and the spread of the cone increases as the rays propagate into the water mass. The image source of the refracted cone is at the apex of the cone which is indicated in Figure 2 at position (ρ', d'). Since we have chosen the positive z direction to be downward into the ocean, d' is a negative number; the slopes of the rays are positive; and the equations for the bounding rays of the cone are

$$\frac{-d'}{\rho_S - \rho'} = m_1, \qquad (17a)$$

$$\frac{-d'}{\rho_S + \Delta\rho_S - \rho'} = m_2. \qquad (17b)$$

Solving for the image point coordinates we obtain

$$\rho' = \rho_S - \frac{m_2 \Delta\rho_S}{m_1 - m_2}, \qquad (18a)$$

$$d' = -\frac{m_1 m_2 \Delta\rho_S}{m_1 - m_2}. \qquad (18b)$$

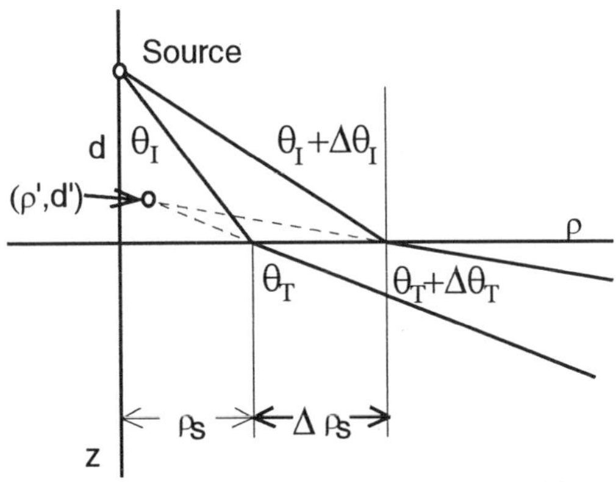

Figure 2 Transmitted Field Image Position

The slopes in (18) can be determined from inspection of Figure 2 and are given by

$$m_1 = \cot\theta_T\,, \qquad m_2 = \cot(\theta_T + \Delta\theta_T) \cong \cot\theta_T - \csc^2\theta_T \Delta\theta_T\,. \qquad (19a,b)$$

In this result $\Delta\theta_T$ is a small angle. Substituting (19) and (11), with help from Snell's law, into (18) we find the image point position to be given by

$$r' = \rho_S\left(1 - \frac{c_1}{c_2}\frac{\sin\theta_T}{\sin\theta_I}\frac{\cos^2\theta_T}{\cos^2\theta_I}\right), \qquad d' = -\rho_S\frac{c_1}{c_2}\left(\frac{\cos^3\theta_T}{\sin\theta_I\cos^2\theta_I}\right). \qquad (20a,b)$$

This gives the image point position in terms of the range to the surface point and the associated angles of incidence and transmission. From the relations

$$\sin\theta_I = \frac{\rho_S}{\sqrt{\rho_S^2 + d^2}}\,, \qquad \sin\theta_T = \frac{\rho - \rho_S}{\sqrt{(\rho - \rho_S)^2 + z^2}}\,, \qquad (21a,b)$$

$$\cos\theta_I = \frac{d}{\sqrt{\rho_S^2 + d^2}}\,, \qquad \cos\theta_T = \frac{z}{\sqrt{(\rho - \rho_S)^2 + z^2}}\,, \qquad (21c,d)$$

we obtain the angles of incidence and transmission in terms of the range to the surface point. This gives all unknown quantities in terms of the surface point range and we obtain that quantity from Snell's law. Substituting into Snell's law we obtain the following fourth order polynomial for the normalized range to the surface point

$$\left(\frac{\rho_S}{d}\right)^4 - 2\frac{\rho}{d}\left(\frac{\rho_S}{d}\right)^3 + \left(\frac{\rho^2}{d^2} + B\left(\frac{c_2^2}{c_1^2}\frac{z^2}{d^2} - 1\right)\right)\left(\frac{\rho_S}{d}\right)^2 + 2B\frac{\rho}{d}\left(\frac{\rho_S}{d}\right) - B\frac{\rho^2}{d^2} = 0\,, \qquad (22a)$$

$$B = \frac{c_1^2}{c_2^2} \Big/ \left(1 - \frac{c_1^2}{c_2}\right). \qquad (22b)$$

Therefore, for a given observation point (ρ,d) at depth, (20), (21) and (22) give the quantities required by (14b) to determine the intensity level at the submarine. From Snell's law we also find that for the angles of incidence such that $\sin\theta_I > c_1/c_2$, the transmission angle has a maximum value of $\pi/2$ and the incident energy is totally reflected. The range to the surface point at which total reflection occurs is found to be $\rho_{S,max} = \sqrt{(B)}d \cong 0.23d$.

A convenient graphical procedure for determining the intensity level at depth in (14b) can be developed in terms of either the submarine parameters or the aircraft parameters. We choose the former case and rewrite (14b) in terms of an additive intensity level $IL_A(\mu Pa)$

$$IL_T(\mu Pa) = \left[SL(AT) - 20Log(d)\right] + IL_A(\mu Pa), \tag{23a}$$

$$IL_A(\mu Pa) = 10Log(\alpha_T) - 20Log\left(\frac{r_T(\rho, z)}{d}\right) + 10Log\left(\frac{c_1|d'|}{c_2 d}\right) + 10Log\left(\frac{I_0(AT)}{I_0(\mu Pa)}\right). \tag{23b}$$

Considering the intensity level, due to an arbitrary aircraft at an arbitrary height, at the point directly beneath the aircraft on the ocean surface, the additive intensity level increments this value to give the intensity level at the submarine. Plots of the additive intensity level are given in Figure 3 in terms of the normalized horizontal range and normalized depth of the submarine. Inspection of Figure 3 shows that the received intensity from an overhead source is larger for shallow submarines but that, when compared to deep submarines, the received intensity falls off rapidly with range. For the case of near surface submarines, e.g. z/d = 1/16, the rapid fall-off begins at $\rho/d \cong 0.25$ which is approximately the range to the point of total reflection. For deep submarines, the fall-off with horizontal range is more gradual and, in fact, at large ranges the received intensity at a deep submarine exceeds the received intensity at a shallow submarine.

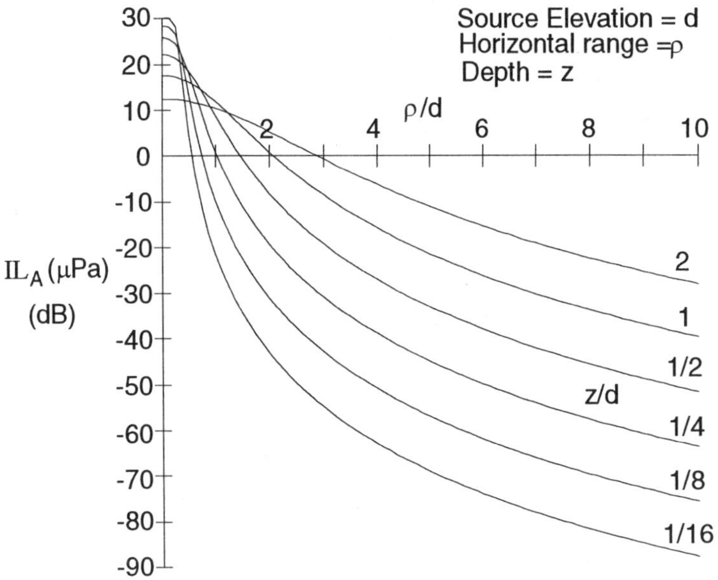

Figure 3 Additive Intensity Level re μPa

Example 3: Consider an airborne source having source level of 120 dB re AT at altitude 100 m and a submarine at 100 m depth. The curve in Figure 3 for z/d = 1 applies and we find that when the aircraft is directly over the submarine the additive intensity is approximately 18 dB re μPa. Since the intensity normally incident on the surface below the aircraft is 120 dB – 20Log(d) = 80 dB, the total received intensity at depth is 98 dB. For an array gain of 10 dB, an ambient noise level of 80 dB and a detection threshold of 10 dB, the signal excess would be 18 dB and the signal would be easily detectable by the submarine. However, at a horizontal range of $\rho = 200$ m = 2d, the additive intensity falls to approximately zero dB and the total received intensity at the submarine falls to 80 dB. Under our assumed scenario the signal excess would be zero dB and the aircraft would not be detectable. However, from the viewpoint of the aircraft crew, for a submarine radiated source level of, say, 130 dB, the intensity level received by an aircraft sonobuoy would be 130 dB – 20Log(200m) ≅ 89 dB. This would give the aircraft a detection advantage of 9 dB over the submarine at 200 m. Furthermore, the aircraft would maintain its advantage out to about 300 m at which point the submarine would no longer be detectable by the aircraft. Of course, these calculations are strongly dependent upon the assumed source levels of the submarine and aircraft and the submarine detection system. In other scenarios the advantage may be with the submarine. ♦

CHAPTER 6
ACOUSTIC PROPAGATION IN THE OCEAN

The effectiveness of ASW operations in countering the threat posed by submarines is dependent upon many factors but among the more formidable are the acoustic propagation conditions. These conditions are largely dependent upon the sound speed variations and the bathymetric structures between the ASW platform and the target. Sound speed variations, discussed in Chapter 5, can cause propagated waves to refract in unwanted directions while intruding bathymetric features can block ray paths that might otherwise connect the target and receiver. When operating in such adverse conditions, the passage of time will sometimes improve, or worsen, the situation since sound speed conditions within any region can change with the watch. Similarly, maneuvering to a new position can sometimes improve, or worsen, the situation since intervening bathymetric conditions change due to platform movements. In all cases, a thorough understanding and ongoing assessment of the propagation conditions are required to conduct effective ASW operations.

The bathymetry of each ocean basin is unique but all share the general characteristics of the world's ocean bathymetry as suggested by the statistical view in Figure 1. This figure shows that the oceans cover approximately 70 percent of the earth's surface with the remaining 30 percent covered by land masses. The distribution of elevations and depths is seen to be bimodal and is biased toward lower elevations and depths with each mode approximately centered, respectively, on the near sea level land masses and the abyssal plane. Continental uprisings have caused the bimodality of this distribution while wind and water erosion coupled with gravity have concentrated the continents near sea level. The extremes in land elevations and ocean depths are both of the order of 10,000 m but, because the continents are concentrated near sea level, the average elevation of the continental land masses is about 1/4 of the average depth of the oceans. Surrounding and near the land masses are the shallow water littorals, sometimes called continental shelves, that are less than approximately 1000 m in depth and that cover approximately 8.5 percent of the earth. Beyond the shallow water littorals are slopes, sometimes called the continental slopes, that drop off into deep water that cover another approximately 7.5 percent of the earth's surface. Beyond the continental slopes is deep water that covers the remaining 54 percent of the earth. Most of the deep water covers rolling abyssal planes that range in depth from approximately 3000 m to approximately 6000 m in depth. However, obtruding sea mounts, some surpassing all continental mountains in size, can penetrate the ocean surface to form islands and submerged chains of sea mounts can also form unseen ocean basins. These basins within the larger ocean basin are generally distinguished by unique acoustic propagation characteristics. Deep ocean trenches that have no continental comparisons extend down to 10,000 m depth and make up a small fraction of a percent of the earth's coverage.

When considering the bathymetry described in Figure 1, we should bear in mind that the vertical scale of the slopes shown is expanded and that actual slopes in the ocean are of the order of 1% and less. This means that, if the oceans were drained and we could hike the abyss, the ocean floor would mostly appear to be a flat plane perforated by continents and an occasional sea mount or sea mount chain. Of course, localized features such as rock outcroppings, depressions and old ship wrecks would also be seen.

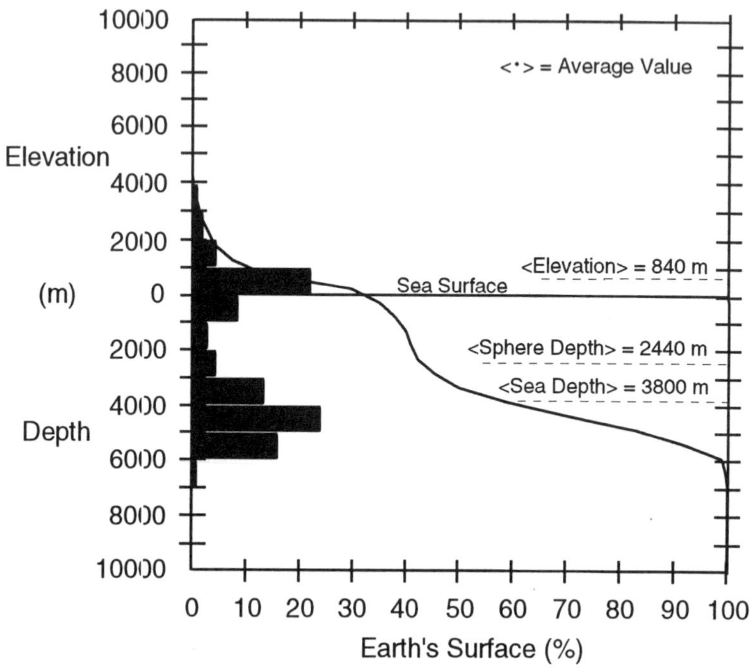

Figure 1 Hypsographic Curve and Frequency Distribution of the Earth's Land Elevations and Sea Depths at 1000 m Intervals

Due to the immensity of the oceans and the variability of environmental conditions from basin-to-basin and within a basin, acoustic propagation conditions can be relatively stable or can be an ever-changing characteristic of the basin. In either case, it is possible to characterize propagation conditions in general terms and any one or several such characterizations can describe the propagation conditions within a particular basin. In the following, we examine some of the more common modes of propagation that are found in ASW scenarios. Our examination is, necessarily, cursory because ocean acoustic propagation is one of the more extensive and complicated topics in ASW sonar. A comprehensive treatment would be a study in itself and, since such treatments are found in other works, our goal in the following is the more modest one of introducing the subject and its methods while examining propagation modes that are useful to ASW sonar. These discussions provide examples of the fundamentals of analyzing acoustic propagation by ray theory and, even though the examples are elementary, they still provide insights into the complexities of more complicated ocean propagation conditions.

6.1 Surface Shadow Zone

During summer-like conditions of surface warming, the sound speed has a negative gradient and decreases into the medium. This results in two systems of downward refracted rays as indicated in Figure 2. The refracted-only rays, indicated in Figure 2a, are delimited by the surface grazing ray which defines the edge of a shadow zone and also defines the maximum range of propagation. The refracted-reflected rays, indicated in Figure 2b, are the rays that can produce the surface interference effect discussed in Section 5.3 and have the same maximum range as the refracted-only rays. The sound speed for this situation can be written in the form

$$c = c_0 + g(z - z_0)\,, \tag{1}$$

where c_0 is the sound speed at the source depth of z_0. The transmission loss along each ray path is obtained from (5.16)

$$TL = 10\log\left[\frac{r}{\cos\theta_0}\left|\cos\theta Z_s - \sin\theta R_s\right|\right], \tag{5.16}$$

where r and z are cylindrical coordinates and R_s and Z_s are their partial derivatives with respect to θ_0 at constant ray path length s.

For the refracted-only rays, the rays follow circular arcs and, we see from Figure 2a, the ray loci are given by

$$r = -\rho(\sin\theta_0 - \sin\theta)\,, \tag{2a}$$

$$z = d - \rho\cos\theta\,, \tag{2b}$$

$$\theta - \theta_0 = s/\rho\,. \tag{2c}$$

The quantity d is the depth of the line of centers, the up-angle θ_0 is a negative angle, down-angles are positive and s is the total distance along the ray path.

The partial derivatives of the ray coordinates are readily found to be

$$R_s = -\rho(\cos\theta_0 - \cos\theta) - P_s(\sin\theta_0 - \sin\theta)\,, \tag{3a}$$

$$Z_s = \rho\sin\theta\Theta_s - P_s\cos\theta\,, \tag{3b}$$

$$\Theta_s = 1 - (\theta - \theta_0)\tan\theta_0\,, \qquad P_s = \rho\tan\theta_0\,. \tag{3c,d}$$

Substituting into the transmission loss expression in (5.16) gives the transmission loss along the ray with launch angle θ_0

$$TL = 10\log\left[\frac{r^2}{\cos^2\theta_0}\right]. \tag{4a}$$

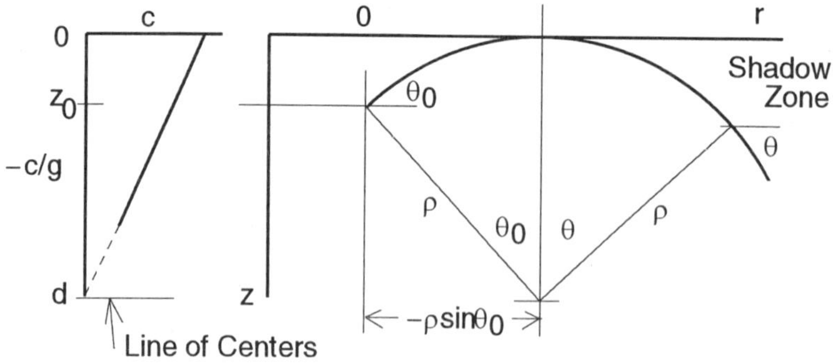

Figure 2a Refracted-Only Rays - Negative Sound Speed Gradient

This expression has the same form as the near field transmission loss in (5.19) for a positive sound speed gradient. An alternative representation in terms of range and depth can be obtained by substituting for $\cos^2\theta_0$ the following expression obtained from (2)

$$\cos^2\theta_0 = \frac{R^2}{\left[(R/2)^2 + \left(1 - Z^2\right)\right]\left[(R/2)^2 + Z^2\right]}, \tag{4b}$$

$$R = \frac{r}{(c/g)}, \qquad Z = \frac{z}{(c/g)}. \tag{4c,d}$$

Along a source depth radial, (4b) reduces to

$$\cos^2\theta_0 = \left[1 + \frac{r^2}{4(c/g)^2}\right]^{-1}, \tag{4e}$$

and the transmission loss along a radial at the source depth is given by

$$TL = 10\text{Log}\left[r^2\left(1 + \frac{r^2}{4(c/g)^2}\right)\right]. \tag{4f}$$

This shows that the transmission loss along the source depth radial is slightly greater than would be found for spherical spreading since the ray travels a longer distance.

For the refracted-reflected rays prior to reflection the ray loci and transmission loss are given by (2) and (4). After reflection the rays follow circular arcs and, we see from Figure 2b, the ray loci are given by

$$r = -\rho\left(\sin\theta_0 - \sin\theta + 2\sin\theta_{ref}\right), \tag{5a}$$

$$z = d - \rho\cos\theta, \tag{5b}$$

$$\theta - \theta_0 = \theta_{ref} - \theta_{inc} + s/\rho, \tag{5c}$$

where θ_{inc} and θ_{ref} are the incident and reflected angles of the ray at the surface; the difference $\theta_{ref} - \theta_{inc}$ is the jump in ray grazing angle between the reflected ray and the incident ray and by the law of reflection we have $\theta_{ref} = -\theta_{inc}$. The quantity s is the total distance along the ray path.

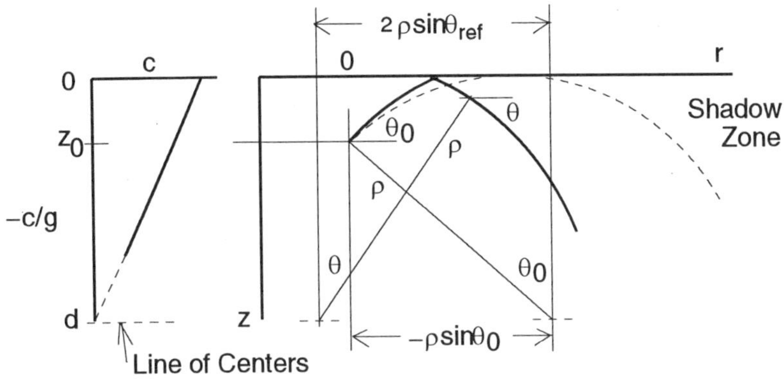

Figure 2b Refracted-Reflected Rays - Negative Sound Speed Gradient

The partial derivatives of the ray coordinates are given by

$$R_s = -\rho(\cos\theta_0 - \cos\theta\Theta_s) - P_s(\sin\theta_0 - \sin\theta) - 2P_s/\sin\theta_{ref}, \tag{6a}$$

$$Z_s = \rho\sin\theta\Theta_s - P_s\cos\theta, \tag{6b}$$

$$\Theta_s = 1 + 2\Theta_{ref,s} - (\theta - \theta_0 - 2\theta_{ref})\tan\theta_0, \tag{6c}$$

$$\Theta_{ref,s} = \frac{\alpha\sin\theta_0}{\sqrt{1-\alpha^2\cos^2\theta_0}}, \qquad \alpha = \frac{d}{d-z_0}. \tag{6d,e}$$

Substituting (6) into the transmission loss expression in (5.16) we obtain

$$TL = 10\mathrm{Log}\left[\frac{r\rho}{\cos^2(\theta_0)}\left|-\sin\theta_0 + \sin\theta + \frac{2\sin\theta\sin\theta_0}{\sin\theta_{ref}}\right|\right]. \tag{7a}$$

Equation (7a) shows that as the refracted-reflected system of rays approach the shadow zone boundary, and the reflection angle vanishes, the transmission loss goes to infinity. Consequently, there is no caustic formed at the boundary by the two systems of rays. Equation (7a) can be recast in terms of the range and depth as was done for the refracted-only rays. However, the algebraic process is tedious and the resulting expressions are complicated. Restricting ourselves to the source depth radial we avoid these problems. The angle of the grazing ray at range is the negative of the launch angle $\theta = -\theta_0$ and the reflection angle is given by

$$\sin\theta_{ref} = \sqrt{1-\alpha^2\cos^2\theta_0}\,, \qquad \alpha = z_0 g / c\,. \tag{7b.c}$$

Collecting results, (7a) gives the transmission loss along the source depth radial in terms of the range and launch angle

$$TL = 10\mathrm{Log}\left[\frac{r}{\cos^2\theta_0} 2\rho \left| -\sin\theta_0 - \frac{\sin^2\theta_0}{\sin\theta_{ref}} \right|\right]. \tag{7d}$$

This can be cast into the range and depth form by reducing (5) to the source depth radial and solving to obtain

$$\cos^2\theta = \frac{R^2}{\left[(2+Z)^2 + R^2/4\right]\left[Z^2 + R^2/4\right]}, \tag{7e}$$

$$R = \frac{r}{(c/g)}, \qquad Z = \frac{z_0}{(c/g)}. \tag{7f,g}$$

Example 1: Consider the case of a source at depth 100m in a negative sound speed gradient of –0.0017 m/sec per m. The ray structures for the refracted-only and refracted-reflected rays are shown in Figure 3a. This figure shows the shadow zone appearing at about 8.5 km at the source depth. Transmission loss curves for the refracted-only and refracted-reflected rays are shown in Figure 3b and are compared to spherical spreading loss. We see that the refracted-only rays propagate essentially as spherical waves with the exception that they terminate abruptly at the shadow zone boundary in contrast to spherical waves. Of course, in practice, measurable sound energy will be found in the shadow zone and it will not be totally dark acoustically. In fact, wave theory shows that weak sound propagates across the boundary that rapidly attenuates into the shadow zone. Additionally, within the context of ray theory, some sound propagates across the boundary because real sound speed profiles deviate from perfect linearity causing some rays to penetrate the shadow zone. Also, sound energy can penetrate the shadow zone if scattering by the surface, bottom or particulate matter in the water column is present. In spite of these deviations from the ray picture in Figure 3, it is found that the attenuation of sound across the shadow zone boundary can be several 10s of dB. The intensity of the refracted-reflected system of rays, in this example, is initially about 46 dB below the intensity of the refracted-only system of rays. This is the loss incurred in propagating to the surface and returning, i.e. in this example $20\mathrm{Log}(2z_0) = 46$ dB. In fact, the loss is slightly more than 46 dB because the focusing of the reflected rays causes the image source to be displaced slightly upward. Over the range from about 120 m to 1200 m the intensities of the two ray systems are seen to be comparable and surface interference effects can occur as discussed in Section 5.3. Beyond about 1200 m the intensity of the refracted-reflected ray system decays and eventually vanishes at the shadow zone boundary. ♦

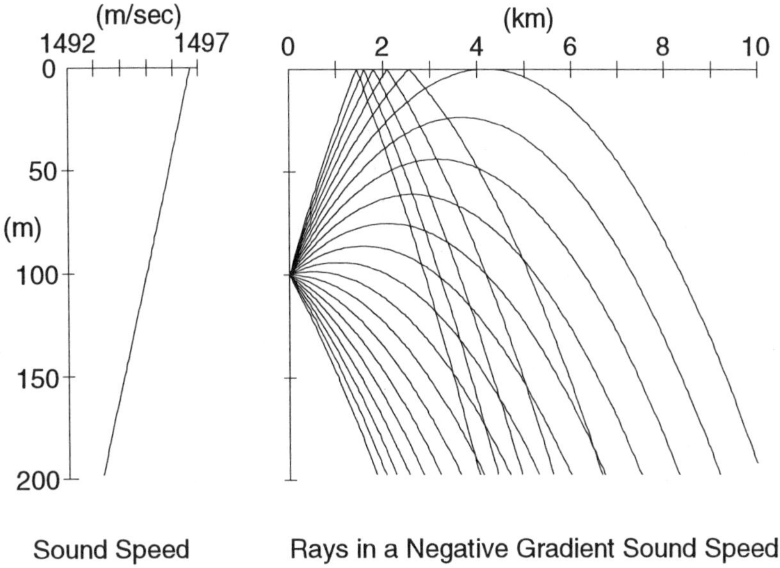

Figure 3a Surface Shadow Zone Formed by Refracted-Only and Refracted-Reflected Rays from a Source at 100 m Depth in a Negative Sound Speed Gradient (g = –0.017 m/sec per m)

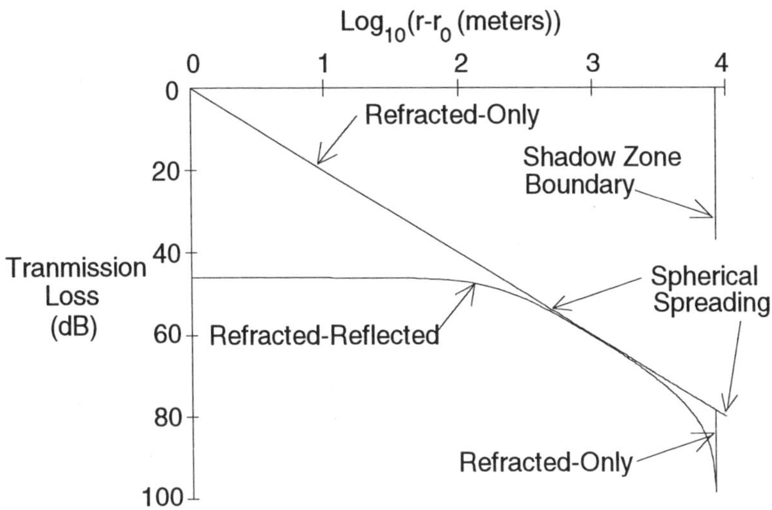

Figure 3b Transmission Loss Along Source Depth Radial at 100 m for Refracted-Only and Refracted-Reflected Rays Compared to Spherical Spreading. (Sound Speed Gradient = –0.017 m/sec per m)

6.2 Surface Duct

During winter-like conditions when surface temperatures increase downward or during mixed turbulent conditions when the surface temperature is constant, the near surface sound speed profile has a positive gradient as shown in Figure 4 and a surface duct is formed. Rays leaving a source in such an environment are refracted upward where they reflect from the surface and the reflected rays are then constrained to a surface duct as they are repeatedly refracted upward and are surface reflected. It's convenient to discuss the down-going and up-going rays, as they leave the source, separately and these ray systems are shown, respectively, in Figures 4a and 4b.

Rays leaving the source follow circular arcs and, we see from Figure 4, the loci for down-going and up-going rays are given by

$$r = \rho\left(\sin\theta_0 + 2N\sin\theta_{ref} - \sin\theta\right), \tag{8a}$$

$$z = \rho\left(\cos\theta - \cos\theta_{ref}\right), \tag{8b}$$

$$\theta = \theta_0 + 2N\theta_{ref} - s/\rho, \tag{8c}$$

where θ_{ref} denotes the angle of reflection at the surface; and N denotes the number of reflections from the surface. To arrive at (8c), we note that the ray path angle is given by $\theta = \theta_{ref} - s_N/\rho$ where s_N represents the ray path length in the Nth cycle. This partial ray path length is given by $s_N = s+\rho(\theta_{inc}-\theta_0)+\rho(N-1)(\theta_{inc}-\theta_{ref})$ where s is the total ray path length and θ_{inc} is the angle of incidence at the surface. We note that $\theta_{inc} = -\theta_{ref}$. Combining expressions gives (8c). If we denote the duct depth by H, then the maximum down and up-angles for rays to remain in the duct are found to be equal for all source positions and is given by

$$\cos\theta_{max} = \frac{(c/g)}{(c/g)+(H-z_0)}. \tag{8d}$$

For the source at the bottom of the duct only the $\theta = 0$ ray is constrained to the duct and the duct ceases to exist. For the source at the surface all rays in the range zero to the maximum angle of reflection, given by $\cos\theta_{max} = 1/(1+Hg/c)$, are constrained to the duct.

The partial derivatives of the ray coordinates in (8) are readily found to be given by

$$R_s = \rho\left(\cos\theta_0 + 2N\cos\theta_{ref}\Theta_{ref,s} - \cos\theta\Theta_s\right) + P_s\left(\sin\theta_0 + 2N\sin\theta_{ref} - \sin\theta\right), \tag{9a}$$

$$Z_s = -\rho\left(\sin\theta\Theta_s - \sin\theta_0\right) + P_s\left(\cos\theta - \cos\theta_0\right), \tag{9b}$$

$$\Theta_{ref,s} = \frac{\alpha\sin\theta_0}{\sqrt{1-\alpha^2\cos^2\theta_0}}, \qquad \alpha = 1 - \frac{z_0}{c/g}. \tag{9d,e}$$

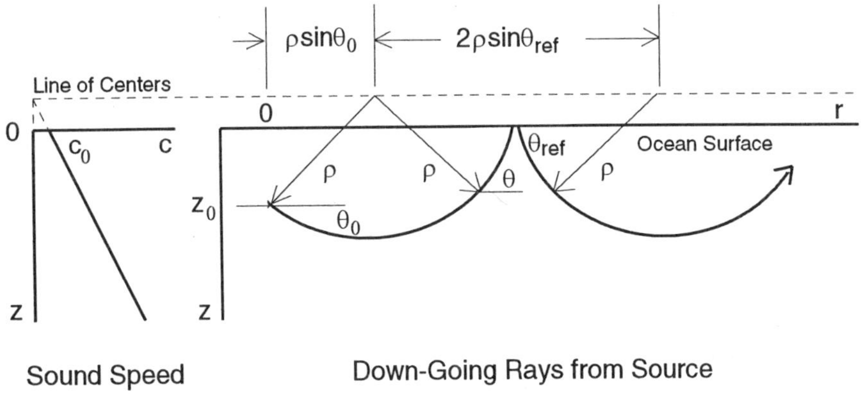

Figure 4a Surface Duct Propagation for Down-Going Rays

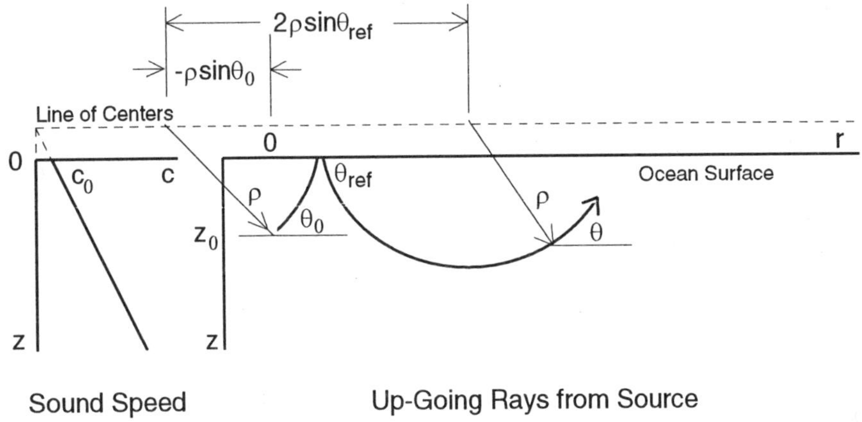

Figure 4b Surface Duct Propagation for Up-Going Rays

Substituting (9) into the transmission loss expression in (5.16) we obtain

$$TL = 10 \text{Log}\left[\frac{r}{\cos^2\theta_0}\left|r\left(1+2N\Theta_0\right)-2N\rho\sin\theta_{Ref}\left(1+2N\Theta_0+\Theta_0^2\right)\right|\right], \quad (10a)$$

$$\Theta_0 = \frac{\sin\theta_0}{\sqrt{1-\alpha^2\cos^2\theta_0}}, \qquad \alpha = 1-\frac{z_0}{c/g}. \quad (10b,c)$$

Caustics occur at the ranges for which the argument of the logarithm vanishes

$$r = 2N\rho\sin\theta_{ref}\left[1+\frac{\Theta_0^2}{1+2N\Theta_0}\right]. \quad (10d)$$

We see that for symmetric down and up launch angles, the range to the caustics of the downward going rays, i.e. $\theta_0 > 0$, is less than the range to the caustics of the upward going rays, i.e. $\theta_0 < 0$. Equation (10d) can be recast into the form

$$r = 2N\rho\sin\theta_{ref} + \frac{\rho\sin\theta_0}{\left(1+\dfrac{1}{2N\Theta_0}\right)}, \qquad (10e)$$

from which we see that for a large number of cycles N >>1 the ranges of the caustics approach

$$r \to 2N\rho\sin\theta_{ref} + \rho\sin\theta_0 . \qquad (10f)$$

Inspection of Figures 4a and 4b show that (10f) asymptotically places the caustics of all rays at the bottom of the ray cycles. This places the caustics mostly in the bottom portion of the duct below the source. However, at shorter ranges (10e) shows that the caustics of the down-going rays slightly precede the bottoms of the cycles and the caustics of the up-going rays slightly follow the bottoms of the cycles. Consequently, the caustics of the rays around zero launch angle, initially, fall slightly in the upper portion of the duct above the source and move lower with range.

A simplified transmission loss expression for surface ducts that is a useful alternative for hand calculations can be readily developed[1]. Since the total power radiated by the source is constrained to the duct, the intensity at range r from the source is given by $I = P/2\pi rH$ where P is the total power radiated by the source and is assumed to be uniformly distributed in depth. Near the source the radiated power is constrained to the angles $\pm\theta_{Max}$ and the radiated intensity is given by

$$I_0 = P\left[2\pi\int_{-\theta_{max}}^{+\theta_{max}}\cos\theta d\theta\right]^{-1} = \frac{P}{4\pi\sin\theta_{max}}, \qquad (11a)$$

where the radiated power is assumed to be uniformly distributed in angle. Combining (8d) and (11a), the transmission loss is found to be

$$TL = -10\text{Log}\left(\frac{I}{I_0}\right) = 10\text{Log}(rr_0), \qquad r_0 = \frac{H}{2\sin\theta_{max}}. \qquad (11b,c)$$

An alternative form of this expression is $TL=10\text{Log}(r_0^2)+10\text{Log}(r/r_0)$. In this form we see that the spreading is spherical out to the range r_0 and is cylindrical beyond that range.

Example 2: Consider a surface duct that is 200 m deep with a source at depth 100m. The sound speed gradient is taken to be 0.0017 m/sec per m. The

[1] Urick, Robert J. *Principles of Underwater Sound for Engineers*, McGraw-Hill, pg. 140.

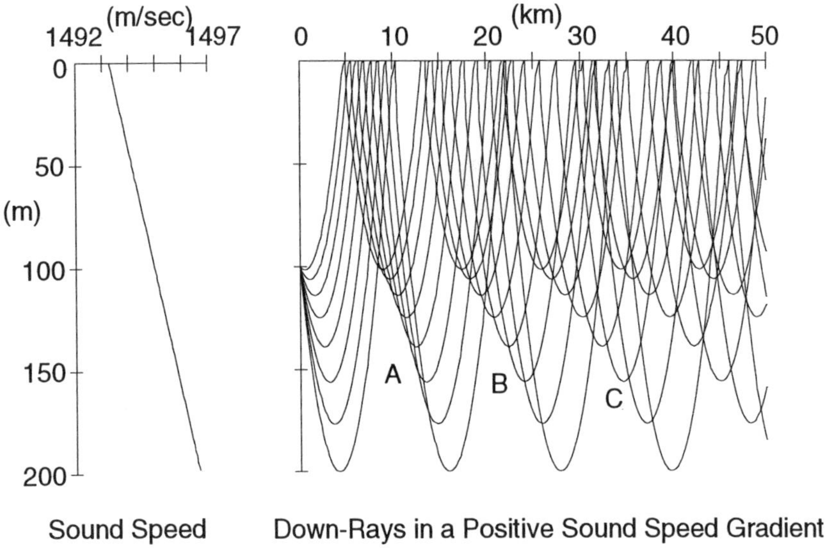

Figure 5a Caustic Structures Formed at A, B, C,··· by Down-Going Rays from a Source at 100 m in a Positive Sound Speed Gradient Environment. Sound Speed Gradient = 0.017 m/sec per m.

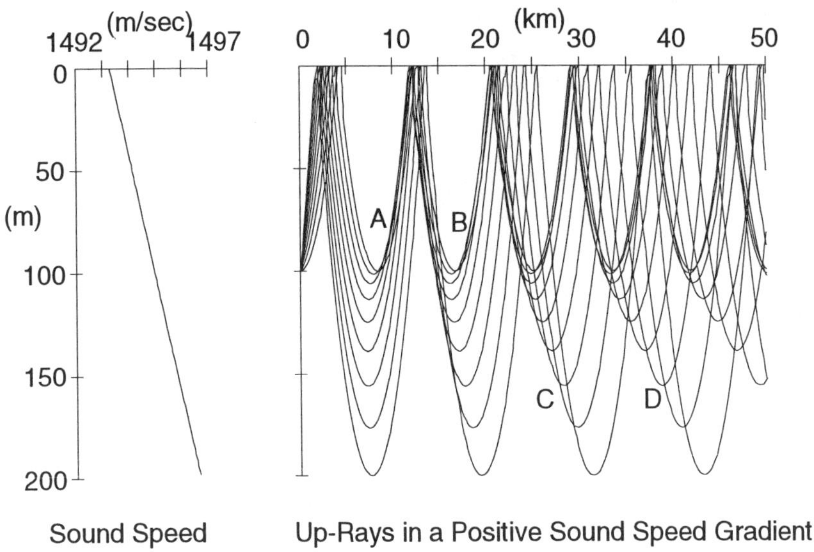

Figure 5b Caustic Structures Formed at A, B, C, D, ··· by Up-Going Rays from a Source at 100 m in a Positive Sound Speed Gradient Environment. Sound Speed Gradient = 0.017 m/sec per m.

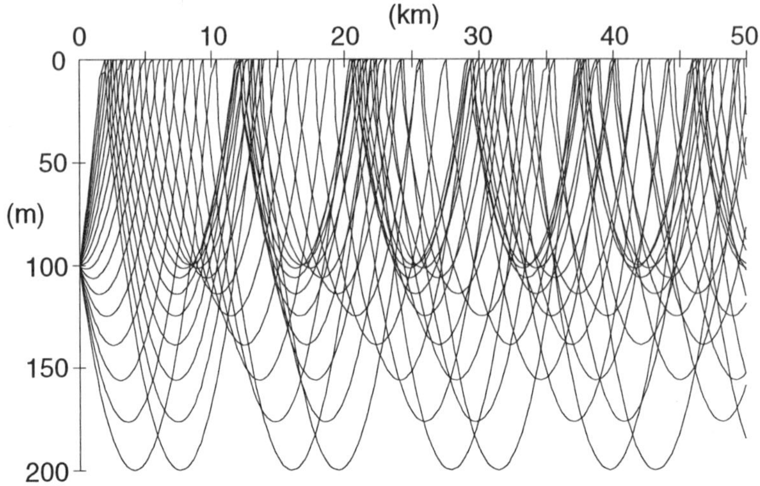

Figure 5c Caustic Structures Formed by Up and Down-Going Rays from a Source at 100 m in a Positive Sound Speed Gradient Environment. Sound Speed Gradient = 0.017 m/sec per m.

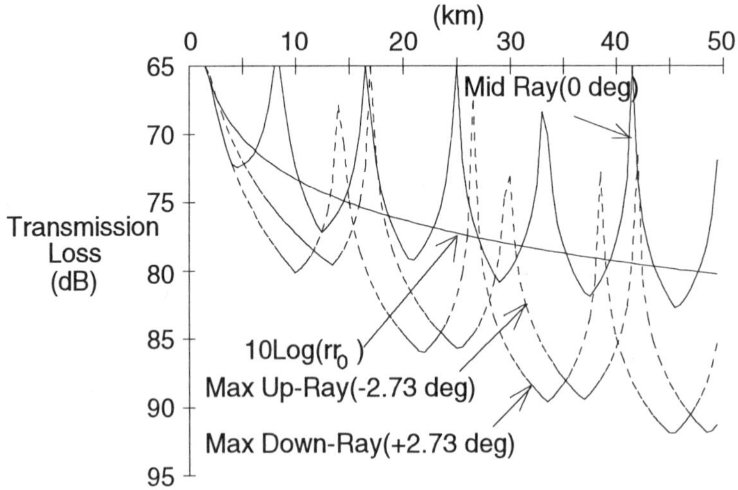

Figure 5d Transmission Loss in a 200 m Surface Duct with Source at a Depth of 100 m. Ray Acoustic Losses are Compared to Sonar Equation Loss Term: $10\text{Log}(rr_0)$, $r_0 = H/(2\sin\theta_{Max})$. H = Duct Depth and θ_{Max} is the Maximum Launch Angle for Trapped Rays.

positive gradient causes the rays to be refracted upward for both downward and upward going rays as shown, respectively, in Figures 4a and 4b. Inspection of the ray trace diagram, in Figure 5a, for down-going rays reveals evidence of caustics in the duct that, at A, start above the source level, descending into the lower half of the duct, and at farther ranges, at B and C, are below the source level in the lower half of the duct. Similarly, inspection of the ray trace diagram for the up-going rays, in Figure 5b, reveals evidence of caustics in the duct, at A and B, in the upper half of the duct and at farther ranges, at C and D, are below the source level in the lower half of the duct. The ray trace diagram, in Figure 5c, shows both down-going and up-going rays and gives some idea of the complexity of propagation in surface ducts. Transmission loss curves for the rays at 0° and the extreme rays at ±2.73°are shown in Figure 5d. The extreme rays graze the bottom of the duct and are confined to the duct. Rays having greater launch angles refract out of the duct and are not trapped. The positions of the caustics in range for these rays are apparent from Figure 5d. Comparison with Figures 5a and 5b indicate that the caustics very quickly approach their asymptotic positions at the bottoms of each ray cycle. A comparison of the ray transmission losses to the sonar equation transmission loss term in (11b,c) indicates that the sonar equation term provides a good mean value. However, the variations of acoustic intensity with range and depth within the duct can be expected to be substantial. ♦

6.3 Duct - Axial Transmissions

A duct is characterized by a water mass layer in which the sound speed profile is concave toward lower values. The sound speed minimum defines the duct axis and rays emanating from a source on the duct axis are repeatedly refracted back toward the duct axis as indicated in Figure 6. Rays with launch angles within a plus and minus range will be constrained to the duct while rays at higher and lower launch angles will escape from the duct. Of major interest to ASW applications, the constrained acoustic energy can propagate over long distances with minimal attenuation. In fact, a common ASW strategy is to place the acoustic source on the duct axis in order to maximize the energy constrained to the duct. This establishes a layer of enhanced target detectability and can inhibit the submariner's choice of operational depths. Ducts of various widths can occur at all depths in the upper ocean but a major permanent duct in the ocean is formed by the main thermocline and deep isothermal layer. The sound speed minimum at their juncture defines the deep sound channel axis. This duct is sometimes referred to as the SOFAR (SOund Fixing And Ranging) channel because, from the times of arrival of pulsed signals at sound receiving stations, the coordinates of a source can be determined by triangulation. The SOFAR channel was used effectively during WWII to locate downed flyers. The downed flyer would release an explosive charge that would sink to the duct axis before detonation. Ducts can occur in all seasons and it occasionally happens that multiple ducts are found at several depths which may necessitate that the ASW crew determine a duct ensonification and/or surveillance schedules. The case for the source and/or receiver on the duct axis is an important special case of duct propagation which is considered in this section. The case for the acoustic source and receiver off the duct axis is discussed in Section 6.4.

The duct boundaries are defined by equal sound speeds at the lower and upper edges of the duct. In the following we adopt the simple case of the linear sound speed profile, as indicated in Figure 6, defined by,

$$c = c_0 + g_U(z - z_0), \quad z \geq z_0 \tag{12a}$$

$$c = c_0 + g_L(z - z_0), \quad z \leq z_0 \tag{12b}$$

where the quantities $g_U < 0$ and $g_L > 0$ denote the sound speed gradients in the upper and lower medium.

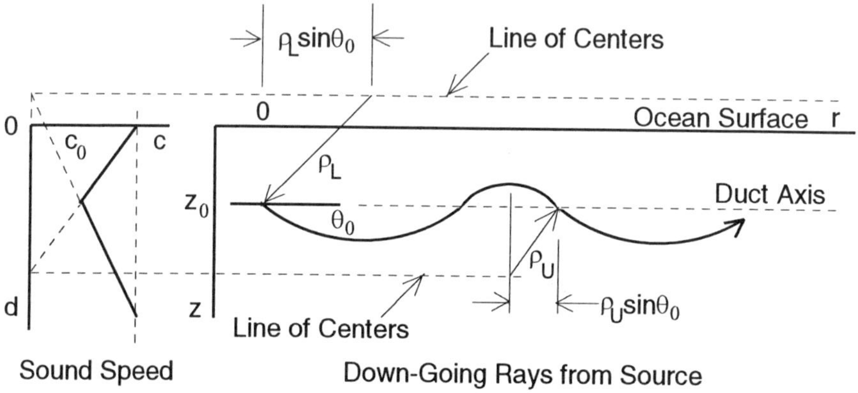

Figure 6a Duct Propagation for Down-Going Rays

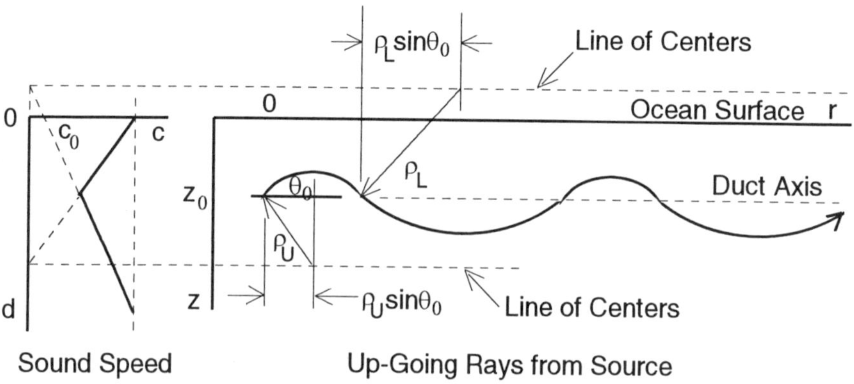

Figure 6b Duct Propagation for Up-Going Rays

It's convenient to discuss the down-going and up-going rays separately. For rays with down-going launch angle and for an even number of axis crossings, i.e. N even and $z > z_0$, we see from Figure 6a that the ray loci are given by

$$r = N(\rho_L - \rho_U)\sin\theta_0 + \rho_L(\sin\theta_0 - \sin\theta), \tag{13a}$$

$$z - z_0 = \rho_L(\cos\theta - \cos\theta_0), \tag{13b}$$

$$\theta = \theta_0\left(1+N\left(1-\frac{\rho_U}{\rho_L}\right)\right)-\frac{s}{\rho_L}. \tag{13c}$$

To arrive at (13c) we observe that $\theta = \theta_0-(s-s_N)/\rho_L$ and $s_N = N(\rho_L-\rho_U)\theta_0$ where s is the total ray path length and s_N is the ray path length up to the N cycle. For an odd number of axis crossings, i.e. N odd and $z < z_0$, the ray loci are given by

$$r = \left((N+1)\rho_L-(N-1)\rho_U\right)\sin\theta_0-\rho_U\left(\sin\theta_0+\sin\theta\right), \tag{13d}$$

$$z-z_0 = \rho_U\left(\cos\theta-\cos\theta_0\right), \tag{13e}$$

$$\theta = \theta_0\left((N+1)\frac{\rho_L}{\rho_U}-N\right)-\frac{s}{\rho_U}. \tag{13f}$$

For (13f) we observe that $\theta = -\theta_0-(s-s_N)/\rho_U$, and $s_N = ((N+1)\rho_L-(N-1)\rho_U)\theta_0$. To get ray loci equations for rays with up-going launch angles we interchange ρ_L and ρ_U in (13).

For the derivatives of the ray loci equations for down-going rays and for N even, i.e. $z > z_0$, we obtain

$$R_s = \left[N(\rho_L-\rho_U)+\rho_L\left(1-\cos\theta_0\cos\theta\Theta-\sin\theta_0\sin\theta\right)\right]/\cos\theta_0, \tag{14a}$$

$$Z_s = -\rho_L\left(\sin\theta\Theta-\tan\theta_0\cos\theta\right), \tag{14b}$$

$$\Theta_s = \left[1+N\left(1-\frac{\rho_U}{\rho_L}\right)\right]+\frac{s\tan\theta_0}{\rho_L}. \tag{14c}$$

For N odd, i.e. $z < z_0$, we obtain the ray loci derivatives

$$R_s = \left[\rho_L(N+1)-\rho_U\left(N+\cos\theta_0\cos\theta\Theta+\sin\theta_0\sin\theta\right)\right]/\cos\theta_0, \tag{14d}$$

$$Z_s = -\rho_U\left(\sin\theta\Theta-\tan\theta_0\cos\theta\right), \tag{14e}$$

$$\Theta_s = \left[(N+1)\frac{\rho_L}{\rho_U}-N\right]+\frac{s\tan\theta_0}{\rho_U}. \tag{14f}$$

Substituting the derivatives in (14) into (5.16) the transmission loss for the down-going rays for N even is found to be given by

$$TL = 10\mathrm{Log}\left[\frac{r}{\cos^2\theta_0}\left|\alpha r-\frac{N}{2}(1+\alpha)(R_L+R_U)\right|\right]. \tag{15a}$$

Similarly, the transmission loss for down-going rays for N odd is found to be given by

$$\mathrm{TL} = 10\mathrm{Log}\left[\frac{\mathrm{r}}{\cos^2\theta_0}\left|\alpha\mathrm{r}\frac{\rho_L}{\rho_U} + \frac{1}{2}\left(1 - \frac{\rho_L}{\rho_U}\right)\left((\mathrm{N}+1)^2 \mathrm{R_L} + \left(\mathrm{N}^2 - 1\right)\mathrm{R_U}\right)\right|\right], \quad (15b)$$

$$\alpha = 1 + \mathrm{N}\left(1 - \frac{\rho_U}{\rho_L}\right), \qquad \mathrm{R_L} = 2\rho_L \sin\theta_0, \qquad \mathrm{R_U} = -2\rho_U \sin\theta_0 .$$

The transmission loss expressions for the up-going rays is obtained by interchanging ρ_L and $-\rho_U$ in (15).

The locations of the caustics for the down-going rays for N even are obtained by setting the argument of the logarithm in (15a) to zero to obtain

$$\mathrm{r} = \frac{\mathrm{N}(1+\alpha)(\mathrm{R_L} + \mathrm{R_U})/2}{\alpha} \xrightarrow{\mathrm{N}\to\infty} \frac{\mathrm{N}}{2}(\mathrm{R_L} + \mathrm{R_U}) + \frac{\mathrm{R_L}}{2} . \quad (16a)$$

The limit expression shows that the caustics for the down-going rays for N even occur at the bottom of each ray cycle. Similarly, the caustic locations for the up-going rays for N even occur at the top of each ray cycle.

The locations of the caustics for the down-going rays for N odd are obtained by setting the argument of the logarithm in (15b) to zero to obtain

$$\mathrm{r} = \frac{(1 - \rho_U/\rho_L)\left((\mathrm{N}+1)^2\mathrm{R_L} + \left(\mathrm{N}^2 - 1\right)\mathrm{R_U}\right)/2}{\alpha} \xrightarrow{\mathrm{N}\to\infty} \left(\frac{\mathrm{N}+1}{2}\right)\mathrm{R_L} + \frac{\mathrm{N}}{2}\mathrm{R_U} . \quad (16b)$$

The limit expression shows that the caustics for the down-going rays for N odd occur at the top of each ray cycle. Similarly, the caustic locations for the up-going rays for N odd occur at the bottom of each ray cycle.

Consequently, sound reception off the duct axis for an axial source is characterized by caustic "hot spots". In contrast, no caustics occur on the duct axis. For down-going and up-going rays, the range to a point on the duct axis can be expressed, in terms of the number of axis crossings, as $(\mathrm{R_L} + \mathrm{R_U})(\mathrm{N}+1)/2$ for N odd and as $(\mathrm{R_L} + \mathrm{R_U})\mathrm{N}/2 + \mathrm{R_{L,U}}$ for N even where the term $\mathrm{R_{L,U}}$ applies, respectively, to down-going and up-going rays. Substituting these values for range into (15) we obtain for both down-going and up-going rays

$$\mathrm{TL} = 10\mathrm{Log}\left[\frac{\mathrm{r}^2}{\cos^2\theta_0}\right], \quad (17)$$

which shows that the spreading law along a ray path between the source and a point on the duct axis is, essentially, spherical spreading. This appears to contradict empirical evidence that axial transmissions within the duct suffer minimal loss. The apparent discrepancy is reconciled by the fact that an infinite number of ray paths connect the source with each axial point. More about this after Example 3.

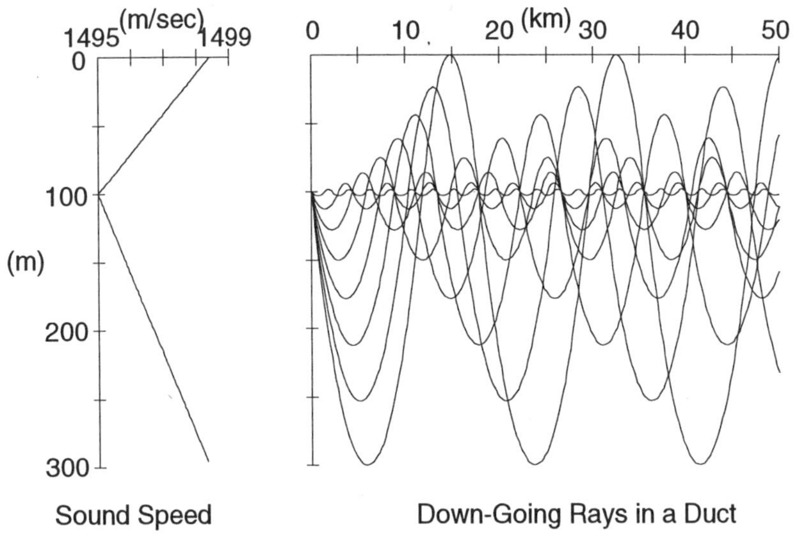

Figure 7a Caustic Structures Formed by Down-going Rays from Source at 100 m in a Bilinear Sound Speed Gradient Environment. Sound Speed Gradients are 0.017 m/sec per m in lower medium and –0.034 m/sec per m in upper medium.

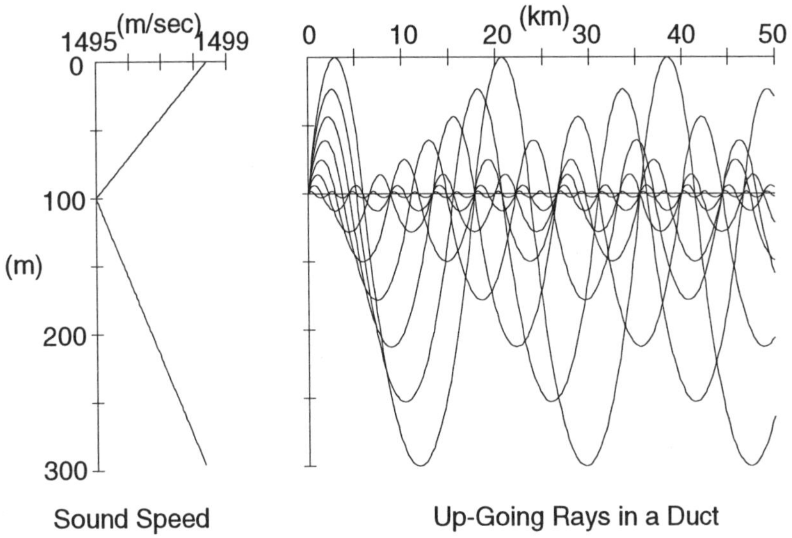

Figure 7b Caustic Structures Formed by Up-going Rays from a Source at 100 m in a Bilinear Sound Speed Gradient Environment. Sound Speed Gradients are 0.017 m/sec per m in lower medium and –0.034 m/sec per m in upper medium.

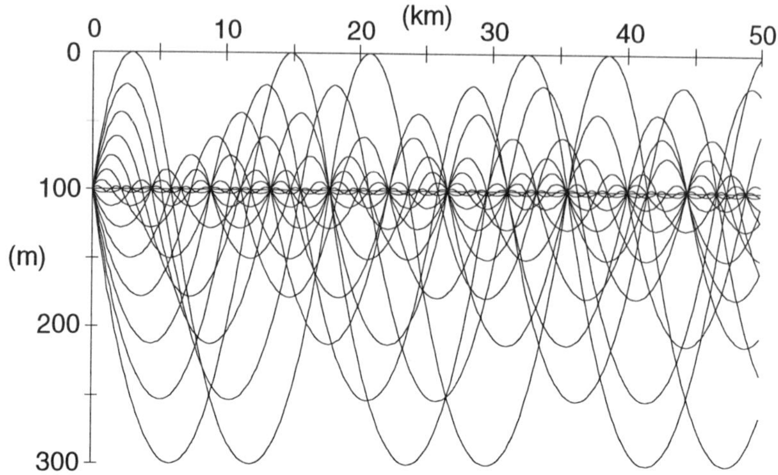

Figure 7c Caustic Structures Formed by Up-Going and Down-Going Rays from a Source at 100 m in a Bilinear Sound Speed Gradient Environment. Sound Speed Gradients are 0.017 m/sec per m in lower medium and –0.034 m/sec per m in upper medium.

Example 3: Consider a duct that is 300 m deep with a source on the duct axis at 100m depth. The sound speed gradients are taken to be 0.0017 m/sec per m in the lower portion of the duct and –0.0034 m/sec per m in the upper portion of the duct. The acoustic rays for both downward and upward going rays are shown, respectively, in Figures 7a and 7b. Inspection of these ray trace diagrams reveals sequences of up-sloping and down-sloping cycle peaks which is where the caustics occur. Since the transmission loss between an axial point and an off-axis caustic is reduced, detections for an axial source/receiver from targets positioned anywhere within the duct are enhanced. Similarly, detections for an off-axis source/receiver from axial targets are enhanced. The ubiquity of caustics within the duct is suggested by the transmission loss curves in Figure 7d. This figure shows the transmission loss along four rays having launch angles that span the duct and compares these losses to the case of spherical spreading losses. Since spherical spreading approximates the loss between two axial points, this figure suggests that detections between an axial point and off-axis points would have higher probabilities than detections between two axial points. This is not the case because, in contrast to the paths between an axial point and an off-axis point, an infinite number of rays paths exist between two axial points which results in reduced losses. In fact, low frequency axial transmissions propagating in the SOFAR channel, have been detected at ranges of several thousand nautical miles.♦

The propagation of signals between points on the duct axis is a special case of ocean propagation that requires additional commentary. As noted above, such signals are found to incur reduced losses because, even though the loss along any one ray path is approximately $10\text{Log}(r^2)$, an infinite number of rays connect any two

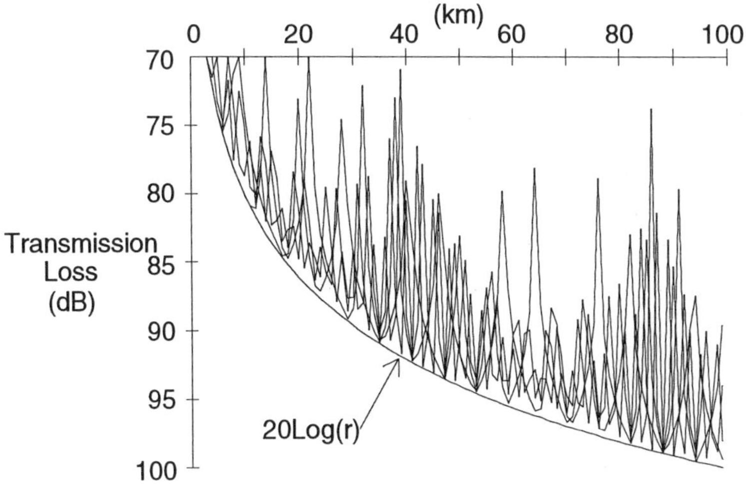

Figure 7d Transmission Loss in a 300 m Duct with Axial Source at 100 m Depth for Rays at −3.86° (Max Up), −1.29°, 1.29° and 3.86° (Max Down). Losses are Compared to Spherical Spreading Loss. Sound Speed Gradients are 0.017 m/sec per m in Lower Duct and −0.034 m/sec per m in Upper Duct.

points on the duct axis and their sum results in a reduced transmission loss. But, in addition, the received signal structure requires commentary because the times of arrival for such signals vary with the launch angle. Since the energy is transmitted over different ray paths having different propagation speeds, the received signal is spread out in time. The propagation times over rays having steep launch angles are less than the propagation times for near axis rays because the steeper ray paths pass through higher sound speed water. However, the axial rays are more numerous and so, for the case of pulsed transmissions, the received signals can be expected to grow in amplitude as the signal is received In contrast, for paths between an axial point and on off-axis point, the arrivals are discrete and are dominated by caustic receptions. These concepts for axial transmissions are especially important for echo ranging and SOFAR transmissions because the reduced transmission loss increases detection ranges substantially but estimates of time of arrival become more problematic.

For the case of an explosive signal having intensity $i(t) = \exp(-\alpha t)$, where α^{-1} gives the time at which the intensity has decayed to exp(−1), the total received signal intensity can be approximated by summing the intensities over all ray paths that connect the source and receiver

$$I(t) = \sum_{n=N(t)}^{1} \sum_{n=-1}^{-N(t)} A_n \exp\left(-\alpha\left(t - t_n\right)\right). \tag{18a}$$

For simplicity we assume a symmetrical duct. The two sums are for, respectively, the down-going and up-going rays. The quantities A_n and t_n represent, respectively, the amplitude and arrival time of the signal components having launch angles θ_n. From (17), the intensity along these rays is $\cos^2\theta_n / r^2 \cong 1/r^2$ where the launch angles are taken to be small. Denoting θ_1 and θ_{-1} as the launch angles of, respectively, the maximum down and up rays, the ray paths having large index numbers are concentrated near the channel axis. The maximum index N(t) in the sum denotes the number of rays that have arrival times up to time t. As t increases and all rays have arrived, this index becomes infinite. At this time, the number of terms ceases to grow and each term in the sum decays exponentially to zero. These mechanisms are the source of the growth and decay of axial transmissions.

We can recast the sum in (18a) as the integral

$$I(t) = \frac{2}{r^2} \int_{\theta_0(t)}^{\theta_{max}} d\theta f_\theta A_\theta \exp\left(-\alpha\left(t - t_\theta\right)\right), \tag{18b}$$

where θ_{max} is the maximum down and up launch angles, $\theta_0(t)$ is the launch angle of the most recent arrival and f_θ is the ray density in the increment $d\theta$. This density is required in the transition to the integral in (18b) because the number of rays in the launch angle increment $d\theta$ increases as θ approaches zero. From the aforementioned relations between the number of axis crossings and range, leading to (17), the number of axis crossings for a ray having launch angle θ can be expressed

$$N = \frac{r}{2(c_0/g)\tan\theta} + C, \tag{19a}$$

where C is a constant term that is different for down-going and up-going rays but is on the order of unity for both cases. The ray density is then found to be

$$f_\theta = -\frac{dN}{d\theta} = \frac{r}{2(c_0/g)\sin^2\theta} \cong \frac{r}{2(c_0/g)\theta^2}. \tag{19b}$$

The time of propagation over a single cycle of a ray having launch angle θ_0 is given by

$$t = \int_0^s \frac{ds}{c} = \frac{1}{g} \int_{-\theta_0}^{\theta_0} \frac{d\theta}{\cos\theta} = \frac{2}{g} \ln\left(\frac{1 + \tan\theta_0/2}{1 - \tan\theta_0/2}\right), \tag{20a}$$

where the variable changes were used: $s = \rho(\theta_0 - \theta)$, $c = c_0 + gz$ and (13b,e). For N ray cycles we obtain for all cases with some manipulation

$$t = \frac{r}{c_0}\ln\left(\frac{1+\tan\theta_0/2}{1-\tan\theta_0/2}\right)/\tan\theta_0 \cong \frac{r}{c_0}\left(1-\frac{\theta_0^2}{6}\right). \tag{20b}$$

The launch angles of the rays arriving at time t and the maximum launch angle are given by

$$\theta_0(t) = \sqrt{6\left(1-\frac{c_0 t}{r}\right)}, \quad \theta_{max} = \sqrt{6\left(1-\frac{c_0 t_{min}}{r}\right)} = \cos^{-1}\left[\frac{1}{1+z_0 g/c_0}\right], \tag{20c,d}$$

where t_{min} is the arrival time of the first ray having the maximum launch angle.

Substituting into (15b) the received signal intensity can be expressed as

$$I(t) = \frac{\exp\left[-\alpha\left(t - r/c_0\right)\right]}{2\left(c_0/g\right)\sqrt{r}}\sqrt{\frac{\alpha}{6c_0}}\int_{\sqrt{\alpha\left(\frac{r}{c_0}-t\right)}}^{\sqrt{\alpha\left(\frac{r}{c_0}-t_{min}\right)}} du\, u^{3/2}\exp(-u). \tag{21}$$

The integral vanishes for $t < t_{min}$, grows as t increases, and becomes a constant for $t > r/c_0$. At this point the leading exponential term causes the intensity to decay to zero. The leading factor shows that, from station-to-station on the duct axis, the amplitude decays as the square root of range. This gives a transmission loss of 5Log(r) as opposed to 20Log(r) for spherical spreading, i.e. for each doubling of range only 1.5 dB is lost as opposed to 6 dB for spherical spreading.

Example 4: We consider a symmetrical duct in which the sound speed gradient is 0.034 m/sec per m and the axial sound speed is 1495 m/sec. Assuming an exponential fall-off in signal intensity with parameter α = 1/6, we numerically integrate (21) to obtain the signal growth curve in Figure 8a for the three ranges shown. Figure 8a shows the time of onset of the received signals and the growth to a maximum intensity at the time r/c_0. However, in practice ducts are not as regular as we have modeled them. The sound speed profile is not fixed. It varies with range and the duct axis meanders. Also the source and receiver may not be exactly on the duct axis. The result is that some rays are not received and, generally speaking, the

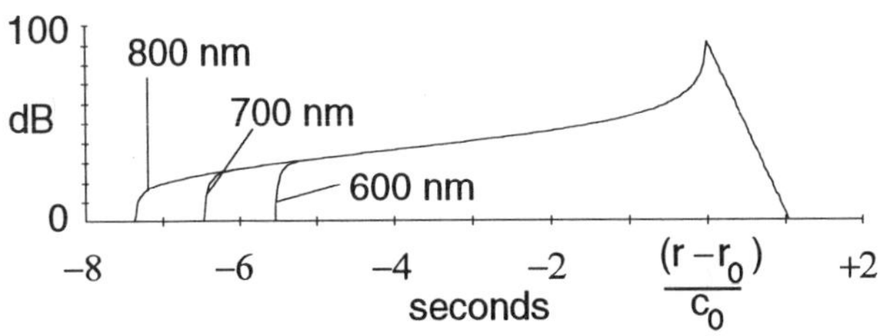

Figure 8a Signal Growth at Three Receiving Stations for Pulsed Axial Duct Transmissions

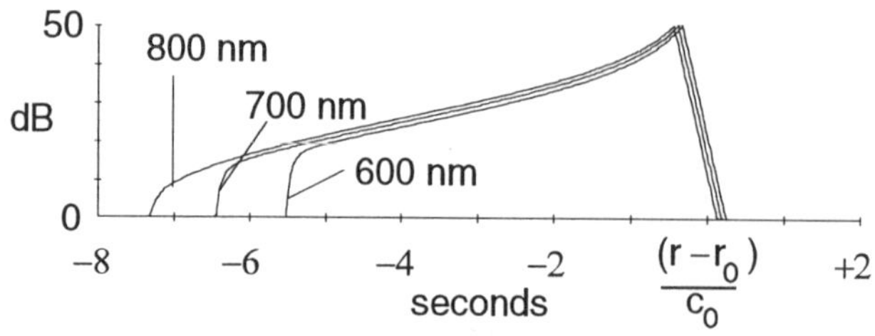

Figure 8b Signal Growth at Three Receiving Stations for Pulsed Axial Duct Transmissions - Excluding < ±3° Rays

low launch angle rays are most likely to be missing. We approximate this situation in Figure 8b by omitting the rays in the arbitrarily chosen range of ±3°. We see that for this case the intensity does not peak as sharply and, more importantly, the arrival time of the peak is earlier than expected by about 0.3 seconds for this example. This bias in arrival times would bias our range estimate by about 0.5 nm. However, since the axial sound speed is variable and unknown over most ranges of interest, this is just one more source of error to be accommodated. ♦

6.4 Convergence Zones

Convergence zone (CZ) propagation is made possible by the presence of the deep sound channel. Under this mode of propagation, transmissions from a source in the upper reaches of the medium are refracted downward by the main thermocline into the deep isothermal layer where they are then refracted back toward the surface where the rays converge at a range of about 30 nm depending upon oceanic conditions. Continuing on, the rays repeat this refraction process and produce convergence zones at about thirty nautical mile intervals from the source. Within each convergence zone the sound intensity is enhanced and received echoes and passive radiations from targets at CZ intervals can yield high confidence detections. Under optimal conditions, convergence zone propagation enables reliable detections to be achieved out to ranges of several hundred nautical miles.

The sound speed profile for CZ propagation is taken to be the linear sound speed profile in (12) and the duct width, as before, is determined by the depths of equal sound speeds at the top and bottom of the duct. For the duct to support useful CZ propagation, the water depth must be sufficient to accommodate the full duct width and more. During summer-like conditions, when the surface temperature and sound speed are a maximum and there is sufficient depth excess, CZs extend to the ocean surface. During winter-like conditions, when the sound speed bends back and the maximum sound speed occurs at depth, or when there isn't sufficient depth excess, CZs pull back from the ocean surface and detection opportunities are reduced.

For a source in the upper medium, as indicated in Figure 9, the acoustic rays are repeatedly refracted toward the deep sound channel axis. The range and

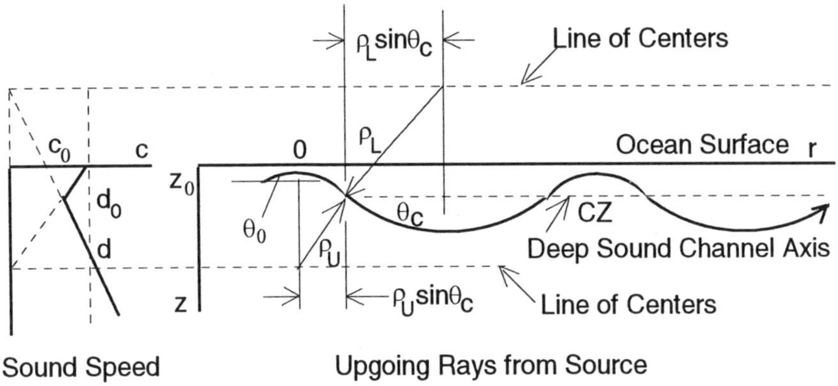

Figure 9 Ray Path for Convergence Zone Propagation

depth of both down-going and up-going rays, for an even number N of axis crossings, are

$$r = \rho_U(\sin\theta_0 - \sin\theta) + N(\rho_L - \rho_U)\sin\theta_c, \tag{22a}$$

$$z - z_0 = \rho_L(\cos\theta - \cos\theta_c), \tag{22b}$$

and, for an odd number N of axis crossings, are

$$r = \rho_U \sin\theta_0 + N(\rho_L - \rho_U)\sin\theta_c - \rho_L \sin\theta, \tag{22c}$$

$$z - z_0 = \rho_U(\cos\theta - \cos\theta_c). \tag{22d}$$

In these relations, the quantity θ_c is the angle at which the rays cross the deep sound channel axis. The relationship between the launch angle and axis crossing angle is found to be

$$\cos\theta_c = \beta\cos\theta_0, \qquad \beta = \left(1 - \frac{d_0 - z_0}{c_0 / g_U}\right), \tag{22e,f}$$

where d_0 is the depth of the axis and z_0 is the depth of the source.

For launch angles in the range $\theta_{0,min} < \theta_0 < \theta_{0,max}$, the acoustic energy is constrained to CZ propagation. For launch angles outside this range, the acoustic energy reflects from the surface on each cycle and a convergence zone for these rays is not formed. The minimum and maximum launch angles are seen to be given by

$$\cos\theta_{0,min} = \left(1 + \frac{z_0}{c / g_U}\right)^{-1}, \tag{22g}$$

$$\cos\theta_{0,max} = \beta^{-1}\cos\theta_{c,max}, \quad \cos\theta_{c,max} = \left(1 - \frac{d - d'}{c / g_L}\right)^{-1}. \tag{22h,i}$$

The extreme rays graze the surface of the ocean and bottom of the deep sound channel, i.e. that depth where the sound is equal to the surface sound speed, and are confined to the deep sound channel. If there isn't sufficient depth excess to accommodate the deep rays, bottom interactions cause the CZs to pull back from the surface and the CZs do not reach the surface.

The derivatives of the range and depth equations, for N even, are found to be

$$R_S = \rho_U\left(\cos\theta_0 - \cos\theta\Theta_s\right) + P_U\left(\sin\theta_0 - \sin\theta\right)$$
$$+N\left(\rho_L - \rho_U\right)\cos\theta_c\Theta_{c,s} + N\left(P_L - P_U\right)\sin\theta_c\,, \quad (23a)$$

$$Z_S = \rho_U\left(-\sin\theta\Theta_s + \sin\theta_c\Theta_{c,s}\right) + P_U\left(\cos\theta - \cos\theta_c\right), \quad (23b)$$

and, for N odd, are found to be

$$R_S = \rho_U\cos\theta_0 + P_U\sin\theta_0 + N\left(\rho_L - \rho_U\right)\cos\theta_c\Theta_{c,s}$$
$$+N\left(P_L - P_U\right)\sin\theta_c - \rho_L\cos\theta\Theta_s - P_L\sin\theta\,, \quad (23c)$$

$$Z_S = \rho_L\left(-\sin\theta\Theta_s + \sin\theta_c\Theta_{c,s}\right) + P_L\left(\cos\theta - \cos\theta_c\right), \quad (23d)$$

$$\Theta_{c,s} = \frac{\beta\sin\theta_0}{\sqrt{1-\alpha^2\cos^2\theta_0}} = \beta\eta\,, \quad (23e)$$

$$P_L = \rho_L\tan\theta_0\,, \qquad P_U = \rho_U\tan\theta_0\,, \quad (23f,g)$$

where the subscript s denotes that partial differentiation is performed holding the ray path constant and the quantity $\Theta_{c,s}$ is the partial derivative of θ_c at constant ray path length.

Substituting into the transmission loss expression in (5.16) we obtain with some manipulation, for N even,

$$TL = 10\text{Log}\left[\frac{r}{\cos^2\theta_0}\left|\alpha r - \frac{N}{2}\left(\alpha+\beta^2\right)\left(R_L + R_U\right)\right|\right], \quad (24a)$$

and, for N odd,

$$TL = 10\text{Log}\left[\frac{r}{\cos^2\theta_0}\left|\alpha r\frac{\rho_U}{\rho_L} - \frac{1}{2}\left(1-\frac{\rho_U}{\rho_L}\right)\beta\left(R_L\left(N-\beta\right)\left(N-\beta^{-1}\right) + R_U\left(N^2-1\right)\right)\right|\right], \quad (24b)$$

$$\alpha = \left(1 + N\beta\left(\frac{\rho_L}{\rho_U} - 1\right)\right). \quad (24c)$$

Equations (24) give the transmission loss along the ray having launch angle θ_0. However, since these rays are deep diving rays, shadow zones are formed between the convergence zones and the matter of greater importance is the transmission loss

$$r = \frac{N\left(\alpha + \beta^2\right)\left(R_L + R_U\right)}{2\alpha} \xrightarrow{N\to\infty} \frac{N}{2}\left(R_L + R_U\right) - \frac{R_U\beta}{2}, \qquad (25a)$$

and, for N odd,

$$r = \frac{\beta\left(1 - \rho_U/\rho_L\right)\left(R_L\left(N-\beta\right)\left(N-\beta^{-1}\right) + R_U\left(N^2-1\right)\right)}{2\alpha} \xrightarrow{N\to\infty} \frac{N}{2}\left(R_L + R_U\right) - \frac{R_U\beta}{2}. \qquad (25b)$$

In the limit expressions, the quantity $-R_U\beta/2$ is the partial cycle range between the source and the first top of the ray cycle for both down-going, i.e. $\beta>0$, and up-going, i.e. $\beta<0$, rays. Consequently, we see that for both down-going and up-going rays, the caustics at range occur at the tops of the ray cycles for the case of N even, and, the caustics occur at the bottom of the ray cycles for the case of N odd. More simply stated, the caustics occur at the turning points of the rays as they refracted back toward the channel axis. A close examination of the limiting expression for N even shows that the caustics occur on the side of the ray turning point nearest the source. This means that the leading edges of the CZs are well defined with sharp onsets and the back sides are less well defined where the intensity decays gradually.

Example 5: We consider a mid-latitude portion of the deep sound channel in which the depth of the channel axis is 1000 m. The axial sound speed is assumed to be 1495 m/sec, the sound speed gradient in the lower medium is taken to be 0.017 m/sec per m and is taken to be −0.034 m/sec per m in the upper medium. The acoustic rays for both downward and upward going rays are shown, respectively, in Figures 10a and 10b. Inspection of these figures reveals evidence of caustics in the regions marked C. The ray trace diagram, in Figure 10c, shows both down-going and up-going rays together and gives some idea of the structure of the CZs. The width of the CZs are seen to be approximately 5% to 10% of the range to the CZ. It follows that at long ranges, say beyond the 10th CZ, the CZs overlap and complete coverage is obtained. However, at these long ranges the caustics are weakened due to propagation irregularities. The depth of the CZs extends from the surface to about the source depth for all ranges. As the source depth is varied, the CZ structure also varies as indicated in Figure 10d. In that figure we see that as the source moves to shallower depths, the CZs become narrower and shallower. Consequently, the best detection performance will be achieved with the deepest source position. ♦

CZ detections can be very effective for locating and tracking submarine targets at great range but this mode of detection requires optimal conditions. At the very minimum, the water mass intervening between the ASW platform and submarine target must support CZ propagation and both platforms must be within each other's CZs. Under these conditions the intensity of active sonar signals will be enhanced at target reflection and the intensity of both active and passive signals will be enhanced at reception leading to high confidence detections. However, the CZ mode of operation requires that the target be detected during its transit through a CZ or during transit of a CZ past the target. Since the CZ width is approximately 10%

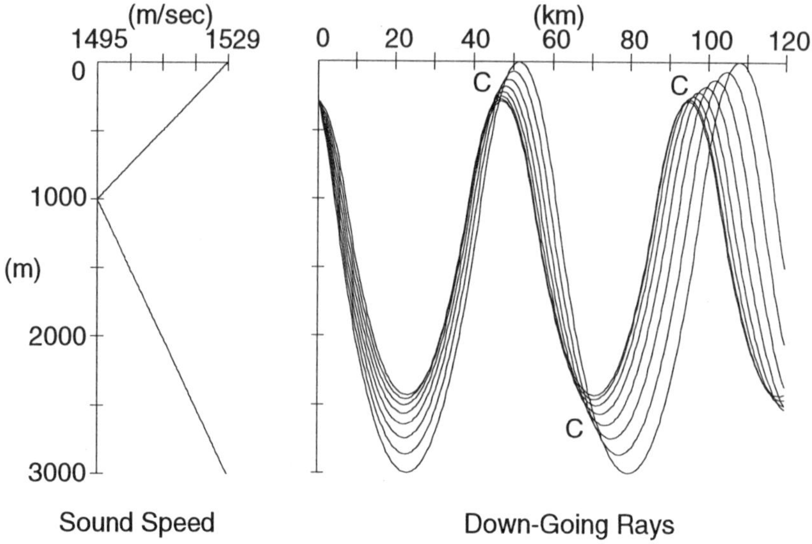

Figure 10a Convergence Zone Propagation for Down-going Rays for a Source Located at 300 m Depth. Sound Speed Gradients in Lower and Upper Media are, respectively, 0.017 m/sec per m and –0.034 m/sec per m.

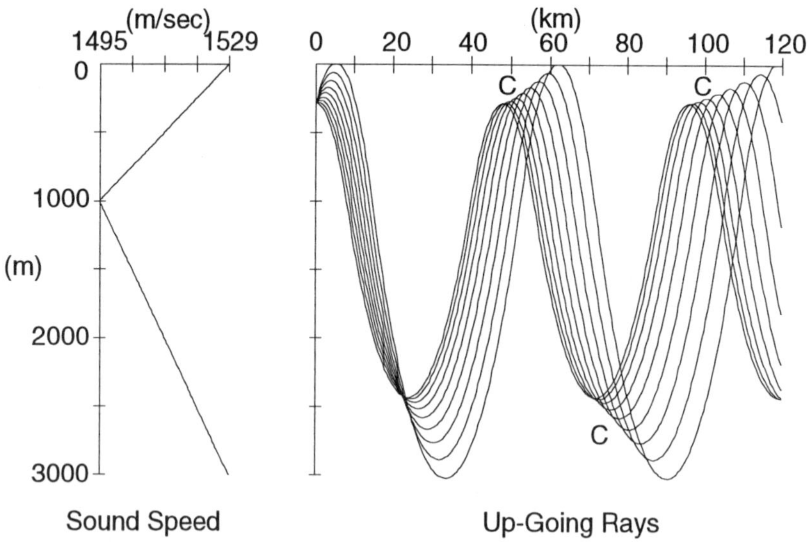

Figure 10b Convergence Zone Propagation for Up-going Rays for a Source Located at 300 m Depth. Sound Speed Gradients in Lower and Upper Media are, respectively, 0.017 m/sec per m and –0.034 m/sec per m.

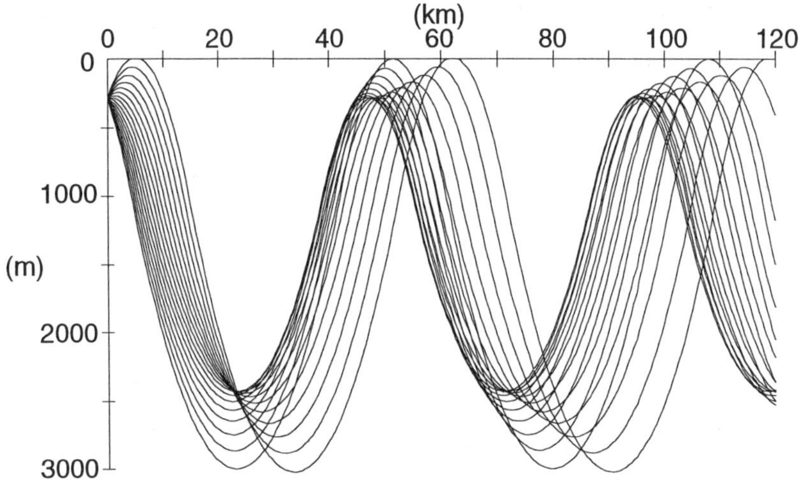

Figure 10c Convergence Zone Propagation for All Rays for a Source Located at 300 m Depth. Sound Speed Gradients in Lower and Upper Media are, respectively, 0.017 m/sec per m and –0.034 m/sec per m.

of the range to the CZ, the CZ detection window will be, generally, less than about an hour for each of the initial CZs. Consequently, in the near CZs the advantage of high confidence detections is weakened by the disadvantage of a low detection rate. At longer ranges this problem is mitigated because CZ propagation has greater area coverage but the signal intensity gains are less pronounced because of the longer ranges and weakened caustics. Under less than optimal conditions CZ propagation may be disrupted or prevented by the intrusion of bottom features, such as seamounts or a lack of depth excess or by unwanted refractions caused by intervening water masses.

6.5 Other Propagation Modes

The modes of propagation discussed in the preceding sections are commonly found in many deep water ASW scenarios in which bottom interactions are not a major factor. However, they do not exhaust the range of possibilities for deep water conditions. For example, currents, eddies and land mass outflows can produce pronounced effects on propagation conditions. These water masses can be warm or cold or of different saline content which would cause additional refraction of acoustic waves upon penetration of their boundaries. Such water masses can provide effective cover for submarine operations either because they prevent viable ray paths from connecting the target and ASW platform or because the ray path refractions are unknown resulting in false target position estimates or incorrect dismissals of the target. Also, while we have discussed several isolated modes of propagation, in practice, it is commonly found that more that one mode of propagation may be occurring. Under these conditions, the ASW crew must decide which mode to emphasize and position the source and/or receiver accordingly.

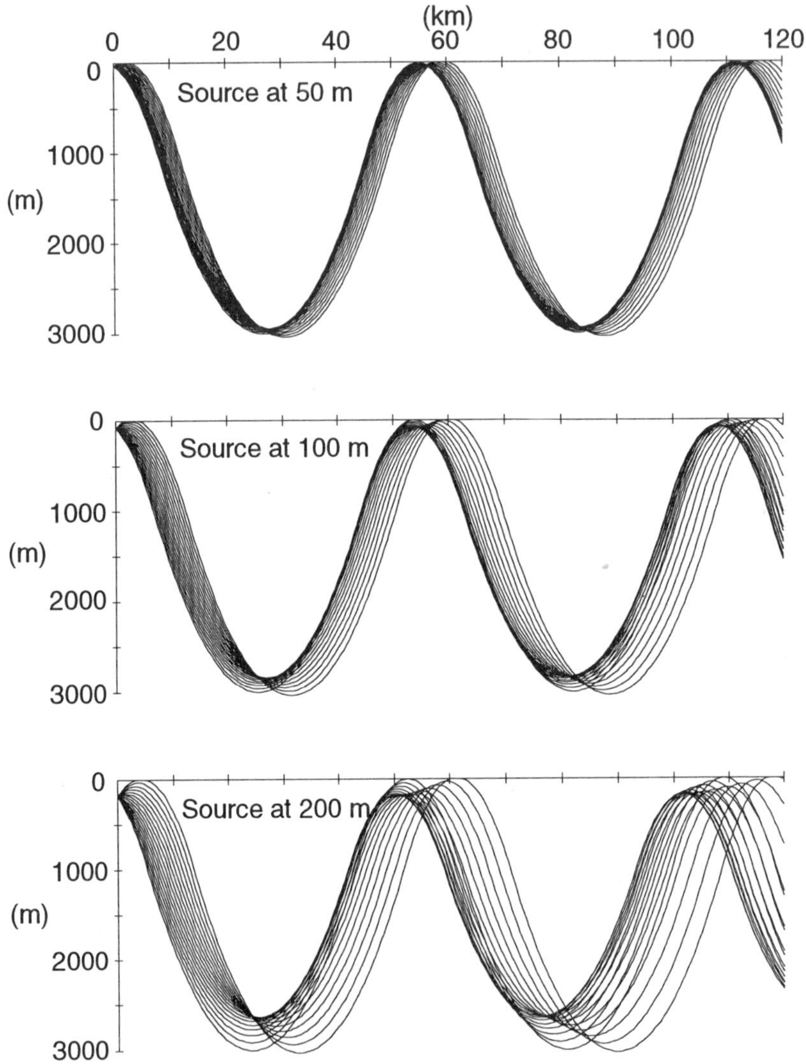

Figure 10d Convergence Zone Propagation for Sources at Three Depths. The Sound Speed Gradients in Lower and Upper Media are, respectively, 0.017 m/sec per m and –0.034 m/sec per m. CZs Extend to the Source Depth and Deeper Sources Cause the Down and Up Going CZs to Separate.

For the important case of shallow water conditions, bottom interactions can produce propagation effects that can be the dominant effect. For example, the surface shadow zone discussed in Section 6.1 can sometimes be circumvented by active bottom-bounce propagation. In this mode of propagation, the source beam is directed toward the bottom and reflected energy penetrates the shadow zone causing targets in the zone to be detected as target reflected energy retraces the bottom-bounce path back to the receiver. In shallow water littorals, multiple bounces from the bottom and surface may be found to be connecting the target and ASW platform. If the intervening bottom is sloping, the problem is even more complicated. For an up-sloping bottom, the ray paths become steeper at each reflection and penetration into the shallower regions of the littoral cannot be achieved. For a down-sloping bottom, the ray paths become more shallow at each reflection and, eventually, as the bottom drops away, the bottom interactions may cease - leaving us with the surface duct propagation discussed in Section 6.2. In the case of CZ propagation, there may not be sufficient depth excess for all transmitted rays to form CZs and bottom reflections will cause the shadow zones between the CZs to fill with acoustic energy giving rise to detection opportunities. Of course, bottom interactions are frequently not an unmixed blessing since they can also give rise to spurious detections. For example, bathymetric returns from beneath the ASW platform are received at arrival times having the periods of the travel times between the source/receiver and bottom and surface. They can be difficult to identify and can have the disconcerting appearance of a close-in target. More commonly, false target detections will be found from distant bottom features such as rock outcroppings and bottom depressions and such detections are abundant in the vicinity of sea mounts and sea chasms. Additionally. these returns will usually be mixed with a strong reverberation background and, in fact, are nothing more than clutter within reverberation. Their correct identification necessitates that estimates of bathymetry, in addition to estimates of the propagation conditions, be available.

CHAPTER 7
TARGET ACOUSTICS

The first contact with submarine targets is generally through their active or passive acoustic signatures which is our primary way of detecting and localizing targets. While detection performance is largely dependent upon the strength of the received signals, detector design is dependent upon the temporal and spectral structure of these signals. Accordingly, for optimal detection performance, we need an understanding of both signal strength and signal structure. In the following, we examine target echo strength and the temporal structure of echoes, which are important for active detection, and target radiation source levels and the spectral structure of these radiations, which are important for passive detection. Actually, knowledge of the structure of received signals is doubly important because, after detection and localization, the information carried by the signals can also be used to classify the targets.

7.1 Scattering of Active Sonar Signals by Targets

When an active sonar signal sweeps over an ASW target, the incident energy is scattered in all directions. This scattering originates from every point of impedance change on and within the target, but the bulk of the scattered energy originates at the surface of most ASW targets. Such targets are nearly acoustically impenetrable and target shape is the primary determinant of echo structure. A receiver can collect the scattered energy from all such ensonified points but, as we will find, the energy that scatters from points on the target surface that lie on specular reflection planes arrive at the receiver in phase and combine coherently to produce a replica of the transmitted signal. However, because of the range extent of the target, the total received signal energy is an elongated version of the transmitted signal.

More specifically, as indicated in Figure 1, an active source at some aspect angle with respect to the target ensonifies the target with an incident field having intensity I_I. A receiver, displaced from the source by the angle θ, intercepts the reflected field having intensity I_R. If the receiver displacement angle is greater than zero, the reception is referred to as bistatic and, if the displacement angle is zero, e.g. when the source and receiver are collocated, the received energy is back scattered and the reception is referred to as monostatic. At each instant, the received signal is found to be the sum of the scattered energy from target surface points that lie on specular planes intersecting the target, as shown. The specular planes are normal to the bisector of the receiver displacement angle and as a result the angles of incidence and reflection are equal as required for specular reflection. The apparent "point" from which the reflected signal appears to originate, from the receiver perspective, is generally found to be

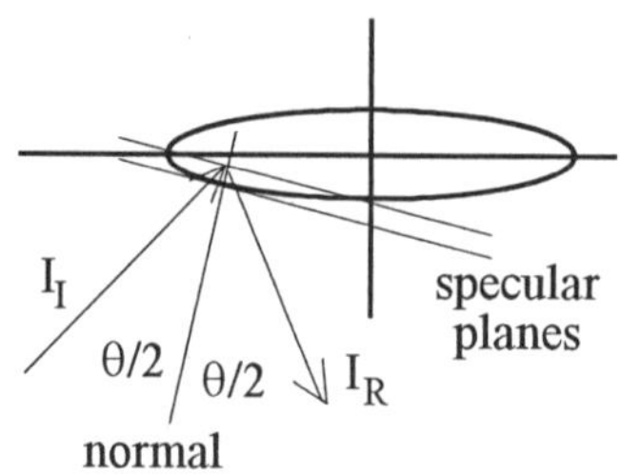

Figure 1 Specular Reflection of a Signal from a Target

within the target, but as the frequency increases we find that this point moves to the surface of the target and the reflection appears to be a reflection from the target surface point that is intersected by the bisector of the receiver displacement angle.

ASW targets are acoustically complicated but, because they mostly share the properties of being massive and/or air-filled vessels, they are nearly acoustically impenetrable. Such targets can be characterized as being either rigid or soft and echo structures from these targets are largely determined by the shape of the target. In Appendices to this chapter we examine the acoustic scattering by such targets. Appendix 7-A develops Kirchoff's integral inversion of the wave equation and, based upon that result, Appendix 7-B examines the general form of scattering by rigid and soft targets. In (7.7-B) we find that the field scattered by these acoustically impenetrable target is given by

$$P_S(\omega,\mathbf{r}) = \pm\frac{1}{4\pi}\int_E d\mathbf{s}'\,\nabla'\left[P_I(\omega,\mathbf{r}')\left(\frac{\exp(-jkr)}{r}\right)\right], \tag{1}$$

where $r = |\mathbf{r}-\mathbf{r}'|$ and the integration is over the ensonified surface of the target. The area element $d\mathbf{s}'$ points outward from the scattering body and ± refers, respectively, to rigid and soft targets. As described in Appendix 7-B, this expression represents bistatic scattering, with monostatic scattering being a special case, but it does not represent the shadow forming field. That field, combined with the direct path field, forms the shadow zone directly behind the target. This is unimportant in much of our analyses because, since forward scattering cannot be resolved from the direct path signal, we exclude the forward scattering region from our discussions of scattering and detection.

A sinusoidal acoustic field incident on the target is given by

$$P_I(\mathbf{r}') = \frac{\exp(-jkr_I)}{r_I}, \tag{2a}$$

where we implicitly assume that the target is within the beam of the source, and the beam is uniform. The range r_I is the distance from the source to a surface element on the target as indicated in Figure 2. We note that, in (2a), there is an implicit factor equal to unit pressure × length. The unit pressure allows for a scaling factor to be

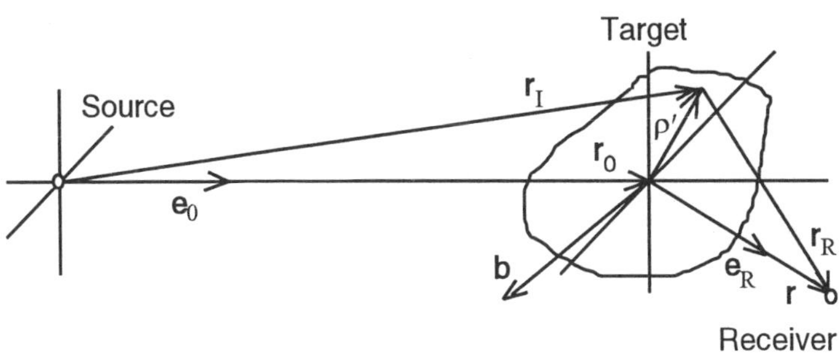

Figure 2 Geometry for Acoustic Scattering by Targets

inserted and the unit length is required for dimensional consistency.

The incident energy is scattered by the target into the receiver located at range r_R. Assuming that the ranges of the source and receiver are large compared to the size of the target we can make the plane wave approximations

$$r_I = r_0 + \mathbf{e}_0 \cdot \rho', \qquad r_R = r - \mathbf{e}_R \cdot \rho', \tag{2b,c}$$

where r_0 and r denote the magnitudes of the vectors $\mathbf{r}_0$ and $\mathbf{r}$, ρ' is the vector locating a point on the target surface and the vectors denoted $\mathbf{e}$ are unit vectors. Substituting (2) into (1), the expression for the scattered field becomes

$$\begin{aligned} P_S(\mathbf{r}) &= \pm \frac{jk}{4\pi}\left(\frac{\exp[-k(r_0+r)]}{r_0 r}\right)\int_E ds' \cdot \nabla'\left[\exp[jk(\mathbf{e}_R - \mathbf{e}_0)\cdot \rho']\right] \\ &= \pm \frac{jk}{4\pi}\left(\frac{\exp[-k(r_0+r)]}{r_0 r}\right)\int_E ds' \cdot (\mathbf{e}_R - \mathbf{e}_0)\exp[jk(\mathbf{e}_R - \mathbf{e}_0)\cdot \rho']. \end{aligned} \tag{3}$$

The vector difference in (3) lies in the plane defined by the unit vectors $\mathbf{e}_R$ and $\mathbf{e}_0$. It bisects the angle between those vectors and defines the orientation of the specular reflection planes.

Denoting the bisector of the unit vectors in (3) as

$$\mathbf{b} = \mathbf{e}_R - \mathbf{e}_0, \tag{4a}$$

we can express $\mathbf{b}$ in spherical coordinates, as indicated in Figure 3,

$$\mathbf{b} = \sin\theta\cos\phi\,\mathbf{i} + \sin\theta\sin\phi\,\mathbf{j} + (1+\cos\theta)\mathbf{k}. \tag{4b}$$

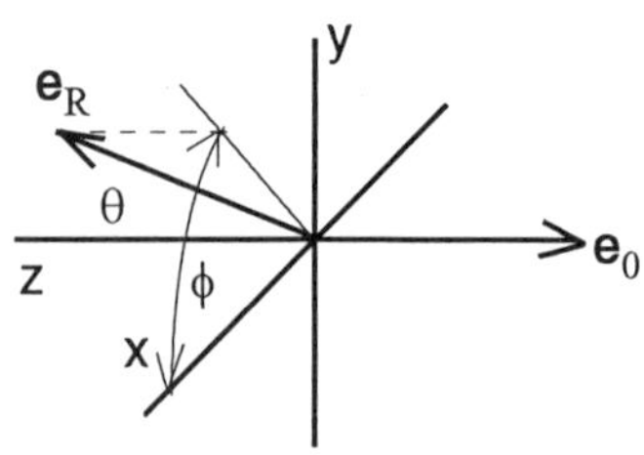

Figure 3 Coordinates of the Unit Vectors

The magnitude of the bisector vector and its expression in terms of the unit bisector vector are readily found to be

$$|\mathbf{b}| = 2\cos\left(\frac{\theta}{2}\right), \qquad \mathbf{b} = 2\cos\left(\frac{\theta}{2}\right)\mathbf{e}_b, \tag{4c,d}$$

where $\mathbf{e}_b$ is the unit vector in the direction of $\mathbf{b}$. The projection of the surface area element vector on the bisector vector is found to be

$$\mathbf{ds}' \cdot \mathbf{b} = 2\cos\left(\frac{\theta}{2}\right) dA_b, \tag{4e}$$

where dA_b is the projection of the target surface element in the direction of the bisector vector and is an area element on a specular reflection plane. Similarly, the projection of the surface location vector on the bisector vector is found to be

$$\mathbf{b}\cdot\rho' = 2\cos\left(\frac{\theta}{2}\right)z_b, \tag{4f}$$

where z_b is the coordinate lying along the bisector vector.

Substituting (4) into (3) we obtain the bistatic scattered field expression

$$P_S(\mathbf{r}) = \pm\frac{jk}{2\pi}\cos\left(\frac{\theta}{2}\right)\left(\frac{\exp[-k(r_0+r)]}{r_0 r}\right)\int_{-\infty}^{a} dz_b\left(\frac{dA_b}{dz_b}\right)\exp\left(j2k\cos\left(\frac{\theta}{2}\right)z_b\right). \tag{5a}$$

The upper limit of integration is the distance to the tangent plane at the point of first contact of the signal with the target and the lower limit extends an infinite distance behind the target. This is an artifice, since A_b becomes constant with zero derivative beyond the ensonified area of the target, and the lower limit is always finite. The back scattering integral is readily found to be

$$P_S(\mathbf{r}) = \pm\frac{jk}{2\pi}\left(\frac{\exp[-k(r_0+r)]}{r_0 r}\right)\int_{-\infty}^{a} dz\left(\frac{dA}{dz}\right)\exp(j2kz). \tag{5b}$$

Equations (5) show that the strength of the field scattered for each surface element is proportional to the rate of change of the area along the direction of the bisector vector and the phase of the received field from the surface element is what is expected from geometrical acoustics for plane waves.

7.2 Target Strength

Target strength is the ratio of the acoustic intensity reflected by the target to the acoustic intensity incident on the target and is computed in decibels

$$TS = 10\mathrm{Log}\left(\frac{I_R}{I_I}\right). \tag{6a}$$

In this expression, the reflected intensity is measured at 1 m from the acoustic center of the target. The acoustic center is the point from which the reflected field, as measured in the far field, appears to emanate. For bistatic sonars, the measurement is made in the direction along which the reflected signal propagates to the bistatic receiver and, as we might expect, for monostatic sonars, the measurement is made in the back scatter direction. Decibel target strengths of submarines and other ASW targets are generally found to be positive in value. This does not mean that the reflected intensity is greater than the incident intensity. It simply means that target sizes are generally large compared to the reference distance of 1 m. Consequently, when the weakly reflected intensity is referred back to the reference distance, it appears to be strong.

Denoting the intensity target strength by lower case

$$ts = \frac{I_R}{I_I}, \tag{6b}$$

the bistatic and monostatic target strength integrals are given, respectively, by

$$ts = \frac{\cos^2(\theta/2)}{\lambda^2}\left|\int_{-\infty}^{a} dz_b\left(\frac{dA_b}{dz_b}\right)\exp\left(j2k\cos\left(\frac{\theta}{2}\right)z_b\right)\right|^2, \tag{7a}$$

$$ts = \frac{1}{\lambda^2}\left|\int_{-\infty}^{a} dz_b\left(\frac{dA_b}{dz_b}\right)\exp(j2kz_b)\right|^2. \tag{7b}$$

An examination of these relations shows that there is no simple relationship by which bistatic target strengths can be obtained from monostatic target strengths for arbitrary targets. While this can be done for certain cases, bistatic target strengths will generally have to be measured or computed for each angle of interest.

Target Strength of a Sphere

Consider the bistatic target strength of a sphere of radius a as shown in Figure 4a. The area of the sphere projected along the z_b axis and its derivative are

$$A_b = \pi\rho^2 = \begin{cases}\pi\left(a^2 - z_b^2\right), & z_b \geq 0\\ \pi a^2, & z_b < 0\end{cases} \tag{8a}$$

$$\frac{dA_b}{dz_b} = \begin{cases}-2\pi z_b, & z_b \geq 0\\ 0. & z_b < 0\end{cases} \tag{8b}$$

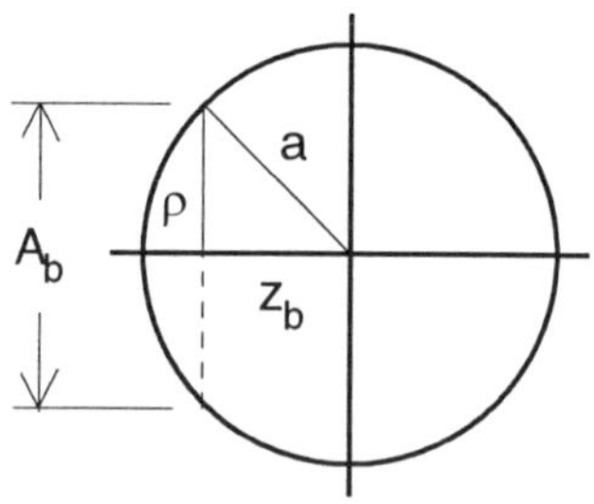

Figure 4a Projected Area Geometry for a Sphere

Substituting into the bistatic target strength relation in (7a) gives the target strength expression

$$ts = k^2\cos^2(\theta/2)\left|\int_0^a dz_b z_b \exp\left(j2k\cos\left(\frac{\theta}{2}\right)z_b\right)\right|^2. \tag{9a}$$

The integral in this expression, which we denote as I, gives the phase of the received signal and can be used to compute the scattered field in (5). It is found to be given by[1]

$$I = \frac{a^2 A}{2B}\exp\left(j(B-\Phi)\right), \qquad B = ka\cos\left(\frac{\theta}{2}\right), \tag{9b,c}$$

---

[1] Abramowitz and Stegun, (4.3.119) and (4.3.123).

$$A = \sqrt{\sin^2(B) + (\cos(B) - \sin(B)/B)^2}\ ,\ \Phi = \tan^{-1}\left(\frac{\cos(B) - \sin(B)/B}{\sin(B)}\right). \quad (9d,e)$$

As an aside, we note that the phase of the integral implies that the apparent reflection point from the sphere is located within the sphere. However, in the limit of high frequencies, the amplitude becomes A = 1 and the phase of the integral becomes $[2ka\cos(\theta/2) - \pi/2]$. Substituting the high frequency limit of the integral into (5), we find that the apparent reflection point from the sphere is the point of specular reflection from the surface of the sphere. Substituting the integral I into the target strength expression in (9a) we obtain

$$ts = \frac{a^2}{4}\left[\sin^2(ka\cos(\theta/2)) + \left(\cos(ka\cos(\theta/2)) - \frac{\sin(ka\cos(\theta/2))}{ka\cos(\theta/2)}\right)^2\right]. \quad (10)$$

The high frequency limit of this target strength expression becomes $ts = a^2/4$. This limit is valid for all meaningful angles since we are restricting the receiver displacement angle to about $|\theta| < 3\pi/4$ to exclude the forward scattering angles. This result shows that in the high frequency limit the bistatic target strength is the same for all bistatic angles and is equal to the back scatter target strength. In fact, denoting the monostatic target strength ts(ka), the bistatic target strength is given by $ts_{bs} = ts(ka\cos(\theta/2))$.

A plot of target strengths for a sphere as a function of frequency and receiver displacement are found from (10) and graphed in Figure 4b. These values are approximate - in the sense of being valid in the high frequency limit. From the graph, we see that, for back scattering, the high frequency limit is reached for wavelengths smaller than about $2\pi a$. For bistatic scattering, the frequency must be greater to reach the high frequency limit. For high frequencies the target strength is given by $TS = 10\text{Log}(a^2/4)$. Thus, for a sphere of 2 m radius, the target strength is found to be zero dB for all frequencies above some limit and all displacement angles excluding the forward scattering direction. Note that the unit of measure for the radius is necessarily the same as the reference distance unit which we have chosen to be 1 m.

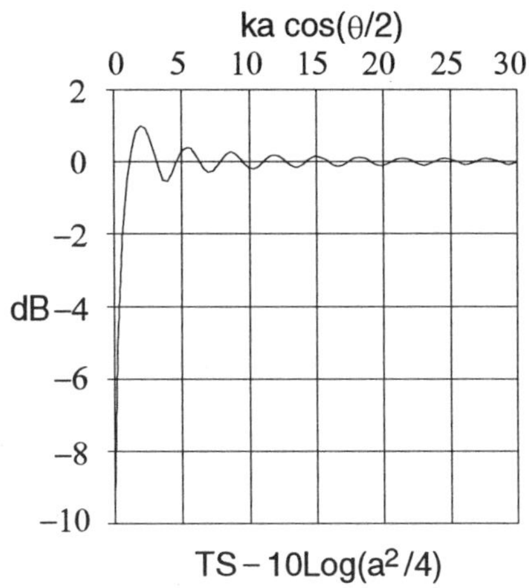

Figure 4b Approximate Target Strength of a Sphere of Radius a vs. Frequency and Receiver Displacement

An alternative viewpoint to determining target strength is also interesting. Since the power intercepted by the sphere is given by $\pi a^2 I_I$ and, assuming the power is reradiated uniformly in all directions, the scattered field has intensity $\pi a^2 I_I/4\pi r^2$. Forming the ratio I_R/I_I, the target strength is again found to be $a^2/4$.

Plots of approximate target strength of a sphere are presented in Figure 4c as functions of receiver displacement angle. These plots show that target strength, in the high frequency limit, is approximately constant for all aspect angles excluding the forward scattering region where target strength vanishes. The forward scattering target strengths must, of course, vanish because there is a shadow zone in the forward scattering region.

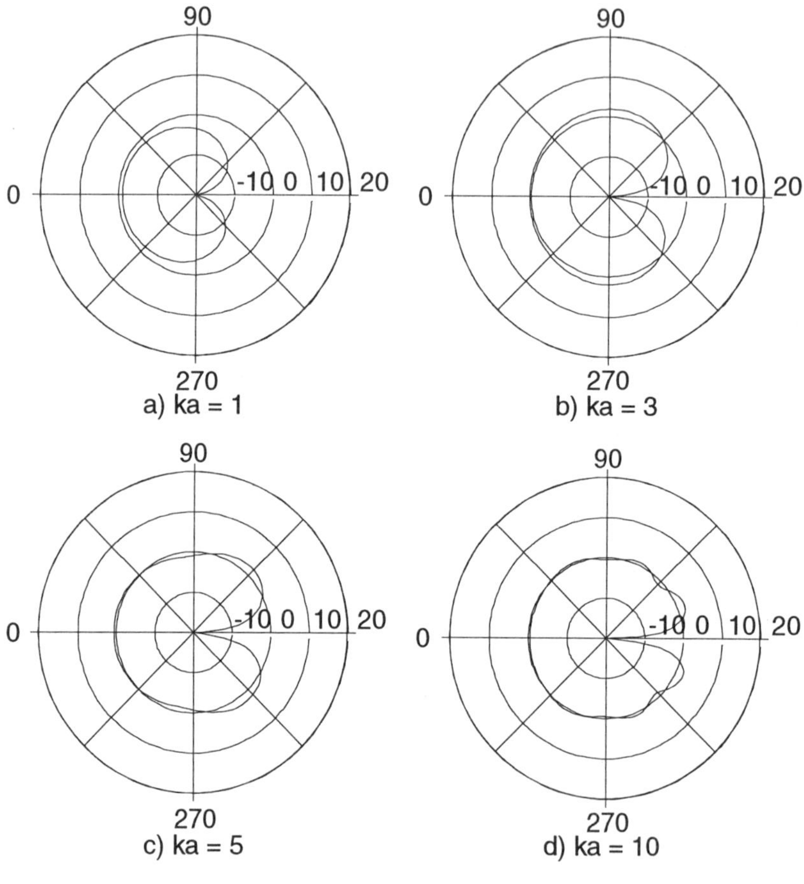

Figure 4c Approximate Target Strength of a Rigid or Soft Sphere of Radius a vs. Receiver Angle. The Plots are Parameterized on ka with 10 dB Increment Scale and Normalized, i.e. Plotted Values = TS – 10Log(a^2/4).

In Appendix 7-C, an expression for the exact target strength of a rigid sphere is determined and plotted. A comparison with the above high frequency approximations shows that the principal difference between the two solutions is the inclusion of the forward scattered field in the exact solution. This field is required to produce the shadow zone. The solutions are also found to differ somewhat in the details of the functional dependence of target strength on frequency and displacement angle.

Target Strength of a Cylinder

Consider the bistatic target strength of a cylinder of radius a and height h as shown in Figure 5. The direction of the incident field is assumed to be normal to the axis of the cylinder and the area of the cylinder projected along the x_b axis and its derivative are

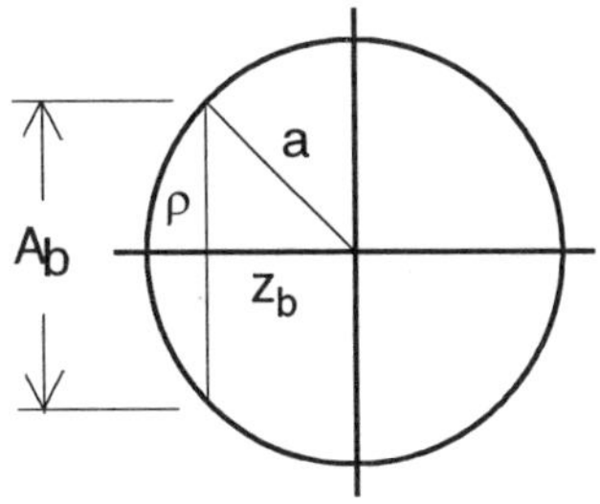

Figure 5 Projected Area for a Cylinder of Radius a and Height h.

$$A_b = 2h\rho = \begin{cases} 2h\sqrt{a^2 - z_b^2}, & z_b \ge 0 \\ 2ha, & z_b < 0 \end{cases} \tag{11a}$$

$$\frac{dA_b}{dz_b} = \begin{cases} \dfrac{-2hz_b}{\sqrt{a^2 - z_b^2}}, & z_b \ge 0 \\ 0. & z_b < 0 \end{cases} \tag{11b}$$

Substituting into the bistatic target strength relation in (7a) gives

$$ts = \frac{4h^2a^2\cos^2(\theta/2)}{\lambda^2}\left|\int_0^1 dz_b z_b \frac{\exp\left(j2ka\cos\left(\frac{\theta}{2}\right)z_b\right)}{\sqrt{1-z_b^2}}\right|^2 . \tag{12}$$

Letting $z_b = \cos\vartheta$ in (12) we obtain for the integral in that relation

$$I = \int_0^{\pi/2} d\vartheta \cos\vartheta \exp(jz\cos\vartheta), \tag{13a}$$

where $z = 2ka\cos(\theta/2)$. Separating (13a) into real and imaginary parts we obtain[2]

$$I = \frac{\pi}{4}\left[\mathbf{E}_1(z) - \mathbf{E}_{-1}(z)\right] + j\frac{\pi}{2}J_1(z), \tag{13b}$$

where $\mathbf{E}_{\pm 1}(z)$ denote Weber functions of first order and $J_1(z)$ denotes a Bessel function of first order. For large ka values we can transform this expression to the form[3]

$$I = j\frac{\pi}{2}H_1^{(1)}(z) = \sqrt{\frac{\pi}{2z}}\exp[j(z-\pi/4)]. \tag{13c}$$

As an aside we note that, substituting this result into (5), the echo return is seen to come from the specular point on the cylinder. Substituting (13) into (12) for the target strength we obtain the target strength expression for the cylinder

---

[2] Gradshteyn and Ryzhik, (3.715-14) and (3.715.17).

[3] Gradshteyn and Ryzhik, (8.583-2), (8.405-1) and (8.451-3).

$$ts = \frac{ah\cos(\theta/2)}{2}\left(\frac{h}{\lambda}\right). \tag{14}$$

This result shows that the scattering in the horizontal plane around the vertical cylinder is nearly uniform but it does have a cardiod pattern. It also shows that the scattering is frequency dependent. This is because the vertical dimension of the cylinder acts as an aperture causing the incident energy to be scattered mostly in the horizontal plane around the cylinder. The effective beam width of the scattered energy is approximately[4] $2\pi/kh$.

To show that our target strength expression is consistent with conservation of energy requirements, we integrate the scattered intensity over the surface of the scattering beam at unit range

$$\frac{2\pi}{kh}\int_{-\pi}^{\pi} d\theta \frac{ah\cos(\theta/2)}{2}\left(\frac{h}{\lambda}\right) = 2ah. \tag{15a}$$

This gives the total power incident on the cylinder as it must. Similarly, to show that our target strength expression is consistent with the target strength for infinitely long cylinders, we note that, for the case of the infinitely long cylinder, the incident energy is not scattered spherically but is scattered cylindrically in the horizontal plane. Assuming the cylinder length in (14) approaches infinity, we multiple by the beam arc at unit range to get the radiated power and renormalize with respect to the vertical extent of the scattered field, i.e. h, to get the cylindrical intensity. The result is the infinite cylinder target strength

$$ts = \frac{a\cos(\theta/2)}{2}. \tag{15b}$$

This indicates that the target strength of a long cylinder is independent of frequency but that is because this is a high frequency limit. In Appendix 7-C we determine the exact target strength of an infinite cylinder and show that (15b) is the high frequency limit. We also show that the angular dependence of target strength is similar to the cardiod pattern in (15b) but the exact scattered field also includes the forward scattered field.

Submarine Target Strength

Target strengths for submarines must be measured for best results. The acoustically impenetrable model, developed in the preceding, is valid for the pressure hull of a submarine but the shapes of pressure hulls and their associated superstructures are not mathematically simple. Fortunately, many modern single-hulled submarines with streamlined Albacore-like designs can be approximated by shapes that are mathematically tractable. However, to the degree that they are doubled-hulled, these approximations break down because the outer hulls of the

---

[4] See (4.12f).

double-hulled portions conceal mathematically complicated shapes and the outer hull itself can also be expected to contribute to the scattered field. Double-hulling is common around hull-mounted sonar arrays, sail-mounted antenna masts, torpedo tubes, cruise missile pods, externally mounted towed arrays, tanks, etc. Unfortunately, submarines that are wholly double-hulled, including the older diesel-electric boats, have even more irregular pressure hulls with surrounding superstructure that defy mathematical representation. Similarly, the scattering from appendages, such as propeller screws and towed array pods, is very difficult to model mathematically.

Granting the difficulty of modeling acoustic scattering by submarines, it is found that, for modern single-hulled submarines, we can do so with sufficient accuracy to determine useful target strength values and to determine the overall functional form of the aspect dependent target strength. Moreover, to the degree that double hulled submarines are similar, in the sense of being long narrow cylindrically shapes bodies, these results also apply to them. To obtain such results we are restricted to frequency domains that are low enough that the signal is scattered from the gross target shape, i.e. the wavelengths are large compared to small irregularities on the target, but high enough that our high frequency approximation is valid, i.e. the wavelengths are small compared to the overall target size. In these approximations we consider only the hull and sail to be the significant scatterers. This will not always be a valid model since the screws are also known to contribute significantly to the echo at some aspects.

From our target strength expression for the doubly convex sphere we observe that in the high frequency limit of (10) the reflected signal comes wholly from the surface specular point and the target strength is given by the product of the half radii of curvatures at the specular point where the curves are assumed to be measured in orthogonal directions. Similarly, for the singly convex cylinder we observe that in the high frequency limit in (14) the target strength is given by the product of the half radii of curvature in the horizontal plane; the ratio of the aperture height to its beam width; and the cardiod scattering pattern. We can use these observations to approximate the target strength of complicated targets such as submarines. We assume that the target strength of a submarine is determined largely by the shape of the pressure hull at the point where the incident signal makes contact. For the doubly convex hull and singly convex sail, the back scattered target strengths are, respectively, approximated by

$$ts_\eta = \frac{R_h R_v}{4}, \qquad ts_\sigma = \frac{R_h H^2}{2\lambda}, \tag{16a,b}$$

where R_h and R_v are the radii of curvature in, respectively, the horizontal and vertical planes, H is the sail height and λ is the acoustic wave length. For narrow band transmissions, we have

$$TS = 10\log\left(ts_\eta + ts_\sigma\right), \tag{16c}$$

where, on average, we assume that the phases of the contributions from the hull and sail are independent. In fact, the phases of these contributions with vary from constructive interference to destructive interference and the target strength will vary about our estimate by as much as +6 dB to −∞ dB depending upon the relative intensities of the reflected components. For sufficiently broad band signals, the contribution from the hull and sail can be resolved and the intensity target strengths need not be combined.

To determine the radii of curvature, we adopt an ellipsoid of revolution as our model for the hull and an elliptical cylinder as our model for the sail. Since the calculations for the radii of curvature are elementary but tedious, they are relegated to Appendix 7-D. From (7.6-D)

$$R_h = \frac{a^2 b^2}{\left(b^2 \cos^2\theta + a^2 \sin^2\theta\right)^{3/2}}, \tag{17a}$$

$$R_v = \frac{a^2}{\sqrt{b^2 \cos^2\theta + a^2 \sin^2\theta}}, \tag{17b}$$

where a and b are, respectively, the semiminor and semimajor axes of the ellipsoid and θ is the relative aspect angle of the sonar system with respect to the submarine with $\theta = 0$ corresponding to bow-on aspect. Note that (17a) applies to both the hull and sail where the appropriate values of a and b are inserted and that (17b) applies only the hull.

Example 1: We estimate the target strength of a modern SSN submarine having hull dimensions of $a_h = 5.5$ m, $b_h = 55$ m and sail dimensions of $a_s = 0.5$ m, $b_s = 2$ m and H = 5 m where we assume that the bulk of the sail is given over to antennae masts and only the air filled portions matter. Substituting (17) into (16) we obtain the decibel target strength plots shown in Figure 6 for four wavelengths. The wavelengths correspond to the approximate frequencies: 3000, 1500, 750 and 375 Hz, which span the range of most ASW operational systems, and are consistent with our requirement on the wavelength being "not-too-big' and "not-too-small."

An examination of Figure 6 shows that in all cases the target strength follows a "butterfly" pattern in which the strongest returns come from beam aspect and weaker returns come from end-on aspects. These figures also indicate that the beam aspect target strengths are dominated by hull returns while the target strengths at all other aspects are dominated by sail returns. For the example shown, the beam aspect target strength is about 30 dB and the quartering and end-on target strengths range down to about, respectively, 15 dB and 11 dB at $\lambda = 0.5$ m to about, respectively, 6 dB and 3 dB at $\lambda = 4$ m. These values are comparable to measured values but the measured target strengths are found to have large variability with aspect. This is to be expected because of constructive and destructive interference between the reflections from the hull and sail and elsewhere.

For a monostatic detection system with source level of 180 dB, a receive array gain of 15 dB operating in an environment of 85 dB noise and cylindrical spreading, the receive signal-to-noise level would be SNR=110+TS−20Log(r). For a quartering aspect target with target strength of 10 dB and a detection threshold of 10 dB, the maximum detection range would be about 158 nm.

At frequencies higher than those shown in Figure 6, the signal wavelengths resolve smaller features on the target and the target begins to appear as a collection of small scatterers distributed over the surface of the target. The factors of an increasing number of scatterers having decreasing target strengths and the complicated phasing between the scattered returns suggest that the target strength would saturate at some maximum level as the frequency increases while maintaining the butterfly patterns shown in Figure 6. This appears to be borne out by measurements reported by Urick[5] for WWII type submarines. ♦

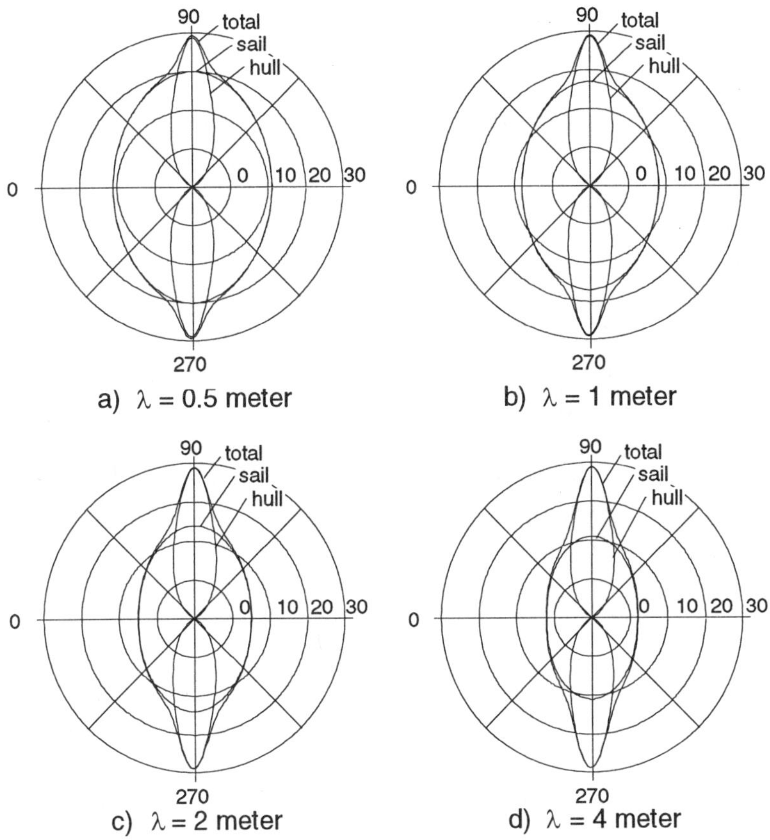

Figure 6 Submarine Target Strength versus Aspect and Frequency. Sail/Hull Target Strengths are shown separately at 10 dB Increments.

7.3 Temporal Structure of Echoes

Echoes received from targets are elongated versions of the transmitted signal because targets are spatially extended. However, the exact temporal structure of echoes depend on the structures of the targets. For a collocated source and receiver, the back scattered received signal, from (5), has the form

---

[5] Urick, Robert J. *Principals of Underwater* Sound for Engineers, McGraw-Hill, 1975,p. 283.

$$P_s(\mathbf{r}) = \pm\frac{jk}{2\pi}\left(\frac{\exp[-2kr_0]}{r_0^2}\right)\int_{-\infty}^{\infty} dz\left(\frac{dA}{dz}\right)\exp(j2kz). \tag{18}$$

The projected target area A is always monotonically increasing with range away from the source/receiver and is always continuous or piece-wise continuous. The area is piece-wise continuous when it contains flat planes such as might be found for, say, a bow-on aspect of a torpedo having a flat sonar array ahead of its warhead. For these cases the projected target area can be written

$$A(z) = \sum_n \Delta A(z_n)u(z_n - z) + A_c(z), \tag{19a}$$

where $\Delta A(z_n)$ is the jump in the area at the point z_n; $u(\cdot)$ denotes the unit step function and the function $A_c(z)$ is the continuous portion of the projected area function. We note that in our choice of coordinate system, the z axis points toward the source/receiver. Differentiating the projected area we obtain

$$\frac{dA(z)}{dz} = -\sum_n \Delta A(z_n)\delta(z_n - z) + \frac{dA_c(z)}{dz}, \tag{19b}$$

where $\delta(\,)$ denotes the Dirac delta function. The derivative of the continuous portion of the projected area is again piece-wise continuous. The positions of the steps in this function are assumed to come at the points z_m, $m = 0,1,2,\cdots,M$, where z_0 and z_M are, respectively, at the leading edge of the target and the shadow line of the target. Substituting (19) into (18)

$$P_s(\mathbf{r}) = \frac{\mp jk}{2\pi r_0^2}\sum_n \Delta A(z_n)\exp(-j2k(r_0 - z_n))$$
$$\pm\frac{jk}{2\pi}\left(\frac{\exp(-2kr_0)}{r_0^2}\right)\sum_{m=1}^{M}\int_{z_{m-1}}^{z_m} dz\frac{dA_c(z)}{dz}\exp(j2kz). \tag{20}$$

The first term represents the sum of echo components from flat planes and the second term represents the sum of echo components from the continuous portions of the projected target area. An important point to note here is that echoes come from both projected area discontinuities and the continuous portions of the projected area but echoes do not come from discontinuities in the derivative of the projected area as some have argued.

Example 2: For the case of a disc of radius a positioned at the origin of coordinates and perpendicular to the z-axis only the jump in the projected area contributes to the echo and we immediately obtain from (20) the back scattered echo

$$P_s(\mathbf{r}) = \frac{\mp jkA}{2\pi r_0^2}\exp(-j2kr_0), \tag{21a}$$

where $A = \pi a^2$ is the disc area. The target strength of the disc is readily found to be given by

$$ts = \left(\frac{A}{\lambda}\right)^2 . \tag{21b}$$

The target strength is seen to increase with frequency.

As might be expected, the field back scattered by a disc can be related to the field that is radiated by a disc projector. For such a projector, we can write the radiated field, from (4.8) and (4.15b), in the form

$$P(\omega, \mathbf{r}) = j\omega\rho_0 V(\omega) A \frac{\exp(-jkr_0)}{r_0}, \tag{22a}$$

where $V(\omega)$ is the spectral density of the face velocity. For a single line component at the frequency ω, we have $V(\omega') = V_0\delta(\omega' - \omega)$ where V_0 is the maximum velocity at the projector face. Fourier transforming (22a) gives the field

$$P(\mathbf{r}) = j\omega\rho_0 V_0 A \frac{\exp(-jkr_0)}{2\pi r_0} . \tag{22b}$$

To relate this result to the field scattered by a disc we remind ourselves that the field incident on the disc was assumed to be the line component in (2a). That field is the source of the "velocity" at the disc face. From Newton's second law in (2.4c-A), we can write the particle velocity at the face of the disc

$$u = \frac{-\nabla P_I(\mathbf{r})}{j\omega\rho_0} = \frac{jk\exp(-jkr_0)}{j\omega\rho_0 r_0} . \tag{22c}$$

For rigid and soft scatterers, the effective face velocities are given by, respectively, $V_0 = \mp u$ in accordance with the boundary conditions given in (7.3-B) and (7.6-B). Substituting (22c) into (22b) gives our scattered field in (21a).♦

Scattering by plane surfaces is not the usual event in ASW sonar. Most targets at most aspects will have projected areas that are continuous or piece-wise continuous and flat reflecting surfaces are avoided in the streamlined design of submarines. For a simple example we consider back scattering by a conical surface.

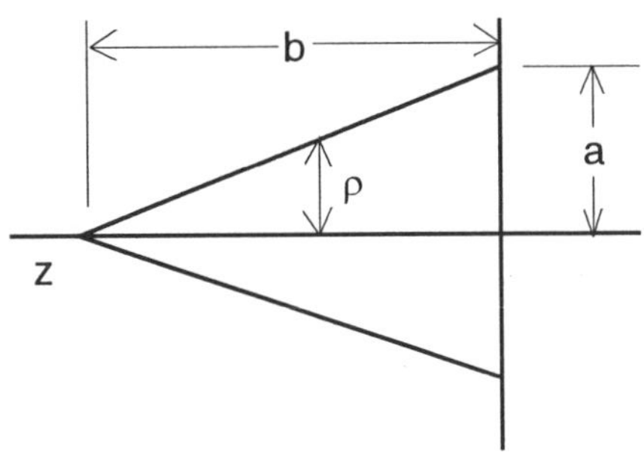

Figure 7 Conical Scattering Surface

Example 3: An acoustic signal is incident from the left on the conical surface depicted in Figure 7. The area and its derivative projected along the z axis are given by

$$A = \pi\rho^2 = \frac{\pi a^2}{b^2}(b-z)^2, \quad 0 \le z \le b \tag{23a}$$

$$\frac{dA}{dz} = -\frac{2\pi a^2}{b^2}(b-z). \quad 0 \le z \le b \tag{23b}$$

Substituting into (20), the back scattered field is given by[6]

$$P_s(\mathbf{r}) = \mp\frac{jka^2}{b^2}\left(\frac{\exp(-2kr_0)}{r_0^2}\right)\int_0^b dz(b-z)\exp(j2kz)$$

$$= \mp jka^2\left(\frac{\exp(-j2k(r_0-b/2))}{2r_0^2}\right)\left[\frac{\sin(kb)}{kb}+\frac{j}{kb}\left(\cos(kb)-\frac{\sin(kb)}{kb}\right)\right]. \tag{24a}$$

This result reveals several interesting properties of the back scattered field. First, if the frequency increases without bound kb $\to \infty$, the back scattered field vanishes because all acoustic energy is reflected away from the back scatter direction. Second, if the length of the cone decreases to the vanishing point b $\to$ 0, the back scattered field becomes the same as the disc scattered field in (21a). Third, the scattered field appears to originate from some point between the tip and base of the cone. For the case of a shallow cone kb $\cong$ 0, the echo appears to originate from the midpoint along the cone. Lastly, the echo is frequency dependent with the target strength

$$ts = \left(\frac{A}{\lambda}\right)^2\left[\frac{\sin^2(kb)}{(kb)^2}+\frac{1}{(kb)^2}\left(\cos(kb)-\frac{\sin(kb)}{kb}\right)^2\right]. \tag{24b}$$

A plot of normalized target strength for a cone as a function of frequency is shown in Figure 8a. We see that the cone target strength saturates very rapidly as frequency increases and that for wavelengths less than about $2\pi b$ the target strength is essentially a constant given by TS = 20Log(A/2πb). For the case of a base radius of 2 m and a cone depth of 2 m the target strength is found to be zero. This behavior deviates markedly from the case of a disc for which the target strength increases with frequency.

The deviation of the cone target strength from the disc target strength can be determined from the relative target strengths of the cone and disc which is found from the ratio of (24b) and (21b)

$$\frac{ts_{cone}}{ts_{disc}} = \left[\frac{\sin^2(kb)}{(kb)^2}+\frac{1}{(kb)^2}\left(\cos(kb)-\frac{\sin(kb)}{kb}\right)^2\right]. \tag{25}$$

A plot of this function is shown in Figure 8b. At very low frequencies the back scattering from the cone is the same as the back scattering from the disc but as

[6] Abramowitz and Stegun, (4.3.119) and (4.3.123).

frequency increases, the relative back scattering from the cone decreases monotonically. This is because the acoustic energy is increasingly reflected away from the cone axis.

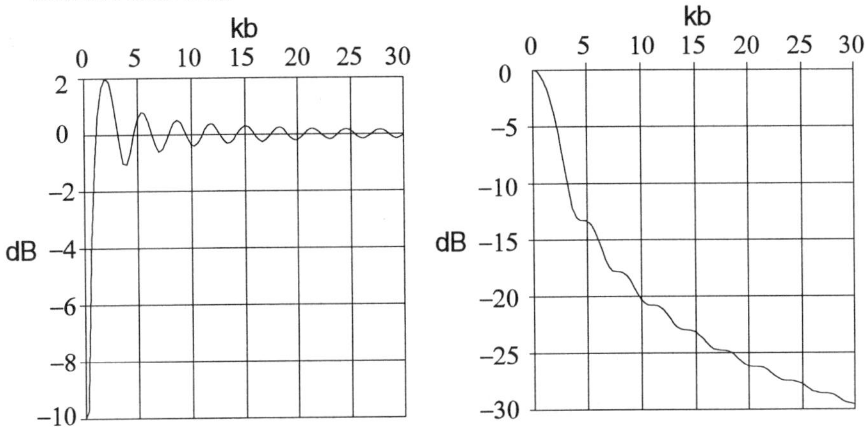

a) TS − 20Log(A/2πb) b) Target Strength: Cone vs. Disc

Figure 8 Back Scattering Target Strength of a Cone of Length b and Base Area A.

7.4 Passive Signatures of Targets

Modern submarines are very quiet and future submarines are expected to be even quieter. Consequently, it is becoming more difficult to detect submarines passively at long ranges. However, such detections still occur when a submarine inadvertently or is operationally required to radiate a burst of acoustic energy and submarines are still passively detected on a routine basis at ranges out to about a 100 nm. Even with modern quieting programs it will probably always be possible for submarines to detect each other at very close ranges. In contrast, most surface craft radiate acoustic energy copiously and submarines use this information to track the world above them. Future ASW platforms are expected to counter this capability with their own quieting programs.

The radiated noise fields from nuclear submarines and advanced capability non-nuclear submarines are closely guarded secrets but the acoustic radiations from older submarines and surface craft are known to come predominately from:

- Propeller Noise;
  - Cavitation,
  - Blade Rate Tonals,
  - Singing,
- Machinery Noise;
  - Gear Trains,
  - Reciprocating Engines,
  - Friction and Turbulence,
- Flow Noise.

Modern submarines are expected to have similar acoustic signatures and similarly for modern surface craft. Brief descriptions of each of these noise types are given in the following.

Propeller Noise

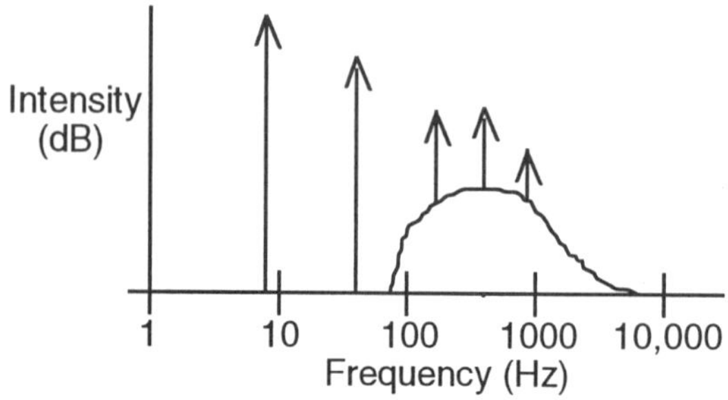

Figure 9a Typical Spectrum of Propeller Noise

Propeller noise is the dominate noise radiated by vessels - when it occurs. For submarines this is not too often because modern propellers are designed to minimize noise radiations and submarine crews drive their boats with an eye to minimizing radiations of all kinds including, especially, propeller noise. Propeller noise is found predominately in the frequency range between 1 Hz to 1000 Hz, as indicated in Figure 9a, with narrow band tonals dominating the first 100 Hz, but also occurring throughout the 1000 Hz band, and broad band cavitation noise dominating the band above 100 Hz. As the speed of the craft increases the broad band noise increases and begins to fill the entire propeller noise band.

Propeller Cavitation:

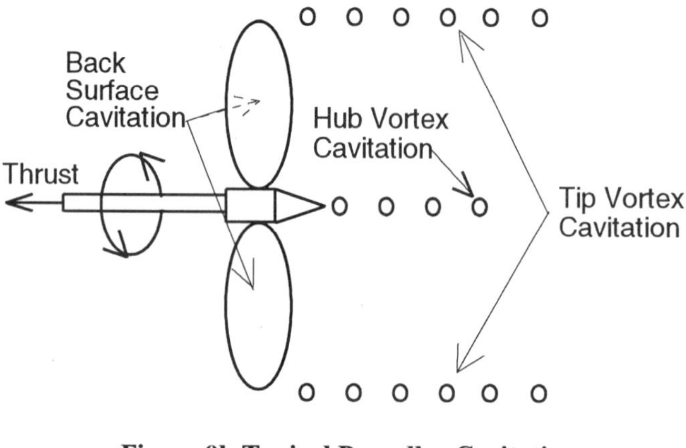

Figure 9b Typical Propeller Cavitation

Cavitation from propellers is generated as either surface cavitation or vortex cavitation and is represented by the broad portion of the spectrum in Figure 9a. As indicated in Figure 9b, surface cavitation is generated mainly on the back (suction) side of a propeller blade and to a lessor degree on the front side of a propeller blade when the thrust is small. Vortex cavitation is generated primarily at the blade tips and at the propeller hub. Surface cavitation from the suction side of the blade causes the loudest noise and vortex cavitation causes lessor noise with hub vortex cavitation being the least noisy. Cavitation from the suction surface of the propeller occurs because the reduced pressure causes dissolved gasses to emerge as small gas bubbles.

These bubbles immediately collapse when the pressure becomes positive and do so with great force thereby releasing significant acoustic energy. The magnitude of the force is great enough to pit the propeller blade when the collapse is against the surface of the propeller. Vortex cavitation is formed when the propeller blades shed vortices. This occurs on the blade trailing edge wherever the blade span changes and is especially strong at the blade tip. If the near hub portion of the blade is providing much of the thrust, cavitation also occurs at the blade hub. The cavitation bubbles caused by vortex shedding are generally filled with vapor as well as gasses and remain in a low pressure environment longer than surface cavitation bubbles causing them to collapse more slowly and to emit less acoustic energy during the collapse. Since all propeller cavitation noise consists of acoustic impulses, it is heard as a broad band noise with a continuous spectrum that has been described as a "hiss". Propeller cavitation is reduced for submarines and torpedoes operating at greater depths where the ambient pressure is greater and cavitation bubbles are less likely to occur. However, when submarines change from shallower to greater depths, propeller cavitation noise can increase as the increasing pressure causes the existing cavitation bubbles to collapse with greater force. Propeller cavitation increases with submarine speed in a stepped shaped curve, as indicated in Figure 9c where the step occurs at a critical speed. Below the critical speed cavitation is virtually non-existent. At the critical speed cavitation increases by several tens of dB and then slowly increases with speed to a saturation level. In contrast, cavitation noise by surface ships increase more linearly with speed and do not exhibit a critical speed.

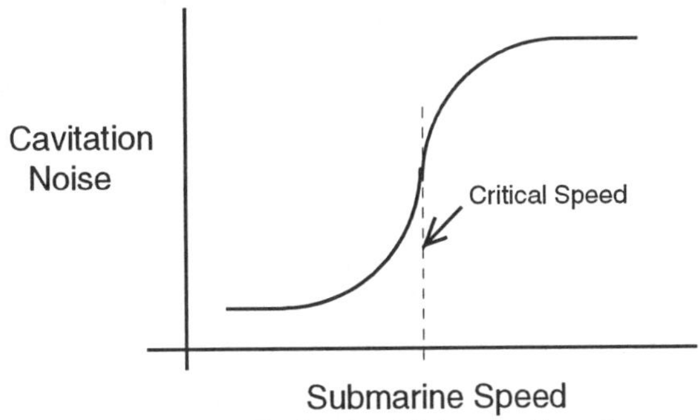

Figure 9c Cavitation Noise vs. Speed

The cavitation noise spectrum generally increases with frequency as f^2 up to some peak level and then decreases with frequency as f^{-2} beyond that frequency. The peak level generally occurs at a frequency above 100 Hz for both submarines and surface craft and the cavitation spectrum in the low frequency region is generally obscured by machinery and flow noises. The frequency at the peak level increases with increasing depth, for submarines, and with decreasing speed for all vessels. The peak level increases with decreasing depth, for submarines, and with increasing speed for all vessels.

Cavitation source levels are usually presented as a spectrum level, i.e. level per 1 Hz, but they are usually measured in bands of width B. The transformation to the spectrum level is

$$SL(1Hz) = SL(B) - 10\log(B)\ . \tag{26}$$

This transformation is obviously exactly correct for white noise but it is also correct for a level falling at –6 dB per octave, i.e. for the upper frequency range of the cavitation spectrum. The cavitation level in the upper frequency range in a band from f_1 to f_2 is given by

$$sl(B) = A\int_{f_1}^{f_2} f^{-2} df = \frac{AB}{f^2}, \tag{27a}$$

where A is a constant, $B = f_2 - f_1$ is the processing bandwidth and $f = \sqrt{(f_1 f_2)}$is the geometric mean of the band edges and is the frequency assigned to the band. For this case we find

$$\frac{sl(1Hz)}{sl(B)} = \frac{1}{B}, \tag{27b}$$

and the transformation from the processing band B to the spectrum band of 1 Hz is as given by (26).

Cavitation measurements on WWII surface ships have lead to the model for cavitation noise[7]

$$SL_c = C + 20 - 20Log(f), \qquad f > 100\text{ Hz} \tag{28a}$$

$$C = 170 + 53Log\left(\frac{v}{10kts}\right), \tag{28b}$$

where v is the ship speed in knots. A similar model is expected to hold for submarines.

Propeller Blade Rate Tonals:

In the absence of cavitation, which is the more usual condition for submarines, blade rate tonals can be heard and are the dominant noise in the 1 Hz to 100 Hz region. Blade rate tonals occur at the frequencies

$$f = mns, \tag{29}$$

where m is an harmonic number, n is the number of blades on the propeller and s is the shaft rate in turns per second. A principal source of blade rate tonals is the fluctuating thrust forces on the propeller blades. Even in a uniform flow, propeller blades flex causing them to emit sound as from a dipole. In the nonuniform flow around the hull, struts and fins of a submarine the flexing is even greater. The direction of sound radiation is primarily in the plane of the propeller blades, i.e. perpendicular to the propeller shaft axis. This is mainly due to the nature of a dipole radiator and partly due to the screening fore and aft provided by the hull and bubble wake. Another source of the blade rate tonals is the blade slap rate at the blade rate ns. It is believed to be caused by the blades passing through vortices created

---

[7] Donald Ross, *Mechanics of Underwater Noise*, Pergamon Press, 1976, (8.30) and (8.31).

Because of irregularities in the propeller blades and the propeller shaft, the blade rate tonals are weakly modulated by a component that has the period of the propeller shaft. The resulting radiated signal has the form

$$p = \left(\sum_{k=-\infty}^{\infty} a_k \exp\left(jk\omega_s t\right) \right) \cos\left(mb\omega_s t\right)$$

$$= \sum_{k=-\infty}^{\infty} \frac{a_k}{2} \left(\exp\left(j(k+mb)\omega_s t\right) + \exp\left(j(k-mb)\omega_s t\right) \right), \qquad (30)$$

where the summation in parentheses is the modulation on the cosine carrier frequency and $\omega_S = 2\pi s$ is the angular shaft rate in radians per second. Since the low order terms, i.e. $k = 0, \pm 1, \cdots$, dominate the modulation, we see that each blade rate tonal is accompanied by a cluster of low level tonals spaced at the shaft rate frequency.

Example 4: For a 5 bladed propeller having a shaft rate of one turn per second, the fourth harmonic will have a blade rate tonal of 20 Hz. Around this tonal we can expect to find low level tonals spaced at 1 Hz intervals. Similar structures will be found for all other harmonics.♦

The blade rate tonals generated by propellers are generally very weak because of their dipole source and the harmonics of the blade rate tonals are even weaker. However, the forces generating these tonals are also transmitted to the submarine hull via the propeller shaft and its thrust bearing and, if the hull is resonant at one of these frequencies, it will vibrate and radiate a much stronger blade rate tonal. Similarly, these tonals can be transmitted to the near portions of the hull through the water and can cause portions of the hull to resonate. These hull vibrations create acoustic radiations that are generally much stronger than the acoustic radiations from the propeller itself.

Propeller Singing:

Propeller singing is a strong tonal radiated in the frequency range from 100 Hz to 1000 Hz. The process is apparently related to vortex shedding at the rear edge of the blades and is dependent upon the shape and cross section of the trailing edge of the blade. If the vortex shedding rate is near a resonance of the blade edge, a strong tone is emitted at the resonance frequency. Propeller singing is a somewhat rare event but, when singing does occur, it is a strong tonal. The singing can be continuous or sporadic and it can occur on only some blades or during only part of the blade cycle. It is dependent upon subtle properties of a propeller and, among apparently identical propellers, some may sing and others will not. For modern submarines with properly designed propellers, singing rarely occurs.

Machinery Noise

Machinery vibrations arise because of imbalances in rotational machinery and impact noise in reciprocating machinery both of which lead to narrow band tonal noise. Machinery vibrations also arise because of friction in bearings and turbulence in pumps and hydraulic lines both of which lead to broad band noise. These vibrations are coupled into the hull of the vessel through the machinery mountings and are radiated into the ocean environment as acoustic noise. Of course every attempt is made to isolate machinery vibrations from the hull on modern submarines but a residual "hull drone" is commonly found.

Imbalances in rotational machinery arise because of static and dynamic imbalances. Static imbalances occur when the center of gravity of the rotating body is off the axis of rotation. However, even if the rotating element is in static balance, it will not be in dynamic balance if the axis of rotation does not coincide with a principal axis of the body. Static and dynamic imbalances produce narrow band tonals in the radiated signature of the vessel with the dominant tonal being the frequency of shaft rotation.

Impact noise is produced by the periodic impact of gears, valves, pistons, etc. The velocity that is transmitted to the hull by the train of impacts can be represented as a periodic train of exponentials given by

$$v_I(t) = \sum_{n=-\infty}^{\infty} u(t - n\tau)\exp(-\alpha(t - n\tau)), \tag{31}$$

where u(·) is the unit step function, α is the exponential damping coefficient and τ is the period of the impulses. Fourier transforming we obtain

$$V_I(\omega) = \sum_{n=-\infty}^{\infty} \frac{\exp(jn\tau\omega)}{(\alpha - j\omega)}. \tag{32a}$$

The series in this expression is periodic with period $\Delta\omega = 2\pi/\tau$. To transform this series, consider the periodic sequence of Dirac delta functions having the period $2\pi/\tau$ which we can represent by a Fourier series

$$\sum_{k=-\infty}^{\infty} \delta\left(\omega - \frac{2\pi k}{\tau}\right) = \sum_{n=-\infty}^{\infty} V_n \exp(jn\omega\tau), \tag{32b}$$

$$V_n = \frac{\tau}{2\pi} \int_0^{2\pi/\tau} \delta(\omega)\exp(-jn\omega\tau)d\omega = \frac{\tau}{2\pi}. \tag{32c}$$

Consequently

$$V_I(\omega) = \frac{2\pi}{\tau(\alpha - j\omega)} \sum_{k=-\infty}^{\infty} \delta\left(\omega - \frac{2\pi k}{\tau}\right). \tag{32d}$$

Multiplying by the transfer function of the hull mountings F(ω) and inverse Fourier transforming we obtain the hull vibration velocity

$$v(t) = \int_{-\infty}^{\infty} d\omega \, \frac{F(\omega)\exp(j\omega t)}{\tau(\alpha - j\omega)} \sum_{k=-\infty}^{\infty} \delta\left(\omega - \frac{2\pi k}{\tau}\right),$$

$$= \sum_{k=-\infty}^{\infty} \frac{F(\omega_k)\exp(j\omega_k t)}{\tau(\alpha - j\omega_k)}. \qquad \omega_k = \frac{2\pi k}{\tau} \tag{33}$$

This series shows that the pulse train is equivalent to a system of tonals separated by $2\pi/\tau$ with powers equal to $|F(\omega_k)|^2/[\tau^2(\alpha^2-\omega_k^2)]$ $k = \pm1, \pm2,\cdots$. The $k = 0$ component is excluded since the hull will not transmit a constant pressure. The dominant component will be the fundamental at frequency $1/\tau$, i.e. at the frequency of the pulse train. If the transfer function of the hull has a resonance near one of the tonals, that tonal will contribute significantly to the acoustic signature of the vessel. The principal sources of impact noise in the signature of a target will be gear noise and reciprocating engine noise.

Machinery/Gear Noise:

Gear noise is dependent primarily upon the power being transmitted by the gears and on the cut of the gears. Obviously, as the power being transmitted increases, the gear noise increases. Similarly, straight cut gears are noisier than helical cut gears and rough cut gears are noisier than polished gears. The repetition rate of gear noise is just the number of teeth on the gear times the turn rate $f = ns$.

Machinery/Reciprocating Engines:

Reciprocating diesel engines and reciprocating compressors emit many noises but the noise due to piston slap has been found to be the greatest. Piston slap occurs because the crank and connecting rod assembly push the piston against one side of the cylinder wall or the other during each cycle of the crank and several slaps will occur during each cycle. Since the slaps are not evenly spaced in time and are not equal in magnitude, the dominant frequency of the slaps will be the turn rate of the crank with harmonics of the crank rate. For the case of multiple cylinders in which the firing rate is staggered in time, harmonics at the firing rate will be emphasized.

Machinery/Friction and Turbulence:

Bearing noise is the result of friction and is usually not a significant source of radiated noise from submarines but, if bearings are worn out, broken, etc., they can generate a significant amount of noise. Ball bearings and roller bearings are found to generate more noise than bushing type bearings which tends to be tonal while bushing noise tends to be broad band. Machinery turbulence arises in hydraulic pumps and plumbing and is broad band. Friction and turbulence noise is generally not a significant generator of noise. It is more serious as a self noise problem in which it serves to mask received signals.

Flow Noise

Flow noise is generated by the flow of water past the hull of the submarine or surface ship. The turbulence at the surface of the hull causes weak sound to be generated, which is usually masked by other noise, but, if the flow causes the hull or other structures such as fins or struts to vibrate, significant amounts of acoustic energy can be radiated. One such noise is the propeller singing discussed above. Flow noise is generally not a significant generator of noise. It is more serious as a self noise problem in which it serves to mask received signals.

7.5 Doppler Shifting of Signal Frequencies

When one or more of the platforms in an active or passive detection scenario is in motion, the radiated and/or the received signals are shifted in frequency. This frequency shift is due to the motion of the platforms and is called the Doppler effect. A classic example of this effect is the decrease in frequency of a locomotive horn signal as it passes from advancing on the observer to receding from the observer. In order for such signals to be properly detected and their parameters to be interpreted, it is important to understand the magnitude and direction of the Doppler shift and to determine the effect of the shift on signal structure. In these endeavors it is convenient to consider initially the case of just one source and one receiver such as would be found in a passive detection scenario. The results of this case are easily extended to the case of a source, target and receiver such as would be found in an active bistatic or monostatic detection scenario.

A source and receiver, as depicted in Figure 10, are each moving along their own track at their own speed and their position vectors are time dependent and denoted $\mathbf{r}_S(\tau)$ and $\mathbf{r}_R(t)$ where τ and t are the times measured, respectively, on the source and receiver clocks.. The signal radiated by the source is denoted $s(\tau)$ and the signal intercepted by the receiver is denoted $s_R(t)$. The source and receiver clocks are assumed to be synchronized and running at the same rates but the time of radiation of a signal wave front and the time of reception of the same wave front are different. The time difference is given by

$$t-\tau=\Delta(t,\tau)\,,\quad \Delta(t,\tau)=\int_{\mathbf{r}_S(\tau)}^{\mathbf{r}_R(t)}\frac{ds}{c(\mathbf{r})}, \tag{34}$$

where $c(\mathbf{r})$ denotes the sound speed along the ray path between the source at the time of radiation and the receiver at the time of interception. The received signal is given by

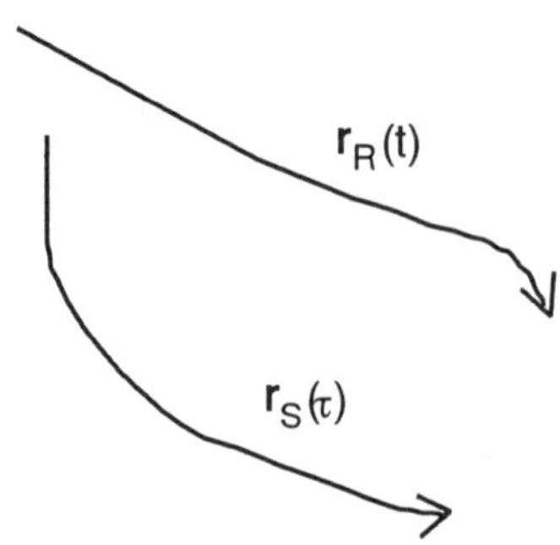

Figure 10 Tracks of Source and Receiver Platforms

$$s_R(t) = s(t - \Delta(t,\tau)). \tag{35a}$$

For the transmit time $\tau = 0$ the reception time can be expressed as $t_{SR} = r_{SR}/c_{SR}$ where r_{SR} is the source to receiver range between the transmission point and interception point and c_{SR} is the effective sound speed between these two points. We note that $\Delta(t_{SR},0) = r_{SR}/c_{SR}$. Expanding $\Delta(t,0)$ about the time t_{SR} we obtain

$$\Delta(t,0) \cong t_{SR} + \frac{d\Delta(t_{SR},0)}{dt}(t - t_{SR}). \tag{35b}$$

Substituting into (35a) gives

$$s_R(t) = s\left((t - t_{SR})\left(1 - \frac{d\Delta(t_{SR},0)}{dt}\right)\right). \tag{35c}$$

This shows that the received signal is a time delayed and time compressed or elongated, i.e. negatively compressed, version of the transmitted signal. The incremental compression/elongation can be expressed as

$$\frac{d\Delta(t_{SR},0)}{dt} = \left.\frac{\partial\Delta(t,0)}{\partial t}\right|_{t=t_{SR}} + \left.\frac{\partial\Delta(t_{SR},\tau)}{\partial\tau}\frac{dt}{d\tau}\right|_{\tau=0}. \tag{36a}$$

From (34) we obtain

$$\frac{dt}{d\tau} = 1 + \frac{\partial\Delta(t,\tau)}{\partial t} + \frac{\partial\Delta(t,\tau)}{\partial\tau}\frac{dt}{d\tau} = \frac{1 + \dfrac{\partial\Delta(t,\tau)}{\partial t}}{1 - \dfrac{\partial\Delta(t,\tau)}{\partial\tau}}. \tag{36b}$$

Substituting (36) into (32c) gives

$$s_R(t) = s\left((t - t_{SR})\left(\frac{1 - \partial_t\Delta(t_{SR},0)}{1 - \partial_\tau\Delta(t_{SR},0)}\right)\right), \tag{37}$$

where ∂_t and ∂_τ denote the partial derivatives with respect to t and τ. It remains to compute these partial derivatives.

From (34) we obtain for the partial derivative with respect to t

$$\left.\partial_t\Delta(t,0)\right|_{t=t_{SR}} = \partial_t \int_{r_S(0)}^{r_R(t_{SR})} \frac{ds}{c(\mathbf{r})} = \lim_{\Delta t \to 0} \frac{1}{\Delta t}\left[\int_{r_S(0)}^{r_R(t_{SR}+\Delta t)} \frac{ds'}{c(\mathbf{r}')} - \int_{r_S(0)}^{r_R(t_{SR})} \frac{ds}{c(\mathbf{r})}\right]. \tag{38a}$$

With reference to Figure 11, the first integral in the right member of this expression is approximated

$$\int_{\mathbf{r}_S(0)}^{\mathbf{r}_R(t_{SR}+\Delta t)} \frac{ds'}{c(\mathbf{r}')} \cong \int_{\mathbf{r}_S(0)}^{\mathbf{r}_R(t_{SR})} \frac{ds}{c(\mathbf{r})} + \left[\mathbf{r}_R(t_{SR}+\Delta t) - \mathbf{r}_R(t_{SR})\right] \cdot \frac{\mathbf{e}_{SR}(\mathbf{r}_R)}{c(\mathbf{r}_R)}, \quad (38b)$$

where the unit vector $\mathbf{e}_{SR}(\mathbf{r}_R)$ points along the ray from source to receiver and is evaluated at the receiver and where $c(\mathbf{r}_R)$ is the sound speed at the receiver. This approximation is valid if the source-to-receiver ray path is nearly unchanged over the time interval t_{SR} to $t_{SR} + \Delta t$. This will always be valid - excluding those situations in which the ray path bifurcates. Substituting (38b) into (38a) we obtain for the partial derivative with respect to t

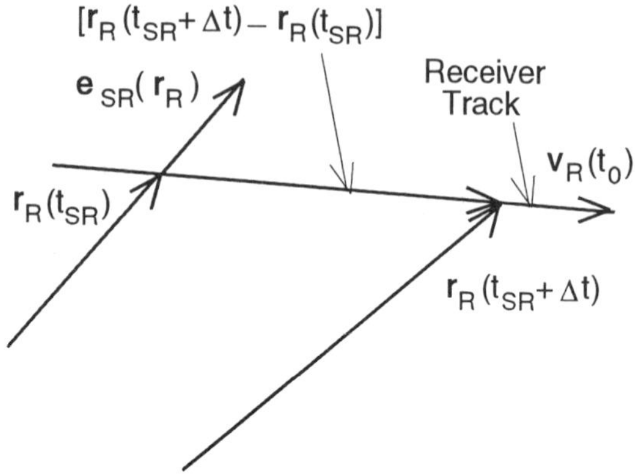

Figure 11 Receiver Track Geometry

$$\partial_t \Delta(t,0)\Big|_{t=t_{SR}} = \frac{\mathbf{v}_R(t_{SR}) \cdot \mathbf{e}_{SR}(\mathbf{r}_R)}{c(\mathbf{r}_R)}. \quad (39a)$$

For the partial derivative with respect to τ we obtain by analogous arguments

$$\partial_\tau \Delta(t_{SR},\tau)\Big|_{\tau=0} = \frac{\mathbf{v}_S(0) \cdot \mathbf{e}_{SR}(\mathbf{r}_S)}{c(\mathbf{r}_S)}, \quad (39b)$$

where the unit vector and sound speed, $\mathbf{e}_{SR}(\mathbf{r}_S)$ and $c(\mathbf{r}_S)$, are evaluated at the point of ray launch and the unit vector points along the ray from source to receiver. Substituting (39) into (37) the received signal has the form

$$s_R(t) = s\left(\alpha_{SR}(t - t_{SR})\right) \qquad \alpha_{SR} = \left(\frac{1 - \mathbf{v}_R(t_{SR}) \cdot \mathbf{e}_{SR}(\mathbf{r}_R) / c(\mathbf{r}_R)}{1 - \mathbf{v}_S(0) \cdot \mathbf{e}_{SR}(\mathbf{r}_S) / c(\mathbf{r}_S)}\right). \quad (40a)$$

We note that there are two sound speed definitions in this expression. One applies to the effective sound speed from source to receiver c_{SR}, i.e. where $t_{SR} = r_{SR}/c_{SR}$, and the other to the sound speeds at the source and receiver, i.e. $c_S = c(\mathbf{r}_S)$ and $c_R = c(\mathbf{r}_R)$. The former determines the source to receiver propagation time and the later determines the Doppler compression at the source and receiver. Moreover, the Doppler compression terms $\mathbf{v}_S(0) \cdot \mathbf{e}_{SR}(\mathbf{r}_S)$ and $\mathbf{v}_R(t_{SR}) \cdot \mathbf{e}_{SR}(\mathbf{r}_R)$ are readily seen to be just the component of platform velocity along the ray path, i.e. the source radial velocity at the time of launch and the receiver radial velocity at the time of reception. Denoting these components as, respectively, v_S and v_R, and noting that they are small compared to the sound speed, the Doppler compression factor can be approximated

$$\alpha_{SR} = \left(1 - (v_R - v_S)/c\right), \tag{40b}$$

where we have also assumed that the local sound speeds at the source and receiver are the same. When the platforms are closing, $v_S > 0$ and $v_R < 0$, the Doppler compression factor is positive and the signal is compressed. Conversely, when the platforms are opening, the Doppler compression factor is negative and the signal is elongated.

The effect of Doppler shifting of frequencies for bistatic and monostatic detection scenarios can be immediately derived from the preceding. For propagation from source to target, (40) describes the situation. For propagation from target to receiver, the target acts as a source, where we imagine that the target "records and retransmits" the signal, sending the signal along to the receiver. Moreover, since the signal is delayed and compressed on each leg of its journey, we can immediately write

$$s_R(t) = s\left(\alpha_{ST}\alpha_{TR}\left(t - t_{ST} - t_{TR}\right)\right), \tag{41a}$$

with Doppler compressions

$$\alpha_{ST} = \left(\frac{1 - \mathbf{v}_T(t_{ST}) \cdot \mathbf{e}_{ST}(\mathbf{r}_T)/c_T}{1 - \mathbf{v}_S(0) \cdot \mathbf{e}_{ST}(\mathbf{r}_S)/c_S}\right), \tag{41b}$$

$$\alpha_{TR} = \left(\frac{1 - \mathbf{v}_R(t_{ST} + t_{TR}) \cdot \mathbf{e}_{TR}(\mathbf{r}_R)/c_R}{1 - \mathbf{v}_T(t_{ST}) \cdot \mathbf{e}_{TR}(\mathbf{r}_T)/c_T}\right). \tag{41c}$$

As in (40b), we can approximate the compound Doppler compression

$$\alpha = \alpha_{ST}\alpha_{TR} = \left(1 - (v_{ST} - v_S + v_R - v_{TR})/c\right), \tag{41d}$$

where v_{ST} and v_{TR} are the target radial velocities along, respectively, the ray from the source and the ray to the receiver. For a monostatic detection scenario the source and receiver are collocated and we have $v_{ST} = -v_{TR}$ and $v_S = -v_R$. For this case (41d) reduces to

$$\alpha = \left(1 - 2(v_T - v_S)/c\right), \tag{41e}$$

where v_T and v_S are the radial velocities directed along the ray from source to target. We see that for monostatic detection, a factor of two is introduced into the Doppler effect due to platform motion. This occurs because the signal is Doppler shifted by the source motion both when it is launched and again when it is received. Similarly, the signal is Doppler shifted by the target motion both when it contacts the signal, i.e. is "recorded", and when it reflects the signal, i.e. when it is "retransmitted".

APPENDIX 7-A: KIRCHOFF'S INVERSION OF THE WAVE EQUATION

The inversion of the Helmholtz equation was developed in Appendix 4-A, using narrow band Green's functions, for solving continuous wave problems in the presence of arbitrary scattering bodies. In the following, we develop Kirchoff's inversion of the wave equation, using broad band Green's functions. The resulting Kirchoff formula gives a solution to the wave equation for broad band signals in the presence of arbitrary scattering bodies.

The wave equation for inviscid media was found in Eq. 2.17-A, to be

$$\nabla^2 p(t,\mathbf{r}) - \frac{1}{c^2}\ddot{p}(t,\mathbf{r}) = -\rho_0 \dot{q}(t,\mathbf{r}), \qquad (1a)$$

with boundary conditions

$$a p(t,\mathbf{r}_S) + b\nabla p(t,\mathbf{r}_S)\cdot \mathbf{e}_S = f(t,\mathbf{r}_S), \qquad (1b)$$

$$p(t_0,\mathbf{r}) = u_0(\mathbf{r}), \qquad \dot{p}(t_0,\mathbf{r}) = v_0(\mathbf{r}). \qquad (1c)$$

The point $\mathbf{r}_S$ is a point on the bounding surface and the condition on this surface in (1b) is inhomogeneous. The boundary conditions in (1c) are the initial conditions of the pressure field.

Kirchoff's formula is obtained by inverting the wave equation in a manner very similar to Helmholtz's inversion of the Helmholtz equation. The Kirchoff formula is found to be more general than Helmholtz's formula and, in fact, Helmholtz's formula is shown to be a special case of Kirchoff's formula. The principal difference between the two formulas is that the Kirchoff formula applies to broad band signals, while the Helmholtz formula applies to narrow band signals.

7.1-A: Green's Functions

A Green's function, $g(t,\mathbf{r}|t',\mathbf{r}')$, is introduced in which t and $\mathbf{r}$ represent an observation time at an observation point within the acoustic field and, similarly, t' and $\mathbf{r}'$ represent a transmission time at a source point within the acoustic field. For simplicity, the clocks at the two points are assumed to be synchronized and to run at the same rate. However, since the observation point and the source point are separated in space, then, by the principle of causality, the observation time of a source event is necessarily later than the source time of the event, i.e. $t > t'$.

The Green's function is defined to be the solution to a wave equation problem that is analogous to that posed in (1). The difference between the two problems is that for the Green's function the source injects a unit volume of fluid at t' into the point at $\mathbf{r}'$, i.e. the Green's function source term $\rho_0 q_g(t)$ is the unit step function and its derivative is the Dirac delta function

$$\rho_0 q_g(t) = u(t) = \begin{cases} 1, & t \geq t' \\ 0, & t < t' \end{cases} \qquad \rho_0 \dot{q}_g(t) = \dot{u}(t) = \delta(t-t'). \qquad (2a,b)$$

In these relations $q_g(t)$ is the Green's function source, the step function u(t) is dimensionless but the implied unit on the step function, contributed by the density, is mass per unit volume. Similarly, the implied unit on the delta function is mass per unit volume per unit time, i.e. the delta function is not dimensionless and provides the per unit time measure.

The Green's function is required to satisfy the wave equation and boundary conditions

$$\nabla^2 g(t,\mathbf{r}|t',\mathbf{r}') - \frac{1}{c^2}\ddot{g}(t,\mathbf{r}|t',\mathbf{r}') = -4\pi\delta(t-t')\delta(\mathbf{r}-\mathbf{r}'), \tag{3a}$$

$$a g(t,\mathbf{r}_S|t',\mathbf{r}') + b\nabla g(t,\mathbf{r}_S|t',\mathbf{r}')\cdot\mathbf{e}_S = 0, \tag{3b}$$

$$g(t,\mathbf{r}|t',\mathbf{r}') = 0, \quad \dot{g}(t,\mathbf{r}|t',\mathbf{r}') = 0, \quad t < t' \tag{3c}$$

where the space derivatives are with respect to the observation point **r** and the vector $\mathbf{r}_S$ denotes a point on the bounding surface. The boundary condition in (3b) is the homogeneous form of the wave equation boundary condition in (1b) and the boundary conditions in (3c) assert that, prior to the impulse at $t = t'$, the Green's function and its derivative vanish everywhere, i.e. the impulse has yet to produce an acoustic field.

The form of the Kirchoff integral solution of the wave equation is relatively simple in concept. Since the Green's function is the solution to the wave equation for a point source, the solution to the original wave equation problem in (1), for the case of a homogeneous surface boundary condition in (1b), can be expected to be a sum of Green's functions weighted by the source strength in (1) where the sum is over the domain of the sources. We show this in the following. For the case of an inhomogeneous surface boundary condition in (1b), the same result is found where, though, the Green's function is required to satisfy an additional condition. Of course, finding the Green's function in the presence of boundaries can be difficult but, fortunately, the solution can be shown to be more simply and advantageously cast in terms of free-field Green's function. Preliminary to this work, we demonstrate the Green's function reciprocity relation and determine the form of the Green's function.

Green's Function Reciprocity Relation

The Green's function reciprocity relation is concerned with interchanges of the source and receiver and is expressed as

$$g(t,\mathbf{r}|t',\mathbf{r}') = g(-t',\mathbf{r}'|-t,\mathbf{r}'). \tag{4}$$

This relation asserts that if the observation point and source point are interchanged, with the times negated, the acoustic field values remain the same. The reason the times are negated is because causality requires that $t > t'$ and reversing signs in the reciprocal relation results in the equivalent relation $-t' > -t$. For example, if an event at **r** at time $t = 10$ is due to an event at **r**′ at time $t' = -5$, then the elapsed time would be t – t′ = 15. Interchanging coordinates, an event at **r′** at time –t′ = 5 would be due to an event at **r** at time –t = –10. The elapsed time would still be –t′ –(–t) = 15. Moreover, adding 5 to both –t′ and –t shows that an event at 10 would be due to an event at –5, i.e. the time axes for the Green's function and the reciprocal Green's function not only have the same unit of measure but they also have the same origin making them identical.

To prove that the Green's function reciprocity relation is valid, we write a wave equation problem for the reciprocal Green's function. This, of course, can be done whether the Green's function reciprocity relation is valid or not. For the reciprocal Green's function solution we assume the same observation point **r**, as in (3), and a source point at **r″**, which gives

$$\nabla^2 g(-t, r|-t'', r'') - \frac{1}{c^2}\ddot{g}(-t, r|-t'', r'') = -4pd(t - t'')d(r - r''), \tag{5a}$$

$$ag(-t, \mathbf{r}_S|-t'', \mathbf{r}'') + b\nabla g(-t, \mathbf{r}_S|-t'', \mathbf{r}'')\cdot \mathbf{e}_S = 0, \tag{5b}$$

$$g(-t, \mathbf{r}|-t'', \mathbf{r}'') = 0, \quad \dot{g}(-t, \mathbf{r}|-t'', \mathbf{r}'') = 0, \quad -t < -t'' \tag{5c}$$

where the space derivatives operate on the observation point and the source points for (3) and (5) are indicated in Figure 1.

To prove the validity of the Green's function reciprocity relation we manipulate as follows: multiply the wave equation in (3a) by the Green's function in (5a); multiply the wave equation in (5a) by the Green's function in (3a); subtract the latter equation from the former equation. Then integrate the difference equation over the observation coordinates. The ranges of integration are over the entire volume and over all times up to the present. The result is (6a) where **dr** = dv = dxdydz and where the integrations in the right member, with the assistance of the delta functions, were immediately performed.

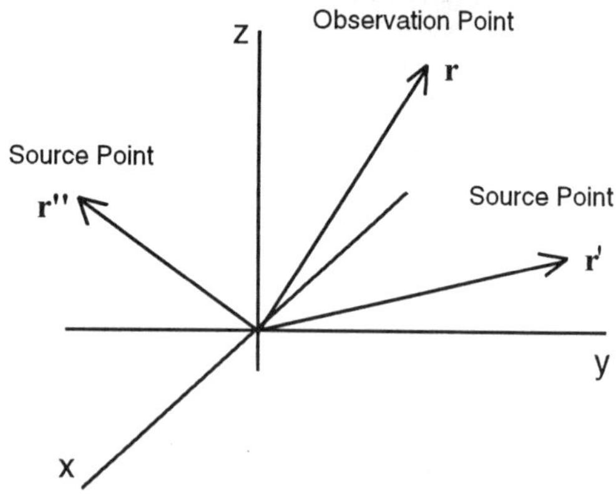

Figure 1 Configuration of Sources and Obseration Point for Green's Functions

$$\int_{-\infty}^{t^+} dt \int_V d\mathbf{r}\left[g(-t, \mathbf{r}|-t'', \mathbf{r}'')\nabla^2 g(t, \mathbf{r}|t', \mathbf{r}') - g(t, \mathbf{r}|t', \mathbf{r}')\nabla^2 g(-t, \mathbf{r}|-t'', \mathbf{r}'')\right]$$

$$-\frac{1}{c^2}\int_{-\infty}^{t^+} dt \int_V d\mathbf{r}\left[g(-t, \mathbf{r}|-t'', \mathbf{r}'')\ddot{g}(t, \mathbf{r}|t', \mathbf{r}') - g(t, \mathbf{r}|t', \mathbf{r}')\ddot{g}(-t, \mathbf{r}|-t'', \mathbf{r}'')\right]$$

$$= -4\pi\left[g(-t', \mathbf{r}'|-t'', \mathbf{r}'') - g(t'', \mathbf{r}''|t', \mathbf{r}')\right]. \tag{6a}$$

The time t^+ is slightly greater than the greater of t or t″ to ensure that all delta function peaks are covered by the integration. If the left member of (6a) can be shown to vanish, the right member would be zero which would demonstrate the validity of the Green's function reciprocity relation. The first integral in (6a) can be transformed to a surface integral by means of Green's Theorem in (4.6-A) and the integrand in the second integral can be transformed using elementary differential properties to give

$$\int_{-\infty}^{t^+} dt \int_S ds \cdot \left[g(-t,\mathbf{r}_S|-t'',\mathbf{r}'')\nabla g(t,\mathbf{r}_S|t',\mathbf{r}') - g(t,\mathbf{r}_S|t',\mathbf{r}')\nabla g(-t,\mathbf{r}_S|-t'',\mathbf{r}'')\right]$$

$$-\frac{1}{c^2}\int_{-\infty}^{t^+} dt \int_V d\mathbf{r}\,\frac{\partial}{\partial t}\left[g(-t,\mathbf{r}|-t'',\mathbf{r}'')\dot{g}(t,\mathbf{r}|t',\mathbf{r}') + g(t,\mathbf{r}|t',\mathbf{r})\dot{g}(-t,\mathbf{r}|-t'',\mathbf{r}'')\right]$$

$$= -4\pi\left[g(-t',\mathbf{r}'|-t'',\mathbf{r}'') - g(t'',\mathbf{r}''|t',\mathbf{r}')\right]. \qquad (6b)$$

The value of the integrand of the first integral in (6b) is found as follows: multiply the surface boundary condition in (3b) by the Green's function on the boundary in (5b); multiply the surface boundary condition in (5b) by the Green's function on the boundary in (3b); subtract the latter from the former and it is found that the integrand of the first integral of (6b) vanishes at all points of the bounding surface. In the second integral of (6b) the time integration can be immediately performed to give

$$\frac{1}{c^2}\int_V d\mathbf{r}\left\{\left[g\left(-t^+,\mathbf{r}|-t'',\mathbf{r}''\right)\dot{g}\left(t^+,\mathbf{r}|t',\mathbf{r}'\right) - g\left(t^+,\mathbf{r}|t',\mathbf{r}'\right)\dot{g}\left(-t^+,\mathbf{r}|-t'',\mathbf{r}''\right)\right]\right.$$

$$\left.-\left[g(\infty,\mathbf{r}|-t'',\mathbf{r}'')\dot{g}(-\infty,\mathbf{r}|t',\mathbf{r}') - g(-\infty,\mathbf{r}|t',\mathbf{r}')\dot{g}(\infty,\mathbf{r}|-t'',\mathbf{r}'')\right]\right\}$$

$$= -\left[g(-t',\mathbf{r}'|-t'',\mathbf{r}'') - g(t'',\mathbf{r}''|t',\mathbf{r}')\right]. \qquad (6c)$$

The terms $g(-t^+,\mathbf{r}|-t'',\mathbf{r}'')$ and $\dot{g}(-t^+,\mathbf{r}|-t'',\mathbf{r}'')$ are zero by the time boundary conditions in (5c) and the terms $g(-\infty,\mathbf{r}|t',\mathbf{r}')$ and $\dot{g}(-\infty,\mathbf{r}|t',\mathbf{r}')$ are zero by the time boundary conditions in (3c). The result is that all terms in the left member of (6a) vanish and the right member shows that the Green's function reciprocity relation in (4) is generally valid.

<u>Form of the Green's Function</u>

The Green's function represents the field due to the point source and the scattering from the surfaces within the medium. We first consider just the field due to the point source, i.e. we determine the free-field solution of (3). For this case the Green's Function is dependent on range only and we can express the homogeneous form of (3) as

$$\frac{1}{r^2}\frac{\partial}{\partial r}\left(\frac{r^2 \partial g}{\partial r}\right)-\frac{1}{c^2}\frac{\partial^2 g}{\partial t^2}=0 \quad \rightarrow \quad \frac{\partial^2 (gr)}{\partial r^2}-\frac{1}{c^2}\frac{\partial^2 (gr)}{\partial t^2}=0 . \tag{7a,b}$$

$$r = |\mathbf{r}-\mathbf{r}'|$$

From (7b) we immediately find the solution

$$g = \frac{u(t-t'-r/c)+v(t-t'+r/c)}{r}, \tag{8}$$

where the first term represents an outgoing wave and the second represents an incoming wave. We choose the first term to represent waves propagating away from the point source. To determine the functional form of u, we integrate (3a) over a vanishingly small sphere of volume $V \rightarrow 0$. This gives

$$\int_V dv\left(\nabla^2 g - \frac{\ddot{g}}{c^2}\right) = -4\pi\delta(t-t')\int_V dv\,\delta(\mathbf{r}-\mathbf{r}') = -4\pi\delta(t-t') . \tag{9a}$$

The first term in the left member can be determined from Gauss' theorem in (4.3-A) to obtain

$$\int_V dv\nabla^2 g = \int_S d\mathbf{s}\cdot\nabla\frac{u(t-t')}{r} = -4\pi u(t-t') , \tag{9b}$$

where $d\mathbf{s}$ points outward from the sphere. The second term in the left member is found to vanish

$$-\frac{1}{c^2}\int_V dv\ddot{g} = -\frac{1}{c^2}\int_V dv\frac{\ddot{u}(t-t')}{r} = 0 . \tag{9c}$$

Collecting terms, we find the functional form

$$u(t-t') = \delta(t-t') . \tag{9d}$$

From the Green's Function representation in (8) we obtain

$$g(t,\mathbf{r}|t',\mathbf{r}') = \frac{\delta(t-t'-r/c)}{r} . \tag{10a}$$

$$r = |\mathbf{r}-\mathbf{r}'|$$

Equation (10a) shows that the field of the point source is an impulse that propagates outward from the source as a spherical shell. The amplitude of the impulse drops off as 1/r as it propagates.

For the case of the total field, which includes the impulse and the field scattered by the bounding surfaces, the Green's function has the form

$$g(t,\mathbf{r}|t',\mathbf{r}') = \frac{\delta(t-t'-r/c)}{r} + h(t,\mathbf{r}|t',\mathbf{r}'), \tag{10b}$$

where h represents the scattered field and contains no singularities. The determination of h for a given scattering problem is generally difficult. However, we will adopt an alternative method using only the free-field Green's function making this problem academic. Our pursuit of the more general Green's function is primarily for the reason of introducing standard Green's function theory.

7.2-A: Integral Solution of the Inhomogeneous Wave Equation

We are now ready to integrate the wave equation in (1) by means of Kirchhoff's method. In the process we want to determine the acoustic field at t,$\mathbf{r}$ in terms of integrals over the Green's function g(t,$\mathbf{r}$|t',$\mathbf{r}'$) weighted by the source density and the boundary conditions in which the integrations are with respect to the source and boundary coordinates. To achieve this we write the wave equation in (1a) in terms of primed coordinates t', $\mathbf{r}'$as shown in (11a). This changes nothing since the wave equation can be expressed in terms of any set of variables. Moreover, it remains true that, when the primed coordinates are within the domain of the active source, the wave equation is inhomogeneous and, when they are exterior to the domain of the active source, the wave equation is homogeneous. The wave equation surface boundary condition in (1b) remains the same but is expressed in terms of primed coordinates in (11b). However, the time boundary conditions in (1c) are made explicit in (11c) which asserts that the acoustic field and its derivative vanish in the distant past, i.e. the source q is activated in the recent past

$$\nabla'^2 p(t',\mathbf{r}') - \frac{1}{c^2}\ddot{p}(t',\mathbf{r}') = -\rho_0\dot{q}(t',\mathbf{r}'), \tag{11a}$$

$$ap(t',\mathbf{r}'_S) + b\nabla' p(t',\mathbf{r}'_S)\cdot\mathbf{e}_S = f(t',\mathbf{r}'_S), \tag{11b}$$

$$p(-\infty,\mathbf{r}') = 0, \qquad \dot{p}(-\infty,\mathbf{r}') = 0. \tag{11c}$$

We similarly rewrite the Green's function wave equation in (3) in terms of new coordinates

$$\nabla'^2 g(-t',\mathbf{r}'|-t,\mathbf{r}) - \frac{1}{c^2}\ddot{g}(-t',\mathbf{r}'|-t,\mathbf{r}) = -4\pi\delta(t'-t)\delta(\mathbf{r}'-\mathbf{r}), \tag{12a}$$

$$ag(-t',\mathbf{r}'_S|-t,\mathbf{r}) + b\nabla' g(-t',\mathbf{r}'_S|-t,\mathbf{r})\cdot\mathbf{e}_S = 0, \tag{12b}$$

$$g(-t',\mathbf{r}'|-t,\mathbf{r}) = 0, \quad \dot{g}(-t',\mathbf{r}'|-t,\mathbf{r}) = 0. \qquad -t' < -t \tag{12c}$$

In (11) and (12) the prime on the space derivatives denotes differentiation with respect to the primed coordinates. Combining (11) with (12) by following procedures analogous to those leading to (6a) to we obtain

$$\int_{-\infty}^{t^+} dt' \int_V d\mathbf{r}' \left[g(-t',\mathbf{r}'|-t,\mathbf{r}) \nabla'^2 p(t',\mathbf{r}') - p(t',\mathbf{r}') \nabla'^2 g(-t',\mathbf{r}'|-t,\mathbf{r}) \right]$$

$$-\frac{1}{c^2} \int_{-\infty}^{t^+} dt' \int_V d\mathbf{r}' \left[g(-t',\mathbf{r}'|-t,\mathbf{r}) \ddot{p}(t',\mathbf{r}') - p(t',\mathbf{r}') \ddot{g}(-t',\mathbf{r}'|-t,\mathbf{r}) \right]$$

$$= -\rho_0 \int_{-\infty}^{t^+} dt' \int_V d\mathbf{r}' g(-t',\mathbf{r}'|-t,\mathbf{r}) \dot{q}(t',\mathbf{r}') + 4\pi p(t,\mathbf{r}) , \quad (13a)$$

where t^+ is larger than the greater of t or t′ to ensure that the integration covers all delta function peaks. Similarly, procedures analogous to those leading to (6b) and (6c) give

$$p(t,\mathbf{r}) = \frac{\rho_0}{4\pi} \int_{-\infty}^{t^+} dt' \int_V dv' g(t,\mathbf{r}|t',\mathbf{r}') \dot{q}(t',\mathbf{r}')$$

$$+ \frac{1}{4\pi} \int_{-\infty}^{t^+} dt' \int_S d\mathbf{s}' \cdot \left[g(t,\mathbf{r}|t',\mathbf{r}') \nabla' p(t',\mathbf{r}') - p(t',\mathbf{r}') \nabla' g(t,\mathbf{r}|t',\mathbf{r}') \right] , \quad (13b)$$

where dv′ = $d\mathbf{r}'$ and $d\mathbf{s}'$ points outward from the volume, i.e. into the scattering body. The volume integral in (13b) represents the pressure field at t,**r** in terms of a sum over the Green's Function weighted by the source distribution. If the Green's function is taken to include the direct path and scattered fields, the volume integral must be the whole solution and the surface integral in (13b) must vanish. Substituting the boundary conditions in (11b) and (12b) into the surface integral we obtain the condition on the Green's function

$$\int_{-\infty}^{t^+} dt' \int_S ds' g(t,\mathbf{r}|t',\mathbf{r}') f(t',\mathbf{r}') = 0 . \quad (14)$$

If the boundary condition for the pressure field is homogeneous, i.e. f(t,**r**)=0, this condition is clearly satisfied. For the inhomogeneous boundary condition, i.e. f(t,**r**)≠0, (14) establishes an additional condition on the Green's function.

Alternatively, we find that an another approach to determining the surface integral in (13b) is more productive. Since all we require of the kernel in (13b) is that it be a Green's function that satisfies the wave equation in (12), and, since the free-field Green's function in (10a) does this, we choose the free-field Green's function for (13b). For this choice the surface integral no longer vanishes but we are left with a useful physical interpretation of (13b). The volume integral over acoustic sources is the direct path field and the surface integral over the scattering surfaces is the scattered field. The sum of the two terms provides a complete solution to the acoustic propagation problem for the case of homogeneous media and for distributed sources ensonifying scattering bodies. All that is required to make this integral formulation a solution is the specification of the source function and the acoustic pressure and its gradient on the surface of the scattering body for all past times up to the present. Moreover, this representation of the field applies to broad band signals in contrast to the narrow band Helmholtz integral formulation in Appendix 4-A.

Substituting the free-field Green's function in (10a) into (13b) and performing the time integration gives the Kirchoff formula

$$p(t,\mathbf{r}) = \frac{\rho_0}{4\pi}\int_V dv' \frac{\dot{q}(t-r/c,\mathbf{r}')}{r} + \frac{1}{4\pi}\int_S d\mathbf{s}' \cdot \left[\frac{2}{r}\left(\nabla' p(t-r/c,\mathbf{r}')\right) - \nabla'\left(\frac{p(t-r/c,\mathbf{r}')}{r}\right)\right]. \qquad (15)$$

The volume integral is the broad band equivalent of the direct path source integral in (4.14-A) and the surface integral is the broad band equivalent of the scattering integral in (4.14-A).

Representing the fields in (15) by their Fourier transforms

$$p(t',\mathbf{r}') = \frac{1}{2\pi}\int_{-\infty}^{\infty} d\omega P(\omega,\mathbf{r}')\exp(j\omega t'), \qquad (16a)$$

$$q(t',\mathbf{r}') = \frac{1}{2\pi}\int_{-\infty}^{\infty} d\omega Q(\omega,\mathbf{r}')\exp(j\omega t'), \qquad (16b)$$

and substituting these Fourier transforms into (15), we find that the amplitude of the Fourier components have the form

$$P(\omega,\mathbf{r}) = \frac{j\omega\rho_0}{4\pi}\int_V dv' Q(\omega,\mathbf{r}')\left(\frac{\exp(-jkr)}{r}\right) + \frac{1}{4\pi}\int_S d\mathbf{s}'\left[\left(\frac{\exp(-jkr)}{r}\right)\nabla' P(\omega,\mathbf{r}') - P(\omega,\mathbf{r}')\nabla'\left(\frac{\exp(-jkr)}{r}\right)\right], \qquad (17)$$

which is the Helmholtz integral formula as in (4.14-A) and which shows that the Helmholtz formula is a special case of the Kirchoff formula.

APPENDIX 7-B: SCATTERING BY RIGID AND SOFT TARGETS

The total acoustic field in the presence of sources and scattering surfaces was found in (7.17-A) in terms of a volume integral over sources and a surface integral over scatterers. We write the total field in that expression in terms of the incident and scattered fields

$$P(\omega,\mathbf{r}) = P_I(\omega,\mathbf{r}) + P_S(\omega,\mathbf{r}), \tag{1}$$

where the incident field is given by the volume integral in (7.17-A). Substituting (1) into the surface integral in (7.17-A), that integral reduces to an expression for the scattered field in terms of two surface integrals over the scattering surfaces. In the first integral the surface values are in terms of the incident field and in the second integral the surface values are in terms of the scattered field. If, in fact, no scatterer bodies were present, the scattered field would vanish which shows that the integral over the incident field values is always zero. Consequently, the acoustic field scattered by the target for an incident continuous wave field is obtained from

$$P_S(\omega,\mathbf{r}) = \frac{1}{4\pi}\int_S d\mathbf{s}'\left[\left(\frac{\exp(-jkr)}{r}\right)\nabla' P_S(\omega,\mathbf{r}) - P_S(\omega,\mathbf{r})\nabla'\left(\frac{\exp(-jkr)}{r}\right)\right], \tag{2}$$

$$r = |\mathbf{r} - \mathbf{r}'|$$

where $d\mathbf{s}'$ points inward into the scattering body. To evaluate this expression we need values for the pressure and its gradient on the scattering surface. Fortunately, for many scatterers of interest, the surface is approximately impenetrable to acoustic energy, either because the surface is rigid or soft, and the surface values for these targets are readily approximated. For such targets it is apparent that the scattered fields are determined solely by the size and shape of the target. The question of whether a target is either rigid or soft is relatively unimportant since we find that the fields differ only in their sign. However, it is clear that modern submarine hulls and mines are approximately rigid while bubbles and the sea surface are approximately soft.

7.1-B: Boundary Conditions and the Scattering Integral

For rigid targets, the normal gradient of the total pressure vanishes at the surface since the surface does not deform under pressure. Consequently, the normal gradient of the scattered field is the negative of the normal gradient of the incident field

$$d\mathbf{s}' \cdot \nabla P_S = -d\mathbf{s}' \cdot \nabla P_I\,. \tag{3a}$$

To determine the scattered pressure on the target surface, we distinguish between the ensonified surface and the shadowed surface and approximate

$$\text{Ensonified Surface:} \qquad P_S = +P_I, \tag{3b}$$

$$\text{Shadowed Surface:} \qquad P_S = -P_I\,. \tag{3c}$$

On the ensonified surface, the scattering is assumed to conform to the case of plane wave reflection by a plane rigid surface while on the shadowed surface the total pressure is assumed to vanish. These approximations are valid for short wavelengths because surface curvatures are small compared to a wavelength and the region of refraction into the shadow zone is small compared to the target size which causes the shadows to be nearly acoustically black. Additionally, for both ensonified and shadowed surfaces we assume that there is no multiple scattering which implies that all surfaces are convex.

Substituting the rigid target boundary conditions into (2) we obtain

$$P_S(\omega,\mathbf{r}) = \frac{1}{4\pi}\int_E ds'\nabla'\left[P_I(\omega,\mathbf{r}')\left(\frac{\exp(-jkr)}{r}\right)\right]$$
$$+\frac{1}{4\pi}\int_S ds'\left[\left(\frac{\exp(-jkr)}{r}\right)\nabla' P_I(\omega,\mathbf{r}') - P_I(\omega,\mathbf{r}')\nabla'\left(\frac{\exp(-jkr)}{r}\right)\right], \qquad (4)$$

where E and S denote integration over the ensonified and shadowed surfaces and where we have reversed the surface element vectors to point inward into the volume, i.e. outward from the scattering surfaces. The first integral gives the scattered field of interest and the second integral can be shown to contribute only to the shadow forming field.

To evaluate the second integral[1], we denote the value of the integrand of the shadowed integral as the vector **A** and, through the use of the Helmholtz equation, we find that its divergence vanishes $\nabla\cdot\mathbf{A}=0$. Consequently, **A** can be expressed as the curl of some other vector $\mathbf{A}=\nabla\times\mathbf{B}$. From Stokes' theorem, derived at the end of this section, the shadowed integral can be expressed as a line integral over **B** in which the line of integration is along the boundary between the ensonified and shadowed surfaces denoted by Γ.

$$\int_S \nabla\times\mathbf{B}\cdot d\mathbf{s} = \oint_\Gamma \mathbf{B}\cdot d\mathbf{r}\,. \qquad (5)$$

Fortunately, we don't need the vector **B** to compute the surface integral. Since we have shown that the integral over the shadowed surface of arbitrary shape can be replaced by the line integral around the shadow line, we can replace the target surface by a surface having the shape of a membrane stretched over the shadow line. The incident pressure and its gradient on this new surface is readily determined and, using the methods for sources in Chapter 4, we see that this new surface acts as an acoustic source that forms a beam in the direction of the incident field. In the short wavelength limit, the beam falls off rapidly in angle away from the direction of the incident field and is present only behind the target. Consequently, the shadow forming field, when combined with the incident field, gives the field in the shadow zone behind the target. Since forward scattering cannot be resolved from the direct

[1] Morse and Feshback, *Methods of Theoretical Physics,* **II**, McGraw-Hill Book Co. 1953, pp. 1551-4.

path signal, we exclude this region from our discussions and, consequently, the shadow forming field is unimportant to our discussions.

For soft targets, the total pressure vanishes on the surface since the surface deforms freely under pressure. Consequently, the scattered field is the negative of the incident field

$$P_S = -P_I \,. \tag{6a}$$

To determine the gradient of the scattered pressure on the soft target surface, we, again, distinguish between the ensonified surface and the shadowed surface and approximate

$$\text{Ensonified Surface:} \qquad \mathbf{ds}' \cdot \nabla' P_S = \mathbf{ds}' \cdot \nabla' P_I \,, \tag{6b}$$

$$\text{Shadowed Surface:} \qquad \mathbf{ds}' \cdot \nabla' P_S = -\mathbf{ds}' \cdot \nabla' P_I \,. \tag{6c}$$

On the ensonified surface, the scattering is assumed to conform to the case of plane wave reflection by a soft plane surface while on the shadowed surface the total suffice motion is assumed to vanish. The validity of these assumptions for short wavelength conditions is similar to the rigid target case. Substituting these boundary conditions into (2) gives an equation that is identical to (4) for rigid targets except that the leading sign on the ensonified integral is negative.

Collecting results we find the field scattered by an acoustically impenetrable target is given by

$$P_S(\omega, \mathbf{r}) = \pm \frac{1}{4\pi} \int_E \mathbf{ds}' \nabla' \left[P_I(\omega, \mathbf{r}') \left(\frac{\exp(-jkr)}{r} \right) \right], \tag{7}$$

$$r = |\mathbf{r} - \mathbf{r}'|$$

where the $\pm$ sign applies, respectively, to a rigid or soft target and the surface element vector points outward from the scattering body. This expression is valid in the high frequency limit and is useful for scattering angles that do not include the forward scattered direction.

7.2-B: Stokes' Theorem

We consider a closed surface S that is bounded by a curve Γ through which a vector field **f** flows as indicated in Figure 1. We also introduce the equations

$$x = x(u, v) \,, \tag{8a}$$

$$y = y(u, v) \,, \tag{8b}$$

$$z = z(u, v) \,, \tag{8c}$$

where u and v range over a set of values. If we fix u and let v vary, these equations

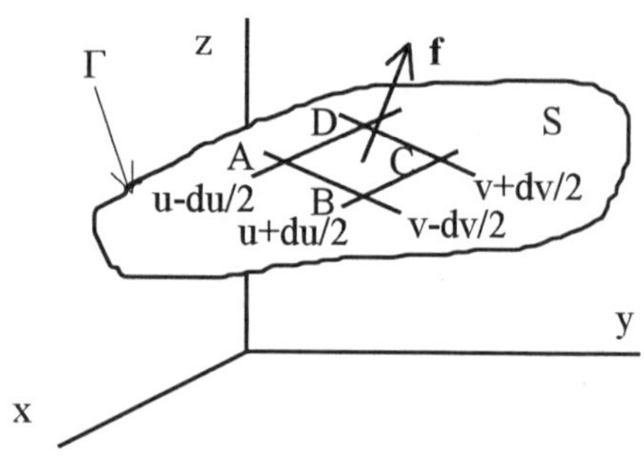

Figure 1 Vector Field Penetrating a Surface Bounded by Closed Curve

define a curve in space. One such curve exists for each u and letting u vary produces a surface. Consequently, we can define the surface by the vector

$$\mathbf{r}(u,v) = x(u,v)\mathbf{e}_x + y(u,v)\mathbf{e}_y + z(u,v)\mathbf{e}_z . \tag{9a}$$

At each point on the surface, we can specify unit vectors lying in the surface by

$$\mathbf{e}_u = \frac{\partial \mathbf{r}}{\partial u}, \qquad \mathbf{e}_v = \frac{\partial \mathbf{r}}{\partial v}. \tag{9b,c}$$

We specify that these vectors are orthogonal and that with the unit vector normal to the surface $\mathbf{e}_n$ these vectors define an orthogonal coordinate system.

Consider the line integral around the surface element ABCD in Figure 1

$$\oint_{ABCD} \mathbf{f}\cdot d\mathbf{r} = \mathbf{f}(u, v-dv)\cdot\mathbf{e}_u du + \mathbf{f}(u+du, v)\mathbf{e}_v dv$$
$$-\mathbf{f}(u, v+dv)\cdot\mathbf{e}_u du - \mathbf{f}(u-du, v)\cdot\mathbf{e}_v dv . \tag{10a}$$

The first term in the right member can be expanded to yield

$$\mathbf{f}(u, v-dv)\cdot\mathbf{e}_u du = \left(\mathbf{f}(u,v) - \frac{d\mathbf{r}_v}{2}\cdot\nabla\mathbf{f}(u,v)\right)\cdot\mathbf{e}_u du$$
$$= \left(\mathbf{f}(u,v) - \frac{\mathbf{e}_v dv}{2}\cdot\nabla\mathbf{f}\right)\cdot\mathbf{e}_u du = \left(\mathbf{f}(u,v) - \frac{\partial\mathbf{f}}{\partial v}\frac{dv}{2}\right)\cdot\mathbf{e}_u du . \tag{10b}$$

Similar expansions hold for the other terms of (10) and combining all such terms gives

$$\oint_{ABCD} \mathbf{f}\cdot d\mathbf{r} = \left[\left(\frac{\partial\mathbf{f}}{\partial u}\right)\cdot\mathbf{e}_v - \left(\frac{\partial\mathbf{f}}{\partial v}\right)\cdot\mathbf{e}_u\right] du dv . \tag{10c}$$

Consider next the quantity

$$(\nabla\times\mathbf{f})\cdot(\mathbf{e}_u\times\mathbf{e}_v) = \left(\mathbf{e}_u\times\frac{\partial\mathbf{f}}{\partial u} + \mathbf{e}_v\times\frac{\partial\mathbf{f}}{\partial v} + \mathbf{e}_n\times\frac{\partial\mathbf{f}}{\partial n}\right)\cdot(\mathbf{e}_u\times\mathbf{e}_v)$$
$$= \left[\left(\mathbf{e}_u\times\frac{\partial\mathbf{f}}{\partial u}\right)\times\mathbf{e}_u\right]\cdot\mathbf{e}_v - \left[\left(\mathbf{e}_v\times\frac{\partial\mathbf{f}}{\partial v}\right)\times\mathbf{e}_v\right]\cdot\mathbf{e}_u$$
$$= \frac{\partial\mathbf{f}}{\partial u}\cdot\mathbf{e}_v - \frac{\partial\mathbf{f}}{\partial v}\cdot\mathbf{e}_u , \tag{10d}$$

where $\nabla\times\mathbf{f}$ is the curl of $\mathbf{f}$ and where we have used the identities $\mathbf{a}\cdot(\mathbf{b}\times\mathbf{c})=(\mathbf{a}\times\mathbf{b})\cdot\mathbf{c}$ and $\mathbf{a}\times(\mathbf{b}\times\mathbf{c})=(\mathbf{a}\cdot\mathbf{c})\mathbf{b}-(\mathbf{a}\cdot\mathbf{b})\mathbf{c}$. Comparing (10c) and (10d) gives the relationship

$$\oint_{ABCD} \mathbf{f} \cdot d\mathbf{r} = (\nabla \times \mathbf{f}) \cdot (\mathbf{e}_u \times \mathbf{e}_v) dudv = (\nabla \times \mathbf{f}) \cdot d\mathbf{s}, \qquad (10e)$$

where $d\mathbf{s}$ is the differential surface element that is normal to the surface.

For a physical interpretation of the curl of a vector, we examine a vector field $\mathbf{f}$ that penetrates the area element $d\mathbf{s}$ and consider just the component of $\mathbf{f}$ that lies on the surface element, i.e. for the component perpendicular to the surface element both members of (10e) vanish. The left member is the net circulation around the surface element and the curl is seen to be the net circulation per unit area. If $\mathbf{f}$ were a displacement, the net circulation would be a net rotation of the area element that would be equivalent to a rigid body rotation. Consequently, for irrotational motion a necessary and sufficient condition is $\nabla \times \mathbf{f} = 0$.

Summing (10e) over the entire enclosed surface in Figure 1, we see the line integrals cancel on adjacent elements and only the line integrals along the bounding curve contribute to the sum while the quantities in the right member of (10e) accumulate giving

$$\oint_{\Gamma} \mathbf{f} \cdot d\mathbf{r} = \int_{S} \nabla \times \mathbf{f} \cdot d\mathbf{s}. \qquad (11)$$

This is Stoke's theorem and is a companion theorem to Gauss' theorem.

APPENDIX 7-C: SCATTERING BY RIGID SPHERES AND CYLINDERS

In Chapter 7 the acoustic fields scattered by rigid and soft spheres and cylinders were determined for the high frequency case. Because of the simplicity of these solutions, they were immediately pressed into the service of providing estimates of target strengths for all frequencies. In the following, we amend this expediency by determining the exact solutions for scattering by rigid spheres and cylinders and show that, while those high frequency solutions do describe the gross features of scattering at all frequencies, they fail to capture the details in the low frequency region as might be expected. Exact solutions for the soft sphere and cylinder can be similarly developed.

7.1-C: Scattering by Rigid Spheres

We assume that a sinusoidal field is incident on a sphere which has the form p(t,**r**) = P(**r**)exp(jωt). For a plane wave propagating from the negative z direction and incident on the sphere from the left as indicated in Figure 7.4a the field can be written[1]

$$P(\mathbf{r}) = \exp(jkr\cos\theta) = \sum_{n=0}^{\infty} j^n(2n+1)P_n(\cos\theta)j_n(kr)\,. \tag{1}$$

Since the incident and scattered fields are dependent on the latitudinal angle but not on the longitudinal angle, the wave equation of interest is

$$\frac{1}{r^2}\frac{\partial}{\partial r}\left(r^2\frac{\partial P(\mathbf{r})}{\partial r}\right) + \frac{1}{r^2\sin\theta}\frac{\partial}{\partial\theta}\left(\sin\theta\frac{\partial P(\mathbf{r})}{\partial\theta}\right) + k^2P(\mathbf{r}) = 0\,, \tag{2}$$

where k = ω/c. By the method of separation of variables we express the field as the product P(**r**) =Q(θ)R(r). Substituting into (2) yields the two equations

$$\frac{1}{Q\sin\theta}\frac{d}{d\theta}\left(\sin\theta\frac{dQ}{d\theta}\right) = \frac{1}{R}\frac{d}{dr}\left(r^2\frac{dR}{dr}\right) + k^2r = C\,, \tag{3a}$$

where both are equal to the same constant C.

In the first equation of (3a) we let z = cosθ and obtain the equation

$$\left(1-z^2\right)Q'' - 2zQ' + CQ = 0\,. \tag{3b}$$

Bounded solutions of this equation are obtained if the constant has the values C = n(n + 1) n = 0,1,2,··· and the solutions are the Legendre functions of first and second

---

[1] Abramowitz and Stegun, (10.1.47).

kind. For solutions with no singularities we choose the Legendre functions of the first kind of degree n denoted $Q = P_n(z)$.[2]

In the second equation of (3a) we let z = kr and obtain the equation

$$z^2R'' + 2zR' + \left(z^2 - n(n+1)\right)R = 0. \tag{3c}$$

The solutions of this equation are spherical Bessel functions of degree n[3]. The spherical Bessel functions of the first and second kind are denoted, respectively, $j_n(z)$ and $y_n(z)$. The spherical Bessel functions of the third kind, also called spherical Hankel functions, are the complex pairs $h_n^{(1,2)}(z) = j_n(z) \pm jy_n(z)$. An examination of the asymptotic forms of these functions shows that, for our time convention, $h_n^{(1)}(z)$ represents incoming waves and $h_n^{(2)}(z)$ represents outgoing waves.

Assembling our results the scattered field can be represented by the expression

$$P_s(\mathbf{r}) = \sum_{n=0}^{\infty} A_n P_n(\cos\theta) h_n^{(2)}(kr), \tag{4}$$

where the coefficients A_n are determined from the boundary conditions.

Since the sphere is rigid, the radial velocity of the acoustic field must vanish at each point on the surface

$$v_I(t,a) + v_S(t,a) = 0, \tag{5a}$$

and from Newton's second law in (2.4c-A) we obtain the boundary condition

$$\lim_{r \to a}\left[-\frac{\partial p(t,r)}{\partial r}\right] = j\omega\rho_0 v(t,a). \tag{5b}$$

Substituting the field expressions from (1) and (4) into (5a) we obtain

$$\sum_{n=0}^{\infty} j^n(2n+1)P_n(\cos\theta)\left.\frac{dj_n(z)}{dz}\right|_{z=ka} + \sum_{n=0}^{\infty} A_n P_n(\cos\theta)\left.\frac{dh_n^{(2)}(z)}{dz}\right|_{z=ka} = 0, \tag{5c}$$

$$A_n = -\left.\frac{j^n(2n+1)dj_n(z)/dz}{dh_n^{(2)}(z)/dz}\right|_{z=ka}. \tag{5d}$$

The derivatives are[4]

$$\frac{dh_n^{(2)}(z)}{dz} = nh_n^{(2)}(z)/z - h_{n+1}^{(2)}(z) = -jD_n\exp(j\delta_n), \tag{6a}$$

which yield

---

[2] Abramowitz and Stegun, (8.1.1).
[3] Abramowitz and Stegun, (10.1.1).
[4] Abramowitz and Stegun, (10.1.22).

$$\frac{dj_n(z)}{dz} = nj_n(z)/z - j_{n+1}(z) = D_n \sin(\delta_n), \tag{6b}$$

$$\frac{dy_n(z)}{dz} = ny_n(z)/z - y_{n+1}(z) = D_n \cos(\delta_n). \tag{6c}$$

Substituting into (5d), the constants are given by

$$A_n = -j^{n+1}(2n+1)\sin(\delta_n)\exp(-j\delta_n), \tag{7a}$$

$$\delta_n = \tan^{-1}\left[\frac{nj_n(ka)/ka - j_{n+1}(ka)}{ny_n(ka)/ka - y_{n+1}(ka)}\right]. \tag{7b}$$

These calculations determine the scattered field around the sphere.

In the far field of the sphere kr >> 0, the scattered field expression can be written[5]

$$P_s(\mathbf{r}) = \frac{1}{kr}\sum_{n=0}^{\infty}(-1)^n(2n+1)\sin(\delta_n)\exp(-j\delta_n)P_n(\cos\theta)\exp(-jkr), \tag{8a}$$

and the target strength is immediately found to be given by

$$ts = \frac{a^2}{4}\left(\frac{4}{(ka)^2}\right)\sum_{n,m=0}^{\infty}(-1)^{n+m}(2n+1)(2m+1)\sin(\delta_n)\sin(\delta_m)$$
$$\times\cos(\delta_n - \delta_m)P_n(\cos\theta)P_m(\cos\theta). \tag{8b}$$

Plots of target strength as functions of frequency and aspect angle are presented in Figures 1 and 2.

The target strength versus ka value plots in Figure 1 show that, as frequency increases, target strength approaches the limit of $10\text{Log}(a^2/4)$ in the back scattering directions. A comparison between the exact target strengths plotted in Figure 1 and the approximate target strengths plotted in Figure 7.4b, as computed from the high frequency approximation in (7.10), show that their overall behavior is the same. Both generally increase with frequency to the same limit and the frequency of their oscillatory behavior decreases with aspect angle. They differ in the fact that the exact values oscillate more rapidly with frequency.

The target strength versus aspect angle plots in Figure 2 show that, in the back scattering directions, target strength approaches a constant with aspect angle as the frequency increases and that constant is $10\text{Log}(a^2/4)$ analogous to the plots in Figure 1. A comparison between the exact aspect dependent target strengths plotted in Figure 2 and the approximate aspect dependent target strengths plotted in Figure 7.4c, as computed from the high frequency approximation in (7.10), show

[5] Abramowitz and Stegun, (10.1.1) and (9.2.4).

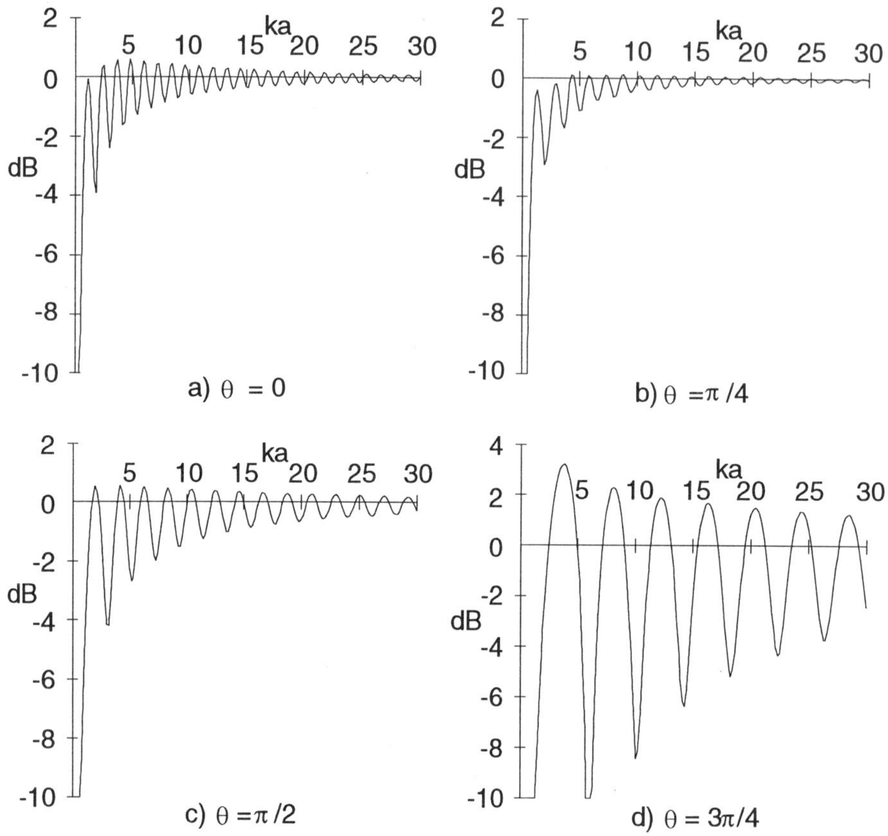

Figure 1 Normalized Target Strength of a Rigid Sphere of Radius a versus ka and Parameterized on Receiver Displacement Angle: Values = TS – 10Log(a^2/4).

that their overall behavior is very similar. Both are generally constant with aspect angle in the back scattering directions and have the approximate value 10Log(a^2/4). However, they differ markedly in the forward scattering direction. This is because the exact target strength plots represent the forward scattered field, which is required to form the shadow zone behind the target by canceling the incident field, while the high frequency approximation to the target strengths, in effect, assume the presence of the shadow zone and do not represent a forward scattered field. This discrepancy between these plots is a small loss because the forward scattered portion of the field is not of great interest in ASW applications.

7.2-C: Scattering by Rigid Cylinders

The solution for the scattering of sound by a rigid cylinder is obtained in a manner similarly to that for the scattering of sound by a rigid sphere. We assume that a sinusoidal field is incident on the sphere having the form p(t,**r**) = P(**r**)exp(jωt). Analogous to the case of the sphere, for a plane wave propagating in the negative z

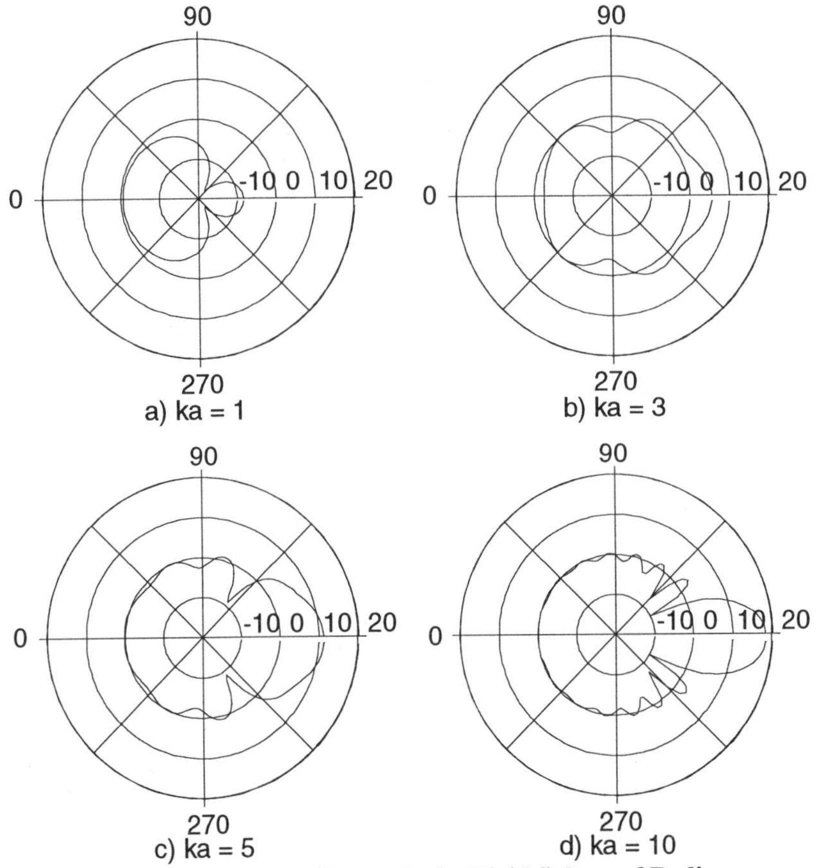

Figure 2 Normalized Target Strength of a Rigid Sphere of Radius a versus Receiver Displacement Angle for four values of ka: Values are equal to TS – 10Log(a^2/4) and the scale increment is 10 dB.

direction and incident on the cylinder from the left, as indicated in Figure 7.5, the field can be written[6]

$$P(\mathbf{r}) = \exp(jkr\cos\theta) = \sum_{n=1}^{\infty} \alpha_n j^n \cos(n\theta) J_n(kr), \tag{9}$$

where $\alpha_n = 1$ for n = 0 and $\alpha_n = 2$ for n = 1,2,$\cdots$ and $j = \sqrt{-1}$. The form of the scattered field for the cylinder is found as for the sphere but the wave equation separates into differential equations for cylinder functions and Bessel functions. The cylinder functions are the cos(nθ) and sin(nθ) functions and we choose the former since our problem is symmetrical in θ. The Bessel functions of the first and second kind are denoted, respectively, $J_n(kr)$ and $Y_n(kr)$. The cylindrical Bessel functions of the third kind, also called Hankel functions, are the complex pairs $H_n^{(1,2)}(z)$ =

[6] Abramowitz and Stegun, (9.1.44) and (9.1.45).

$J_n(z) \pm jY_n(z)$. An examination of the asymptotic forms of these functions shows that, for our time convention, $H_n^{(1)}(z)$ represents incoming waves and $H_n^{(2)}(z)$ represents outgoing waves. Assembling our results the scattered field can be represented by

$$P_s(\mathbf{r}) = \sum_{n=0}^{\infty} A_n \cos(n\theta) H_n^{(2)}(kr) . \tag{10a}$$

where the coefficients A_n are determined from the boundary conditions.

Since the cylinder is rigid, the radial velocity of the acoustic field must vanish at each point on the surface. Performing calculations as for the case of the sphere, the coefficients for the cylinder are found to be given by

$$A_n = -\left.\frac{\alpha_n j^n dJ_n(z)/dz}{dH_n^{(2)}(z)/dz}\right|_{z=ka} . \tag{10b}$$

The derivatives in this expression can be written[7]

$$\frac{dH_n^{(2)}(z)}{dz} = nH_n^{(2)}(z)/z - H_{n+1}^{(2)}(z) = -jD_n \exp(j\delta_n), \tag{11a}$$

which yield

$$\frac{dJ_n(z)}{dz} = nJ_n(z)/z - J_{n+1}(z) = D_n \sin(\delta_n), \tag{11b}$$

$$\frac{dY_n(z)}{dz} = nY_n(z)/z - Y_{n+1}(z) = D_n \cos(\delta_n). \tag{11c}$$

Substituting into (10b), the constants are given by

$$A_n = -\alpha_n j^{n+1} \sin(\delta_n) \exp(-j\delta_n), \tag{12a}$$

$$\delta_n = \tan^{-1}\left[\frac{nJ_n(ka)/ka - J_{n+1}(ka)}{nY_n(ka)/ka - Y_{n+1}(ka)}\right]. \tag{12b}$$

These calculations determine the scattered field around the cylinder.

In the far field of the cylinder kr >> 0, the scattered field expression can be written[8]

$$P_s(\mathbf{r}) = \sqrt{\frac{2}{\pi kr}} \sum_{n=0}^{\infty} (-1)^n \alpha_n \sin(\delta_n) \exp(-j\delta_n) \cos(n\theta) \exp(-jkr), \tag{13a}$$

and the target strength is immediately found to be given by

---

[7] Abramowitz and Stegun, (9.1.27).

[8] Abramowitz and Stegun, (10.1.1) and (9.2.4).

$$ts = \frac{a\cos(\theta/2)}{2}\left(\frac{4}{\pi ka\cos(\theta/2)}\right)\sum_{n,m=0}^{\infty}(-1)^{n+m}\alpha_n\alpha_m \sin(\delta_n)\sin(\delta_m)$$
$$\times\cos(\delta_n-\delta_m)\cos(n\theta)\cos(m\theta)\,, \qquad (13b)$$

where the leading factor was found to be the target strength of the infinite cylinder in the high frequency limit in (7.15b). Plots of target strength as functions of frequency and aspect angle are presented in Figures 3 and 4.

The target strength versus ka value plots in Figure 3 show that, as frequency increases, target strength approaches the limit of 10Log(acos(θ/2)/2). A comparison of the exact target strengths versus frequency plotted in Figure 3 and the constant target strength versus frequency given by the high frequency approximation in (7.15b) show that they have the same limit. But, whereas the high frequency approximation gives only the limiting value, the exact target strength values reveal the oscillatory behavior of target strength at low frequencies.

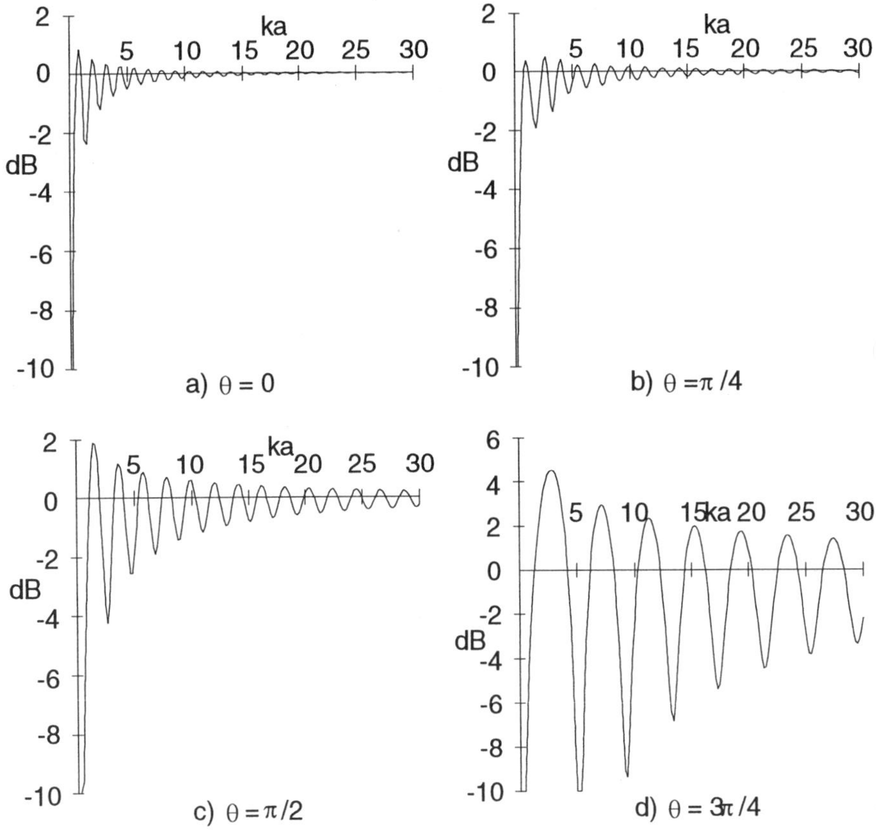

Figure 3 Normalized Target Strength of a Rigid Cylinder of Radius a versus ka Parameterized on Receiver Displacement Angle: Values=TS−10Log(acos(θ/2)/2).

The target strength versus aspect angle plots in Figure 4 show that, in the back scattering directions, target strength approaches a constant with aspect angle as the frequency increases and that constant is 10Log(acos(Θ/2)/2) which is consistent with the plots in Figure 3. A comparison between the exact aspect dependent target strengths plotted in Figure 4 and the approximate aspect dependent target strengths, as computed from the high frequency approximation in (7.15b), show that their overall behavior is similar. Both follow a cosine-like law in the back scattering directions and have the approximate value 10Log(acos(Θ/2)/2). However, as for the case of the sphere, they differ markedly in the forward scattering direction. This is, again, because the exact target strength plots represent the forward scattered field, which is required to form the shadow zone behind the target by canceling the incident field, while the high frequency approximation to the target strengths effectively assume the presence of the shadow zone and do not represent a forward scattered field.

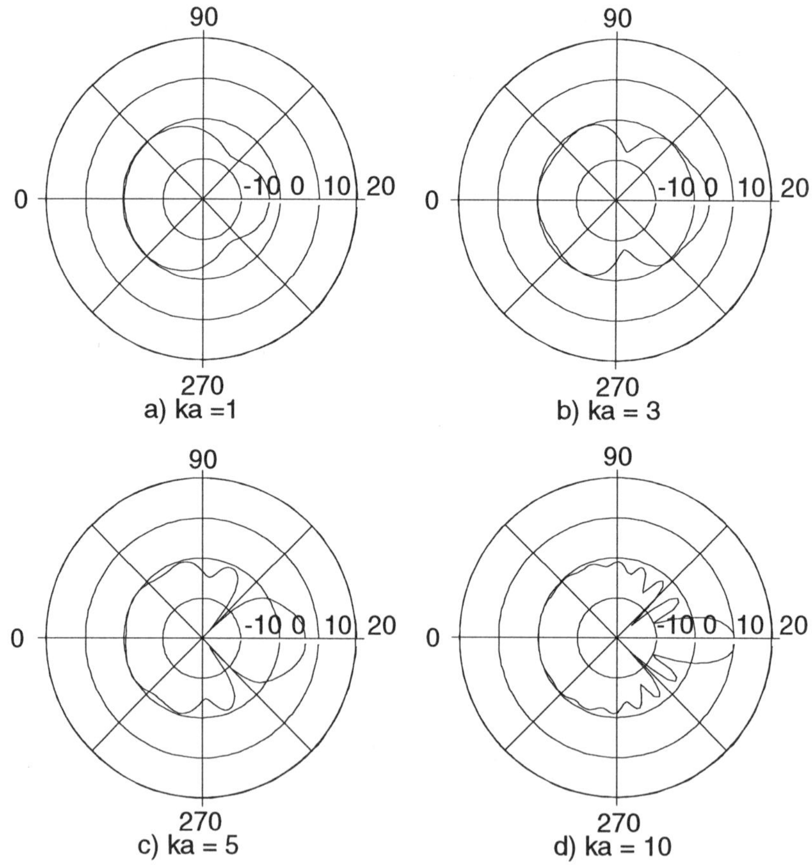

Figure 4 Normalized Target Strength of a Rigid Cylinder of Radius a vs. Receiver Displacement Angle for four values of ka: Plotted Values are equal to TS – 10Log(a/2) dB and the scale increment is 10 dB.

APPENDIX 7-D ELLIPSOID OF REVOLUTION RADII OF CURVATURE

In the following we examine the radii of curvature of an ellipsoid of revolution for the purpose of determining the target strength of an acoustically impenetrable target having ellipsoidal shape. The ellipsoid is shown in plan view in Figure 1. An acoustic plane wave is assumed to be incident from the left and the ellipsoid is assumed to be rotated through the arbitrary angle θ. The required radii of curvature are for horizontal and vertical cuts through the body at the point (x_0,z_0) which is the point of first contact of the acoustic wave.

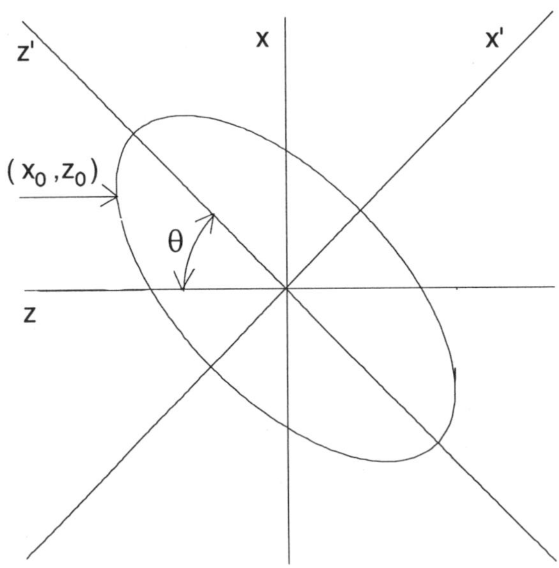

Figure 1 Ellipsoid of Revolution Geometry Showing Point of Tangential Contact for a Plane Wave Incident from the Left

The ellipsoid of revolution can be expressed in the primed coordinate system as

$$\frac{x'^2}{a^2}+\frac{y'^2}{a^2}+\frac{z'^2}{b^2}=1, \tag{1}$$

where a is the semiminor axis and b is the semimajor axis and the y' coordinate is directed downward into the page. Transforming to the unprimed coordinate system

$$\begin{aligned} z' &= z\cos\theta + x\sin\theta, \\ x' &= z\sin\theta - x\cos\theta, \\ y' &= y, \end{aligned} \tag{2}$$

the ellipsoid can be expressed in the $y = 0$ plane as

$$x^2A - 2xzB + z^2C + \frac{y^2}{a^2} = 1, \tag{3a}$$

$$A = \left(\frac{\cos^2\theta}{a^2} + \frac{\sin^2\theta}{b^2}\right), \tag{3b}$$

$$B = \cos\theta\,\sin\theta\left(\frac{1}{a^2} - \frac{1}{b^2}\right), \tag{3c}$$

$$C = \left(\frac{\sin^2\theta}{a^2} + \frac{\cos^2\theta}{b^2}\right). \tag{3d}$$

To find the point of tangential contact with the plane wave, we set y = 0 in (3a), solve for z and differentiate with respect to x. Setting the derivative to zero gives

$$x_0 = \frac{B}{\sqrt{A\left(AC - B^2\right)}} = \frac{\cos\theta\,\sin\theta\left(b^2 - a^2\right)}{\sqrt{b^2\cos^2\theta + a^2\sin^2\theta}}. \tag{4a}$$

Substituting into (3a), gives

$$z_0 = \sqrt{\frac{A}{AC - B^2}} = \sqrt{b^2\cos^2\theta + a^2\sin^2\theta}\,. \tag{4b}$$

These manipulations determine the point of first contact of the plane wave on the ellipsoid of revolution. We next turn to determining the radii of curvature through this point.

The radii of curvature in the horizontal and vertical planes are found from the relations

$$R_h = \left|\frac{d^2z}{dx^2}\right|^{-1}, \qquad R_v = \left|\frac{d^2z}{dy^2}\right|^{-1}, \tag{5a,b}$$

evaluated at the point of first contact. Solving (3a) for z, performing the differentiations and evaluating at the contact point gives

$$R_h = \sqrt{\frac{AC - B^2}{A^3}} = \frac{a^2b^2}{\left(b^2\cos^2\theta + a^2\sin^2\theta\right)^{3/2}}, \tag{6a}$$

$$R_v = a^2\sqrt{\frac{AC - B^2}{A}} = \frac{a^2}{\sqrt{b^2\cos^2\theta + a^2\sin^2\theta}}. \tag{6b}$$

These expressions are used in Chapter 7 to determine submarine target strengths.

CHAPTER 8
NOISE AND REVERBERATION IN THE OCEAN

Noise and reverberation have the effect of reducing detection performance when they find their way into sonar receiving systems which usually happens. They are among the most significant detriments to sonar operations to be found in the ocean environment. Only acoustic propagation conditions that disrupt or extinguish received signals can be more serious for sonar operations. Noise and reverberation can appear to the detection system as false targets or they can mask target returns during detection processing. In the former case they have the effect of raising the system false alarm rate while in the latter case they have the effect of lowering the system detection rate. Reverberation returns are especially pernicious to detection processing because these returns, being caused by the transmitted signal, are highly correlated with that signal and are prone to generating false alarms. Otherwise, noise and reverberation are similar in the sense that both are random processes but differ in the sense of having differing statistical parameters.

Noise in the oceans comes from a variety of sources and propagates to the receiver within frequency bands that are appropriate to each source. In general terms, ocean noise consists of ambient noise, ownship noise and other identifiable noise sources. The sources of ambient noise and their approximate frequency bands include

1. Quasi-Hydrostatic Forces (0 to 1 Hz),
2. Ocean Turbulence (1 to 20 Hz),
3. Shipping Traffic (20 to 500 Hz),
4. Sea Surface (500 to 50,000 Hz),
5. Thermal Noise (50,000 to 100,000 Hz).

Quasi-hydrostatic forces are the slowly varying forces caused by tides, surface waves and seismic activity. The acoustic pressures caused by these forces are very large but fortunately they are out of the frequency band of importance to virtually all ASW systems. Similarly, the acoustic noise attributed to ocean turbulence is also out of the frequency band of interest for most ASW systems. In contrast, shipping noise and sea surface noise are the predominate sources of noise to most ASW sonar systems and thermal noise is a source of interference to very high frequency, i.e. short range, ASW sonar systems.

The source to receiver propagation paths for shipping noise, as for the other low frequency noises, are predominately in the horizontal direction, while the propagation paths for sea surface noise are predominately from the overhead direction. The propagation paths for thermal noise come from all directions. In the deep ocean, spectrum levels for shipping noise range from about 60 to 90 dB re μPa per Hz, depending on the proximity of distant shipping lanes, and spectrum levels for surface noise range from about 40 to 75 dB re μPa per Hz, depending on sea state. While these levels are fairly typical for all deep water oceans of the world, in shallow water and coastal regions, noise levels can vary markedly from these values depending on the absence or proximity of harbors, shipping lanes, storm activity, etc. Spectrum levels for thermal noise range from about 20 to 30 dB re μPa per Hz and this is relatively consistent throughout the worlds oceans. All of these spectrum levels are, of course, frequency dependent.

Ownship noise can also be a significant source of noise that can dominate ambient noise and discussions of it are found in Section 7.4. Other identifiable noises include ships, biologics, rain squalls and storms that are nearby, i.e. all acoustic radiations that would normally be included in the ambient noise level if they were not identifiable which usually means they are near to the ASW system.

Reverberation is acoustic scattering that occurs within the ocean as a result of active transmissions. These returns consist of acoustic energy that has been scattered from objects that are distributed within the ocean water-mass and acoustic energy that has been scattered from the surface and bottom of the ocean water-mass. It is frequently seen on sonar displays as a decaying signal following an active transmission that has a noise-like quality. For a receiver that is collocated with the source, reverberation consists of back scattered returns but for bistatic receivers reverberation can consist of laterally scattered and forward scattered acoustic energy depending on the location of the receiver vis-à-vis the target and source. Perhaps our most personal experience with reverberation is the rumbling that follows a thunder clap in atmospheric acoustics. In that case we act as a bistatic receiver for the lightning bolt source.

8.1 Statistics of Noise and Reverberation

Both noise and reverberation are random processes and, as such, all we know about them is contained in their probability distributions. Unfortunately, the particular distributions to be attributed to these processes remains a controversial issue but, for short time intervals, many investigators agree that both processes are best described by stationary Gaussian distributions. For longer time intervals, investigations have shown that both noise and reverberation are no longer suitably represented as stationary Gaussian processes and alternative distributions must be developed. Two approaches for developing such distributions have been pursued. On the one hand these processes have been assumed to remain stationary for long time intervals and non-Gaussian distributions have been postulated to describe them, e.g. the Weibull distribution[1]. On the other hand these processes have been assumed to be nonstationary over long time intervals and time varying distributions have been postulated to describe them, e.g. a Gaussian distribution with time varying parameters. This approach is preferred in the following and another application of this approach is presented in Appendix 10-B.

Both noise and reverberation are best represented as zero mean nonstationary Gaussian processes and, consequently, the statistics of their complex envelopes are represented by the well-known nth order complex Gaussian probability density function, from (8.16-A),

$$p(\mathbf{z}) = \frac{1}{\pi^{n}|\mathbf{C}|}\exp\left(-\mathbf{z}^{*T}\mathbf{C}^{-1}\mathbf{z}\right), \tag{1a}$$

[1] Tests by the author and others have indicated that the Weibull distribution does not provide a satisfactory description of either noise or reverberation.

$$\mathbf{z} = \begin{bmatrix} z_1 & \cdots & z_n \end{bmatrix}^T, \qquad \mathbf{C} = E\left(\mathbf{z}\mathbf{z}^{*T}\right), \tag{1b,c}$$

where $\mathbf{z}$ is a vector of complex envelope samples, the superscript asterisk denotes complex conjugate, the superscript T denotes transpose, $\mathbf{C}$ is the covariance matrix, and $|\mathbf{C}|$ means determinant of $\mathbf{C}$. The envelope samples can be viewed as having been collected at different times and/or at different points within the field, i.e. at different sensors. For our present purposes it is sufficient to note that the noise and reverberation fields are completely specified by the elements of the covariance matrix $\mathbf{C}$ which we are assuming is time dependent. Of course, since noise and reverberation are readily distinguished by sonar operators, it is expected that their covariance matrices differ markedly and this, in fact, is the case.

Additionally, noise and reverberation are frequently assumed to be temporally and spatially first order Markoff processes. For these processes, samples collected at particular times and from particular sensors shed light only on the next samples to be collected, or on samples collected from adjacent sensors. For such processes a sufficient description of the field is given by the second order Gaussian joint probability density function. With the second order joint probability density function known, the conditional probability function can be determined and this function represents all that we know about the next envelope sample to be collected given the current envelope sample. In this case the probability density function is completely specified by the variances $\mathbf{E}(|z|^2)$, $\mathbf{E}(|z'|^2)$ and the covariance $\mathbf{E}(zz'^*)$ where the primes denotes a different sample time or sensor and $\mathbf{E}$ is expectation. In this case, the covariance matrix $\mathbf{C}$ is reduced to a 2×2 matrix with the variances as diagonal elements and the covariances as the off-diagonal elements. The variances are just the intensity or power of the process and the covariance indicates how well the signals at two points are correlated. Also, the variances are given by the covariances in which the temporal and spatial differences are zero.

The form of the variances in terms of sample times and sensors are given by

$$\sigma^2 = \mathbf{E}\left[|z(t,\mathbf{r})|^2\right], \qquad \sigma'^2 = \mathbf{E}\left[|z(t',\mathbf{r}')|^2\right], \tag{2a,b}$$

and the covariance by

$$C(t,\mathbf{r},t',\mathbf{r}') = \mathbf{E}\left[z(t,\mathbf{r})z^*(t',\mathbf{r}')\right], \tag{3a}$$

or, equivalently,

$$C(t,\mathbf{r},\tau,\mathbf{u}) = \mathbf{E}\left[z(t,\mathbf{r})z^*(t-\tau,\mathbf{r}-\mathbf{u})\right], \tag{3b}$$

where $\tau = t - t'$ and $\mathbf{u} = \mathbf{r} - \mathbf{r}'$. In the equivalent form of the covariance function the parameters τ and $\mathbf{u}$ are, respectively, the temporal and spatial displacements from the point t, $\mathbf{r}$.

The variances and covariance are seen to be generally dependent upon time and space but several special cases can be delineated. If the variance and covariance functions are independent of when they were determined, i.e. both are independent of t, the process is stationary and the covariance can be written as

$$C(\mathbf{r},\tau,\mathbf{u}) = \mathbf{E}\left[z(t,\mathbf{r})z^*(t-\tau,\mathbf{r}-\mathbf{u})\right]. \qquad (4a)$$

Similarly, if the variance and covariance functions are independent of where they were determined, i.e. both are independent of **r**, the process is homogeneous and the covariance elements can be written as

$$C(t,\tau,\mathbf{u}) = \mathbf{E}\left[z(t,\mathbf{r})z^*(t-\tau,\mathbf{r}-\mathbf{u})\right]. \qquad (4b)$$

Accordingly, for a stationary and homogeneous process, the variance is constant over all time and space and the covariance is represented by

$$C(\tau,\mathbf{u}) = \mathbf{E}\left[z(t,\mathbf{r})z^*(t-\tau,\mathbf{r}-\mathbf{u})\right]. \qquad (4c)$$

The autocovariance of the process is the covariance of a single sensor output. For notational simplicity the autocovariance function is written

$$C(\tau) = C(\tau,\mathbf{0}) = \mathbf{E}\left[z(t,\mathbf{r})z^*(t-\tau,\mathbf{r})\right]. \qquad (4d)$$

Similarly, the cross covariance, which is the covariance of the output of two sensors, at the same time which, for notational simplicity, is written

$$C(\mathbf{u}) = C(0,\mathbf{u}) = \mathbf{E}\left[z(t,\mathbf{r})z^*(t,\mathbf{r}-\mathbf{u})\right]. \qquad (4e)$$

Even though noise and reverberation are, in general, neither stationary nor homogeneous, the stationary homogeneous representation proves to be useful over short time intervals, such as a signal processing interval, and over small distances, such as the interarray distance between sensor elements.

For stationary homogeneous processes, the Fourier transform pair for the covariance function is found to be useful and is given by

$$C(\tau,\mathbf{u}) = \frac{1}{2\pi}\int_{-\infty}^{\infty} Q(\omega,\mathbf{u})\exp(j\omega\tau)d\omega\ , \qquad (5a)$$

$$Q(\omega,\mathbf{u}) = \int_{-\infty}^{\infty} C(\tau,\mathbf{u})\exp(-j\omega\tau)d\tau, \qquad (5b)$$

where the cross spectrum $Q(\omega,\mathbf{u})$ is the Fourier transform of the cross covariance $C(\tau,\mathbf{u})$.

For those cases in which the stationary and homogeneous covariance function is dependent only on the magnitude of the displacement and not on the direction of displacement, the field is isotropic, as well as homogeneous, and written as

$$C(\tau, u) = \mathbf{E}\left[z(t, \mathbf{r}) z^*(t - \tau, \mathbf{r} - \mathbf{u})\right], \tag{6}$$

where u is the magnitude of **u**.

With these preliminaries completed we turn to the development of models for the representation of the intensity and covariance functions for both noise and reverberation in the ocean. As a final thought on terminology, we note that, strictly speaking, the above functions are correlation functions instead of covariance functions but for zero mean processes these functions are the same.

8.2 Noise in the Ocean

Noise in the ocean comes from all directions into the receiving system and the intensity of the noise is generally dependent upon the direction from which it comes. However, if we consider, momentarily, the noise from just one direction **e′**, with reference to Figure 1, the noise at a sensor at position **r′** = **r** − **u** is the time delayed noise received at a sensor positioned at **r** and this can be expressed as

$$n(t, \mathbf{r}') = n(t, \mathbf{r} - \mathbf{u}) = n\left(t - \frac{\mathbf{e}' \cdot \mathbf{u}}{c}, \mathbf{r}\right). \tag{7}$$

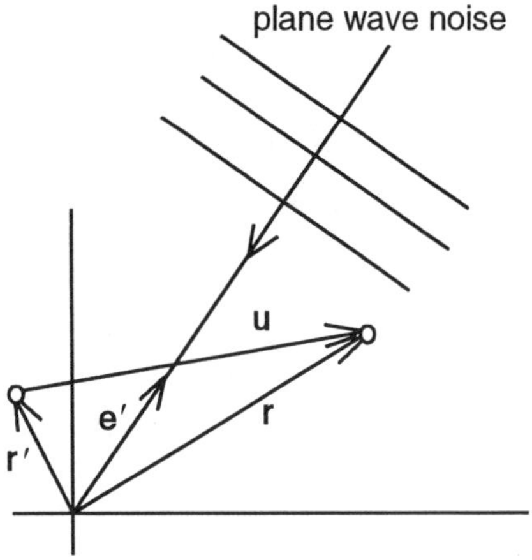

Figure 1 Noise Reception

Following Cox,[2] this expression relates the noise at the two sensor positions that are indicated in Figure 1. The unit vector **e′** is pointing in the direction opposed to the propagating noise and **e′·u**/c is the time delay between the two sensors along the direction **e′**. We see that the noise at **r′** is just the noise at **r** received at an earlier time. If the noise from the direction **e′** is stationary and homogeneous, then from (4c) the cross covariance can be expressed as

$$\begin{aligned} C(\tau, \mathbf{u}, \mathbf{e}') &= E\left[n(t, \mathbf{r}) n^*\left(t - \tau - \frac{\mathbf{e}' \cdot \mathbf{u}}{c}, \mathbf{r}\right)\right], \\ &= C\left(\tau + \frac{\mathbf{e}' \cdot \mathbf{u}}{c}, \mathbf{e}'\right). \end{aligned} \tag{8}$$

This expression asserts that the cross covariance between **r** and **r′** is just the autocovariance function at **r** time shifted by **e′·u**/c where we are explicitly calling out

---

[2] Cox, Henry, "Spatial correlation in arbitrary noise fields with application to ambient sea noise," J. Acoust. Soc. Am., **54**, No. 5, 1289-1301, 1973.

the dependence of the function on the direction of the noise. Fourier transforming the cross covariance function we obtain the cross spectrum function

$$Q(\omega,\mathbf{u},\mathbf{e}') = \int_{-\infty}^{\infty} C\left(\tau + \frac{\mathbf{e}'\cdot\mathbf{u}}{c}, \mathbf{e}'\right)\exp(-j\omega\tau)d\tau\,,$$

$$= Q(\omega,\mathbf{e}')\exp\left(j\frac{\omega\mathbf{e}'\cdot\mathbf{u}}{c}\right), \tag{9a}$$

where

$$Q(\omega,\mathbf{e}') = \int_{-\infty}^{\infty} C(\tau,\mathbf{e}')\exp(-j\omega\tau)d\tau\,. \tag{9b}$$

By the Weiner-Kinchine theorem in (2.22f), (9b) represents the two sided spectral density function at $\mathbf{r}$ of the noise from the direction $\mathbf{e}'$ and (9a) represents the cross spectrum of the noise from $\mathbf{e}'$ at the two sensor positions. Integrating (9) over all directions gives the cross spectral density of the noise at the two sensors for the noise from all directions

$$Q(\omega,\mathbf{u}) = \frac{1}{4\pi}\int_0^{2\pi}\int_0^{\pi} Q(\omega,\mathbf{u},\theta',\phi')\sin(\theta')d\theta'd\phi'\,. \tag{10}$$

In this expression the spherical coordinate angles of $\mathbf{e}'$ have been made explicit in the spectral density of the noise field.

Substituting (9) into (10), we obtain the cross spectral density of the noise field

$$Q(\omega,\mathbf{u}) = \frac{1}{4\pi}\int_0^{\pi}\int_0^{2\pi} Q(\omega,\theta',\phi')\sin(\theta')$$

$$\times\exp\left(j\frac{\omega u}{c}(\sin\theta\sin\theta'\cos(\phi'-\phi)+\cos\theta\cos\theta')\right)d\phi'd\theta'\,, \tag{11a}$$

where the vectors were expressed in coordinate form

$$\mathbf{e}' = \sin\theta'\cos\phi'\mathbf{e}_x + \sin\theta'\sin\phi'\mathbf{e}_y + \cos\theta'\mathbf{e}_z\,, \tag{11b}$$

$$\mathbf{u} = \sin\theta\cos\phi\mathbf{e}_x + \sin\theta\sin\phi\mathbf{e}_y + \cos\theta\mathbf{e}_z\,. \tag{11c}$$

To evaluate the integral in (11a) we must determine the angular dependence of the noise field represented by $Q(\omega,\theta',\phi')$ for each frequency. This can be done experimentally or by means of a theoretical model. For both it is found that the noise field for many cases is approximately uniform in all directions. While the deviations from uniformity are generally important, for those instances in which they are absent or in which they can be ignored, the noise field can be taken to be constant in all directions and equal to Q_0. In this case the integral over ϕ' can be immediately performed to yield[3]

$$Q(\omega,\mathbf{u}) = \frac{Q_0}{2}\int_0^{\pi} J_0\left(\frac{\omega u}{c}\sin\theta\sin\theta'\right)\sin(\theta')\exp\left(j\frac{\omega u}{c}(\cos\theta\cos\theta')\right)d\theta' \,. \quad (12a)$$

With simple coordinate transformations, this result can be put into the form[4]

$$\begin{aligned} Q(\omega,\mathbf{u}) &= Q_0\int_0^{\pi/2} J_0\left(\frac{\omega u}{c}\sin\theta\sin\theta'\right)\sin(\theta')\cos\left(\frac{\omega u}{c}(\cos\theta\cos\theta')\right)d\theta' \\ &= Q_0\left(\frac{\sin(\omega u/c)}{\omega u/c}\right). \end{aligned} \quad (12b)$$

This is the cross spectral density function between two sensor elements separated by the distance u at the frequency ω. This result shows that in the event that the noise field is homogeneous and spatially uniform, it is isotopic. We also see that the noise effectively decorrelates for argument values $\omega u/c \geq \pi$ or equivalently $u \geq \lambda/2$. This result ensures that the noise from receive array elements that are separated by a half wave length or more is uncorrelated which gives rise to the half wave length rule for receive array design.

Noise in the ocean is generally not stationary over time periods greater than several minutes or, at most, hours and is not homogeneous over spatial distances greater than several meters or, at most, 10s of meters. Consequently, while our preceding results are useful for signal processing and array design they are not generally applicable to the wider ocean environment. From that larger perspective we find that noise levels vary markedly both temporally and spatially but, as might be expected, their measured values are independent and the noise field statistics are adequately described by first order nonstationary Gaussian distributions in which the intensity level is time varying and required to be known. We refer the reader to the literature for these values and additional discussions of ocean noise. Fortunately, the literature is rich in commentary and results concerning ocean noise and, in particular, we mention the excellent summary of much of the earlier material by Urick[5] concerning, especially, noise intensity variations in the ocean.

8.3 Reverberation in the Ocean

Reverberation is an unavoidable concomitant of active sonar operations in most propagation environments. Objects within the ocean that scatter sound are varied and ubiquitous but, principally, they include biologics, especially biologics with gas filled swim bladders, entrained air bubbles caused by ocean surface

---

[3] Abramowitz and Stegun, (9.1.21).
[4] Gradshteyn and Rhyzhik, (6.688-2).
[5] Urick, Robert. J., *Principles of Underwater Sound*, McGraw-Hill Book Co. 1975, Chapter 7.

dynamics and the irregular shapes of the ocean surface and bottom as well as the irregular structure and variety of materials within the bottom. Because the sources of reverberation can be separated in this manner, reverberation is usually classified as

1. Volume,
2. Surface ,
3. Bottom.

Reverberant-like returns are also sometimes found from ship wakes that contain entrained gas bubbles that are created by the ship propellers and hull, as part of the dynamics of ship motion through the water. But, since ship wakes are localized and shallow, they are not usually a significant source of reverberation in modern ASW operations.

For volume reverberation the strongest sources of scattering within the medium are biologics with gas filled swim bladders, which are mostly fish, and other biologics and objects that are distributed sufficiently densely to cause significant scattering even though their acoustic impedances might differ only slightly from that of the medium and are individually weak scatterers. Scattering by swim bladders is especially strong when the incident signal causes them to resonate. This occurs at frequencies for which the wavelengths are large compared to the bubble sizes while at shorter wavelengths scattering is due to the whole organism. Scattering also occurs, theoretically speaking, wherever there is a change in acoustic impedance within the medium[6] and, in principle, within all regions in which density, temperature and salinity are changing. However, it is found that those regions in which ρc is changing because of changes in the microstructure do not produce significant scattering and are not, in fact, considered to be sources of volume reverberation.

Scattering by the ocean surface is complicated by the fact that at least two mechanisms can be involved. At low sea states the dominant mechanism is the irregular shape of the surface. At high sea states a layer of entrained air bubbles and distributions of bubble plumes immediately beneath the surface partially screens the surface from the acoustic signal and these air bubbles become the dominant source of scattering. Surface scattering is also enhanced or diminished depending upon whether the sound speed profile structure directs energy onto the surface. For example, in a surface duct environment surface reverberation can be strong but in a deep duct environment it could be nonexistent. Also, the movement of the surface causes a small spreading of the frequency band of the scattered energy.

Scattering by the ocean bottom is complicated by, not only the irregular shape of the bottom, but also by, since acoustic energy penetrates the bottom, the irregular shapes of the sublayers that make up the bottom and the irregularities in the distributions of materials within the bottom. It is also enhanced or diminished depending upon whether the sound speed profile structure directs energy onto the bottom. For example, in a convergence zone environment, very strong bottom reverberation can be generated by grazing angle rays at the bottom but, fortunately,

---

[6]See Rayleigh reflection and transmission laws in (5.30) and (5.32).

this reverberation is only received from the deep regions between CZs and for signals that are not too long the returns from the CZs are free of bottom reverberation. For a discussion of the effect of bottom reverberation in CZ environments on signal design see Appendix 12-B. In all environments that produce bottom scattering, on any given transmission, the scattering will be into, predominately, one to several discrete directions depending upon the bottom structure. As the source and receiver move, though, new bottom structures are encountered and the scattering directions change. An average over all such scattered fields gives the bottom scattering "beam pattern". The shape of this pattern is dependent upon the bottom roughness, i.e. the size of the bottom irregularities as compared to the acoustic wave length. If the irregularities are small compared to the wave length, i.e. the smooth bottom low frequency limit, the scattering pattern will approximate specular reflection with the bulk of the energy scattered (reflected) in the specular direction and the remaining energy scattered in all other directions. If the irregularities are large compared to a wave length, i.e. the rough bottom high frequency limit, the scattering pattern will approximate uniform scattering in all directions with no discernible specular component. Beyond this, it's difficult to form a general characterization of bottom reverberation. However, a safe one might include a statement to the effect that bottom reverberation, when it occurs, will probably consist of strong returns having great variability.

The statistics of reverberation share similarities with the statistics of noise. Reverberation statistics over short time periods and spatial extents, as for oceanic noise, can be modeled as a stationary homogeneous second order Gaussian process. For longer time periods and spatial extents, similarly to oceanic noise, reverberation statistics must be viewed as nonstationary and nonhomogenous but, because all sample values become statistically independent under these conditions, reverberation can be adequately modeled as a first order Gaussian process. Reverberation and noise statistics differ only in their covariance properties and in the fact that reverberation is correlated with the transmitted signal. For most scenarios we are only concerned with the time dependence of the variance of reverberation, i.e. its time dependent intensity. In the following analyses we present simple models for the intensity of volume and surface reverberation along the lines of Urick[7] which lead to his results. Our surface reverberation model can be applied to the case of bottom reverberation - to the degree that bottom reverberation can be viewed as scattering from a moderately rough planar bottom. Unfortunately, when volume reverberation is encountered, this is rarely the case. In these instances bottom reverberation can be strong and highly variable due to the topography and the variety of materials in subterranean layers

In the reverberation analyses that are presented, all reverberant scattering is assumed to be relatively weak compared to the incident signal. Consequently, the scattering caused by scattering, i.e. secondary scattering as opposed to primary scattering, is insignificant. This fact reduces the complexity of our analysis of reverberation since we need to deal only with the primary scattering. This implies that in all cases the strength of the scattering varies as the strength of the incident

---

[7] Urick, Robert J., *Principles of Underwater Sound*, McGraw-Hill Book Co., 1975, Chapter 8.

signal which means that the signal source level has little bearing on detection performance in a reverberation limited environment such as it would have in an ambient noise limited environment.

Reverberation Scattering Strength and Scattering Cross Section

Reverberation scattering strength is defined as the decibel level of intensity scattering strength

$$S(\mathbf{r}, \mathbf{e}_I, \mathbf{e}_S) = 10\text{Log}(s(\mathbf{r}, \mathbf{e}_I, \mathbf{e}_S)), \quad (13a)$$

where

$$s(\mathbf{r}, \mathbf{e}_I, \mathbf{e}_S) = \frac{I_S(\mathbf{r}, \mathbf{e}_S)}{I_I(\mathbf{r}, \mathbf{e}_I)}, \quad (13b)$$

is the intensity scattering strength. In these quantities, $I_S(\mathbf{r},\mathbf{e}_S)$ is the intensity of the scattered wave in the direction $\mathbf{e}_S$ measured at 1 m from the scattering region at $\mathbf{r}$ and $I_I(\mathbf{r},\mathbf{e}_I)$ is the intensity of the wave incident on the scattering region at $\mathbf{r}$ in the direction $\mathbf{e}_I$. Additionally, the scattered intensity $I_S(\mathbf{r},\mathbf{e}_S)$ is a density measure of intensity, i.e. per unit volume for volume reverberation and per unit area for surface and bottom reverberation. This means that the incident and scattered intensities are somewhat different and, we see that in the logarithm argument of (13a) there is an implied factor of unit volume for volume reverberation and an implied factor of unit area for surface and bottom reverberation. However, in (13b) the implied factor does not appear and the intensity scattering strength is per unit volume or per unit area. Reverberation scattering strength is seen to be very similar to target strength but it differs from target strength in being a scattering density. We have made explicit the position and directional dependence of scattering strength in (13) but frequently we find that it is independent of one or both of these parameters.

Reverberation scattering strength is also sometimes measured as a scattering cross section which, in analogy to radar operations, is given by

$$m(\mathbf{r}, \mathbf{e}_I) I_I(\mathbf{r}, \mathbf{e}_I) = \int_A da I_S(\mathbf{r}, \mathbf{e}_S). \quad (14a)$$

In this expression, the cross section $m(\mathbf{r},\mathbf{e}_I)$ is the effective cross sectional area of the target region per unit volume or per unit area which is dependent upon the scattering region at $\mathbf{r}$ and the aspect of the incident wave $\mathbf{e}_I$. The left member is the effective total power intercepted by the scattering region per unit volume or per unit area and the right member is the total power scattered by the scattering region per unit volume or per unit area where the integral is over the surface of a unit sphere for volume scattering or a unit hemisphere for surface scattering. We can recast this expression in the form

$$m(\mathbf{r}, \mathbf{e}_I) = \frac{I_S(\mathbf{r}, \mathbf{e}_S)}{I_I(\mathbf{r}, \mathbf{e}_I)} \frac{\int_A da I_S(\mathbf{r}, \mathbf{e}_S)}{I_S(\mathbf{r}, \mathbf{e}_S)}, \quad (14b)$$

or, equivalently,

$$S(\mathbf{r},\mathbf{e}_I,\mathbf{e}_S) = 10\mathrm{Log}\left(\frac{m(\mathbf{r},\mathbf{e}_I)I_S(\mathbf{r},\mathbf{e}_S)}{\int_A da I_S(\mathbf{r},\mathbf{e}_S)}\right). \tag{15a}$$

In this expression the integral over the unit sphere or hemisphere is the power scattered per unit volume or unit area. If the scattering is omnidirectional, i.e. $I_S(\mathbf{r},\mathbf{e}_S) = I_S(\mathbf{r})$, then

$$S(\mathbf{r},\mathbf{e}_I) = 10\mathrm{Log}\left(\frac{m(\mathbf{r},\mathbf{e}_I)}{A}\right), \tag{15b}$$

where A is the area of a unit sphere for volume scattering, i.e. 4π, or the area of a unit hemisphere for surface and bottom scattering, i.e. 2π. As in (13a), in (15b) there is an implied factor of unit volume or unit area.

Volume Reverberation

A source transmits acoustic energy into the medium and echoes are received at the receive array. The echoes include scattering by the medium as well as target echoes and their relative strengths must be determined. We assume that the source and receiver are collocated and a signal having duration T is transmitted at $t = 0$. As the signal propagates away from the source it encounters scatterers within the medium and a small part of the signal acoustic energy is scattered in all directions. A portion of this scattered energy is scattered back into the receiver which is received as backscattered reverberation.

If we consider the signal to be a sequence of segments each having the transmit time t' where $0 \le t' \le T$, then, as shown in Figure 2, at time t each of these segments will have propagated to the range $c(t - t')$ where c is the sound speed. It's clear that at this time the leading edge of the signal will be at range ct and the trailing edge will be at $c(t - T)$. The backscattered reverberation, at time t, that is due to the segment transmitted at t', comes from half of its total propagation range which is $c(t - t')/2$. Since t' is in the interval 0 to T, reverberation will be received from the range interval ct/2 to $c(t - T)/2$ at time t. This means that, at each time t, reverberation is received from a range interval of length cT/2. This sets a lower limit on our ability to resolve scattering features.

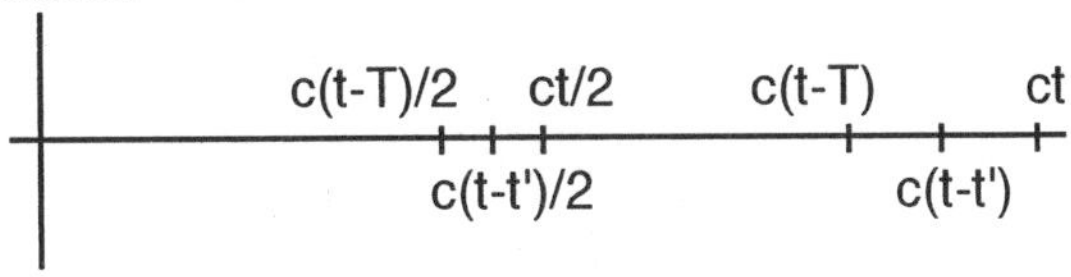

Figure 2 Signal Propagation - Ranges and Times

The intensity of the reverberation at 1 m from each scattering point due to the signal segment that was transmitted at t' is given by

$$\frac{I_0 s_V B_S(\theta' - \theta_S, \phi' - \phi_S)dv}{r'^2}. \tag{16a}$$

In this expression, I_0 is the intensity of the transmitted signal which is assumed to be constant over the duration of the signal. The quantity s_V is the intensity scattering strength per unit volume which is assumed to be constant throughout the medium. B_S is the source beam pattern where θ_S and ϕ_S are the beam axis directions and θ' and ϕ' are the directions to the scattering point. We take θ to be the vertical angle and ϕ to be the azimuthal angle. The range $r' = c(t-t')/2$ is the distance to a scattering region. The differential dv is the volume element containing the scattering point and spherical spreading is assumed.

The intensity of the reverberation at the receiver due to the signal segment that was transmitted at t′ and scattered from each volume element is given by

$$\frac{I_0 s_V B_S(\theta' - \theta_S, \phi' - \phi_S) B_R(\theta' - \theta_R, \phi' - \phi_R) dv}{r'^4}, \tag{16b}$$

where B_R is the receive beam pattern with its angles defined as for the transmit beam and spherical propagation is assumed for the return path.

Integrating (16b) over range and both angles gives the total back scattered reverberation intensity

$$I(t) = I_0 s_V \int_{r_1}^{r_2} dr' \frac{1}{r'^2} \int_0^{\pi} d\theta' \int_0^{2\pi} d\phi' \sin\theta' B_S(\theta' - \theta_S, \phi' - \phi_S) B_R(\theta' - \theta_R, \phi' - \phi_R). \tag{17a}$$

In this expression the range integration limits encompass the scattering volume and are given by

$$r_2 = ct/2, \tag{17b}$$

$$r_1 = c(t - T)/2. \tag{17c}$$

These integration limits implicitly assume $t > T$ which ignores reverberation received during the transmission of the signal. Since a collocated receiver is blanked out during the period $t < T$ this is unimportant.

The integral over the beam patterns in (17) can be written

$$\begin{aligned} \Psi(\theta_S, \phi_S, \theta_R, \phi_R) &= \int_0^{\pi} d\theta' \int_0^{2\pi} d\phi' \sin\theta' B_S(\theta' - \theta_S, \phi' - \phi_S) B_R(\theta' - \theta_R, \phi' - \phi_R) \\ &= \int_S \frac{ds'}{r'^2} B_S(\theta' - \theta_S, \phi' - \phi_S) B_R(\theta' - \theta_R, \phi' - \phi_R) \\ &= \int_{4\pi} d\Omega' B_S(\theta' - \theta_S, \phi' - \phi_S) B_R(\theta' - \theta_R, \phi' - \phi_R), \end{aligned} \tag{18a}$$

where ds′ is the surface element at range r′ and dΩ is the increment of solid angle. This shows that Ψ is, effectively, the beam weighted solid angle that subtends the beam intersection. For example, if the beam intersection of B_S and B_R is nonzero over the solid angle Θ and zero elsewhere, then

$$\Psi(\theta_S, \phi_S, \theta_R, \phi_R) \cong \Theta B_S(\theta'' - \theta_S, \phi'' - \phi_S) B_R(\theta'' - \theta_R, \phi'' - \phi_R), \quad (18b)$$

where the double primed angles are coordinates pointing to a point within the solid angle Θ. If both beams are aligned and pointing in the direction of the scattering region, (18b) reduces to $\Psi = \Theta$. Equation (18) also shows that Ψ is not a function of the range r' and is a constant within the range integral in (17).

Performing the range integration in (17) gives

$$I_R = \frac{I_0}{r^4} s_V \left[(r_2 - r_1)(r^2 \Psi) \right], \quad (19a)$$

where $r = \sqrt{(r_1 r_2)}$ is the geometric mean range to the scattering region. The first factor in brackets in the length of the scattering region where $r_2 - r_1 = cT/2$, and the second factor is the geometrical mean cross sectional area of the scattering region. The product of these factors is the volume of the scattering region. Taking the logarithm of (19a), the volume reverberation level for the sonar equation is given by

$$RL_V = 10\mathrm{Log}(I_0) - 40\mathrm{Log}(r) + S_V + 10\mathrm{Log}(V), \quad (19b)$$

where $S_V = 10\mathrm{Log}(s_V)$ is the volume scattering strength and $V = [(cT/2)(r^2\psi)]$ is the volume of the scattering region For our assumed spherical spreading environment, we see that the reverberant volume grows as r^2 while the angular cross section Ψ is constant. Accordingly, (19b) shows that volume reverberation decays with range as $20\mathrm{Log}(r) = 40\mathrm{Log}(r) - 20\mathrm{Log}(r)$. But, since target echoes decay as $40\mathrm{Log}(r)$ in a spherical spreading environment, distant echoes can easily be masked by volume reverberation.

In the event that bubbles are the dominant source of scattering in the medium and the distribution of bubbles in the medium per unit bubble radius and per unit volume is denoted $n(a,\mathbf{r})$, where a is the bubble radius and $\mathbf{r}$ denotes the position within the medium, then we have the cross section per unit volume

$$m(\mathbf{r}) = \frac{1}{2\pi} \int_0^\infty d\omega P(\omega) \int_0^\infty da \sigma_S(a, \omega) n(a, \mathbf{r}), \quad (20)$$

where $P(\omega)$ is the normalized intensity spectrum of the signal and $\sigma_S(a,\omega)$ is the scattering cross section for a single bubble of radius a and incident signal of angular frequency ω. From (8.14-B) the scattering cross section for an ideal single bubble is given by

$$\sigma_S(a,\omega) = \frac{4\pi a^2}{\left(\omega_R^2/\omega^2 - 1\right)^2 + \delta^2}, \tag{21a}$$

$$\omega_R^2 = 3\gamma p/\rho_0 a^2, \qquad \delta = ka. \tag{21b,c}$$

The quantity ω_R is the angular resonant frequency of the bubble, where γ is the ratio of specific heats for the gas within the bubble (γ = 1.4 for air), p is the ambient pressure and ρ_0 is the density of water. For an actual bubble, small quantities are added to δ which slightly reduce the bubble cross section but do not change the resonance frequency[8].

In the event that the signal is a sinusoid having angular frequency ω_0, its normalized power spectrum is $P(\omega) = \delta(\omega - \omega_0)$. If the bubbles are all the same size with radius a_0, then the distribution of bubbles is given by $n(a,\mathbf{r}) = N(a_0,\mathbf{r})\delta(a - a_0)$ where $N(a_0,\mathbf{r})$ is the number of bubbles per unit volume. Substitution into (20) gives the cross section for the medium

$$m(\mathbf{r}) = \frac{N(a_0,\mathbf{r})4\pi a_0^2}{\left(\omega_{0,R}^2/\omega^2 - 1\right)^2 + \delta_0^2}, \tag{22a}$$

where the zero subscripts indicate that the quantities are to be evaluated for bubble radius a_0. If the signal frequency is the resonance frequency, this expression reduces to

$$m(\mathbf{r}) = \frac{N(a_0,\mathbf{r})4\pi a_0^2}{\delta_{0,R}^2}, \tag{22b}$$

where the subscript R denotes the quantities are evaluated at the resonance frequency in (21b) for the bubble radius a_0. Since gas bubbles are the greatest contributor to volume reverberation, it is advantageous to operate at wave lengths that are very large compared to the bubbles expected to be encountered, i.e. at wave lengths much larger than the resonance wave length which requires very low frequency sonar systems. However, even at very low frequencies, swim bladders of large fish will be found to resonate.

The distribution of biologics in the medium is somewhat complicated and not completely understood. However, certain features of their distribution have been determined by investigators[9]. They find that biologics are found to be horizontally stratified within the ocean and are largely confined to a layer that rises and falls that is referred to as the deep scattering layer (DSL). Generally speaking, the DSL extends from approximately 180 m to 900 m in depth during the day and migrates to

---

[8] See Appendix 8-B.

[9] Urick, Robert J., *Principles of Underwater Sound for Engineers*, McGraw-Hill Book Co. 1975, Chapter 8.

near surface depths at night. The upward migration occurs at sunset and the downward migration occurs at sunrise and the layer depth is essentially constant at other times. These depth changes appear to maintain a constant level of luminosity within the DSL which is not surprising behavior for marine biologics.

For a downward looking sonar system, backscattering strength is found to be both depth dependent and frequency dependent. However, at relatively high frequencies, say above 20 kHz, backscattering strength is found to be only depth dependent. This depth dependence is such that the back scattering strength is about $S_V \cong -70$ dB at the top of the DSL and decreases to about $S_V \cong -80$ dB at the bottom of the DSL, i.e. it decreases at about −5 dB per 300 m increase in depth. At lower frequencies, say below 10 kHz, the back scattering strength is strongly frequency selective in which the returns at the various frequencies come from different depths. This indicates that fish with swim bladders are partitioned in layers within the DSL and their swim bladders are resonating at frequencies below 10 kHz. Further, it is found that the frequencies of the back scattered returns are lower during the night and greater during the day. This is to be expected because during the night the fish are at shallower depths, where the ambient pressure on their swim bladders is lower, and during the day they are at deeper depths, where the ambient pressure on their swim bladders is greater.

Surface Reverberation

A source transmits acoustic energy into the medium and echoes are received at the receive array. The echoes include scattering by the surface and bottom of the medium as well as target echoes and their relative strengths must be determined. We assume that a source and receiver are collocated and a signal having duration T is transmitted at t = 0. As the signal propagates away from the source it encounters scatterers on the surface which cause a part of the signal acoustic energy to be scattered. A portion of this scattered energy is scattered back into the receiver which is received as backscattered reverberation. We compute the backscattering strength and its rate of decay.

The signal contacts the surface as indicated in Figure 3. At time t/2 the signal segment transmitted at time t′, where $0 \le t' \le T$, contacts the surface at range c(t − t′)/2. Similarly, the leading edge of the signal contacts the surface at range ct/2 and the trailing edge contacts the surface at range c(t − T)/2. At time t/2 the intensity of the scattering at 1 m from a surface scattering point is given by

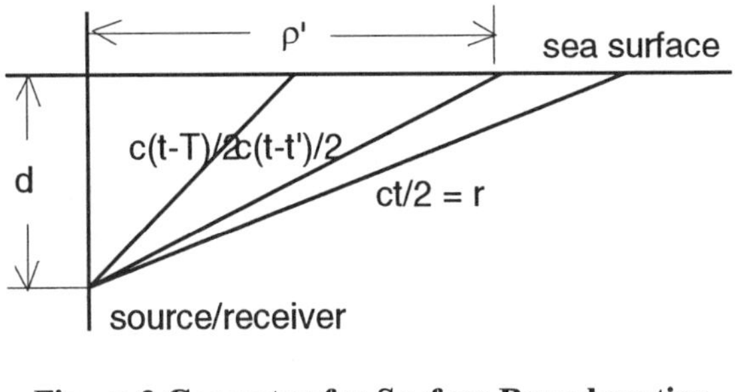

Figure 3 Geometry for Surface Reverberation

$$\frac{I_0 s_S B_S\left(\theta' - \theta_S, \phi' - \phi_S\right) ds}{r'^2}. \tag{23a}$$

In this expression, I_0 is the intensity of the transmitted signal which is assumed to be constant over the duration of the signal. The quantity s_S is the intensity scattering strength per unit area which is assumed to be constant over the sea surface and all angles of incidence. B_S is the source beam pattern where θ_S and ϕ_S are the beam axis directions and θ' and ϕ' are the directions to the scattering point. We take θ to be the vertical angle and ϕ to be the azimuthal angle. The range $r' = c(t-t')/2$ is the distance to the scattering point at t'. The differential ds is the surface area element containing the scattering point and spherical spreading is assumed.

At time t the intensity of the reverberation at the receiver due to the signal segment that was transmitted at t' and scattered from each surface element is given by

$$\frac{I_0 s_S B_S(\theta' - \theta_S, \phi' - \phi_S) B_R(\theta' - \theta_R, \phi' - \phi_R) ds}{r'^4}, \tag{23b}$$

where B_R is the receive beam pattern with its angles defined as for the transmit beam and spherical propagation is assumed for the return path.

Integrating (23b) over the surface scattering region gives

$$r(t) = I_0 s_S \int_{\rho_1}^{\rho_2} d\rho' \frac{\rho'}{r'^4} \int_0^{2\pi} d\phi' B_S(\theta' - \theta_S, \phi' - \phi_S) B_R(\theta' - \theta_R, \phi' - \phi_R), \tag{24a}$$

where the range integration limits are given by

$$\rho_2 = \sqrt{\left(\frac{ct}{2}\right)^2 - d^2}, \quad \frac{2d}{c} \le t \tag{24b}$$

$$\rho_1 = \begin{cases} 0, & \frac{2d}{c} \le t < T + \frac{2d}{c} \\ \sqrt{\left(\frac{c(t-T)}{2}\right)^2 - d^2} . & T + \frac{2d}{c} \le t \end{cases} \tag{24c}$$

The integration of (24) is complicated because of the interdependence between ρ', r' and θ'. However, for short pulses we can approximate

$$r' = \frac{c(t-t')}{2} \xrightarrow{t>>T} \frac{ct}{2} = r, \tag{25a}$$

which implies

$$\theta' = \sin^{-1}\left(\frac{d}{r'}\right) \xrightarrow{t>>T} \sin^{-1}\left(\frac{d}{\sqrt{d^2 + (ct/2)^2}}\right) = \theta . \tag{25b}$$

Therefore, for short pulses, the integral over beams in (24) becomes

$$\Psi(\theta-\theta_S,\theta-\theta_R) \cong \int_0^{2\pi} d\phi' B_S(\theta-\theta_S,\phi'-\phi_S) B_R(\theta-\theta_R,\phi'-\phi_R), \qquad (26a)$$

which is independent of r′ and ρ′. Moreover, this quantity is, effectively, the beam weighted beam width that subtends the beam intersection. For example, if the transmit beam is uniform in the horizontal plane, i.e. the source is a vertical array, and the receive beam is uniform in the vertical plane, i.e. the receiver is a horizontal array, we have

$$\Psi(\theta-\theta_S) = B_S(\theta-\theta_S) \int_0^{2\pi} d\phi' B_R(\phi'-\phi_R) \cong B_S(\theta-\theta_S)\Phi_R. \qquad (26b)$$

If the transmit beam is pointing at the surface scattering points, this becomes $\Psi = \Phi_R$. Note, though, that for all cases the maximum value of Ψ is 2π.

The integral in (24) becomes for the short pulse case

$$r(t) = I_0 s_S \Psi \int_{\rho_1}^{\rho_2} d\rho' \frac{\rho'}{\left(d^2+\rho'^2\right)^2} = \frac{I_0}{r_1^2 r_2^2} s_S \left[\pi\left(\rho_2^2-\rho_1^2\right)\left(\frac{\Psi}{2\pi}\right)\right], \qquad (27)$$

where the leading factor in brackets is the area of the surface that the source is capable of ensonifying and the second factor is the fraction of that area that is jointly "seen" by the transmit and receive beam patterns. The product of these two factors is the area of the surface from which scattering is received. The range attenuation is given by the product of the squared slant ranges to the leading and trailing edges of the ensonified area.

The surface ranges ρ_1 and ρ_2 are time dependent as seen in (24c) and, because the surface range ρ_1 is piece-wise continuous, we can write the reverberation expression in (27) in the forms

$$r(t) = \frac{I_0 s_S}{d^2 r_2^2}\left[\pi\rho_2^2\left(\frac{\Psi}{2\pi}\right)\right], \qquad \frac{2d}{c} \le t < T + \frac{2d}{c} \qquad (28a)$$

$$r(t) = \frac{I_0 s_S}{r^4}\left[\pi\left(\rho_2^2-\rho_1^2\right)\left(\frac{\Psi}{2\pi}\right)\right], \qquad T + \frac{2d}{c} < t < \infty \qquad (28b)$$

where $r = \sqrt{(r_1 r_2)}$. The first form is for the onset of reverberation growth and the second form is for the follow-on fall-off of reverberation.

In (28) the surface scattering mechanism is defined by the surface intensity scattering coefficient s_S. Unfortunately, surface scattering is a complicated phenomenon in which more than one mechanism plays a role. However, for a given wind speed, i.e. sea state, the empirical relation has been established that depends only upon grazing angle, wind speed and signal frequency [10]

$$S_S = 3.3\beta \mathrm{Log}\left(\frac{\theta}{30}\right) - 42.4\mathrm{Log}(\beta) + 2.6\,, \qquad \beta = 158\left(vf^{1/3}\right)^{-0.58}\,, \tag{29}$$

where θ is the grazing angle in degrees, v is the wind speed in knots and f is the signal frequency in Hz.

Surface reverberation can be a serious detriment to detection performance. Since the length of the surface scattering region is a constant, i.e. $\rho_2 - \rho_1 = cT/2$, we can easily show that the scattering area on the surface grows linearly with range. Consequently, in our assumed spherically spreading environment, surface reverberation decays as 30Log(r) = 40Log(r) – 10Log(r). This is a faster decay rate than the 20Log(r) rate for volume reverberation but it is still a slower decay rate than the 40Log(r) rate for target echoes. Consequently, as for volume reverberation, distant echoes can easily be masked by surface reverberation. However, whereas there is hope for mitigating volume reverberation through the use of low frequency sonar systems, the same is not necessarily true for surface reverberation because sea surface features can be found that are greater in size than any realistically achievable sonar wavelength. Therefore, it appears that the most effective way of mitigating surface reverberation is through the use of directionality. In simple terms, this prescription calls for the sonar system to direct its acoustic radiations away from the sea surface. Unfortunately, this cannot always be done because submarine targets are frequently near the surface and become more interesting as they approach the surface.

---

[10] Chapman, R. P. and J. H. Harris, "Surface Backscattering Strength Measured with Explosive Sound Sources," J. Acoust. Soc. Am., **34**, 1596 (1962).

APPENDIX 8-A: STOCHASTIC ANALYTIC SIGNALS

Noise and reverberation in the ocean are real random signals but, upon their reception by modern ASW sonar systems, equivalent analytic signals are typically developed for follow-on signal and data processing, e.g. beam forming, detection analysis, etc. The design and structure of these processing algorithms are based, in part, upon the statistics of the signal and in this appendix we develop models for those statistics. For real in-water ocean noise and reverberation, the problem is relatively straight-forward with many investigators agreeing that, over the short term, these signals are Gaussian distributed and stationary. Since this distribution is well-known, it poses no problem. However, for stationary analytic signals the probability distribution is not so well known but the distribution for these signals is contained within a distribution developed by Goodman[1]. This distribution has the form that he calls the multivariate complex Gaussian distribution.

Following Goodman, we develop the multivariate complex Gaussian distribution for stationary analytic signals in the following. In general terms, Goodman's method transforms the multivariate Gaussian probability density function for real signals into the multivariate complex probability density function for complex signals. In these developments we assume that all signals are zero mean and, therefore, their statistics are defined by their covariance functions. We start our investigation by determining the covariance functions for real and analytic signals.

8.1-A: Covariance Functions of Real and Analytic Signals

One of the advantages of the analytic signal is the simplicity with which it represents narrow band signals in terms of the quadrature components of the signals. Because of this, some signal and data processing algorithms are structured to operate on the narrow band quadrature components of signals while others are structured to operate on broad band analytic signals. In the interests of generality we discuss both cases. Looking ahead, we find that, with the proper covariance function, the same multivariate complex probability density function represents the statistics of both cases.

Broad Band Analytic Signals

Analytic noise signals have the form, from (2.4-B),

$$z(t) = x(t) + jy(t), \tag{1a}$$

where the left member is the analytic signal and the right member consists of the real signal and its Hilbert transform. These components have the representation, from (2.2-B) and (2.5-B),

---

[1] Goodman, N. R., "Statistical Analysis Based on a Certain Multivariate Complex Gaussian Distribution (An Introduction)," Ann. Math. Statistics, **34**, 1963.

$$x(t) = \frac{1}{2\pi} \int_{-\infty}^{\infty} F(\omega) \exp(j\omega t) d\omega \,, \tag{1b}$$

$$y(t) = \frac{-j}{2\pi} \int_{-\infty}^{\infty} \mathrm{sgn}(\omega) F(\omega) \exp(j\omega t) d\omega \,. \tag{1c}$$

$$\begin{aligned} \mathrm{sgn}(\omega) &= +1 \qquad \omega \geq 0 \\ &= -1 \qquad \omega < 0 \end{aligned}$$

Inverse transforming (1b), we note that the Fourier transform of the real signal has the covariance

$$\begin{aligned} \mathbf{E}\left[F(\omega) F^*(\omega')\right] &= \int_{-\infty}^{\infty} dt' R(t - t') \exp(j\omega' t') \int_{-\infty}^{\infty} dt \exp(-j\omega t), \\ &= 2\pi P(\omega') \delta(\omega - \omega'), \end{aligned} \tag{2}$$

where R(t–t′) = **E**[x(t)x(t′)] and P(ω) is the two sided spectrum. The last equality was obtained with the help of the Wiener-Kinchine theorem in (2.22) and the delta function representation. With this result we find the correlation functions of the real and imaginary parts of the analytic signal to be given by

$$\mathbf{E}\left[x(t)x(t')\right] = \mathbf{E}\left[y(t)y(t')\right] = \frac{1}{2\pi} \int_{-\infty}^{\infty} P(\omega) \cos(\omega(t - t')) d\omega = R(t' - t), \tag{3a}$$

$$\begin{aligned} \mathbf{E}\left[x(t)y(t')\right] &= \frac{1}{2\pi} \int_{-\infty}^{\infty} j\,\mathrm{sgn}(\omega) P(\omega) \exp\left[j\omega(t - t')\right] d\omega \\ &= \frac{1}{2\pi} \int_{-\infty}^{\infty} P(\omega) \sin\left[\omega(t' - t)\right] d\omega = R'(t - t'). \end{aligned} \tag{3b}$$

In these derivations the evenness of P(ω) and the oddness of sgn(ω) led to the final results. We observe that R(τ) has its maximum value at τ = 0 and is symmetric about that point. In contrast, R′(τ) vanishes at τ = 0 and is anti-symmetric about that point, i.e. R′(τ) = – R′(–τ).

Combining these results we find that the analytic signal has the covariance function

$$\begin{aligned} \mathbf{E}\left[z(t) z^*(t')\right] &= \mathbf{E}\left[x(t)x(t') + y(t)y(t') + j\left(x(t)y(t') - y(t)x(t')\right)\right] \\ &= 2\left[R(t - t') - jR'(t - t')\right] = C(t - t'), \end{aligned} \tag{4a}$$

$$\mathbf{E}\left[z(t) z(t')\right] = \mathbf{E}\left[x(t)x(t') - y(t)y(t') + j\left(x(t)y(t') + y(t)x(t')\right)\right] = 0. \tag{4b}$$

Accordingly, if we have a set of signal samples we can construct sample vectors in terms of the real signal values or the complex analytic signal values

$$\mathbf{w} = \left(x_1, y_1, x_2, y_2, \cdots, x_N, y_N\right)^T, \qquad \mathbf{z} = \left(z_1, z_2, \cdots, z_N\right)^T, \tag{5a,b}$$

where $z_i = x_i + jy_i$. The covariance matrices of these vectors are found to have the form

$$\mathbf{R} = \mathbf{E}\left(\mathbf{w}\mathbf{w}^T\right) = \mathbf{E}\left[\begin{pmatrix} x_i x_k & x_i y_k \\ y_i x_k & y_i y_k \end{pmatrix}\right] = \begin{cases} \dfrac{1}{2}\begin{bmatrix} \sigma_a^2 & 0 \\ 0 & \sigma_a^2 \end{bmatrix}, & i = k \\ \dfrac{1}{2}\begin{bmatrix} \alpha_{i,k} & -\beta_{i,k} \\ \beta_{i,k} & \alpha_{i,k} \end{bmatrix}, & i \neq k \end{cases}, \tag{6a}$$

$$\mathbf{C} = \mathbf{E}\left(\mathbf{z}\mathbf{z}^T\right) = \mathbf{E}\left[\left(z_i z_k\right)\right] = \begin{cases} \sigma_a^2, & i = k \\ \alpha_{ik} + j\beta_{ik}, & i \neq k \end{cases}, \tag{6b}$$

in which i denotes the row of the (x_i,y_i) pair in the vector $\mathbf{w}$ and k denotes the column of the (x_k,y_k) pair in the vector $\mathbf{w}^T$. Also, $\sigma_a{}^2$ is the variance of the analytic signal and $\alpha_{i,k}$ and $\beta_{i,k}$ are, respectively, the auto and cross covariances of the real signal and its Hilbert transform. Since the cross covariance is anti-symmetric, i.e. $\mathbf{E}(x_i y_k) = \beta_{i,k} = -\beta_{k,i}$, then $\mathbf{R}$ is symmetric, i.e. $\mathbf{R}$ is equal to its transpose $\mathbf{R} = \mathbf{R}^T$, and $\mathbf{C}$ is Hermitian, i.e. $\mathbf{C}$ is equal to its conjugate transpose $\mathbf{C} = \mathbf{C}^{*T}$.

<u>Narrow Band Analytic Signals</u>

For narrow band analytic signals, from (2.14) and (2.19-B), the analytic signal has the form

$$z(t) = \left[x_{nb}(t) + jy_{nb}(t)\right]\exp(j\omega_0 t), \tag{7a}$$

where $x_{nb}(t)$ and $y_{nb}(t)$ are the narrow band quadrature components. Expanding the right member, the broad band quadrature components are obtained

$$x(t) = x_{nb}(t)\cos(\omega_0 t) - y_{nb}(t)\sin(\omega_0 t), \tag{7b}$$

$$y(t) = y_{nb}(t)\cos(\omega_0 t) + x_{nb}(t)\sin(\omega_0 t). \tag{7c}$$

Solving these equations gives the narrow band components in terms of the broad band components.

$$x_{nb}(t) = x(t)\cos(\omega_0 t) + y(t)\sin(\omega_0 t), \tag{7d}$$

$$y_{nb}(t) = y(t)\cos(\omega_0 t) - x(t)\sin(\omega_0 t). \tag{7e}$$

The correlation functions for the narrow band components are readily obtained

$$\begin{aligned} R_{nb}(t - t') &= E\left[x_{nb}(t)x_{nb}(t')\right] = E\left[y_{nb}(t)y_{nb}(t')\right] \\ &= R(t - t')\cos(\omega_0(t - t')) - R'(t - t')\sin(\omega_0(t - t')), \end{aligned} \tag{8a}$$

$$R'_{nb}(t-t') = E\left[x_{nb}(t)y_{nb}(t')\right]$$
$$= R'(t-t')\cos(\omega_0(t-t')) - R(t-t')\sin(\omega_0(t-t')). \qquad (8b)$$

These relations show that $R_{nb}(t-t')$ is a maximum at $t = t'$ and is symmetric about that point and that $R'_{nb}(t-t')$ vanishes at $t = t'$ and is antisymmetric about that point. Accordingly, the covariance functions of the quadrature components of the narrow band analytic signal in (9) have the same form as the components of the broad band analytic signal in (3). Consequently, a sample set of narrow band quadrature components $\mathbf{w} = (x_{nb,1}, y_{nb,1}, \cdots, x_{nb,N}, y_{nb,N})^T$ and the corresponding analytic signal values $\mathbf{z} = (z_{nb,1}, \cdots, z_{nb,N})^T$ have covariance matrices with the same form as (6). Therefore, in either the broad band form or the narrow band form of both the real and analytic signals the covariance matrices of the information components of the signal have the same form.

8.2-A: Probability Density Functions for the Analytic Signal

The multivariate probability density function for the real signal is well-known and can be written

$$p(\mathbf{w}) = \frac{\exp\left(-\frac{1}{2}\mathbf{w}^T\mathbf{R}^{-1}\mathbf{w}\right)}{\sqrt{(2\pi)^{2N}|\mathbf{R}|}}, \qquad (9)$$

where $\mathbf{R}$ is the covariance matrix and $|\mathbf{R}|$ denotes the determinant of the covariance matrix and N is the number of xy pairs. The multivariate complex Gaussian density function is obtained by transforming (9) to the complex form. In particular, we need to show that the isomorphic matrices in (6) are such that $\frac{1}{2}\mathbf{w}^T\mathbf{R}^{-1}\mathbf{w} = \mathbf{z}^T\mathbf{C}^{-1}\mathbf{z}$ and $2^{2N}|\mathbf{R}| = |\mathbf{C}|^2$. We start with a short diversion into isomorphisms.

Comparing the covariance matrices for the real and analytic signals in (6) we see that if either the real or complex matrix is known that the other can be immediately written, i.e. there is a one-to-one correspondence between these matrices. In more mathematical terms, for the simplest case, there is an isomorphism between a complex number and the 2×2 matrix

$$(\alpha + j\beta) \cong \begin{bmatrix} \alpha & -\beta \\ \beta & \alpha \end{bmatrix}, \qquad (10a)$$

where $\cong$ denotes an isomorphic relation. The isomorphism is clearly maintained under addition and scalar multiplication. Not so clearly, it is also maintained under matrix multiplication, i.e. if two complex numbers are multiplied, the result is isomorphic to the product of their isomorphic matrices. For example, the expanded product

$$(\alpha_1 + j\beta_1)(\alpha_2 + j\beta_2) \cong \begin{bmatrix} \alpha_1 & -\beta_1 \\ \beta_1 & \alpha_1 \end{bmatrix}\begin{bmatrix} \alpha_2 & -\beta_2 \\ \beta_2 & \alpha_2 \end{bmatrix}, \qquad (10b)$$

$$\left(\alpha_1\alpha_2 - \beta_1\beta_2\right) + j\left(\alpha_1\beta_2 + \alpha_2\beta_1\right) \cong \begin{bmatrix} \left(\alpha_1\alpha_2 - \beta_1\beta_2\right) & -\left(\alpha_1\beta_2 + \alpha_2\beta_1\right) \\ \left(\alpha_1\beta_2 + \alpha_2\beta_1\right) & \left(\alpha_1\alpha_2 - \beta_1\beta_2\right) \end{bmatrix}, \quad (10c)$$

retains the isomorphism. Similarly, for higher order matices, matrix multiplication retains the isomorphism. This is clearly true for matrix addition and scalar multiplication and it is readily shown to be true under matrix multiplication. For example, extending the previous calculations shows that there is an isomorphism between the 2×2 complex matrix and the 4×4 real matrix

$$\begin{bmatrix} \left(\alpha_{11} + j\beta_{11}\right) & \left(\alpha_{12} + j\beta_{12}\right) \\ \left(\alpha_{21} + j\beta_{21}\right) & \left(\alpha_{22} + j\beta_{22}\right) \end{bmatrix} \cong \begin{bmatrix} \begin{pmatrix} \alpha_{11} & -\beta_{11} \\ \beta_{11} & \alpha_{11} \end{pmatrix} & \begin{pmatrix} \alpha_{12} & -\beta_{12} \\ \beta_{12} & \alpha_{12} \end{pmatrix} \\ \begin{pmatrix} \alpha_{21} & -\beta_{21} \\ \beta_{21} & \alpha_{21} \end{pmatrix} & \begin{pmatrix} \alpha_{221} & -\beta_{22} \\ \beta_{22} & \alpha_{22} \end{pmatrix} \end{bmatrix}. \quad (10d)$$

In fact, extending these results we find that if we have two matrices and their isomorphisms, e.g. $\mathbf{A} \cong \mathbf{C}$ and $\mathbf{B} \cong \mathbf{D}$, their product remains isomorphic, i.e $\mathbf{AB} \cong \mathbf{CD}$. With these preliminaries completed we turn to the first half of Goodman's analysis.

Show $\frac{1}{2}\mathbf{w}^T\mathbf{R}^{-1}\mathbf{w} = \mathbf{z}^T\mathbf{C}^{-1}\mathbf{z}$

We first show that, for the isomorphic matrices $\mathbf{R}$ and $\mathbf{C}$, if $\mathbf{C}$ is nonsingular then $\mathbf{R}$ is nonsingular, in which case their inverses exist, and the inverse of $\mathbf{R}$ is isomorphic to the inverse of $\mathbf{C}$, i.e. $\mathbf{R}^{-1} \cong \mathbf{C}^{-1}$.

Since $\mathbf{C}$ is nonsingular then $\mathbf{C}^{-1}$ exists and we have $\mathbf{CC}^{-1} = \mathbf{I}_N$ where $\mathbf{I}_N$ is the Nth order identity matrix having ones along the diagonal and zeros elsewhere. Denote the form of $\mathbf{C}^{-1}$ as

$$\mathbf{C}^{-1} = \left[\alpha^{ik} + j\beta^{ik}\right]. \quad (11a)$$

This matrix is isomorphic to a real matrix. Denote the isomorphism of $\mathbf{C}^{-1}$ as S. Then, by our preliminary discussion, $\mathbf{RS}$ is isomorphic to $\mathbf{CC}^{-1}$ and, therefore, $\mathbf{RS}$ is also isomorphic to $\mathbf{I}_N$. But since we have the isomorphism $\begin{bmatrix} 1 & 0 \\ 0 & 1 \end{bmatrix} \cong 1$ then there is an isomorphism between the identity matrices $\mathbf{I}_{2N} \cong \mathbf{I}_N$. Consequently, $\mathbf{RS} = \mathbf{I}_{2N}$ and $\mathbf{S} = \mathbf{R}^{-1}$ which has the form

$$\mathbf{R}^{-1} = \begin{bmatrix} \alpha^{ik} & -\beta^{ik} \\ \beta^{ik} & \alpha^{ik} \end{bmatrix}, \quad (11b)$$

proving our assertion. This result works in both directions, i.e. if $\mathbf{R}$ is nonsingular then $\mathbf{C}$ is nonsingular; their inverses exist and they are isomorphic, i.e. $\mathbf{C}^{-1} \cong \mathbf{R}^{-1}$.

For the matrices in (6) we have the isomorphism $2\mathbf{R} \cong \mathbf{C}$. Given the preceding, we have $\mathbf{R}^{-1}/2 \cong \mathbf{C}^{-1}$ and we can write

$$\frac{1}{2}\mathbf{w}\mathbf{R}^{-1}\mathbf{w} = \sum_{i,k=1}^{N} \begin{bmatrix} x_i & y_i \end{bmatrix} \begin{bmatrix} \alpha^{ik} & -\beta^{ik} \\ \beta^{ik} & \alpha^{ik} \end{bmatrix} \begin{bmatrix} x_k \\ y_k \end{bmatrix},$$

$$= \sum_{i,k=1}^{N} \alpha^{ik}(x_i x_k + y_i y_k) - \beta^{ik}(x_i y_k + y_i x_k). \tag{12a}$$

$$\mathbf{z}\mathbf{C}^{-1}\mathbf{z} = \sum_{i,k=1}^{N} (x_i - jy_i)(\alpha^{ik} + j\beta^{ik})(x_i + jy_i),$$

$$= \sum_{i,k=1}^{N} \alpha^{ik}(x_i x_k + y_i y_k) - \beta^{ik}(x_i y_k + y_i x_k). \tag{12b}$$

Therefore, $\frac{1}{2}\mathbf{w}\mathbf{R}^{-1}\mathbf{w} = \mathbf{z}\mathbf{C}^{-1}\mathbf{z}$.

Show $2^{2N}|\mathbf{R}| = |\mathbf{C}|^2$

We next show that, for the isomorphic matrices **R** and **C**, that their determinants satisfy the required relation. For this development, analytic signals permit a small restriction on the generality of Goodman's result.

Since **C** is Hermitian, there exists a unitary matrix **U**, i.e. $\mathbf{U}^{-1} = \mathbf{U}^{*T}$, that transforms **C** to a real diagonal matrix[2]. To determine that matrix we solve the eigenvalue problem

$$\mathbf{C}\mathbf{u} = \lambda\mathbf{u} \rightarrow |\mathbf{C} - \lambda \mathrm{I}| = 0, \tag{13a}$$

where λ is an eigenvalue and **u** is an eigenvector which are determined from the characteristic equation $|\mathbf{C}-\lambda\mathbf{I}| = 0$. For an Nth order matrix **C** there will be N such values λ_n and $\mathbf{u}_n$ where $n = 1,\cdots,N$. These solutions can be assembled in matrix form

$$\mathbf{C}\mathbf{U} = \mathbf{C}[\mathbf{u}_1,\cdots,\mathbf{u}_N] = [\lambda_1\mathbf{u}_1,\cdots,\lambda_N\mathbf{u}_N] = \mathbf{U}\mathbf{D}_N, \tag{13b}$$

equivalently,

$$\mathbf{U}^{-1}\mathbf{C}\mathbf{U} = \mathbf{D}_N, \tag{13c}$$

where $\mathbf{D}_N$ is a real Nth order diagonal matrix with the eigenvalues along the diagonal. If we perform the conjugate transpose operations to (13c), we obtain

$$\left(\mathbf{U}^{-1}\mathbf{C}\mathbf{U}\right)^{*T} = \mathbf{U}^{*T}\mathbf{C}\left(\mathbf{U}^{-1}\right)^{*T} = \mathbf{D}_N, \tag{14a}$$

and see that **U** is unitary, i.e. $\mathbf{U}^{-1} = \mathbf{U}^{*T}$.

Since **U** is unitary, its inverse exists and is isomorphic to a real orthogonal matrix which we denote $\mathbf{M}^{-1}$, i.e. $\mathbf{U}^{-1} \cong \mathbf{M}^{-1}$. More explicitly

---

[2] Birchoff, G. and S. MacLane, *A Survey of Modern Algebra*, Macmillan Co. 1971, 3d Ed. pg. 267, Theorem 26.

$$\mathbf{U}^{-1} = \mathbf{U}^{*T} = \left[a_{i,k} + jb_{i,k}\right]^{*T} = \left[a_{k,i} - jb_{k,i}\right] \cong$$

$$\begin{bmatrix} a_{k,i} & b_{k,i} \\ -b_{k,i} & a_{k,i} \end{bmatrix} = \begin{bmatrix} a_{i,k} & -b_{i,k} \\ b_{i,k} & a_{i,k} \end{bmatrix}^T = \mathbf{M}^T = \mathbf{M}^{-1} . \quad (14b)$$

Consequently, **M** is orthogonal.

Then, for the isomorphic matrices R and C in (6), we have the isomorphism $\mathbf{U}^{-1}\mathbf{C}\mathbf{U} \cong 2\mathbf{M}^{-1}\mathbf{R}\mathbf{M}$ where **U** and **M** are, respectively, unitary and orthogonal matices. Consequently, from (13c), we have the isomorphism

$$\mathbf{D}_N = [\lambda_n] \cong \begin{bmatrix} \lambda_n & 0 \\ 0 & \lambda_n \end{bmatrix} = \mathbf{D}_{2N} . \quad (14c)$$

where $2\mathbf{M}^{-1}\mathbf{R}\mathbf{M} = \mathbf{D}_{2N}$.

To show that $2^{2N}|\mathbf{R}| = |\mathbf{C}|^2$ for the isomorphic matrices in (6) we first transform 2**R** and **C** to diagonal matrices by means of the isomorphic pair **U** and **M** which are, respectively, unitary and orthogonal. The determinants of these matrices are

$$|2\mathbf{R}| = \prod_{n=1}^{N} \begin{vmatrix} \lambda_n & 0 \\ 0 & \lambda_n \end{vmatrix} = \prod_{n=1}^{N} \lambda_n^2 , \quad |\mathbf{C}| = \prod_{n=1}^{N} \lambda_n . \quad (15a,b)$$

Therefore,

$$|2\mathbf{R}| = 2^{2N}|\mathbf{R}| = |\mathbf{C}|^2 . \quad (15c)$$

Collecting results, the probability density function in (9) becomes

$$p(\mathbf{z}) = \frac{\exp\left(-\mathbf{z}^T\mathbf{C}^{-1}\mathbf{z}\right)}{\pi^N|\mathbf{C}|} . \quad (16)$$

APPENDIX 8-B: SCATTERING AND ABSORPTION BY BUBBLES

When acoustic energy in incident upon a field of bubbles, such as may be found in a school of fish possessing swim bladders or the layer of entrained air bubbles below an agitated sea surface, the bubbles compress and decompress in response to the incident wave. If the incident wave length is large compared to the typical bubble size, the bubbles compress and decompress uniformly and this action causes a spherical wave to be radiated away from each bubble. If the incident wave length is small compared to the typical bubble size, the situation is more complicated and the scattered wave must be analyzed by the methods used in Appendix 7-C. Here we consider only the long wave length case. If bubble compression and decompression is reversible, the bubbles are considered to be ideal bubbles and the conversion of incident energy into scattered energy incurs no loss. For real bubbles this is not the case and losses are incurred. The losses typically arise because the compression and decompression of the gas within the bubbles is not adiabatic and as a result net work is performed on the bubbles. These and other mechanisms convert some of the incident energy into heat and cause the bubble motion to lag the driving force of the incident wave. Consequently, the incident energy is partly scattered and partly absorbed. Also, if the separation of the bubbles is small compared to the acoustic wavelength, the bulk modulus of the bubble filled medium will be less than the bulk modulus of water and the sound speed will be decreased. This can result in slower propagation speeds and refraction of the acoustic wave as it passes through the bubble field. This can be an important effect, but, in the following, we restrict our examination to the scattering and absorption of acoustic energy by bubbles. In fact, in the following, we restrict our examination to the scattering and absorption by a single bubble which we extend to the scattering and absorption by a field of bubbles in Section 8.3. Our approach follows that of the National Defense Research Committee[1] whose work during WWII lead to solutions to the problems of scattering and absorption by bubbles.

8.1-B: Scattering by an Ideal Bubble

A plane CW acoustic field is assumed to be incident upon an ideal bubble with pressure and intensity which can be written

$$p = A\exp(j\omega t), \qquad I = \frac{|A|^2}{2\rho_0 c}. \tag{1a,b}$$

The scattered pressure field and intensity at a distance from the bubble can be written

$$p_S = \frac{B}{r}\exp(j(\omega t - kr)), \qquad I_S = \frac{|B|^2}{2\rho_0 c r_2}. \tag{2a,b}$$

The scattering cross section of the bubble is then given by

---

[1] *Physics of Sound in the Sea*, Summary Technical Report of Division 6, NDRC Vol. 8, Chapter 28, Washington D.C., 1946.

$$\sigma_S = 4\pi|B|^2/|A|^2, \tag{3}$$

where B is unknown.

To determine B we match the total exterior acoustic pressure to the pressure field within the bubble at the boundary. The field within the bubble follows the driving incident field and is given by

$$p_I = A_I \exp(j\omega t), \tag{4}$$

where A_I is the unknown amplitude of the interior field. At the surface of the bubble the total pressure then satisfies the relation

$$p + p_S = p_I \quad \rightarrow \quad A + \frac{B}{a}\exp(-jka) = A_I, \tag{5a,b}$$

where a is the radius of the bubble. This gives a relation for B in terms of the unknown A_I which requires a second relation for its determination.

The radial velocity of the bubble is continuous across the boundary but we must distinguish between the radial motion of the bubble and the to-and-fro motion caused by the incident wave. Only the former is important to the scattering process. Consequently, the radial particle velocity exterior to the bubble is obtained from Newton's second law and the scattered field

$$\rho_0 \frac{\partial v_r}{\partial t} = -\frac{\partial p_S}{\partial r}. \tag{6a}$$

Substituting from (2a) and integrating gives the radial velocity

$$v_R = -\frac{jB}{\omega\rho a^2}\exp(j\omega t). \tag{6b}$$

Since the bubble volume is $V = 4\pi a^3/3$, the surface velocity of the bubble can be expressed

$$v_R = \frac{1}{4\pi a^2}\frac{\partial V}{\partial t}. \tag{7a}$$

The rate of change of volume in this expression is related to the rate of change of pressure within the volume but the relationship must satisfy the laws of thermodynamics. Since the volume change is very rapid, we assume, initially, that the change occurs under adiabatic conditions. For this case the pressure and volume satisfy the relation pV^γ = constant where γ is the ratio of the specific heat at constant pressure to the specific heat at constant volume. Differentiating this relation we obtain

$$\frac{1}{p_I}\frac{\partial p_I}{\partial t} + \frac{\gamma}{V}\frac{\partial V}{\partial t} = 0. \tag{7b}$$

From (4) and (7b) we obtain the expression for the volume velocity of the bubble

$$\frac{\partial V}{\partial t} = -\frac{j\omega V A_I}{p\gamma}\exp(j\omega t), \tag{7c}$$

and from (7a) we obtain the radial velocity of the bubble surface

$$v_R = -\frac{j\omega a A_I}{3p\gamma}\exp(j\omega t). \tag{7d}$$

Equating this with (6b) gives the amplitude of the pressure within the bubble in terms of the scattered amplitude

$$A_I = \frac{3B\gamma p}{\omega^2 \rho_0 a^3}. \tag{7e}$$

Substituting this into (5b) gives the relation for the scattered field amplitude

$$\frac{B}{A} = \frac{a}{\left(\omega_R^2/\omega^2 - 1\right) + jka}, \qquad \omega_R^2 = \frac{3\gamma p}{\rho_0 a^2}, \tag{8}$$

where $k = \omega/c$. The scattering cross section of the bubble is found from (3) and (8) to be

$$\sigma_S = \frac{4\pi a^2}{\left(\frac{\omega_R^2}{\omega^2} - 1\right)^2 + (ka)^2}. \tag{9}$$

Since the scattering cross section peaks at the angular frequency $\omega = \omega_R$, this frequency is called the angular resonance frequency.

8.2-B: Scattering by an Actual Bubble

For the case of an actual bubble, compression and decompression incurs losses that cause energy to be converted into heat which results in the internal pressure lagging the external driving pressure and the bubble surface velocity lagging that of an ideal bubble. Consequently, for an actual bubble we need to replace (5) and (7d), respectively, by

$$p + p_S = p_I - \alpha v_R \quad \rightarrow \quad A + \frac{B}{a}\exp(-jka) = A_I + \frac{\alpha B}{\omega \rho_0 a^2}, \tag{10a,b}$$

$$v_R = -\frac{j\omega a A_I}{3p\gamma}\exp(j(\omega t - \beta)). \tag{11}$$

In these expressions, α and β are small constants. The quantity αv_R introduces a balancing pressure, i.e. when the bubble is being compressed $v_R < 0$, the internal pressure lags the external pressure by the amount αv_R and when the bubble is expanding $v_R > 0$ the external pressure lags the internal pressure by the amount αv_R. Similarly, the quantity β introduces a lag in the bubble surface velocity. Proceeding as before we find in place of (7e) the amplitude of the pressure within the bubble to be

$$A_I = \frac{3B\gamma p}{\omega^2 \rho_0 a^3} \exp(j\alpha), \tag{12}$$

and in place of (9) the scattered field amplitude is given by

$$\frac{B}{A} = \frac{a}{\left(\omega_R^2/\omega^2 - 1\right) + j\delta}, \tag{13a}$$

$$\omega_R^2 = \frac{3\gamma p}{\rho_0 a^2}, \qquad \delta = \frac{\omega a}{c} + \frac{\alpha}{\omega \rho_0 a} + \beta \frac{\omega_R^2}{\omega^2}. \tag{13b,c}$$

Consequently, the scattering cross section for an actual bubble is found to be

$$\sigma_S = \frac{4\pi a^2}{\left(\dfrac{\omega_R^2}{\omega^2} - 1\right)^2 + \delta^2}. \tag{14}$$

We see that the resonance frequency for an actual bubble is the same as for an ideal bubble but the resonance peak is smaller in magnitude. However, since α and β are small the difference in magnitude is small.

8.3-B: Absorption by an Actual Bubble

When an acoustic wave passes over a bubble, part of the incident energy is scattered by the bubble and part is absorbed by the bubble and converted to heat. Both effects serve to extinguish the incident acoustic wave and we define the extinction cross section to include the effects of both scattering and absorption

$$\sigma_E = \frac{P_E}{I_0}, \tag{15}$$

where P_E is the total power extinguished from the incident wave. This quantity is given by the expression

$$P_E = \left\langle \mathrm{Re}(p)\, \mathrm{Re}\left(\frac{\partial V}{\partial t}\right) \right\rangle, \tag{16a}$$

where the first factor is the pressure of the incident wave and the second factor is the volume velocity of the bubble and both factors are the real physical values. Also, the brackets $\langle\rangle$ denote time average.

Expanding (16a) using (1a) and also (7c) and (7e) we obtain

$$P_E = \frac{1}{4}\left\langle \left(A\exp(j\omega t) + A^*\exp(-j\omega t)\right)\left(-j\frac{4\pi B}{\omega\rho_0}\exp(j\omega t) + j\frac{4\pi B^*}{\omega\rho_0}\exp(-j\omega t)\right)\right\rangle . \quad (16b)$$

From (13a) we see that B has the form AF where A is the amplitude of the incident wave and F is the right member of (13a). Expanding (16b) and setting the time averages over $\exp(\pm j2\omega t)$ to zero, we obtain

$$P_E = -\frac{2\pi|A|^2 F_I}{\omega\rho_0} = \frac{2\pi|A|^2}{\omega\rho_0}\left[\frac{a\delta}{\left(\omega_R^2/\omega^2 - 1\right)^2 + \delta^2}\right], \quad (17)$$

where F_I is the imaginary part of F. Substituting into (15), the extinction cross section can be written

$$\sigma_E = \frac{\delta}{ka}\left[\frac{4\pi a^2}{\left(\omega_R^2/\omega^2 - 1\right)^2 + \delta^2}\right]. \quad (18)$$

The extinguished power is the sum of the scattered and absorbed power and therefore the extinction cross section can be expressed

$$\sigma_E = \sigma_S + \sigma_A , \quad (19)$$

where σ_A is the absorption cross section which is given by

$$\sigma_A = \left(\frac{\delta}{ka} - 1\right)\left[\frac{4\pi a^2}{\left(\omega_R^2/\omega^2 - 1\right)^2 + \delta^2}\right]. \quad (20)$$

For $\delta = ka$, i.e. if we reduce (18) and (20) to the case of an ideal bubble, we see that that the absorption cross section vanishes and the total extinguished power is due to scattering.

CHAPTER 9
DETECTION FUNDAMENTALS

Target detection is the process of determining the presence of a target echo in an active sonar return or a target radiation signature in a passive sonar interception. The target portion of the received data can be masked by reverberation and/or background noise and sophisticated signal and data processing techniques are required to detect the presence of the target signal. Detection theory provides methods for determining optimal detection processing structures and for determining the performance of those optimal detection processing structures.

Optimal detection techniques balance the need for detecting targets with certainty against the need for an acceptable false alarm rate. Target detectability is, of course, required for the ASW mission to be accomplished but, at the same time, a low false alarm rate is required to minimize the burden on the system. The detectability of the target is found to be directly dependent upon the strength of the received target signal, i.e. the echo return or intercepted target radiation, as compared to the received reverberation and/or noise level. The false alarm rate is found to be primarily dependent upon the background reverberation and/or noise level. The sonar system is designed to achieve a level of detection performance, which is characterized by a minimum probability of detecting a standard target within a standard environment and by the maximum allowable probability of false alarm. The minimum required probability of detection and the maximum acceptable false alarm rate, that the system must meet, are the specified detection performance parameters. The standard target is described by the expected target strength for active systems and by the expected target source level and frequency band for passive systems. The standard environment is specified by the system figure of merit (FOM), which is a measure of the maximum detection range of the system within a given background of reverberation and/or noise. For a specified environment, the FOM determines the detection zones within the environment and the maximum target range at which the design detection and false alarm probabilities can be met. In summary form, the preceding comments define the detection problem but they are in need some elaboration to make them fully comprehensible. This is done in the following.

9.1 Active and Passive Detection Concepts

Active and passive detection systems have similar functional structures but differ somewhat in the details of how these functions are implemented. In general terms both active and passive systems process received beam data as indicated in Figure 1. The acoustic beam data x(t) are filtered to the frequency bands of interest to reject as much noise as possible at the outset which produces beam-band data for each channel of data which, for the nth frequency channel of data, is denoted $y_n(t)$. The processing for channel n beam-band data consists of signal compression via optimal likelihood ratio techniques to produce the detector output $\lambda_n(t)$. Follow-on post processing may be performed to enhance the signal-to-noise ratio to produce the enhanced detector output $\lambda'_n(t)$. The resultant output is compared to the detection

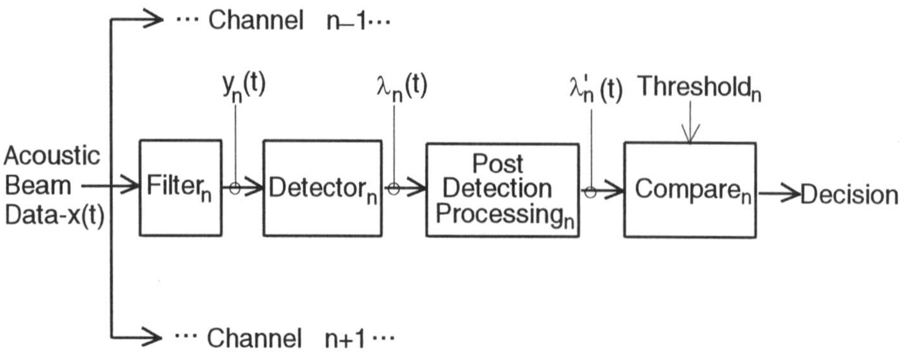

Figure 1 A General Detection System

threshold which leads to a target present or target absent decision. For active detection systems the selected frequency bands will include the bands of the transmitted signal with guard bands to accommodate Doppler shifting that may be caused by the target and receiver. If a single waveform is transmitted, only the band of that waveform is processed but if multiple waveforms are transmitted then multiple frequency bands will be selected. For passive data, the selected frequency bands range from broad band to narrow band depending upon the search strategy. Broad band data can include the full range of expected frequencies from all targets while narrow band data may encompass only a single line component from a specific target. Intermediate bands can also be selected that would include other characteristic signals of the targets of interest and, in all cases, the selected bands will generally include guard bands to accommodate frequency shifting that may occur due to Doppler shifting of the signal that is caused by both the target and receiver motion.

The detector for the active system is configured to search for signal returns that match the transmitted signal, i.e. have the same spectral band and time structure, but with unknown time of arrival, unknown acoustic channel effects and unknown target reflection effects. The unknown time of arrival is accommodated by sequentially searching the received data in time and converting time of arrival to target range via the relation r = ct/2 where c is the speed of sound and t is the receive time for the signal which is the time required for a round trip to the target and back. The ranging clock starts running at the start of transmission and ends at the time corresponding to the maximum detection range. Since the receive processing is generally "blanked" during the transmission period, the target search usually starts at the end of the transmission. The Doppler compression on the target reflected signal is generally not known but this is accommodated by searching over a range of possible Doppler bands. For each beam-band and Doppler channel the detector signal-to-noise levels can be estimated by forming the ratio of the detector output level to the detector background noise level. The background noise level is estimated by averaging the detector output over time in which a split window averager, as indicated in Figure 2, generally performs the averaging process. The duration of the split exceeds the expected compressed echo duration to ensure that target echo data does not contaminate the estimate of the background noise level. Additionally, ranking and censoring strategies can be employed to reject outliers, due to target and reverberation intruders, in the averaging windows. With or without post processing the detector output is compared to a threshold for the purpose of either facilitating post processing or for detection analysis. If post processing is employed, its output is compared to a detection threshold for detection analysis. Post processing is generally implemented to compensate for acoustic channel effects and target reflection effects. Each of these effects can result in a spread of the signal energy in both time and frequency which generally degrades the signal compression techniques and follow-on detection performance. For the case of single and multi-waveform transmissions, post detection averaging over energy can be performed to accumulate waveform energies. For the case of multi-waveform transmissions various post detection cross-waveform combining techniques can be employed to further reduce the effects of signal spreading and to enhance signal-to-noise ratio.

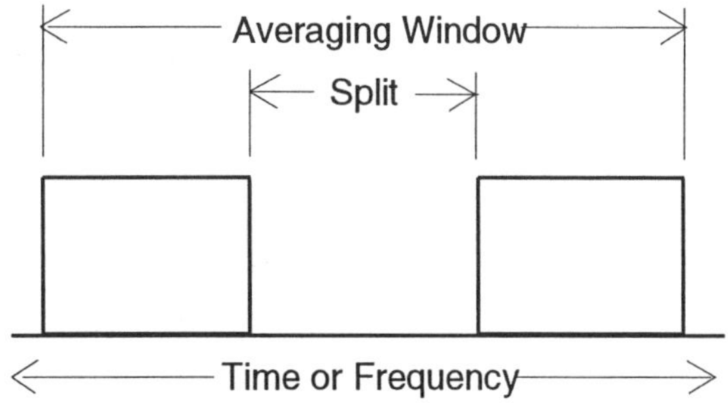

Figure 2 Split Window Averager

The detector for the passive system is configured to search for signals of approximately known spectral band but of unknown structure. It performs this function by continuously searching the spectral bands of interest for power levels that exceed the background level. Time averaging is typically performed over short time periods of a few minutes and/or over long time periods of up to an hour or more. Short time averaging is required to ensure that the signal remains in the processing band during the integration period because, due to the changing geometry between the target and receiver, the intercepted target signature can drift from beam-to-beam and from band-to-band during the processing period. Long time averaging is required to raise a weak target signal out of the background noise because, in the integration process, target signature energy accumulates only marginally more quickly than noise energy. To determine the received signal-to-noise ratio, the time averaged spectra from several frequency channels can be further averaged over frequency to obtain measures of the background noise level. The frequency averaging is usually performed with a split window averager, as indicated in Figure 2, in which the split exceeds the expected signal spectral width to ensure that target signature energy does not contaminate the background noise level estimate. Ranking and censoring strategies can also be employed to reject outliers in the averaging windows. In addition, post processing can be employed to combine signals from different frequency bands or from different beams.

For both active and passive detection systems, the background signal-to-reverberation and/or noise level of the detection system output is monitored, and the detection threshold is set at a level above the background. When the detection system output exceeds the detection threshold, a detection decision is triggered. The performance of the detection system is, of course, designed to be optimal and dependent upon its structure but sometimes they do not rise to this level for various reasons. While active and passive detectors are generally structured as optimal likelihood ratio detectors, some environments do not offer the conditions required for optimal performance by these systems and post processing is required to recover some of the lost performance. The post processors are generally based upon heuristic considerations and serve as an ad-hoc add-on to the likelihood ratio detector for a particular environment. When this occurs the overall system is generally known to be an improvement over the likelihood ratio detector within the environment of interest, but it is generally not known to be an optimal detection system, i.e. other post processors may produce a better system. Also, while modern ASW systems ideally search all beams of data over all bands of interest, the substantial data processing load that is incurred frequently results in suboptimal compromises in coverage.

As an historical aside concerning terminology, the detection function, which is the entire function depicted in Figure 1, is a generalization of an older concept of the detector function, which is shown as a single element in the overall detection function in Figure 1. During the early days of radio, the detector performed the entire detection function by extracting the signal information content, i.e. the modulation, from the carrier signal which it displayed as an audio signal. In modern target detection theory the modulation is still extracted from the signal, either within or prior to the detector, but the importance of the "audio" content of the resulting signal depends solely on the fact of its presence or absence. We determine its presence or absence by compressing the signal energy into a localized pulse and comparing its signal-to-noise ratio to a threshold. The principal function of the detector in Figure 1 is to perform this signal compression. Follow-on post-processing can also be performed on the data, after the detector, to enhance the signal-to-noise of the detected signal and the sum total of these functions, jointly, is referred to as the target detection function.

9.2 Detection Theory

Detection theory is concerned with the detection of signals masked by background interference. For the detection system in Figure 1 the signals are target echoes or target radiations and the received beam data x(t) consists of one or more target signals plus added background reverberation and/or noise or is just background reverberation and/or noise. For both cases the received data is always random and cannot be predicted but the set of possible receptions, i.e. windows of data, being of finite duration and constrained to a finite amplitude range, can be specified or at least imagined to be specified, i.e. for a sampling system there will be a finite number of samples in which each sample is quantized to one of a finite number of levels. The set of possible receptions, defined by the possible number of unique sample sets, constrains the population of received data and an omniscient observer could, in principle, partition the data into a signal plus reverberation and/or noise population and a reverberation and/or noise only population. In the absence of an omniscient observer, the detection system must decide whether the received data came from the signal plus reverberation and/or noise population or from the reverberation and/or noise only population. Effectively, the detection system must try to duplicate the partitioning of the omniscient observer and it uses a criterion against which the received data is tested to do so. Depending upon the criterion selected, the detection system partitioning will be more or less correct. Since to the nonomniscient observer, i.e. the detector, the data are random, but with presumed known statistics, an optimal criterion partitions the data such that data samples having a higher degree of probability of being from signal plus reverberation and/or noise are placed in the signal plus reverberation and/or noise population, i.e. a target detection is decided, and data having a higher degree of probability of being from reverberation and/or noise only are placed in the reverberation and/or noise only population, i.e. no target present is decided. Each decision will have an associated probability of being correct which implies that the decisions reached by the detection system will not always be correct and, in fact, two kinds of errors can arise. The false alarm error, referred to as an error of the first kind, occurs when a target is, in fact, not present but a target present decision is reached. The missed detection error, referred to as an error of the second kind, occurs when a target is, in fact, present but a no target decision is reached. Errors of the first kind are generally more expensive than errors of the second kind because, whereas a missed detection requires no action and can be rectified during the next interrogation period, resources must be committed to deal with every false detection.

Denoting the hypothesis that a target is present as H_1 and the hypothesis that no target is present as H_0, the probabilities of correct decisions and errors are represented by the conditional probabilities

$$\begin{aligned} P_D &= P(\text{correct detection} \mid H_1), \\ P_{FD} &= P(\text{false dismissal} \mid H_1), \\ P_{FA} &= P(\text{false alarm} \mid H_0), \\ P_{CD} &= P(\text{correct dismissal} \mid H_0). \end{aligned}$$

These conditional probabilities are related by the expressions

$$P_D = 1 - P_{FD}, \tag{1a}$$

$$P_{FA} = 1 - P_{CD}, \tag{1b}$$

where statisticians refer to P_D as the power of the test and to P_{FA} as the size of the test.

Neyman-Pearson / Likelihood Ratio Criterion

For sonar (and radar) applications the Neyman-Pearson criterion partitions the data such that P_D is maximized for a given maximum acceptable P_{FA}. This criterion is shown below to be equivalent to the likelihood ratio test for partitioning the data, and, in fact, many criteria can be shown to be equivalent to the likelihood ratio test.

The likelihood ratio test is a simple mechanical procedure for reaching detection decisions. For each set of received time sampled beam-band data

$$\mathbf{y} = (y_1, y_2, \cdots, y_m, \cdots y_M)^T, \tag{2}$$

where the individual data samples y_m are collected together as a vector with superscript T denoting transpose, the likelihood ratio is computed and a test of the likelihood ratio value is performed to decide whether the data comes from a target, with additive reverberation and/or noise, or if it is reverberation and/or noise only. The test criterion partitions the data and is based upon the statistics of the data. The conditional probability density functions for the data populations are denoted $p(\mathbf{x}|H_1)$ and $p(\mathbf{x}|H_0)$ where the argument $\mathbf{x}$ is an n-tuple of scalar variables that represent beam-band samples. The likelihood functions of the received data are obtained by substituting data vector values, from (2), into these functions giving the random values $p(\mathbf{y}|H_1)$ and $p(\mathbf{y}|H_0)$[1]. When the data comes from a target, with additive reverberation and/or noise, the likelihood $p(\mathbf{y}|H_1)$ is expected to have a larger value on average than for the case of the reverberation and/or noise only likelihood $p(\mathbf{y}|H_0)$. Conversely, when the data is reverberation and/or noise only, the likelihood $p(\mathbf{y}|H_0)$ is expected to have a larger value than the target plus reverberation and/or noise likelihood $p(\mathbf{y}|H_1)$. Accordingly, the likelihood ratio, $p(\mathbf{y}|H_1)/p(\mathbf{y}|H_0)$, is expected to be large for target data and to be small for reverberation and/or noise only data which implies the likelihood ratio test

$$\frac{p(\mathbf{y}|H_1)}{p(\mathbf{y}|H_0)} \mathop{\gtrless}_{H_0}^{H_1} \Lambda_T, \tag{3a}$$

where the H_1 decision is made, if the likelihood ratio exceeds the threshold Λ_T, and the H_0 decision is made if the likelihood ratio fails to exceed the threshold. An attractive test because it is simple but, to be an appropriate test for the sonar problem, it must be shown to be equivalent to the Neyman-Pearson criterion.

Since the population of receive beam-band data and their corresponding likelihood ratios are, in principle, known, all likelihood ratios can be ordered in an ascending sequence

$$0 \le \Lambda_1 \le \Lambda_2 \le \cdots \le \Lambda_k \le \cdots \le \Lambda_N < \cdots \infty. \tag{3b}$$

The data can be partitioned at the kth sample such that all receive beam-band data having likelihood ratios greater than or equal to Λ_k give rise to the target present decision and all samples having likelihood ratios less than Λ_k give rise to the no-target decision. This gives two sets of data that lead to the following decisions

---

[1] Note that the probability density function, with a particular n-tuple $\mathbf{x}$ substituted, is a real deterministic value while the likelihood function, defined on the population of possible returns $\mathbf{y}$, is a real random variable.

$$\{y_k, y_{k+1}, \cdots, y_N\} \rightarrow H_1, \tag{4a}$$

$$\{y_1, y_2, \cdots, y_{k-1}\} \rightarrow H_0, \tag{4b}$$

where the data is subscripted in accordance with its likelihood ratio and there are N unique data sets. The likelihood ratio test becomes, with Λ_k the decision threshold,

$$\frac{p(\mathbf{y}_n|H_1)}{p(\mathbf{y}_n|H_0)} \underset{H_0}{\overset{H_1}{\gtrless}} \Lambda_k. \quad n = 1,2,3\ldots,N \tag{5}$$

The probability of detection for this test is the probability of receiving one of the data sets in (4a), assuming that there is a target, and the probability of false alarm for this test is, similarly, the probability of receiving one of the same data sets but assuming that there is no target. These probabilities are given by

$$P_D = \sum_{j=k}^{N} p(\mathbf{y}_j|H_1)\Delta\mathbf{y}_j, \tag{6a}$$

$$P_{FA} = \sum_{j=k}^{N} p(\mathbf{y}_j|H_0)\Delta\mathbf{y}_j, \tag{6b}$$

where $\Delta\mathbf{y}_j$ denotes the product of quantizing bin sizes in the digitizing process.

To ensure that the likelihood ratio criterion satisfies the Neyman-Pearson criterion, the threshold at which the data is partitioned Λ_k is chosen to include a maximum amount of data in the signal plus noise decision set, defined by (4a), which maximizes the probability of detection in (6a), while keeping the P_{FA} in (6b) below a maximum acceptable false alarm rate. (Note that, if all data were included in the target present sample set, both P_D and P_{FA} would be unity and, conversely, if all data were included in the no-target sample set, both would vanish.) This partitioning of the data satisfies the Neyman-Pearson criterion of maximizing P_D while minimizing P_{FA}. However, since the data can be partitioned in other ways that don't conform to the likelihood ratio partitioning, it must be determined whether or not an alternative test, assumed to also satisfy the Neyman-Pearson criterion, might yield a higher probability of detection for the same false alarm rate. In mathematical terminology, the likelihood ratio test has been shown to be sufficient to formally satisfy the Neyman-Pearson criterion but the question remains to be answered as to whether it is it a necessary test to satisfy the Neyman-Pearson criterion, i.e. might an alternative test also formally satisfy the Neyman-Pearson criterion but with higher P_D for a given P_{FA}?

To show that the likelihood ratio test is both sufficient and necessary, modify the data partition in (4) by moving one or more samples from the no-target data set to the target present data set and one or more samples from the target present data set to the no-target data set. The new probabilities of detection and false alarm can be written

$$P'_D = P_D + \left(\sum_m \Lambda_m p(y_m|H_0)\Delta y_m - \sum_n \Lambda_n p(y_n|H_0)\Delta y_n \right), \tag{7a}$$

$$P'_{FA} = P_{FA} + \left(\sum_m p(y_m|H_0)\Delta y_m - \sum_n p(y_n|H_0)\Delta y_n \right), \tag{7b}$$

where P_D and P_{FA} are given by (6) and where the mth data samples were added to the target present data set (removed from the no-target data set) and the nth data samples were added to the no-target data set (removed from the target present data set). Note that each of the Λ_m likelihood ratios are less than each of the Λ_n likelihood ratios. Three different cases can arise in this transfer of data:

1. The term in parentheses in (7a) is greater than zero which necessarily implies that the term in parentheses in (7b) is also greater than zero. For this case

   $$P'_D \geq P_D,$$

   $$P'_{FA} \geq P_{FA}.$$

   The new partition raises the probability of detection but it also raises the probability of false alarm which would then exceed the maximum acceptable probability of false alarm and would violate the Neyman-Pearson criterion.
2. The term in parentheses in (7b) is less than zero which necessarily implies that the term in parentheses in (7a) is also less than zero. For this case

   $$P'_D \leq P_D,$$

   $$P'_{FA} \leq P_{FA}.$$

   The new partition lowers the probability of false alarm but it also lowers the probability of detection which would then not be the minimum probability of detection for the chosen maximum acceptable probability of false alarm and would violate the Neyman-Pearson criterion.
3. The term in parentheses in (7a) is less than zero and the term in parentheses in (7b) is greater than zero. For this case

   $$P'_D \leq P_D,$$

   $$P'_{FA} \geq P_{FA}.$$

   The new partition not only lowers the probability of detection but it also raises the probability of false alarm. On both counts this partition violates the Neyman-Pearson criterion.

It follows that no other test can meet the Neyman-Pearson criterion and exceed the likelihood ratio test performance. Therefore partitioning the data via the likelihood ratio test is not only a sufficient requirement for meeting the Neyman-Pearson criterion but it is also a necessary requirement for meeting the Neyman-Pearson criterion.

Receiver Operating Characteristics (ROC)

The receiver operating characteristics (ROC) of the detection system summarize the detection performance of the system in a set of mathematical curves. The output of the detection system is, in principle, the likelihood ratio or an equivalent statistic but, for the case of sonar (and radar) detection systems, the output may include suboptimal post processing. Here, we assume that the output is a likelihood ratio statistic but, in principle, ROC curves can in all cases be developed. While the detection system, consisting of a detector and post processor, was indicated in Figure 1, it is instructive to examine a "blackbox" view of the detection process in terms of input and output waveforms as depicted in Figure 3.

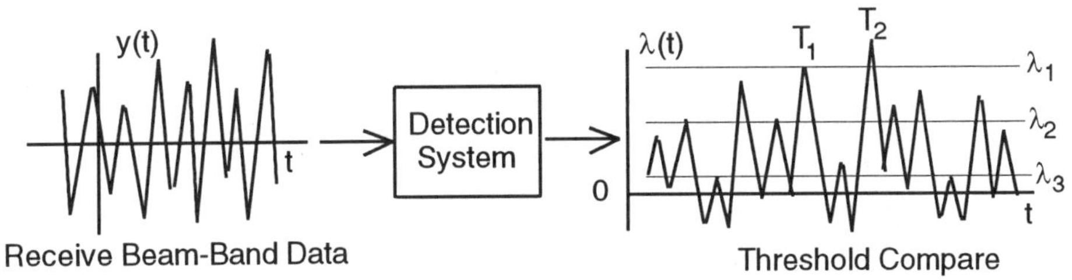

Figure 3 Detection System Input-Output

In Figure 3, the received beam-band data y(t) is processed by the detection system to produce the detection system output $\lambda(t)$ which is the likelihood ratio statistic or an equivalent statistic. The time sequence of computed output values are compared to a background noise threshold λ_T and exceedances trigger target present detection decisions. The choice of detector threshold is determined by the Neyman-Pearson criterion. In Figure 3 the threshold λ_1 would produce one detection decision, threshold λ_2 would produce 6 detection decisions and threshold λ_3 would produce 11 detection decisions. Not all of these decisions will necessarily be correct. If the peaks T_1 and T_2 were the only peaks that, in fact, correspond to targets, then threshold λ_1 would detect target T_2 with no false alarms but with one false dismissal; threshold λ_2 would detect targets T_1 and T_2 with 4 false alarms; and threshold λ_3 would detect T_1 and T_2 with 9 false alarms.

The probability of detection and probability of false alarm corresponding to these thresholds are found from the expressions

$$P_D = \int_{\lambda_T}^{\infty} p(\lambda|H_1)d\lambda\,, \qquad P_{FA} = \int_{\lambda_T}^{\infty} p(\lambda|H_0)d\lambda\,, \tag{8a,b}$$

where the conditional probability density functions, $p(\lambda|H_1)$ and $p(\lambda|H_0)$, are for the detection system output statistic $\lambda(t)$. Note that the probabilities in (8) are the equivalent of the probabilities in (6) but are expressed in terms of detection system output probabilities instead of input beam-band data probabilities and, in accordance with common practice, are expressed in terms of integrals instead of sums.

The detection performance expressions in (8) are parametric representations of the receiver operating characteristics (ROC) of the detection system, with parameter λ_T. The ROC curves are symbolically depicted in Figure 4 as a set of graphs of P_D versus P_{FA}. From (8) all graphs in the figure are seen to range between the coordinates (0,0) to (1,1) as the threshold λ_T varies from infinity to zero. This occurs because, at very high thresholds, no data trigger detection decisions while, at very low thresholds, all data trigger detection decisions. The slopes of the ROC curves are given by

$$\frac{dP_D}{dP_{FA}} = \frac{-p(\lambda_T|H_1)}{-p(\lambda_T|H_0)} = f(\lambda_T) = \Lambda_T \geq 0, \tag{9}$$

which shows that all curves increase steadily in the ascent from (0,0) to (1,1). The curves are necessarily concave upward in the ascent because, at large threshold values, the slope is greater than unity since the probability of a threshold exceedance being from a target is greater than the probability of the threshold exceedance being from noise only. Similarly, at low threshold values the slope is less than unity since the probability of a threshold exceedance being from noise is greater than the probability of the threshold exceedance being from a target. Equation (9) shows that the slope of the ROC curves at each operating point is equal to the likelihood ratio threshold Λ_T. However, since the detection statistic $\lambda(t)$ defines an equivalent statistic, the threshold of operational importance is the detector threshold λ_T. Of course, both thresholds are related by (9).

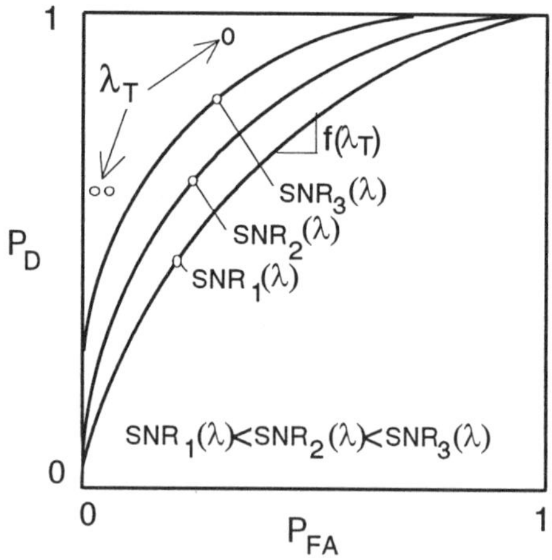

Figure 4 Receiver Operating Characteristics

The probability of detection hypothesizes a target present which produces a received signal level plus background noise and the probability of false alarm hypothesizes a received background noise level. It follows that each curve in the family of ROC curves can be expected to be parameterized by the signal and noise levels. In fact, it is generally found, that the parameter is the signal-to-noise ratio at the detection system output and the ROC graphs are in ascending order of increasing signal-to-noise ratio of the matched filter output which we denote SNR(λ). The order is ascending because an increase in SNR(λ) raises the probability of detection of the target for any given false alarm rate. The signal-to-noise ratio at the detection system output is called the detection index d and is given by

$$d = \mathrm{SNR}(\lambda) = \frac{\left[E(\lambda|H_1) - E(\lambda|H_0)\right]^2}{\mathrm{Var}(\lambda|H_0)}. \tag{10}$$

The detection index is related to the deflection signal-to-noise ratio of the output, denoted DSNR(λ), by the relation

$$\mathrm{DSNR}(\lambda) = \sqrt{d}\,. \tag{11}$$

The deflection ratio has historical interest but it has been largely replaced as a measure of detection performance by the detection index.

The family of ROC curves in Figure 4 define the performance of the detection system. By specifying a minimum acceptable detection rate and a maximum acceptable false alarm rate, the system operating point is defined and the required minimum detection index d of the system is specified by the ROC curve passing through the operating point. From this detection index, the minimum in-water detection threshold DT can be determined, which equals the minimum in-water signal-to-noise ratio needed to achieve the required detection performance. This detection threshold determines the maximum system figure of merit, as indicated in (2.4), or, equivalently, the maximum operational detection range of the system. Lastly, the detection system threshold λ_T corresponding to the selected ROC operating point is given by (8).

9.3 Summary

A summary of the elements of ASW sonar detection theory is encapsulated in Figure 5. This ASW sonar detection theory "map" implicitly assumes that the detector is a likelihood ratio detector but we have indicated that post processing might take place after the likelihood ratio detector - with appropriate changes to the figure. The figure can be viewed from a system operational viewpoint, a system test and evaluation viewpoint or a system design viewpoint.

From a system operational viewpoint, the system FOM, possible targets and the operational environment are set. These parameters jointly determine the "footprint" within the operational environment of the system, from (2.3), and the detection threshold, from (2.4). The detection threshold, in turn, specifies the detection index which, from the ROC curves, determines a range of possible P_D and P_{FA} values. A maximum P_{FA} value is selected which dictates the minimum P_D at which the system will operate and the matched filter threshold is found from (8). If the resulting P_D is not great enough, a higher P_{FA} must be tolerated. In brief terms, the tactical problem is to adjust the ROC operating point to achieve the mission objectives within the constraints of the system figure of merit.

From a system test and evaluation viewpoint, operationally important parts of the ROC curves must be determined. This entails performing detection studies on targets having known acoustic properties at known positions in acoustically known environments. From detection histories, P_D and P_{FA} can be measured versus received signal-to-noise ratios which, effectively, give operating points on the ROC curves. FOM can be estimated from maximum detection ranges at which P_D and P_{FA} are acceptable. Jointly, these values establish standards for system validation. Of course, other measures of system performance are also performed.

From the system design viewpoint, a minimum P_D, maximum P_{FA} and minimum FOM are set. The task of the designer, in broad terms, is to produce a system concept that meets these requirements. Of course, other requirements, including cost considerations, will also need be met.

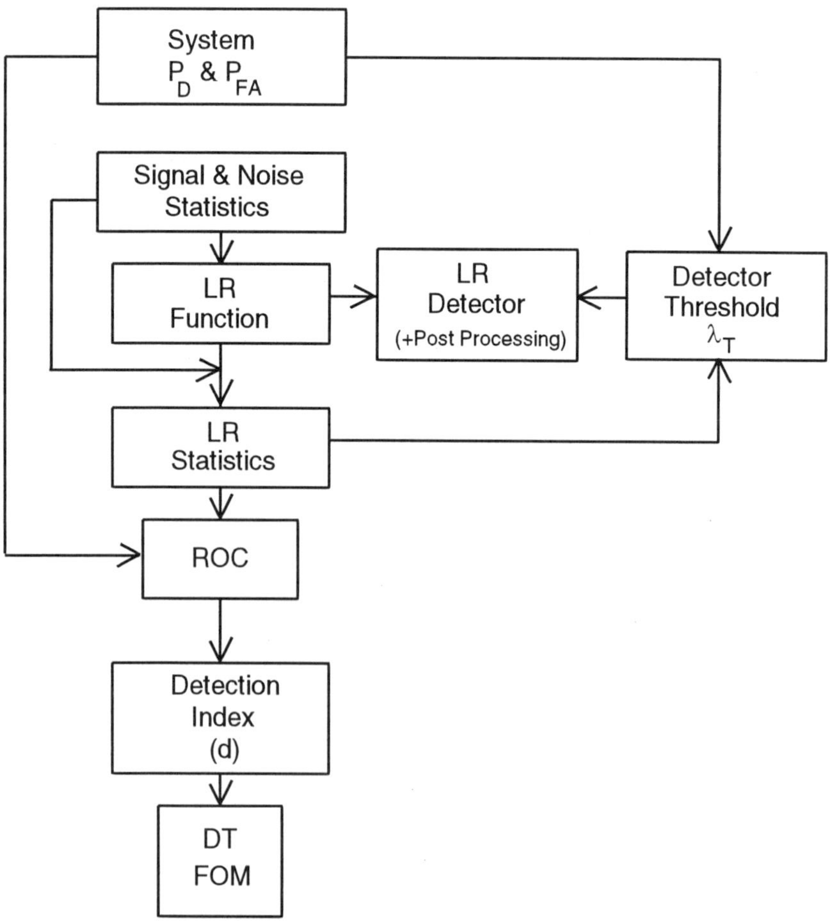

Figure 5 Detection Theory "Map" for ASW Sonar Systems

CHAPTER 10
ACTIVE DETECTION

Active detection systems search for targets by transmitting signals and listening for target echoes. The transmitted signals are designed to enhance detection performance, especially in reverberant environments, and to provide target disposition and classification information. Target parameters that are commonly estimated from echo returns include range, range rate, bearing and target highlight information. Other parameters of interest, that are more difficult to obtain and less commonly estimated, include target strength, depth and speed. A typical time sequence of returns acquired from a transmission might appear as indicated in the stylized representation in Figure 1. This detector output is generally normalized by a constant false alarm rate (CFAR) split window average to produce a level background for comparison with the detector threshold, but the CFAR averager that passes target detections would also pass reverberation and noise "spikes" and other false target detections. Reverberation and echoes from prominent bottom features are the principal sources of false target detections, when the system is bottom reverberation limited, but, as indicated in Figure 1, active sonar systems are usually reverberation limited only during part of the reception cycle and are noise limited during other parts of the reception cycle.

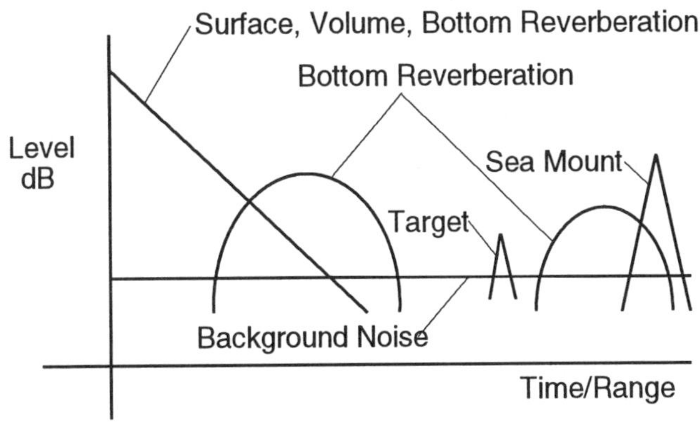

Figure 1 Target Echo versus Reverberation and Noise

Surface reverberation is strongly dependent upon the sea state and typically appears early-on in the received data with a decay rate of 30Log(r) but, in an upward refracting environment with a surface duct, the rate of decay can be reduced as propagation shifts to cylindrical spreading from spherical spreading. In a CZ environment, surface reverberation might reappear at regular intervals as the transmission sequentially visits the surface CZs.

Volume reverberation is largely dependent upon biologics and these reverberant returns depend upon the distribution of biologics within the water column. The initial volume reverberation typically decays at the rate of 20Log(r) but, if the biologics are contained within a surface duct or deep duct, the rate of decay can be prolonged as the propagation transitions to cylindrical spreading from spherical spreading.

Bottom reverberation when it occurs, is usually very strong and can be the dominant energy over much of the received data. Frequently, it is initially generated by the bottom directly below the source, from which a discrete set of bathymetric returns can be obtained due to vertical bottom-to-surface bounces from beneath the source ship, followed by a decay at the rate of 30Log(r). However, in shallow water or downward refracting environments, the rate of decay can be prolonged as the propagation transitions to cylindrical spreading from spherical spreading. In CZ environments, bottom reverberation can reappear periodically as the transmission reencounters the bottom between CZs as suggested by Figure 1. Additionally, certain bottom features such as sea mounts, rock outcroppings, canyon walls, etc. can give rise to significant singular echoes that can appear target-like.

Since all of the reverberant returns are necessarily correlated with the transmitted signal, the detection system can be expected to report numerous detections in a reverberant environment because reverberation tends to produce more "spikes" within the detector than ambient noise does. The estimated target parameters from all such false detections are the primary discriminants for separating false target echoes from correct detections and for classifying detections as target or non-target. For example, a sequence of detections that are associated with a significant bathymetric feature and that do not link to form a credible submarine track (and do not link with existing target tracks) and that have large target strengths would likely be dismissed as a non-submarine target.

10.1 Signal and Noise Statistics

The signal transmitted by the active system is represented by the analytic form

$$s_a(t) = m(t)\exp\left(j\omega_0 t\right), \tag{1a}$$

where m(t) is a complex modulation, ω_0 is the angular carrier frequency and, without loss of generality, the transmitted signal is assumed to have an average in-water intensity of unity at one meter from the source, i.e. a source level of 0 dB. The received signal is given by

$$r_a(t) = Am\left(\alpha\left(t - t_d\right)\right)\exp\left[j\alpha\omega\left(t - t_d\right) - \theta\right], \tag{1b}$$

where A is a real constant that denotes an amplitude change due to attenuation and reflections, α is the Doppler compression factor, t_d denotes a time delay due to propagation to the target and return and θ is a phase change due to propagation and reflections. For both the transmitted and received signals the time t is measured from the start of transmission. Other distortions, such as frequency dependent absorption and frequency dependent target reflection effects, could, in principle, be included but this is normally not done because both are assumed to be nearly constant over the signal bandwidth and to be included in (1b).

The energy flux density of the received analytic signal, without Doppler compression, is

$$E_a = \int_0^T dt\left|r_a\left(t + t_d\right)\right|^2 = A^2\int_0^T dt\left|m(t)\right|^2 = A^2 T, \tag{1c}$$

in which the unit of intensity is implicit and A^2 is assumed to carry that unit. Moreover, the quantity $10\log(A^2)$ is the received echo level in the sonar equation, i.e.

$10\text{Log}(A^2) = \text{SL}-2\text{TL}+\text{TS}$ from (2.1). Furthermore, as a reminder, we note that half of the analytic signal energy is carried by the real signal and half by the Hilbert transform of the real signal as indicated in (2.15-B). See Appendix 10-A for comments concerning signal design for CZ environments.

Example 1[1]: Consider an active source having source level of SL = 160 dB re μPa at 1 meter. . The transmitted intensity would be $I = I_0 \times 10^{16} = 6.67\times10^{-7}$ watts/cm^2 = 6.67 ergs/sec-cm^2 at one meter from the source and the rms signal sound pressure would be $|p| = \sqrt{(\rho c I)} = 10^3$ dyne/cm^2 = 10^8 μPa. If the transmitted signal had a duration of, say, 10 seconds, then the energy flux density of the signal would be E = 66.7 ergs/cm^2. Assuming the sonar system is operating in a deep duct environment, with 10Log(r) transmission loss, against a submarine target at 30 nm range, the two way transmission loss would be 2TL = 20Log(30 nm) = 95 dB. For a target strength of TS = 20 dB, and no time spreading of the signal, the received target echo level would be $10\text{Log}(A^2/I_0\rho c) = 85$ dB and the intensity of the analytic echo level would be $A^2/\rho c = 2I_0\times10^{8.4} = 33.5\times10^{-15}$ watts/cm^2 = 33.5×10^{-8} ergs/sec-cm^2. The sound pressure amplitude would be $A = 2.24\times10^{-1}$ dyne/cm^2 = 2.24×10^4 μPa and the analytic energy flux density would be $E_a = 33.5\times10^{-7}$ ergs/cm^2. However, if time spreading occurs, the received echo level, sound pressure and energy flux density could be more or less than the received signal, depending on the multipath interference effects, but the received signal would have a longer duration than the transmitted signal. ♦

For each received set of N data samples, we hypothesize that the data consists of either signal plus noise or noise only. Explicitly, the hypotheses: H_1 = target echo plus noise, and H_0 = noise only, are represented by the relations

$$H_1\colon\ \mathbf{y} = \mathbf{r}_a + \mathbf{n}_a, \tag{2a}$$

$$H_0\colon\ \mathbf{y} = \mathbf{n}_a, \tag{2b}$$

where $\mathbf{y} = (y_1, y_2, \cdots, y_N)^T$ represents the data vector of beam-band time samples. The quantity $\mathbf{r}_a$ is the vector of received analytic signal time samples, which is known to within the parameters: A, θ, α and t_d; and the quantity $\mathbf{n}_a$ represents the vector of analytic Gaussian zero mean noise time samples which are assumed to be statistically independent. We denote the unknown receive signal parameters by the vector

$$\Phi = \left(A, \theta, \alpha, t_d\right)^T, \tag{3}$$

where superscript T denotes transpose. The conditional means and variances of the received beam-band data for both hypotheses are given by

---

[1] In Example 1 we are implicitly assuming that the signal amplitude that drives the source would produce, in lieu of the transducer source level, an acoustic pressure of 1 μPa over the 10 second duration at one yard from the source. However, in operational systems, the signals that drive the transducers are tapered at each end to avoid "shocking" the transducer elements. Accordingly, the amplitude of the driving signal is adjusted to give the desired energy flux density at one yard from the source.

$$\mathbf{E}(\mathbf{y}|H_1,\Phi) = \mathbf{r}_a(\Phi), \quad \mathbf{E}(\mathbf{y}|H_0) = 0, \tag{4a,b}$$

$$\mathrm{Var}(\mathbf{y}|H_1,\Phi) = \mathbf{E}\left((\mathbf{y}-\mathbf{r}_a)\left(\mathbf{y}^*-\mathbf{r}_a^*\right)^T |H_1,\Phi\right) = 2\sigma^2 I_N, \tag{4c}$$

$$\mathrm{Var}(\mathbf{y}|H_0) = \mathbf{E}(\mathbf{y}\mathbf{y}^{*T}|H_0) = 2\sigma^2 I_N. \tag{4d}$$

The moments for the signal present hypothesis are conditioned upon the received signal parameters Φ; the quantity σ^2 is the power of the real and imaginary components of the analytic noise samples; and I_N denotes the identity matrix of order N. Note that, while the assumption of stationary noise with constant power σ^2 is generally adequate for the noise background, for ocean noise and reverberation it is a valid assumption only over limited time periods which are presumed to include the received signal duration. See Appendix 10-B for comments on nonstationary noise.

The probability density functions for the received beam-band analytic data under the two hypotheses are given by, from (8.16-A),

$$p(\mathbf{y}|H_1,\Phi) = \frac{\exp\left[-\frac{1}{2\sigma^2}\left(\mathbf{y}^* - \mathbf{r}_a^*(\Phi)\right)^T\left(\mathbf{y}-\mathbf{r}_a(\Phi)\right)\right]}{\left(2\pi\sigma^2\right)^N}, \tag{5a}$$

$$p(\mathbf{y}|H_0) = \frac{\exp\left[-\frac{1}{2\sigma^2}\mathbf{y}^{*T}\mathbf{y}\right]}{\left(2\pi\sigma^2\right)^N}. \tag{5b}$$

These probability density functions repesent all that is known about the received beam-band data and are the basis for partitioning the data into suspected target detections and noise only decisions. To perform the partitioning, it is expected that the optimal detector would perform a search for that signal. It is shown below that this is accomplished by correlating the received data against a replica of the transmitted signal but the exact correlation procedures are dependent upon the how much information is available concerning the signal parameters in (3).

10.2 Likelihood Ratio Detector

Likelihood Ratio Function

The likelihood ratio test is found from (8.5) and (5) to be

$$\rho(\mathbf{y}|\Phi) = \exp\left[\frac{1}{2\sigma^2}\left(2\mathrm{Re}\left[\mathbf{y}^{*T}\mathbf{r}_a(\Phi)\right] - \mathbf{r}_a^{*T}(\Phi)\mathbf{r}_a(\Phi)\right)\right] \underset{H_0}{\overset{H_1}{\gtrless}} \rho_T, \tag{6}$$

for a particular sample data set. The sample rate f_S on the beam-band data in (6) is equal to the signal bandwidth W

$$f_S = W = \frac{1}{\delta t}, \tag{7a}$$

where δt is the time interval between samples[2]. Therefore the analytic noise power can be expressed

$$2\sigma^2 = N_a W = \frac{N_a}{\delta t}, \tag{7b}$$

where N_a is the one sided spectral density of the analytic noise, assumed to be constant over the signal band, which is twice the one sided spectral density of the real noise. Substituting into (6) and taking the Riemann limit $\delta t \to 0$, the likelihood ratio test statistic can be expressed in terms of the integral

$$\rho(y|\Phi) = \exp\left[\frac{2\text{Re}}{N_a}\int_{t_d}^{T+t_d} y^*(t) r_a(t,\Phi)dt - \frac{E_a}{N_a}\right] \mathop{\gtrless}_{H_0}^{H_1} \rho_T, \tag{8}$$

where y and r_a are time dependent signals and not vectors of samples. Equation (8) represents the likelihood ratio receiver as a correlation between the data and the replica signal which is how the detector searches for the known signal. The additional term represents the analytic signal-to-noise ratio.

Two cases of analysis are pursued in the following in which the signal parameters Φ are, on the one hand, assumed to be known and, on the other hand, assumed to be unknown random variables. The completely known signal case is an important classical application of detection theory that serves as a reference for the case of unknown signal parameters but its direct applicability to ASW sonar scenarios is limited.

Likelihood Ratio Statistics - Known Signal

For the case of a known signal, the signal-to-noise ratio term in (8) is a known constant and the likelihood ratio receiver can be simplified. We note that, since the exponential function increases monotonically, the logarithm of (8) also increases monotonically and the log-likelihood ratio test, given by

$$\lambda(y|\Phi) = \text{Re}\int_{t_d}^{T+t_d} y^*(t) r_a(t,\Phi)dt \mathop{\gtrless}_{H_0}^{H_1} \lambda_T, \tag{9}$$

---

[2] The Nyquist criterion for real signals, $f_S = 2\times$(Band Width), is satisfied since basebanding the analytic signal gives two real quadrature component functions having band widths W/2. The required sample rate on each is W and since there are two components the effective sample rate is 2W, i.e. there are sufficient samples to separately reproduce the analytic signal, its real part and its imaginary part.

which is equivalent to the likelihood ratio test in (8) where, since the term involving the analytic signal-to-noise ratio is a known constant that is independent of data, it was absorbed within the left member of (9). The thresholds for the likelihood ratio and the log-likelihood ratio are related by the expression: $\exp[(2\lambda_T - E_a)/N_a] = \rho_T$.

The resulting log-likelihood detector from (9) consists of the sum of Gaussian random variables and, consequently, the log-likelihood ratio is also a Gaussian random variable with conditional means

$$\mathbf{E}_1(\lambda \mid \Phi) = \mathbf{E}_1 \mathrm{Re} \int_{t_d}^{T+t_d} \left(r_a^* + n_a^*\right) r_a dt = E_a \,, \tag{10a}$$

$$\mathbf{E}_0(\lambda \mid \Phi) = \mathbf{E}_0 \mathrm{Re} \int_{t_d}^{T+t_d} n_a^* r_a dt = 0 \,, \tag{10b}$$

where $\mathbf{E}_1$ and $\mathbf{E}_0$ are the expectation operators conditioned, respectively, on H_1 and H_0. The expected output of the correlator is seen to vanish when the signal is absent, since the received signal and noise are uncorrelated, but when the signal is present the correlator output is expected to be the energy of the analytic signal E_a.

The variances of the log likelihood detector for both the H_1 and H_0 hypotheses are the same and given by

$$\mathrm{Var}(\lambda) = \mathbf{E}\left[\mathrm{Re} \int_{t_d}^{T+t_d} dt n_a^* r_a\right]^2 = \mathbf{E}\left[\frac{1}{2} \int_{t_d}^{T+t_d} dt \left(n_a^* r_a + n_a r_a^*\right)\right]^2$$

$$= \frac{1}{4}\left[\int_{t_d}^{T+t_d} dt_1 \int_{t_d}^{T+t_d} dt_2 E\left(n_{a,1}^* n_{a,2}\right) r_{a,1} r_{a,2}^* + \int_{t_d}^{T+t_d} dt_1 \int_{t_d}^{T+t_d} dt_2 E\left(n_{a,1} n_{a,2}^*\right) r_{a,1}^* r_{a,2}\right], \tag{10c}$$

where integer subscripts reference the corresponding time variables t_1 and t_2. The expectations, $E(n_{a,1}{}^* n_{a,2})$ and $E(n_{a,1} n_{a,2}{}^*)$, are obtained from correlation function in (8.4a-A) and that correlation function is the Dirac delta function in (2.22i,j). The expectations $E(n_{a,1}{}^* n_{a,2}{}^*)$ and $E(n_{a,1} n_{a,2})$ vanish in accordance with (8.4b-A). Equation (10c) becomes

$$\mathrm{Var}(\lambda) = \frac{1}{4}\left[\frac{N_a}{2} \int_{t_d}^{T+t_d} dt_1 \int_{t_d}^{T+t_d} dt_2 \left[\delta(t_1 - t_2) - \frac{j}{\pi(t_1 - t_2)}\right] r_{a,1} r_{a,2}^* \right.$$

$$\left. + \frac{N_a}{2} \int_{t_d}^{T+t_d} dt_1 \int_{t_d}^{T+t_d} dt_2 \left[\delta(t_1 - t_2) + \frac{j}{\pi(t_1 - t_2)}\right] r_{a,1}^* r_{a,2}\right]$$

$$= \frac{1}{4}\left[\frac{N_a}{2}\int_{t_d}^{T+t_d} dt\left(2|r_a|^2 - j\left(r_a\hat{r}_a^* - r_a^*\hat{r}_a\right)\right)\right],$$

$$= \frac{1}{4}\left[N_aE_a - jN_a\int_{t_d}^{T+t_d} dtj\left(r^2+\hat{r}^2\right)\right] = \frac{1}{4}\left[N_aE_a + N_aE_a\right] = \frac{N_aE_a}{2}, \quad (10d)$$

Since the average detector output with signal is E_a and the average output vanishes when no signal is present, the detection index from (9.10), i.e. the signal-to-noise ratio at the detector output, is

$$d = \frac{2E_a}{N_a} = \frac{2E}{N_0}. \quad (10e)$$

The quantity E/N_0 is just the in-water energy flux density signal-to-noise ratio. This expression shows that measurements at the detector output give a means of estimating the noise spectral density which is required to implement the detector.

ROC Curves - Known Signal

Since the log likelihood ratio receiver is a Gaussian random variable with the means and variances in (10), the conditional probabilities of the log likelihood ratio receiver are

$$p(\lambda|H_1) = \sqrt{\frac{1}{\pi N_aE_a}}\exp\left[-\frac{1}{N_aE_a}(\lambda - E_a)^2\right], \quad (11a)$$

$$p(\lambda|H_0) = \sqrt{\frac{1}{\pi N_aE_a}}\exp\left[-\frac{\lambda^2}{N_aE_a}\right]. \quad (11b)$$

Substituting (11) into the expressions for the probability of detection and false alarm, which are obtained from (9.8), gives the receiver operating characteristics

$$P_D = \frac{1}{\sqrt{2\pi}}\int_{z_T-\sqrt{d}}^{\infty} dz\,\exp\left(-z^2/2\right), \quad (12a)$$

$$P_{FA} = \frac{1}{\sqrt{2\pi}}\int_{z_T}^{\infty} dz\,\exp\left(-z^2/2\right), \quad (12b)$$

where $z_T = \lambda_T/(N_aE_a/2)^{1/2}$ is the normalized threshold of the log-likelihood ratio detector in which the normalization is the square root of the background noise of the detector output. The normalized threshold is determined from the maximum allowable false alarm rate in (12b) and the detector threshold λ_T is determined from the known, i.e. estimated, background noise level which in a reverberant environment can be expected to fluctuate. The detection index d, in the lower limit of the probability of detection integral in (12a), is determined by the minimum required probability of detection.

The family of ROC curves for the known signal case, which are determined from the detection and false alarm probabilities in (12), are shown in Figure 2a. This form of the ROC plot is the classic[3] form which is excellent for illustrating the relations among the variables but it is not very useful for sonar applications because most design false alarm probabilities are found on the extreme left hand end of the probability of false alarm axis and detection indices greater than 16 are of interest. A more useful presentation[4] of the ROC curves is obtained by replotting the curves on probability coordinates. Such a graph is shown in Figure 2b.

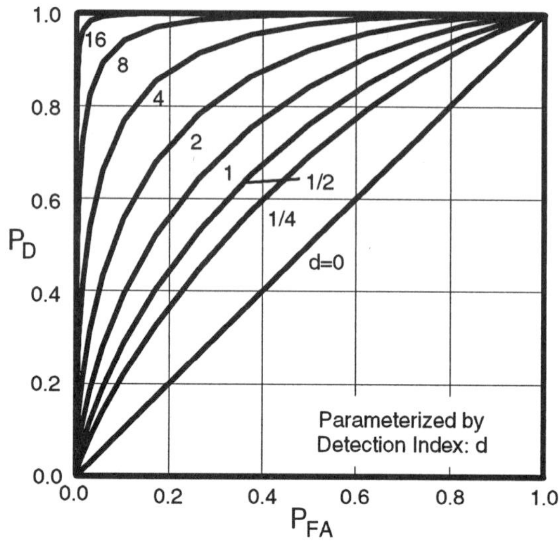

Figure 2a ROC for Known Signals

Example 2: For the ROC operating point, P_D = 50% and P_{FA} = 0.0001%, the required detection index is d = 22.4 = 13.5 dB. The lower limit on the probability of false alarm integral in Eq. 12b is 4.75342[5] which puts the detector threshold that factor above the background noise level, i.e. the corresponding detector threshold would be $\lambda_T = 4.75342\times\sqrt{(N_a E_a/2)}$. For the submarine target detection scenario in Example 1, the received in-water analytic energy flux density was found to be $E_a = 33.5\times10^{-7}$ erg/cm^2. If the noise level were, say, N_0 = 75 dB re µPa per Hz, the spectral intensity level of the in-water analytic noise would be $N_a = 2\times6.67\times10^{-23}\times10^{7.5}$ watt/cm^2-Hz = 42.2×10^{-9} erg/cm^2 and the noise at the detector output would be $\sqrt{\mathrm{Var}\lambda} = \sqrt{(N_a E_a/2)} = 2.66\times10^{-7}$ erg/cm^2 giving a signal-to-noise ratio at the detector output of 22 dB which is in excess of the signal-to-noise ratio required to achieve a probability of detection of 50%. The detector threshold would be $\lambda_T = 4.75342\times2.66\times10^{-7}$ erg/cm^2 = 12.6×10^{-7} erg/cm^2. This shows that the detector threshold lies intermediate to the detector output when signal plus noise is present and when noise only is present and, therefore, the signal would be detected and, as indicated, with a probability in excess of the minimum required probability. ♦

---

[3] Peterson, W. W., T. G. Birdsall, and W. C. Fox, "The Theory of Signal Detectability," IRE Trans. IT, PGIT-4, 171ff, Sept. 1954.

[4] See footnote 3.

[5] Abramowitz and Stegun, Table 26.6.

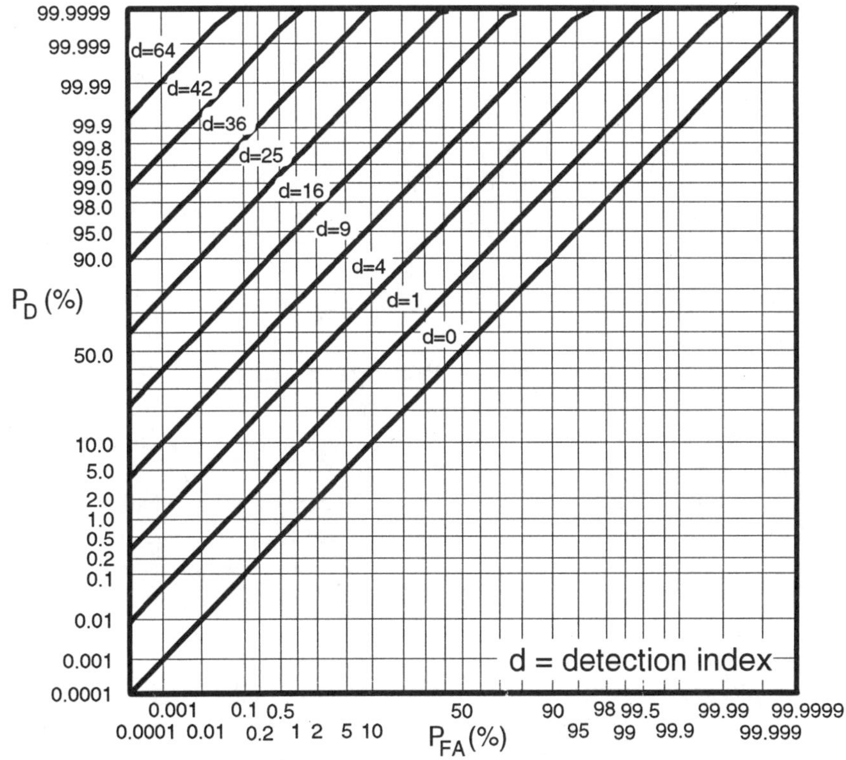

Figure 2b ROC for Known Signal on Probability Coordinates

Detection Threshold (DT) - Known Signal

The relationship between the detection index and detection threshold can be found from (10f). Reexpressing that equation in terms of the power of the physical signal gives

$$d = \frac{2E}{N_0} = \frac{2TS}{N_0}, \tag{13a}$$

where T is the duration of the signal and S denotes the signal intensity. The detection threshold is then given by

$$DT = 10\text{Log}\left(\frac{S}{N_0}\right) = 10\log\left(\frac{d}{2T}\right). \tag{13b}$$

Note that, since d is dimensionless, the argument of the right member of (13b) contains an implicit divisor having the unit of frequency which is balanced in the central member of those equations by another implicit divisor having the unit of frequency. This makes the argument of the central member to be the ratio of signal intensity to the noise intensity in a band of unit frequency. If the noise spectral density has the unit of intensity per Hz., the unit of the implicit multiplier is necessarily one Hertz. In this case care must be taken to express the signal duration T in seconds.

Example 3: For the submarine target detection scenario in Examples 1 and 2, the required detection index was found to be d = 22.4. The corresponding detection threshold is, from (13b), DT = 0.5 dB. Since the received echo level was 85 dB and the background noise level was taken to be 75 dB, the received signal-to-noise ratio would be 10 dB which is far in excess of the detection threshold.♦

Likelihood Ratio Statistics - Unknown Signal Parameters

The likelihood ratio statistic in (8) is dependent upon the signal amplitude attenuation A, phase shift θ, Doppler compression α, and time delay t_d. These parameters represent the effects of propagation to and from the target including reflections from the ocean surface and bottom, as well as the target, and, in general, each is an unknown random variable. In principle, these parameters, with the possible exception of phase, carry information concerning the target which, when estimated, can shed light on the target disposition and operations. In practice, estimates of amplitude attenuation can reveal target strength, which is a useful target classifier, but only if propagation conditions are known. Since propagation losses are frequently known only inaccurately, target strength estimates are generally unreliable. Moreover, while phase estimates not only suffer the same and worse inaccuracies, their principal disadvantage is that they have yet to be shown to carry useful target information. Consequently, while phase change estimates may be an element of some future system, they are not part of any operational ASW system in current use. Doppler compression of the signal and time delay are another matter, since they reveal target range rate and range, and can generally be estimated with sufficient accuracy to help localize and track the target.

In so far as amplitude and phase are assumed to be devoid of useful target information, the log-likelihood ratio receiver is structured to be optimal "on the average", i.e. for the average received amplitude level and phase, and the log-likelihood ratio receiver performs no estimates of amplitude and phase. Note, however, that post-detection estimates of amplitude are routinely performed for those cases in which it is judged that such estimates can reveal useful target information. In contrast, Doppler compression and time delay estimates are relatively insensitive to propagation effects, i.e. their standard deviations are generally small compared to their means, and estimates of Doppler compression and time delay are structured into the receiver. Because of the different treatment of these signal parameters, it is convenient to examine the dependence of the log-likelihood ratio receiver on amplitude and phase separately from the dependence on Doppler shifting and time delay.

The amplitude of received signals fluctuate from reception-to-reception and the statistics of these fluctuations depend principally upon the multipath structure and sound speed variations within the environment. Of course, in a static situation no reception-to-reception fluctuations would occur but, since the target and source/receiver are generally in motion (and also the medium), fluctuations are caused, largely, by the interference patterns between multipaths and, somewhat, by the divergence and convergence of rays due to random sound speed variations. While other effects can also cause fluctuations, e.g. variable absorption over changing paths, these effects are of lesser importance. The degree of fluctuations depends upon the acoustic path. Over reliable acoustic path (RAP) ranges the received waveform will be relatively constant in amplitude from reception-to-reception and the amplitude can be modeled statistically as a constant having a magnitude that is known, in so far as the propagation loss is known, and otherwise must be estimated. Over fading acoustic path (FAP) ranges, the changing environment causes constructive and destructive interference of the received waveform from reception-to-reception. In the extreme case of many multipaths, the received waveform can be approximated by the sum of a number of transmitted waveforms having variable amplitudes and phases. The sum signal is an approximately Gaussian distributed random variable, in so far as the Central Limit Theorem holds, and the waveform amplitude is inferred to be Rayleigh distributed in which the mean amplitude level is known if the multipath structure is known and otherwise must be estimated.

The phases of the received signals also fluctuate from reception-to-reception within the ocean environment because of the changing path structure. Moreover, since total phase change depends upon the number of refractions and reflections within the medium and also upon target reflection characteristics, then, even for the shortest RAP ranges the received signal phase can be significantly shifted by an unknown amount. Consequently, the phase must be considered to be wholly unknown and is assumed to be uniformly distributed over 2π radians for both constant and fading amplitude cases.

Assuming known Doppler shift and time delay, momentarily, and averaging over the amplitude attenuation and phase shift, the likelihood ratio in (8) can be written

$$\rho(y|\alpha, t_d) = \int dA \int d\theta \ \rho(y|A,\theta,\alpha,t_d) p(A) p(\theta), \tag{14}$$

where the integrand consists of the conditional likelihood ratio in (8) and probability density functions are for the random amplitude and phase. This is equivalent to the likelihood ratio obtained from (5) in which the probability density function conditioned on the signal in (5a) is averaged over amplitudes and phase. In this averaging process the probability density function for the noise only case in (5b) is independent of the signal parameters.

In the evaluation of (14), the two cases of amplitude fluctuations of interest are modeled as constant known amplitude and Rayleigh fading amplitude with known mean intensity. The probability density functions for each case are:

Constant Known Amplitude;

$$p(A) = \delta(A - A_0), \qquad 0 \le A < \infty \tag{15a}$$

Rayleigh Fading Amplitude;

$$p(A) = \frac{2A}{A_0^2}\exp\left(\frac{-A^2}{A_0^2}\right), \quad 0 \le A < \infty \tag{15b}$$

where A_0 denotes, in the first case, a known constant mean amplitude and, in the second case, a known mean squared amplitude, i.e. $\mathbf{E}A^2 = A_0^2$.

In both amplitude fluctuation cases the phase is assumed to be uniformly distributed over 2π radians

$$p(\theta) = \frac{1}{2\pi}. \qquad 0 \le \theta \le 2\pi \tag{15c}$$

It is convenient to average the likelihood ratio in (14) over the random phase first. Rewriting the likelihood ratio explicitly in terms of amplitude and phase, the average over phase gives the expression

$$\Lambda(y|A,\alpha,t_d) = \frac{1}{2\pi}\int_0^{2\pi} d\theta \, \exp\left[\frac{2A}{N_a}\lambda(y|\alpha,t_d)\cos(\theta-\phi) - \frac{A^2T}{N_a}\right], \tag{16a}$$

where

$$\lambda(y|\alpha,t_d) = \left|\int_{t_d}^{T+t_d} y^*(t)s_a(\alpha(t-t_d))dt\right|, \tag{16b}$$

$$s_a(t) = m(t)\exp[j\omega t]. \tag{16c}$$

In (16a) the quantity ϕ is the phase of the correlation integral and the energy of the analytic signal is expressed as A^2T. Equation (16b) defines λ which is the magnitude of the correlation integral which is also called the matched filter. In this integral, the data is correlated against the Doppler shifted and time delayed transmit replica which is defined in (16c) and which is a Doppler shifted and time delayed version of the transmitted signal in (2a). Since the integral in (16a) is over 2π, the phase ϕ does not matter and the integral is seen to be the modified Bessel function of zeroth order[6], denoted $I_0(\cdot)$, which immediately reduces (16a) to

$$\Lambda(y|A,\alpha,t_d) = \exp\left(-\frac{A^2T}{N_a}\right)I_0\left(\frac{2A\lambda(y|\alpha,t_d)}{N_a}\right). \tag{17}$$

To determine the optimal likelihood ratio receiver for known constant signal amplitude and Rayleigh fading signal amplitudes, the expression in (17) is averaged over amplitude for both cases in the following sections.

[6] Abramowitz and Stegun, (9.6.16).

Likelihood Ratio Statistics - Constant Known Amplitude and Unknown Phase

The average of the likelihood ratio in (17) over random amplitudes, using the amplitude density function for a constant amplitude in (15a), leads immediately to the likelihood ratio receiver

$$\rho(y|\alpha, t_d) = \exp\left(-\frac{A_0^2 T}{N_a}\right) I_0\left(\frac{2A_0\lambda(y|\alpha, t_d)}{N_a}\right) \underset{H_0}{\overset{H_1}{\gtrless}} \rho_T . \tag{18a}$$

In this expression everything is constant and known with only the correlation integral λ being data dependent. Moreover, since the modified Bessel function increases monotonically with its argument, an equivalent statistic is given by the correlation envelope raised to an arbitrary power. It is mathematically convenient to choose the squared envelope giving the equivalent statistic

$$\lambda^2(y|\alpha, t_d) \underset{H_0}{\overset{H_1}{\gtrless}} \lambda_T^2 . \tag{18b}$$

In (18b), the new threshold $\lambda^2{}_T$ is related to the old threshold ρ_T by the expression $\exp(-A^2{}_0T/2)I_0(2A_0\lambda_T/N_a) = \rho_T$. Also of interest is the independence of (18b) on the amplitude A_0. This parameter is needed only to relate the thresholds λ_T and ρ_T.

To this point the likelihood ratio test has been determined under the assumption that λ has been fixed in value by an assumed data sample y(t). To determine the performance of the likelihood ratio receiver for all possible data samples, we consider λ to be a random variable and determine its statistics. Writing the equivalent statistic in the form

$$\lambda^2 = X^2 + Y^2 , \tag{19a}$$

$$X = \mathrm{Re} \int_{t_d}^{T+t_d} y^*(t) s_a(t - t_d) dt , \tag{19b}$$

$$Y = \mathrm{Im} \int_{t_d}^{T+t_d} y^*(t) s_a(t - t_d) dt . \tag{19c}$$

The quantities X and Y are the real and imaginary components of the complex correlation between the received data and the Doppler shifted and time delayed replica of the tranmitted signal.

As in the correlator for the known signal case in (9), the quadrature components of the complex correlator in (19) are sums of Gaussian random variables and are again Gaussian. From inspection of (10a) and (10b), the conditional means of (19b) and (19c) are immediately seen to be

$$\mathbf{E}_1(X|\Phi_0) = A_0 T\cos\phi, \qquad \mathbf{E}_1(Y|\Phi_0) = A_0 T\sin\phi, \tag{20a,b}$$

$$\mathbf{E}_0(X) = 0, \qquad \mathbf{E}_0(Y) = 0, \tag{20c,d}$$

where $\Phi_0 = (A_0,\alpha,t_d)^T$ and in which each quantity in (20a,b) contains the implied mean intensity of unity for the transmit replica and the terms represent the inner product between the received signal and the Doppler shifted and time delayed transmit replica. From further inspection of (10c) and (10d) the conditional variances are equal and given by

$$\mathrm{Var}(X) = \frac{N_a T}{2}, \qquad \mathrm{Var}(Y) = \frac{N_a T}{2}. \tag{20e,f}$$

Each variance contains the mean intensity of unity for the transmit replica and represents the product between the noise spectral power density and the replica energy flux. Moreover, the conditional covariances $\mathbf{E}_1(XY)$ and $\mathbf{E}_0(XY)$ both vanish which show that X and Y are independent Gaussian random variables.

The joint density function between X and Y, conditioned on target present and the unknown phase, is

$$p_1(X,Y|\phi) = \frac{1}{\pi N_a T}\exp\left\{\frac{-1}{N_a T}\left[(X - A_0 T\cos\phi)^2 + (Y - A_0 T\sin\phi)^2\right]\right\}. \tag{21}$$

We transform the joint probability density function in (21) to (λ,ϕ_0) coordinates by the transformation

$$X = \lambda\cos\phi_0, \qquad Y = \lambda\sin\phi_0, \tag{22a,b}$$

where λ is the correlator output defined in (16b) and ϕ_0 is an arbitrary random phase. The transformation of the probability density function is obtained from the relationship for the transformation of variables within an integral

$$\begin{aligned}\int p_{X,Y}(X,Y|\phi)dXdY &= \int p_{X,Y}(X=\lambda\cos\phi_0, Y=\lambda\sin\phi_0|\phi)|\det J|d\lambda d\phi_0 \\ &= \int p_{\lambda,\phi_0}(\lambda,\phi_0|\phi)d\lambda d\phi_0 .\end{aligned} \tag{22c}$$

In this transformation the quantity $|\det J|$ is the absolute value of the determinant of the Jacobian matrix

$$J = \begin{bmatrix} \dfrac{\partial X}{\partial\lambda} & \dfrac{\partial Y}{\partial\lambda} \\ \dfrac{\partial X}{\partial\phi_0} & \dfrac{\partial Y}{\partial\phi_0} \end{bmatrix}. \tag{22d}$$

It is readily found that $\det(J) = \lambda$. Making the appropriate substitutions gives the joint probability density function for λ and ϕ_0

$$p_{\lambda,\phi_0}(\lambda,\phi_0|H_1,\Phi_0,\phi)=\frac{\lambda}{\pi N_a T}\exp\left[\frac{-1}{N_a T}\left(\lambda_2+(A_0T)^2-2A_0T\lambda\cos(\phi-\phi_0)\right)\right]. \quad (23)$$

Since the phase ϕ_0 is arbitrary, we define it to be uniformly distributed over 2π and average to get the marginal distribution for λ, as in (17),

$$p_\lambda(\lambda|H_1)=\frac{2\lambda}{N_a T}\exp\left[\frac{-\left(\lambda^2+(A_0T)^2\right)}{N_a T}\right]I_0\left(\frac{2A_0\lambda}{N_a}\right). \quad (24a)$$

This is the Rician probability density function for the envelope of the correlator output and is a generalization of the Rayleigh probability density function. Note that the envelope probability density function is independent of the correlator phase ϕ, i.e. it does not appear in (24a).

To determine the probability density function for the squared envelope, make the variable change $u=\lambda^2$ in (24a). From the transformation relations in (22c) and (22d) we obtain

$$p(u)=\frac{p_\lambda(\lambda=\sqrt{u})}{2\sqrt{u}}=\frac{1}{N_a T}\exp\left[\frac{-\left(u+(A_0T)^2\right)}{N_a T}\right]I_0\left(\frac{2A_0\sqrt{u}}{N_a}\right), \quad (24b)$$

which is the probability density function for the envelope squared.

ROC Curves - Constant Known Amplitude and Unknown Phase

The probability of detection for the squared envelope of the correlator output is given by

$$P_D=\frac{1}{N_a T}\int_{u_T}^{\infty}\exp\left[\frac{-\left(u+(A_0T)^2\right)}{N_a T}\right]I_0\left(\frac{2A_0\sqrt{u}}{N_a}\right)du, \quad (25a)$$

where $u_T=\lambda_T^2$ is the detector threshold for the envelope squared. Making the variable change $u=z^2(N_aT/2)$ in (25a) gives the expression

$$P_D=\int_{z_T}^{\infty}dz\, z\exp\left[-\frac{1}{2}\left(z^2+\frac{2E_a}{N_a}\right)\right]I_0\left(z\sqrt{\frac{2E_a}{N_a}}\right), \quad (25b)$$

where the threshold z_T is the envelope threshold normalized, i.e. $z_T=\lambda_T/\sqrt{(N_aT/2)}$. The result in (25b) is the Marcum Q-Function. The probability of false alarm follows immediately

$$P_{FA} = \int_{z_T}^{\infty} z \exp\left[\frac{-z^2}{2}\right] dz = \exp\left(-\frac{z_T^2}{2}\right). \tag{25c}$$

The ROC curves for the case of a received signal having known amplitude but unknown phase are given by parametric plots of (25b) and (25c) which are presented in Figure 3. For comparison purposes, the ROC curves for the case of a wholly known received signal are also presented in Figure 3. These curves show that the penalty paid for not knowing the received signal phase is a lowered probability of detection at all false alarm rates. However, while the relative decrease in detection probability is negligible at high signal-to-noise ratios, it is substantial at low signal-to-noise ratios. For the important case of intermediate detection probabilities at, say the 50% probability level, the reduction remains significant. For example, at a probability of detection of 50% and probability of false alarm of 10^{-8} for the case of known signal, the probability of detection is reduced to approximately 37% if the phase is unknown.

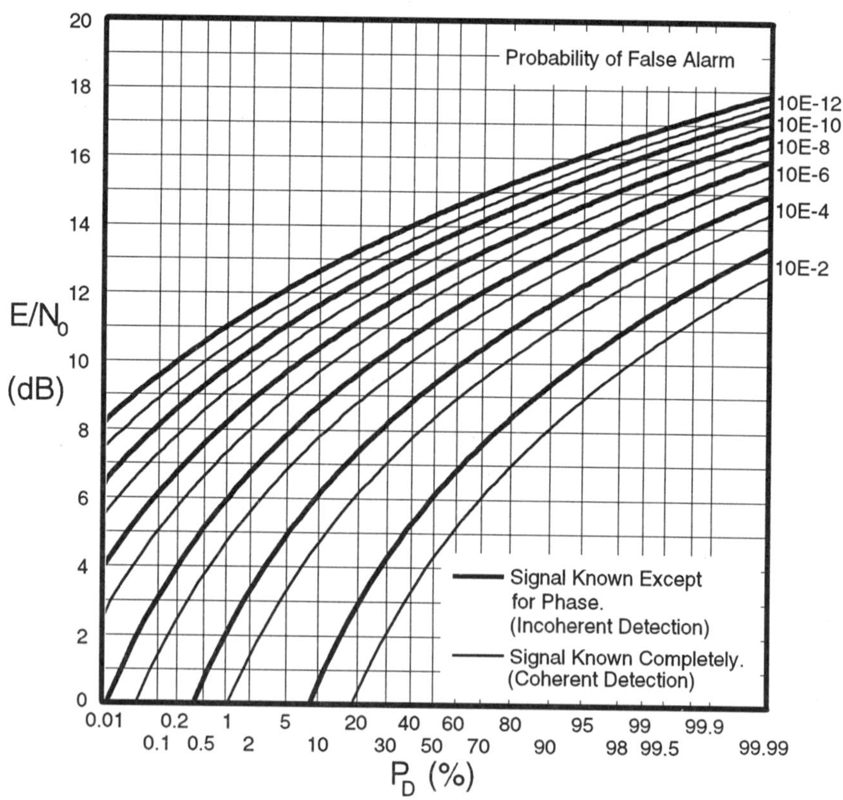

Figure 3 Comparison of ROC for an Unknown Phase Signal and a Known Phase Signal - both with Known Amplitude

Detection Threshold (DT) - Constant Known Amplitude and Unknown Phase

Since the ROC curves in Figure 3 are plotted directly in terms of the in-water energy signal-to-noise ratio the detection threshold is immediately found from the relation

$$DT = 10\log\left(\frac{S}{N_0}\right) = 10\log\left(\frac{E}{TN_0}\right). \tag{26}$$

Analogous to (13b), if the noise spectral density unit is intensity per Hz., there is necessarily an implied factor of 1 Hz in the denominator of both members. Moreover, since the signal intensity must have the same intensity unit as the noise, the energy and time units are constrained to be compatible with the relation ST = E.

Example 4: For the detection operating point P_D = 50%, P_{FA} = 0.0001%, the required in-water energy signal-to-noise ratio is approximately 11.3 dB. This is greater than the in-water energy signal-to-noise ratio for the known signal case of 10.5 dB. Extending the submarine target detection scenario in Examples 1-3, to the detector configured for unknown phase, the detection threshold, from (26), is DT = $10\text{Log}(10^{1.13}/T)$ = 1.3 dB. Compared to the case of wholly known signal, the case of detection of a signal with unknown phase over RAP ranges requires a received echo level of about 0.8 dB higher for comparable detection performance. Of course, for the assumed echo level of 10 dB, both detectors would handily detect the signal. ♦

The detection index in (9.10) for the squared envelope of the correlator output can be determined from the squared envelope statistics in (24b). Substituting $u = \lambda^2$ in that equation, the nth moments of the squared envelope for the signal-plus-noise hypothesis are

$$E_1\lambda^{2n} = \frac{2}{N_aT}\int_0^\infty d\lambda\, \lambda^{2n+1} \exp\left(-\frac{\lambda^2}{N_aT} - \frac{A_0^2T}{N_a}\right) I_0\left(\lambda\frac{2A_0}{N_a}\right). \tag{27}$$

This immediately yields[7]

$$E_1\lambda^{2n} = (N_aT)_n \exp\left(-\frac{A_0^2T}{N_a}\right)\Gamma(n+1)M\left(n+1,1,\frac{A_0^2T}{N_a}\right), \tag{28a}$$

where $\Gamma(\cdot)$ denotes the Gamma Function and $M(\cdot,\cdot,\cdot)$ denotes the Confluent Hypergeometric Function. The nth order moment of the envelope for the noise only hypothesis follows immediately

$$E_0\lambda^{2n} = (N_aT)^n\Gamma(n+1). \tag{28b}$$

Expanding (28a) for high and low signal-to-noise ratio[8] cases, i.e. exact closed form solutions are not available, we find the first moments to be

[7] Abramowitz and Stegun (11.4.28).

[8] Abramowitz and Stegun (13.1.2) and (13.5.1).

$$\mathbf{E}_1\lambda^2 = A_0^2T^2, \qquad \text{High SNR} \tag{29a}$$

$$\mathbf{E}_1\lambda^2 = N_aT\left(1+\frac{A_0^2T}{N_a}\right). \qquad \text{Low SNR} \tag{29b}$$

Similarly, the first moment and variance of the squared envelope under the noise only hypothesis are

$$\mathbf{E}_0\lambda^2 = N_aT, \qquad \mathrm{Var}_0\lambda^2 = \left(N_aT\right)^2. \tag{29c,d}$$

The detection index is computed directly from (29). For both high and low signal-to-noise cases we obtain

$$d = \left(\frac{A_0^2T}{N_a}\right)^2 = \left(\frac{E_a}{N_a}\right)^2 = \left(\frac{E_0}{N_0}\right)^2. \qquad \text{High and Low SNR} \tag{29e}$$

Example 5: For submarine detection scenario in Examples 1-4, the expected detector output when the signal is present is $\mathbf{E}_1\lambda^2 = E_a = 33.5\times10^{-6}$ erg-sec/cm^2 while the expected detector output and standard deviation when noise only is present is $\mathbf{E}_0\lambda^2 = \sqrt{\mathrm{Var}_0\lambda^2} = 4.22\times10^{-7}$ erg-sec/cm^2. This gives a detector signal-to-noise output of 38 dB which is substantially in excess of the 22.6 dB detection index required for the detector operating point as found from (29e). The detector threshold is given by $\lambda^2_T = z^2_T(N_aT/2)$. Since $z^2_T = -2\times\ln P_{FA} = 27.6$, the envelope squared threshold is $\lambda^2_T = 5.8\times10^{-6}$ erg-sec/cm^2 which is intermediate to the signal present and noise only cases. ♦

Likelihood Ratio Statistics - Rayleigh Fading Amplitude and Unknown Phase

The average of the likelihood ratio in (17) over random amplitudes using the amplitude density function in (15b) for Rayleigh fading amplitudes leads immediately to

$$\Lambda(y|\alpha, t_{1d}) = \int_0^\infty dA\, \frac{2A}{A_0^2}\exp\left[-A^2\left(\frac{1}{A_0^2}+\frac{T}{N_a}\right)\right] I_0\left(\frac{2A\lambda(y|\alpha,t_d)}{N_a}\right). \tag{30a}$$

This can be evaluated[9] to yield

$$\Lambda(y|\alpha, t_d) = \left(\frac{N_a}{N_a + A_0^2T}\right)\exp\left[\frac{A_0^2\lambda^2(y|\alpha,t_d)}{N_a\left(N_a+A_0^2T\right)}\right] \underset{H_0}{\overset{H_1}{\gtrless}} \Lambda_T. \tag{30b}$$

In (30b) only the $\lambda^2(y|\alpha,t_d)$ term is data dependent, with the remaining terms being constants where λ is defined in (16b). Since the likelihood ratio is monotonic in λ^2, then an equivalent statistic is

---

[9] Abramowitz and Stegun, (11.4.28), (9.6.3) and (13.6.12).

$$\lambda^2(y|\omega_D,t_d) \underset{H_0}{\overset{H_1}{\gtrless}} \lambda_T^2, \tag{31a}$$

with the threshold relation

$$\left(\frac{N_a}{N_a + A_0^2 T}\right)\exp\left[\frac{A_0^2\lambda_T^2}{N_a\left(N_a + A_0^2 T\right)}\right] = \Lambda_T. \tag{31b}$$

Note that the known mean squared envelope is not required for the likelihood ratio detector in (31a) and is only required for the threshold transformation in (31b).

<u>ROC Curves - Rayleigh Fading Amplitude and Unknown Phase</u>

The statistics of the squared envelope were determined in (24c) for the case of a constant known amplitude. The probability of detection, i.e. the Marcum Q-Function, and false alarm for the squared envelope detector are repeated from (25b) and (25c) in which the amplitude is made explicit

$$P_D(A) = \int_{z_T}^{\infty} z\exp\left[\frac{-1}{2}\left(z^2 + \frac{2A^2T}{N_a}\right)\right] I_0\left(z\sqrt{\frac{2A^2T}{N_a}}\right)dz, \tag{32a}$$

$$P_{FA} = \int_{z_T}^{\infty} z\exp\left[\frac{-z^2}{2}\right]dz = \exp\left(-\frac{z_T^2}{2}\right). \tag{32b}$$

While for the case of known amplitude and unknown phase the dependence of the squared envelope detector statistics on that signal amplitude was deterministic, for the case of Rayleigh fading amplitude and unknown phase the relationship between the squared envelope detector and the signal amplitude is probabilistic. However, the false alarm probabilities and thresholds for both cases are the same.

Averaging over amplitudes in (32a) can be performed in the same manner as the integral in (30a) was evaluated. The resulting integral for the probability of detection has the same form as the integral for the probability of false alarm in (32b) and, since it is elementary, it immediately yields

$$P_D = \exp\left(\frac{-z_T^2}{2}\left(1 + \frac{A_0^2 T}{N_a}\right)^{-1}\right). \tag{32c}$$

Combining (32b) and (32c) gives

$$P_D = P_{FA}^Z, \qquad Z = \left(1 + \frac{\overline{E}_a}{N_a}\right)^{-1}, \tag{33}$$

where the quantity $\overline{E}_a = A_0^2 T$ is the average energy of the analytic signal and the ratio can be expressed in terms of the ratio of the average energy of the real signal, i.e. $\overline{E}_a / N_a = \overline{E}_0 / N_0$.

The ROC curves for the case of unknown signal parameters is shown in Figure 4 in which the ROC curves for the case of known signal are also plotted for comparison purposes. Notice that at low average signal-to-noise ratios the signal amplitude fluctuations are an advantage. Weak signals can be expected to occasionally fluctuate above the detection threshold producing a detection. However, at high average signal-to-noise ratios the signal amplitude fluctuations degrade detection performance because strong signals occasionally drop below the detection threshold.

Detection Threshold (DT) - Rayleigh Fading Amplitude and Unknown Phase

The detection index for the case of Rayleigh fading is very similar to the case of constant known amplitude and the steps leading to it are the same as those in (26), (27) and (28) in which the known signal energy is replaced by the known

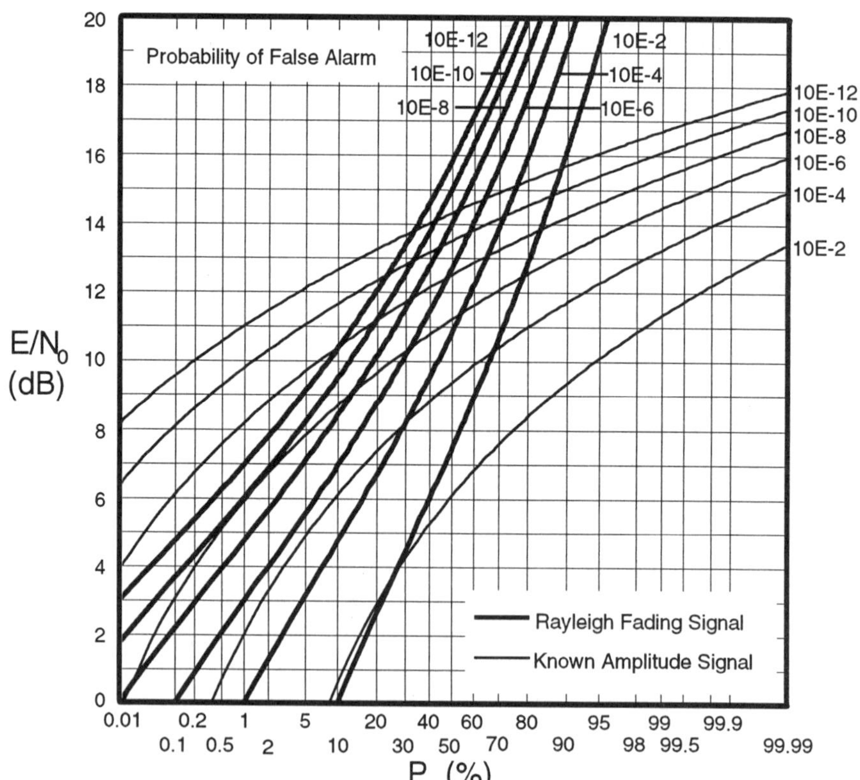

Figure 4 Comparison of Receiver Operating Characteristics for a Rayleigh Fading Signal and a Known Amplitude Signal - both with Unknown Phase

average energy. The final result is

$$d = \left(\frac{\overline{E}_a}{N_a}\right)^2, \tag{34a}$$

and it differs from the detection index for the constant amplitude case only in the detail of being dependent on the average energy instead of the known energy.

The detection threshold for the Rayleigh fading case is given by

$$DT = 10\log\left(\frac{\overline{S}_0}{N_0}\right) = 10\log\left(\frac{1}{T}\left(\frac{\overline{E}_0}{N_0}\right)\right), \tag{34b}$$

and is similarly dependent on the average intensity and energy.

Example 6: For the detection operating point P_D = 50%, P_{FA} = 0.0001%, the required in-water energy signal-to-noise ratio is 12.8 dB from Figure 4. This exceeds the required in-water signal-to-noise ratio for the cases of known signal and signal with unknown phase. Extending the submarine target detection scenario in Examples 1-4, to the detector configured for unknown amplitude and phase, the detection threshold, from (34b), is DT = $10\log(10^{1.28}/T)$ = 2.8 dB which exceeds the detection threshold for the unknown phase case by 1.5 dB and, compared to the case of wholly known signal, the case of detection of a signal with Rayleigh fading amplitude and unknown phase over RAP ranges requires a received echo level of 2.3 dB higher for comparable detection performance. The expected detector output when the signal is present and when noise only is present is the same as in Example 4 with the difference that the signal energy is the expected energy. As for Example 4 the detector signal-to-noise output is 38 dB which is far in excess of the 22.6 dB required for the detector operating point. The detector threshold is also the same as for the detector configured for a constant amplitude and unknown phase signal. ♦

Likelihood Ratio - Unknown Doppler and Time Delay

The development of the likelihood ratio detector from (8) in the above sections proceeded under the assumption that the Doppler shift impressed upon the signal and the time delay were known while the amplitude attenuation and phase shift were treated variously as known, or estimable, parameters and unknown random parameters with known statistics. In the immediately following we revisit (8) and consider the fact that, for active ASW operations, the Doppler shift and time delay of the received signal are unknown.

The Doppler compressions for all expected targets fall between a maximum and a minimum range of Doppler compressions, the time delays for the search area fall between zero and a maximum that corresponds to the maximum detection range. Since the peak of the likelihood ratio detector output at, say, the −3 dB points has a width in both Doppler and time delay dimensions, it resolves both the Doppler range and time delay range into a discrete number of Doppler compressions and time delays. Each of these Doppler-time delay bins is inspected for a detection and when found the corresponding Doppler and time delay become estimates of the target range and range rate. This effectively replaces the hypotheses in (2) by a set of hypotheses

$$H_{j,k}:\ \mathbf{y} = \mathbf{r}_a\left(\alpha_j, t_{d,k}\right) + \mathbf{n}_a, \quad -J \le j \le J,\ \ k = 0,\cdots,K \tag{35a}$$

$$H_0:\ \mathbf{y} = \mathbf{n}_a, \tag{35b}$$

where j and k denote Doppler and time delay bins and the likelihood ratio test in (8) becomes

$$\rho\left(y \mid A, \theta, \alpha_j, t_{d,k}\right) \underset{H_0}{\overset{H_{j,k}}{\gtrless}} \rho_T. \tag{35c}$$

All else following (8) proceeds as before with the difference that the assumed known Doppler compression and time delay are now given explicit values associated with the Doppler-time delay bin being examined. This yields the important benefit of providing a Doppler and time delay estimate for each detection.

10.3 System P_D and P_{FA}

Active sonar systems usually display detection data in range-azimuth maps for each transmission. Detections that link-up with previous detections and that produce meaningful target tracks are given close attention and the means of examining these detections for Doppler structure, target structure highlights and target strength are frequently provided to the operators. The detection maps are usually earth fixed and can be overlaid on bathymetric maps to keep the operators properly oriented in their examination of the data.

To ensure that the operators have quality data for examination, the probability of detection is set to accommodate the weakest of submarine targets while the probability of false alarm is set to minimize the number of spurious detections. The selected operating point specifies the single-cell detection probability, which is an appropriate operational probability of detection since the target occupies only one detection cell, or at most a small cluster of detection cells, on any one transmission. However, the probability of false alarm that is of operational interest is the number of false alarms produced by all detection cells in some period of time that is typically a period of, say, hours or days. The operational false alarm rate is related to the single-cell false alarm rate through the number of detection cells and the interrogation period. For example, if there are N_C detection cells, including range, bearing and Doppler, and an interrogation rate of R_T, i.e. rate of active transmissions per unit time, then, since the operational false alarm rate is the event of one or more false detections from the system within the period of time T, the situation can be represented by the expression

$$P\left(N \ge 1 \text{ detections in period } T \mid H_0\right) = 1 - P\left(N = 0 \text{ detections in period } T \mid H_0\right). \tag{36a}$$

The left member of (36a) is the operational false alarm rate and, assuming that the correct misses in all cells in the right member are independent events, (36a) can be rewritten as

$$P_{OFA} = 1 - (1 - P_{FA})^{N_C R_T T}, \tag{36b}$$

where P_{OFA} is the operational probability of false alarm and P_{FA} is the single cell probability of false alarm. Assuming the value P_{FA} is a small quantity, (36b) can be approximated

$$P_{FA} \cong \frac{P_{OFA}}{(N_C R_T T)}. \tag{36c}$$

Example 6: If the system probability of false alarm is $P_{OFA} = 1$ false detection in a period of one day, and the number of detection cells is $N_C = 10^4$, and the interrogation rate is $R_T = 100$ transmissions per day, the required single cell false alarm probability is $P_{FA} = 10^{-6}$. This choice of single-cell probability of false alarm has the effect of setting the threshold of the detector as indicated in (8-8b).♦

For active systems there is an interdependence between the probabilities of detection and false alarm. For example we can write the relations

$$P_{OFA} + A_O \xrightarrow{(36c)} P_{FA} \xrightarrow{(25c)} z_T, \tag{37a}$$

$$P_D + z_T \xrightarrow{(25b)} DT + A_O \rightarrow FOM. \tag{37b}$$

The first relation asserts that choosing an operational probability of false alarm and an operations area A_O, then with (36c) the single cell probability of false alarm is determined. Similarly, this single cell probability of false alarm and (25c) determine the detector threshold. Moving to the second relation we find that choosing a probability of detection, with the previously determined detector threshold, (25b) determines the detection threshold which, with the operations area, determines the FOM from either (2.3) or (2.4a,b). (This operations area concept, of course, encompasses the geometric area, the propagation conditions, the noise background and likely targets to be found.) If the FOM as determined by the process in (37b) does not meet or exceed the system FOM, then the system will not meet its operating point within the operations area. To determine the discrepancy in the operating point, we run (37b) backwards,

$$FOM + A_O \rightarrow DT + z_T \xrightarrow{(25b)} P_D. \tag{37c}$$

The system FOM with the chosen operations area leads to the detection threshold with which, in conjunction with the detector threshold from (37a), (25b) leads to the probability of detection. This probability of detection with the probability of false alarm define the system operating point within the operations area. Since this probability of detection applies primarily at the outer limits of the operations area and since it is never graphically evident without extensive testing, the resulting operating point for the system is generally accepted as an operational trade-off.

APPENDIX 10-A THE "CZ RULE" FOR ACTIVE SONAR WAVEFORM DESIGN

Active sonar detection performance in a convergence zone (CZ) environment is enhanced for targets within the surface CZs because of the increased signal intensity within those regions. Targets outside of the surface CZs are generally not detected. This phenomenon is especially evident for broadband high range resolution sonar systems and operational procedures are sometimes developed in which only contacts received from within the surface CZ are viewed as possible targets and detections obtained from intermediate ranges are rejected as clutter from the bottom.

For such operational procedures to be useful, the returns from the CZs and the returns from the intermediate bottom scattering regions must be separated. This will be the case for sufficiently short duration signals but, for longer duration signals, reverberation from the intermediate bottom scattering region can intrude into the returns from within the surface CZ. This phenomenon is especially important when dealing with long duration wavetrains consisting of several waveforms occupying the same frequency band. However, in the interests of simplicity, the following discussions focus on the phenomenon of bottom reverberation in CZ environments for a single waveform.

A CZ environment is modeled as shown in Figure 1. The surface CZs and bottom scattering regions have range extents of, respectively, S and B, and are spaced in range with range interval R. The surface CZs are displaced from the bottom scattering regions by R/2. Target detections are assumed to be received from the surface CZs but only bottom backscattering is received from the bottom scattering region.

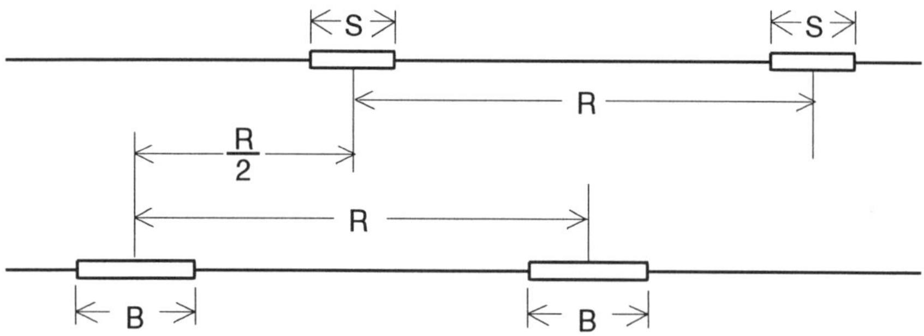

Figure 1 Surface CZs and Bottom Scattering Regions

A single waveform is assumed to be transmitted and the time at which echo returns at a particular frequency are received are indicated in the time-frequency plot in Figure 2. The transmitted waveform is assumed to be a linear frequency modulated waveform with increasing frequency but the exact structure of the frequency sweep is unimportant and, in fact, any broadband waveform could be chosen with no substantial change to the results. Also, implicit in the analysis is the assumption that the medium depth is sufficiently less than the CZ intervals and that the problem can be considered to be only range dependent. Including the depth dependence would change the analysis somewhat but it would not change the principles involved. In fact, in practice, the times of arrival of the signal components are determined empirically and not from models.

The nth CZ at range R_n is assumed to be of primary interest. Bottom reverberation is assumed to be received prior to and following any target returns from the

CZ at range R_n. The reverberation has a broadband continuous spectrum and produces an increase in the matched filter noise floor in addition to that caused by the ambient noise level.

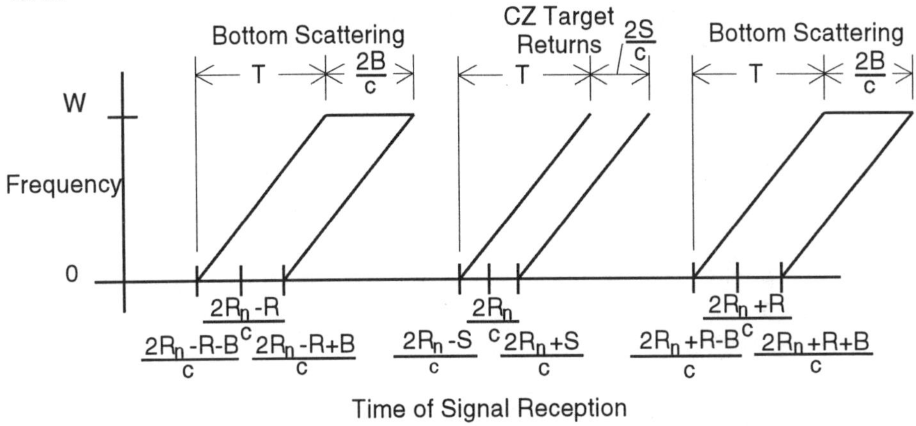

Figure 2 Time-Frequency Structure of Backscattered Waveform

From Figure 2 we see that, if the transmitted waveform has sufficient duration, the tail of the reverberation from the uprange bottom scattering region can encroach into a target return from the leading edge of the CZ. Similarly, a target return from the trailing edge of the CZ can be intruded upon by the scattering from the downrange bottom scattering region. For each respective case the conditions that will ensure separation of the target return from the reverberation are given by

$$\frac{(2R_n - R - B)}{c} + T + \frac{2B}{c} \leq \frac{(2R_n - S)}{c}, \tag{1a}$$

$$\frac{(2R_n - S)}{c} + T + \frac{2S}{c} \leq \frac{(2R_n + R - B)}{c}. \tag{1b}$$

From (1a) and (1b)

$$\frac{cT}{2} \leq \frac{(R - S - B)}{2}. \tag{1c}$$

This expression asserts that the duration of the waveform must be restricted such that half the spatial extent of the waveform does not exceed the range interval between the edges of the surface CZ and the bottom scattering region. This simply states that the onset of an echo from the leading edge of a CZ must arrive after the last of the reverberation from the back edge of the uprange bottom scattering region. Similarly, the trailing edge of an echo from the back edge of the CZ must arrive before the onset of reverberation from the downrange bottom reverberation region. This result represents the basic CZ rule for waveform design in CZ environments. In practice, the ranges in (1c) are determined from empirical times of arrival of signal and reverberation and these measured times can be used to determine the optimal waveform duration.

The basic CZ rule is conservative in requiring that no reverberation be allowed to encroach on the target return. However, in many cases the onset and trailing edges of reverberation are weak. By allowing some reverberation to interfere with the target return, the waveform duration, i.e. energy, can be increased which will increase the target returned energy. As long as the target returned energy increases faster than the reverberation energy we obtain a processing gain. The optimal waveform duration would produce a maximum signal-to-noise-plus-reverberation ratio (SNRR). This SNRR criterion leads to a modified CZ rule.

Assuming that the waveform duration has been increased to the point of violating the basic CZ rule, i.e. $(R - S - B) \le cT$, bottom reverberation intrudes into the target return interval for a leading edge target as shown in Figure 3. The broadband reverberation spectral density is assumed to have the level R_0 and the total reverberation energy processed by the matched filter is the product of the spectral density and the time-bandwidth area of the reverberation. This area is just the time-frequency area of the intrusive reverberation that is indicated in Figure 3. Accordingly, the total reverberation energy processed by the matched filter is given by

$$\frac{R_0}{2}\left(1 - \frac{(R - S - B)}{cT}\right)^2 WT. \qquad cT \ge (R - S + B) \tag{2}$$

Since the matched filter noise suppression factor is the reciprocal time-bandwidth product 1/(TW), the reverberation noise that is output by the matched filter is given by

$$\frac{R_0}{2}\left(1 - \frac{(R - S - B)}{cT}\right)^2 . \qquad cT \ge (R - S + B) \tag{3}$$

Consequently, the SNRR at the matched filter output is given by

$$\mathrm{SNRR} = \left(\frac{PT}{N_0}\right)\left(1 + \frac{R_0}{2N_0}\left(1 - \frac{(R - S - B)}{cT}\right)^2\right)^{-1}, \qquad cT \ge (R - S + B) \tag{4}$$

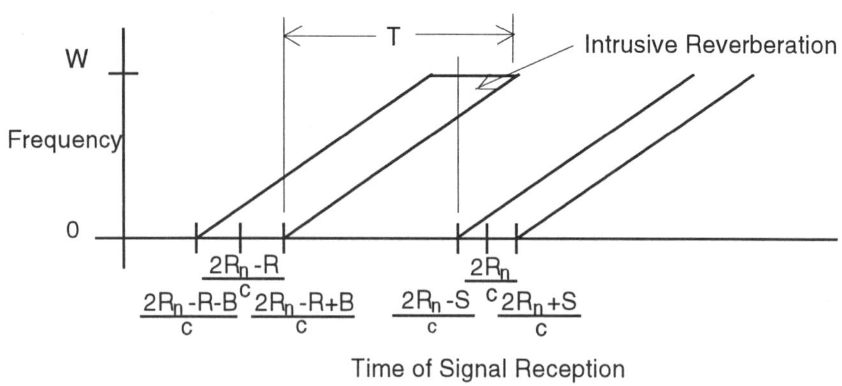

Figure 3 Reverberation Intrusion into Target Return

where P denotes the signal power. The leading factor represents the signal-to-noise ratio which shows a linear increase with signal duration. The denominator represents the effect of the reverberation intrusion. The magnitude of the effect is dependent upon the reverberation-to-noise ratio R_0/N_0.

The maximum SNRR is found by setting the first derivative of (4) with respect to signal duration to zero. Performing this operation we obtain the modified CZ rule

$$cT \le \frac{R-S-B}{\frac{2}{3}\left[1+\frac{1}{2}\sqrt{1-\frac{6}{R_0/N_0}}\right]}. \tag{5a}$$

We see that there are no real solutions unless $R_0/N_0 \ge 6$, i.e. 7.7 dB, which implies that for weaker reverberation there are no constraints on signal duration. For the case of stronger reverberation, i.e. $R_0/N_0 > 6$, (5a) can be approximated

$$cT \le \frac{(R-S-B)}{\left(1-\frac{1}{R_0/N_0}\right)}. \tag{5b}$$

The approximate form of the CZ rule indicates that for a reverberation-to-noise ratio of, say, 10 dB that the waveform duration could be extended by about 10% over the duration given by the basic CZ rule. These results are for a target in the leading edge of the CZ but identical results are readily obtained for a target in the trailing edge of the CZ. They ensure that targets within the leading and trailing edges of the CZ will not be negatively impacted by bottom reverberation returns and that targets located at all other intermediate points within the CZ will suffer no impact by bottom reverberation.

The CZ rule provides a ready guideline for designing waveforms for use in CZ environments. However the rule is based upon a simple model of the propagation channel and is necessarily an approximate guideline. The principle benefits to be derived from the CZ rule lie in its offering of an estimate of when to apply the rule; and in its provision of an estimate of what the maximum waveform duration should be. However, the values given by the CZ rule will no doubt require adjustments based upon either in-situ measurements or sophisticated propagation model estimates.

APPENDIX 10-B: FALSE ALARMS IN NONSTATIONARY PROCESSES

Casual observation shows that both reverberation and background noise are usually nonstationary processes in which the nonstationarity is most obviously seen as a time varying power. Since both are arguably Gaussian processes at each time and are independent from time-to-time, an appropriate nonstationary model for the statistics of reverberation and noise is a Gaussian distribution with time varying power and time-to-time independence. With respect to our discussions of detection in Chapter 10, since, over the duration of a matched filter window, the time varying power is constant, or very nearly so, all the results in Chapter 10 remain valid within this model of reverberation and noise. Therefore, the matched filter output remains a Chi-squared distributed process that is independent from time-to-time, but it has a time varying mean and we cannot normalize the matched filter output with a constant as was done in Chapter 10. Instead, we need a measure of the time varying mean of the matched filter output, which we usually obtain with a sliding split window averager as discussed in Chapter 9. Normalizing the matched filter output with the estimated time varying mean gives a matched filter output that is expected to have a constant false alarm rate (CFAR), i.e. the normalized output is expected to be a stationary Chi-squared distributed process with constant mean. In this case, the procedures for thresholding and for detecting threshold exceedances discussed in Chapter 10 remain valid. If the estimated means of the matched filter output were, in fact, the correct mean values, the normalized values would certainly be Chi-squared distributed with constant mean and the false alarm rate would be as obtained from (10.25c) and (10.32b). Equivalently, we can write those false alarm expressions as

$$P_{FA} = \exp(-T)\,. \tag{1}$$

In this expression, $T = \lambda^2/N_aT$ is the normalized threshold on the matched filter output where unity corresponds to the background noise level. A plot of the false alarm rate in (1) is shown in Figure 1 along with a typical false alarm rate curve that might be measured in a reverberant environment[1]. We see that if the mean of the matched filter output is stationary and known, as implied by (1), a false alarm probability of, say, 10^{-2} can be achieved with a threshold setting of about T = 4. In practice, though, we find that

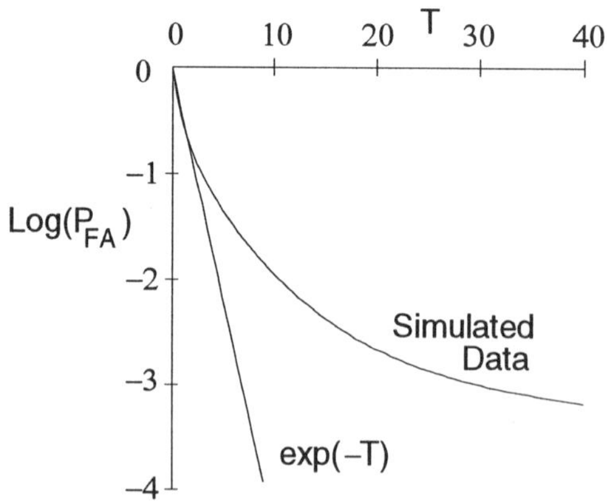

Figure 1 Probability of False Alarm for the Ideal Case exp(−T) and for Simulated Nonstationary Data. T = Threshold.

[1] The P_{FA} values that "might be measured in a reverberant environment" are, in fact, simulations of actual measured false alarm rates and are more fully explained in the following.

the achievement of a 10^{-2} false alarm probability in a reverberant environment requires a threshold of approximately double that level. For lower false alarm rates the discrepancy is even greater. The need for a higher threshold in a reverberant environment to hold down the false alarm rate causes lower level targets to be missed. To see why this happens we need to determine the false alarm rate, as in (1), but for a reverberant environment.

We let the output of the unnormalized matched filter be represented by the random process $\lambda(t)$. Summing over M samples in a split window averager we obtain the estimate of the mean

$$\mu(t) = \frac{1}{M} \sum_{m=1}^{M} \lambda(t_m) . \tag{2}$$

This estimate of the mean is the CFAR normalizer which gives the normalized output

$$z(t) = \frac{\lambda(t)}{\mu(t)} . \tag{3}$$

The normalized output will, ideally, have a constant mean. Comparisons of the normalized output to a constant threshold would then produce detections with a constant false alarm rate in accordance with (1). In reality, the situation is more complicated and less benign as is suggested by the discrepancy in Figure 1.

The probability of false alarm is given by

$$P_{FA} = \int_{T}^{\infty} dz p(z) , \tag{4}$$

where p(z) is the probability density function of the normalized output and T is the threshold applied to that output. The function p(z) is the density function of all the normalized matched filter output for the transmission cycle of interest. This function can be readily determined at the end of each transmission cycle by forming a histogram of the normalized matched filter output values. Such a procedure could be performed in a reverberant environment for each transmission and used to compute the probability of false alarm in (4). The result obtained would usually be a false alarm rate curve similar to the simulated data curve in Figure 1. This would serve to document the discrepancy between theory and practice but it would not explain it nor would it give any insight into methods of correcting it. For that we turn to analysis.

The probability density function of the normalized output in (3) is given by[2]

$$p(z) = \int_{0}^{\infty} dx x p_{\lambda}(zx) p_{\mu}(x) , \tag{5}$$

---

[2]Papoulis, A., *Probability, Random Variables and Stochastic Processes*, McGraw-Hill Book Co. 1965, (7.21).

where $p_\lambda(\cdot)$and $p_\mu(\cdot)$ are the probability density functions of, respectively, the unnormalized matched filter output and the CFAR normalizer. Our immediate task is to determine these probability density functions.

10.1-B: Determination of $p_\lambda(x)$

The probability density function $p_\lambda(x)$ is the density function of all the matched filter outputs prior to normalization. Each of the output samples of the matched filter is assumed to satisfy the nonstationary Chi-squared density function

$$p(\lambda, t) = \frac{\exp(-\lambda / m(t))}{m(t)}, \tag{6}$$

where $m(t) = 2\sigma^2(t)$ is the time varying mean of the matched filter output and $\sigma^2(t)$ is the nonstationary power of the quadrature components of the matched filter output.

Denoting the histogram of the unnormalized matched filter output as $p(\lambda)$, which implies stationarity of the data, the stationary distribution is given by the average of the nonstationary probability density function over the transmission cycle of duration τ

$$p(\lambda) = \frac{1}{\tau}\int_0^\tau p(\lambda, t)dt. \tag{7a}$$

We can transform this to a more useful form by first replacing the integral in (7a) by its Riemann sum

$$p(\lambda) = \frac{\Delta t}{\tau}\sum_{n=0}^{N} p(\lambda, m(t_n)), \tag{7b}$$

where Δt is the sample interval. Grouping the terms that have equal probability, i.e. the terms that fall within bins around the mean values m_j where $j = 0,1,2,\cdots J$, the sum can be rewritten

$$p(\lambda) = \frac{\Delta t}{\tau}\sum_{j=0}^{J} N_j p(\lambda, m_j), \tag{7c}$$

where N_j is the number of "hits" within the jth bin. Since $N = \tau/\Delta t$ is the total number of matched filter output samples, this can be written

$$p(\lambda) = \sum_{j=0}^{J} P_j p(\lambda, m_j), \tag{7d}$$

where $P_j = N_j/N$ is the probability that the mean value is m_j. Multiplying and dividing by Δm and going to the Riemann limit we obtain

$$p(\lambda) = \int_0^\infty p(m)p(\lambda, m)dm, \tag{7e}$$

where p(m) is the probability density function of the mean of the matched filter output. Substituting the nonstationary Chi-squared density function from (6) gives

$$p(\lambda) = \int_0^\infty p(m) \frac{\exp(-\lambda / m)}{m} dm. \tag{8}$$

This gives an integral equation for the probability density function of the mean of the matched filter output. Actually, since (6) is a density function conditioned on m, (8) could have been written immediately but such an approach would be less convincing and provide less insight.

The integral in (8) is an integral equation for the density function of the matched filter output means, i.e. p(m). It is a Fredholm integral equation of the first kind. It can be solved exactly, if somewhat tediously, but for our purposes a simple alternative approach has been found to be effective. We let the probability density function of the mean be the discrete distribution

$$p(m) = \sum_{n=1}^{N} P_n \delta(m - m_n), \tag{9a}$$

$$0 \le P_n \le 1, \qquad \sum_{n=1}^{N} P_n = 1, \tag{9b,c}$$

where P_n are probabilities which satisfy (9b,c) and m_n $n = 1,2,\cdots,N$ denotes a discrete set of means. This assumption implies that the reverberation power is discrete and comes from a number of scattering types. For example, for the case of bottom reverberation, the returns may come from various regions of, say, rock and sand from various ranges. Similarly, noise sources may include nearby shipping noises and distant shipping noises depending on receive array orientation. However, even if the reverberation and noise powers varied continuously, we could still approximate the mean levels of the matched filter output by a step function. As long as the number of steps are not excessive, this can be and has been an effective approach

Substituting (9) into (8) we obtain the probability density function for the unnormalized output

$$p(\lambda) = \sum_{n=1}^{N} P_n \frac{\exp(-\lambda / m_n)}{m_n}. \tag{10}$$

If we "fit" the right member to the data produced left member we can obtain estimates of the probabilities P_n and the mean levels m_n. This can be done via a least mean square fit subject to the constraints in (9b,c).

10.2-B: Determination of $p_\mu(x)$

The probability density functions of the CFAR averager terms $\mu_m = \lambda_m/M$ in (2), where μ_m represents one term in the series in (2), are related to the probability density function of the matched filter output terms by

$$p(\mu_m)d\mu_m = p(\lambda_m)d\lambda_m . \tag{11a}$$

Solving, with the aid of (10), gives

$$p(\mu_m) = M \sum_{n=1}^{N} \frac{P_n \exp(-\mu_m M / m_n)}{m_n} . \tag{11b}$$

This expression implicitly assumes that in every split window of data, each component has the probability P_n of occurring. This will be the case if the reverberation power is a random process but, in an environment of slowly vary reverberant power, only one or two components of reverberation will appear in the sum at any time. The effects of this are more fully illustrated in Section 10.4.

The characteristic functions of the CFAR averager terms are found from elementary methods to be

$$\phi_m(\omega) = M \int_0^\infty d\mu_m \exp(j\omega\mu_m) \sum_{n=1}^{N} \frac{P_n \exp(-\mu_m M / m_n)}{m_n} = \sum_{n=1}^{N} \frac{P_n}{(1 - jm_n\omega / M)} . \tag{11c}$$

Since the matched filter output samples are independent, the probability density function of the CFAR averager is

$$p(\mu) = \frac{1}{2\pi} \int_{-\infty}^{\infty} d\omega \exp(-j\omega\mu) \left[\sum_{n=1}^{N} \frac{P_n}{(1 - jm_n\omega / M)} \right]^M . \tag{12}$$

10.3-B: Probability of False Alarm

Substituting (12) and (10) into (5) gives the probability density function of the normalized matched filter output

$$\begin{aligned} p(z) &= \frac{1}{2\pi} \int_0^\infty d\mu\mu \left[\sum_{n=1}^{N} \frac{P_n \exp(-z\mu / m_n)}{m_n} \right] \int_{-\infty}^{\infty} d\omega \exp(-j\omega\mu) \left[\sum_{k=1}^{N} \frac{P_k}{(1 - jm_k\omega / M)} \right]^M \\ &= \frac{1}{2\pi} \sum_{n=1}^{N} \frac{P_n}{m_n} \int_{-\infty}^{\infty} d\omega \left(\frac{1}{z / m_n + j\omega} \right)^2 \left[\sum_{k=1}^{N} \frac{P_k}{(1 - jm_k\omega / M)} \right]^M , \end{aligned} \tag{13}$$

where the integration over μ was performed by elementary methods. From this expression we can immediately write an expression for the probability of false alarm for the normalized output, i.e. by combining (4) and (13) we obtain

$$P_{FA} = \frac{1}{2\pi}\sum_{n=1}^{N} P_n \int_{-\infty}^{\infty} d\omega \left(\frac{1}{T/m_n + j\omega}\right)\left[\sum_{k=1}^{N} \frac{P_k}{(1 - jm_k\omega / M)}\right]^M . \tag{14}$$

We can perform this integration as a contour integral in the complex ω plane. The contour encircles the upper half plane and the only pole within the contour is at $\omega = jT/m_n$. Symbolically we can write this integration as

$$\oint_C = \int_{-\infty}^{\infty} + \oint_\Gamma = 2\pi j F(jT/m_n), \tag{15}$$

where C is the contour encircling the upper half plane, the integral from $-\infty$ to ∞ is the integral in (14), Γ is the contour on the half circle of C in the upper half plane and F is the Cauchy integrand in (14). Since the integral over Γ is seen to vanish., we obtain

$$P_{FA} = \sum_{n=1}^{N} P_n \left[\sum_{k=1}^{N} \frac{P_k}{1 + Tm_k / Mm_n}\right]^M . \tag{16}$$

The gives the false alarm probability in terms of the known probabilities of the components of the unnormalized matched filter output and the mean values of those components. These probabilities and means can be determined from reverberation data using one of the algorithms discussed above. In fact, such values were used to produce the curve "that might be measured in a reverberant environment" of P_{FA} in Figure 1. The parameters used to generate that curve are given in Table 1.

Table 1 Parameters for Simulated Data

n	1	2	3	4	5
Log(P_n)	−0.130	−0.591	−2.90	−3.10	−3.74
m_n	1	4	32	64	256

If there were only one component in (16), i.e. the stationary case, that expression reduces to

$$P_{FA} = \left[\frac{1}{1 + T/M}\right]^M \xrightarrow{M\to\infty} \exp(-T), \tag{17}$$

which shows that for the case of stationary matched filter output, increasing the number of samples in the CFAR averager produces an output false alarm rate that approaches the ideal case. In fact, for an environment in which the matched filter mean is slowly varying, i.e. there is a small number of power components within the split window averager, the samples within the split window averager and within the split have nominally the same mean and the false alarm rate approaches the limit in (17). This is the allure of the split window averager. However, in many reverberant environments the mean values can vary rapidly because of acoustic “clutter” or “glint”, i.e. strong target-like receptions that appear randomly within the reverberation background. These receptions are, effectively, targets but they are due to bottom or surface features and are not submarine targets and are of no interest. Because of this, the achievable false alarm performance of the split window averager can fall short of the ideal in (1).

For the case of a reverberant environment or a nonstationary noise environment having random power, when the number of matched filter output samples in the CFAR averager increase without bound, the probability of false alarm goes to

$$P_{FA} \xrightarrow{M\to\infty} \sum_{n=1}^{N} P_n \exp\left(-T\langle m\rangle / m_n\right), \tag{18a}$$

where <m> is the average

$$\langle m\rangle = \sum_{n=1}^{N} P_n m_n . \tag{18b}$$

Graphs of probability of false alarm are shown if Figure 2 for the infinite degrees of freedom case in (18), the case of 10 degrees of freedom in (16) and the stationary case in (1) where the parameters in Table 1 were used. These graphs show that increasing the number of samples in the CFAR normalizer for the case of a nonstationary process will not substantially improve detection performance. In fact, for the case of a nonstationary process having random mean, the false alarm rate will never approximate the ideal case in (1).

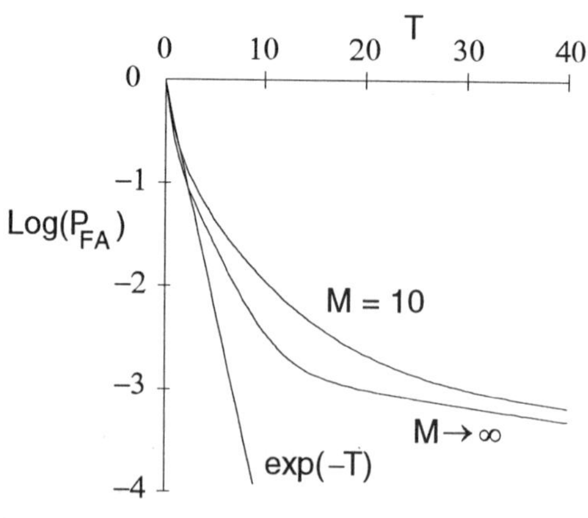

Figure 2 Probability of False Alarm for the Ideal Case and for Simulated Nonstationary Data. T = Detection Threshold. M = DOF.

If we distinguish between nonstationary processes having a slowly varying mean, i.e. quasi-stationary processes, and those having a random mean, we see that the split window averager does not achieve the same performance in both environments. For quasi-stationary processes, the performance can be expected to approach the ideal case but for random mean processes the performance is degraded. To illustrate this we turn to an example using simulated data.

10.4-B: Random versus Quasi-Stationary Mean Power - An Example

In the preceding we found that it is necessary to distinguish between nonstationary processes having random mean power and nonstationary processes having quasi-stationary mean power. In the former situation, the false alarm performance is degraded from the stationary case while in the latter situation the false alarm performance can approach the ideal stationary case. These results are not surprising but it is instructive to see an example of just what they mean.

In the following we simulate the detection process in two different environments and compare the performance of the split window CFAR normalizer for each. The first process is assumed to be for a matched filter output with a random mean and the second is assumed to be for a matched filter output with a quasi-stationary mean. Graphs of these means for both processes are shown in Figure 1.

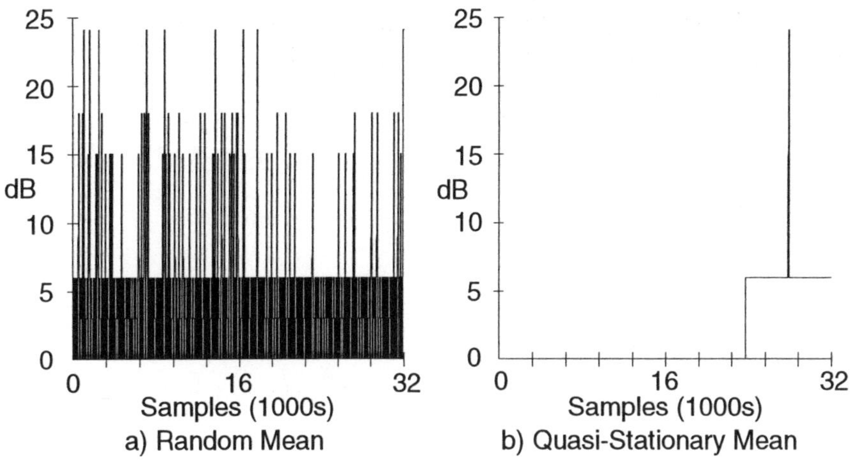

Figure 3 Mean of Nonstationary Matched Filter Output

The total number of samples in both cases is 32000 and both sets of samples are identical. By this we mean that there is a one-to-one correspondence between the two sample sets but their orderings in time are different. The values of the means and their frequency of occurrence for both sample sets were constructed using the parameters in Table 1. For the random mean case the order of the values of the mean was chosen randomly resulting in the graph in Figure 3a while for the quasi-stationary case the order was chosen arbitrarily as indicated in the graph in Figure 3b. In both cases the magnitudes and frequency of occurrence of the samples are consistent with the values in Table 1.

Both sample sets of mean powers were used to construct simulated unnormalized matched filter outputs that are displayed in Figure 4. The procedure followed is based upon the observation that, from (6), the unnormalized matched filter output values must satisfy the relation $m \times p(\lambda) = \exp(-\lambda/m)$. Accordingly, the values in Figure 3 are obtained from the relation $\lambda = -m \times \ln(\alpha)$ where m is a mean value shown in Figure 1, and α is a uniformly distributed random value from the interval 0 to 1. The relatively high levels of clutter in the random mean case as compared to the quasi-stationary case in Figure 2 is readily apparent. Since both sets of samples are identical, except for ordering, they have the same distribution $p(\lambda)$ that is described by (7) and (8).

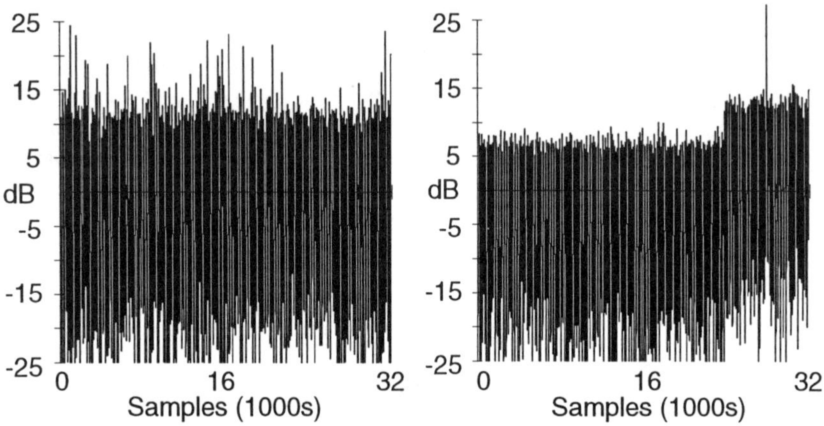

a) Random Mean b) Quasi-Stationary Mean

Figure 4 Simulated Nonstationary Matched Filter Output

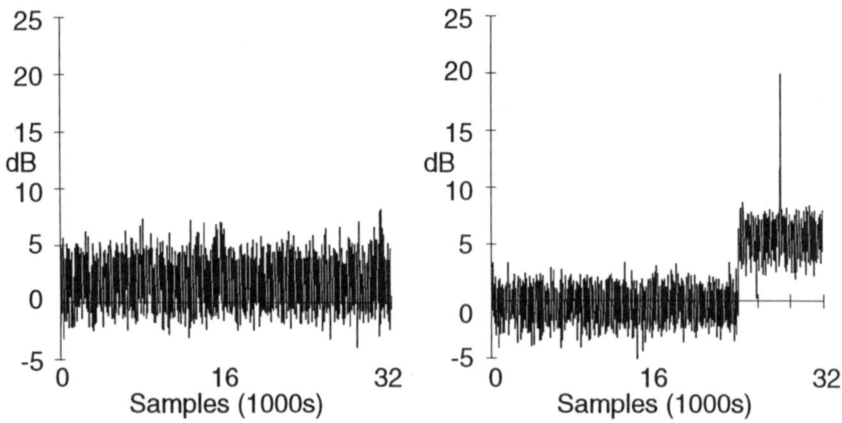

a) Random Mean b) Quasi-Stationary Mean

Figure 5 Split Window Average of Nonstationary Matched Filter Output

A split window averager was used to estimate the means of the unnormalized matched filter output in Figure 4 in accordance with (2) and the normalized matched filter output is shown in Figure 5. For both cases the split window averager consisted of 20 samples with 10 samples in each side window with a single sample in the "split". Of course, for real targets the echo would require a split of more than a single sample but this is unimportant for our simulation. The sum over the split window averager was performed as a two step operation. The first sum obtained the gross mean in the side windows. The second sum omitted, i.e. censored, all sample values that exceeded the gross mean by a factor of ten. This effectively assumes a detection threshold of 10 dB and excludes all samples from the side window sum that are possible targets that would bias the noise estimate. The variance of the estimated mean in Figure 5 for the random mean case is seen to be approximately twice that of the quasi-stationary case for which the 2 sigma value is approximately the expected value of 3 dB that a stationary process would exhibit.

The censored mean of the unnormalized matched filter output was used as the CFAR normalizer to produce the normalized matched filter output graphs shown in Figure 6. Inspection of the graphs shows that both have approximately constant means but that the variance of the random mean case is much greater than the variance of the quasi-stationary case. Effectively, the CFAR normalizer has passed randomly occurring clutter "spikes" of the random mean case as potential targets as it should.

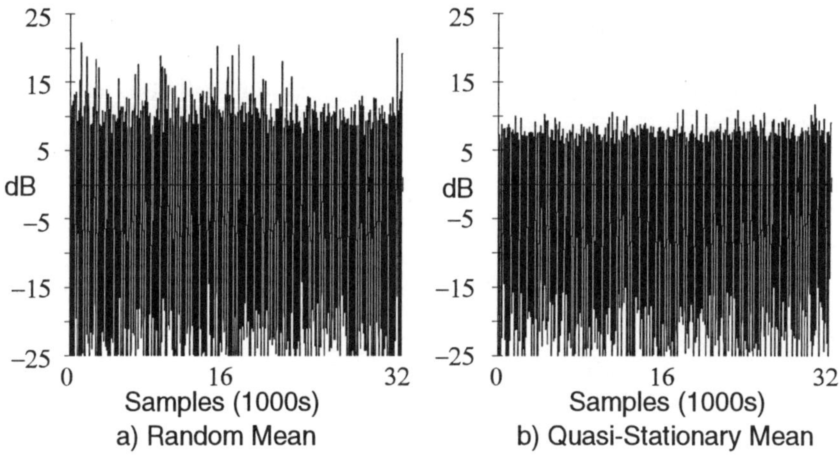

Figure 6 CFAR Normalized Matched Filter Output

The probabilities of false alarm for the random mean and quasi-stationary cases were determined from the data in the graphs in Figure 6 and are shown plotted in Figure 7 along with the stationary false alarm case. We see that the false alarm performance of the random mean case is degraded from the stationary case while the performance of the quasi-stationary case approximates that of the stationary case. Consequently, this example suggests that CFAR normalization by the split window average can be an optimal detection technique but that in environments which produce much clutter the performance can be expected to be degraded. A simplistic generalization might be that the performance of the split window averager is optimal in surface and volume reverberation environments, and all noise environments, but is suboptimal in bottom reverberation environments. Of course, this generalization requires testing and is, no doubt, too simple to be universally valid but it might prove useful as a rule of thumb. The question remains, "Are there mean estimation procedures that are optimal in all environments?" No answer to this question is given here but, until an affirmative answer is given, the split window averager is the means estimator of choice.

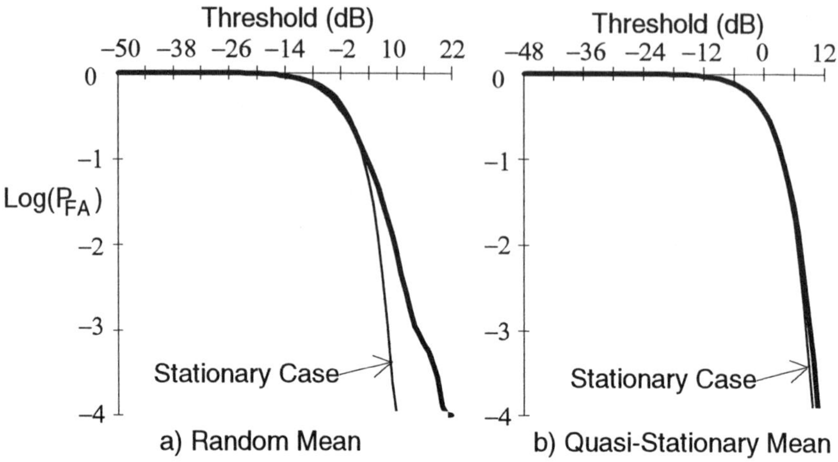

Figure 7 Probability of False Alarm for Nonstationary Matched Filter Output

CHAPTER 11
PASSIVE DETECTION

Passive detection systems search for target radiated signals that include periodic components, which can be caused by machinery; noise-like components, which can be propeller cavitation noise; and transients, which can arise from events such as the opening and closing of torpedo ports and missile hatches. The periodic components can be narrow band tonals or can be modulated signals having several harmonically related narrow band tonals. The noise-like components can be narrow band or broad band and the transients are short duration broad band signals. The temporal structures of target radiated fields are essentially unknown to the ASW receiver but their spectral content exhibits characteristic frequency signatures that can be used to identify target class and, in some cases, individual targets. To capitalize on this information, passive sonar receivers perform spectral decomposition of the received acoustic data. This serves the dual functions of revealing the frequency signature of the detected target, which provides target classification information to the system, and identifying and removing out-of-band noise, which enhances target detection. As indicated in Figure 1, selected search bands of frequencies, that span the expected targets spectral regions, are examined for excess energy that is attributable to targets and which can be processed either as a single analysis band or as several spectrally resolved analysis bands. The selected search bands cover the general regions in which targets radiations are expected to be found and their resolution into analysis bands optimizes the capture of the tonal components of the targets signatures within the search bands. Time averages within analysis bands accumulate energy for detection processing.

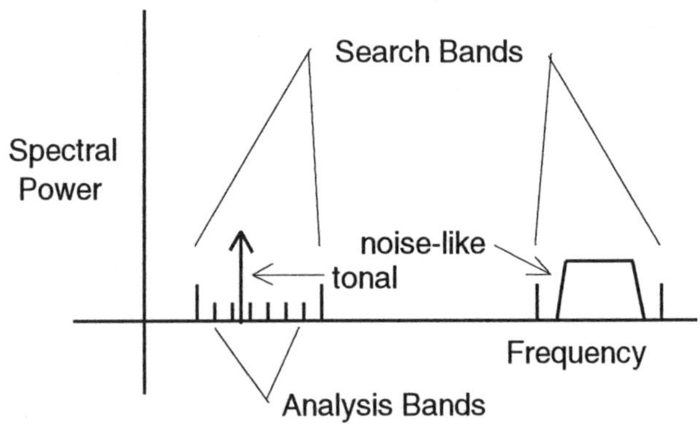

Figure 1 Passive Spectral Features

The analytic form of target radiated signals, that are expected to be found within a search band, is modeled similarly to active received signals

$$r_a(t) = m(t)\exp(j\omega t)\,. \tag{1}$$

However, whereas active received signals are generated by active sources and are known to within an attenuation factor, a Doppler compression factor and a phase shift, passive received signals are known only in their spectral shape. This shape can be revealed by Fourier transformation of the passive received signal which, from (2.17), can be expressed

$$r_a(t_n) = \sum_{m=-\infty}^{\infty} Y_{n,m} \exp(j\omega_m t_n)\,, \tag{2a}$$

$$Y_{n,m} = \frac{1}{T_W} \int_{t_n - T_W/2}^{t_n + T_W/2} r_a(t) \exp(-j\omega_m t)\,dt\,, \qquad \omega_m = \frac{2\pi m}{T_W}\,. \tag{2b,c}$$

The quantity $Y_{n,m}$ is the mth Fourier coefficient for the data in the nth time window of data. The time t_n is the time at the center of the data window of length T_W and is the time associated with the nth spectrum of data. Each of the spectral components has the associated frequency bin width of $1/T_W$, which is the analysis bandwidth, and which shows that narrow band analysis requires long duration windows of data. The total duration of all windows of data to be averaged is T and the total number of windows is $N = T/T_W$. The power of the data, from (2.18), is

$$P_n = 2 \sum_{m=1}^{\infty} \left|Y_{n,m}\right|^2 . \tag{2d}$$

The collection of target spectral power coefficients $|Y_{n,m}|^2$ reveal the shape of the target signature at each time t_n but, of course, the target signature, when received in the acoustic data, comes with a background of noise and must be inferred.

Since samples of the received data and their Fourier coefficients are both equivalent representations of the signal, the likelihood ratio detector can be determined in terms of either but it is useful in the following to perform the analysis in terms of Fourier coefficients. The passive detection hypotheses for target plus noise H_1 and for noise only H_0, expressed in terms of Fourier coefficients, are

$$H_1 : \mathbf{Y} = \mathbf{S} + \mathbf{N}\,, \tag{3a}$$

$$H_0 : \mathbf{Y} = \mathbf{N}\,. \tag{3b}$$

The data vector for temporal averaging $\mathbf{Y}^T = (\mathbf{Y}_1^T,\cdots,\mathbf{Y}_m^T,\cdots,\mathbf{Y}_M^T)$, where the superscript T denotes transpose, is represented as a vector of vectors in which each mth sub-vector represents the time sequence of Fourier coefficients for the mth frequency channel. The time sequence is obtained from Fourier transforms of N consecutive windows of data. The mth sub-vector contains the elements $\mathbf{Y}_m^T = (\mathbf{Y}_{1,m},\cdots,\mathbf{Y}_{n,m},\cdots,\mathbf{Y}_{N,m})$, where the first subscript of the elements represents the nth window of data and the second subscript represents the mth frequency channel of the data. Similar definitions apply to the signal vector **S** and noise vector **N**. The number of complex Fourier coefficients N in each sub-vector is equal to the time-bandwidth product of the analysis band data, i.e. $N = T/T_W = T\Delta f$ where T is the total duration of the data and Δf is the analysis band width of the Fourier transform. Similarly, the number M of analysis bands in each spectrum is the time-bandwidth product of the window of data, i.e. $M = W/\Delta f = T_W W$ where W is the width of the search band. Combining terms gives the total number of Fourier coefficients, $NM = TW$, in the data vector. For statistically independent data, twice the total number of coefficients is termed the degrees-of-freedom of the data since the data is complex.

Example 1: During surveillance operations, a search for tonals expected to be within, say, a search band of W = 100 Hz would perform initial filtering to select the 100 Hz band from the in-water data. The data stream from the filter would be sampled at a rate equal to the filter bandwidth, i.e. f_S = 100 Hz, and would be resolved into analysis bands that are narrow enough to exclude a maximum amount of noise and wide enough to contain the tonals and their excursions. Choosing the analysis bandwidth $\Delta f = 0.1$ Hz, the data window duration required to produce this resolution is $T_W = 10$ seconds and the number of frequency bins is $M = T_W W = 1000$. The total number of time samples in each data window is $f_S T_W = 1000$. The number of data windows selected for averaging is essentially arbitrary but is subject to the trade-offs discussed in the following. For, say, a total averaging time T of twenty minutes, the total number of data windows is N = 120 and the total number of Fourier components to be processed is NM = 120,000. The number of degrees-of-freedom for independent data is 240,000. ♦

Example 2: A search for a transit, that is expected to occupy a band of W = 500 Hz, would perform initial filtering to select the 500 Hz search band from the in-water data. The data stream from the filter would be sampled at the bandwidth rate, i.e. f_S = 500 Hz, and the band would not be further resolved. Each data sample would, effectively, be a window of data of T_W = 0.002 seconds duration. The number of frequency bins would be M =1 and the total number of time samples in each data window would be $f_S T_W = 1$. The number of data windows selected for averaging and display might be N = 1 if the transit is expected to be a singular event. If the transient is expected to have longer duration, either because of the nature of its generation or because of spreading due to propagation, more windows of data could be averaged. In this instance, the total number of complex data samples processed would be NM = 1 and the number of degrees-of-freedom would be 2. ♦

A caveat is in order at this point. The preceding discussion of resolving search bands via Fourier transformation implicitly assumed that the data windows are contiguous and non-overlapping and extract unweighted windowed data that are statistically independent from window to window. In fact, in operational systems it is necessary to weight the data, i.e. by effectively using windows tapered at each end to weight the data, in order to reduce the effect of spectral energy created by the window itself due to its onset and decay[1]. However, since the window taper also reduces the impact of the data at the window edges, the windows are commonly overlapped by 25%, 50% or 75% which has the effect of inducing a statistical dependence between the windows of data. Additionally, the spectral energy created by even the best windows adds noise to all analysis bands across the spectrum and, to accurately measure the true background noise level, an effective analysis bandwidth, that is wider than the true bandwidth and which corresponds to a shorter effective data window duration, is assigned to each analysis band. The effective bandwidth is termed the "equivalent noise bandwidth" and is the appropriate analysis bandwidth. However, for "good" windows[2] the data windows can be overlapped, say up to 50%,

---

[1] See Appendix 4-C.

[2] Harris, J. Fredric, "On the Use of Windows for Harmonic Analysis with the Discrete Fourier Transform", Proc. IEEE, **66**, 1, January 1978.

with the window-to-window data approximating the property of statistical independence. Moreover, the effective data window duration T_W that corresponds to the equivalent noise bandwidth is approximately equal to the duration of equivalent contiguous non-overlapping windows. Hence, the assumption of contiguous windows of independent data is nearly correct and adequate for discussing the principles of passive detection even for overlapping windows of data. For more exacting analyses, the overlap of the windows must be accounted for which leads to required modifications to the number of degrees-of-freedom parameter.

11.1 Signal and Noise Statistics

Under the assumption that the Fourier coefficients of both the received signal and the background noise are represented by sums of independent zero mean Gaussian signal samples in (2a), the Fourier coefficients are also similarly distributed. Each Fourier coefficient is complex, $Y_{n,m} = (U_{n,m} + jV_{n,m})$, and the power of the real and imaginary components are each, under the signal plus noise hypothesis, $\sigma^2_S + \sigma^2$ and, under the noise only hypothesis σ^2, i.e. σ^2_S denotes the signal power and σ^2 denotes the noise power of the real and imaginary components. The covariance matrices of the Fourier components are then given by the conditional MN × MN matrices

$$E(\mathbf{YY}^{*T}|H_1) = \begin{bmatrix} 2\left(\sigma^2_{S,1} + \sigma^2_1\right)I_N & 0 \\ 0 & 2\left(\sigma^2_{S,M} + \sigma^2_M\right)I_N \end{bmatrix}_M , \tag{4a}$$

$$E(\mathbf{YY}^{*T}|H_0) = \begin{bmatrix} 2\sigma^2_1 I_N & 0 \\ 0 & 2\sigma^2_M I_N \end{bmatrix}_M , \tag{4b}$$

where the quantities, $2\sigma^2_{S,m}$ and $2\sigma^2_m$, are the powers of the Fourier components of, respectively, the received analytic signal and the analytic background noise within the mth analysis band. The quantity I_N denotes the identity matrix of order N and the symbol $[\ ]_M$ denotes an M × M matrix. Note that the matrix is block diagonal and only the "corner" elements of the matrix are shown.

Since the covariance matrices in (4) are diagonal matrices, their inverses and determinants are easily written. With these quantities the probability density functions for the received beam-band analytic data under the two hypotheses, from (8.16-A), are

$$p(\mathbf{Y}|H_1) = \frac{\exp\left[-\sum_{m=1}^{M} \frac{\mathbf{Y}_m^{*T}\mathbf{Y}_m}{2\left(\sigma^2_{S,m} + \sigma^2_m\right)}\right]}{(\pi)^N\left(\prod_{m=1}^{M} 2\left(\sigma^2_{S,m} + \sigma^2_m\right)\right)^N}, \quad p(\mathbf{Y}|H_0) = \frac{\exp\left[-\sum_{m=1}^{M} \frac{\mathbf{Y}_m^{*T}\mathbf{Y}_m}{2\sigma^2_m}\right]}{(\pi)^N\left(\prod_{m=1}^{M} 2\sigma^2_m\right)^N}. \tag{5a,b}$$

These probability density functions represent the total information available for each beam-band data set and are the basis for detection processing. They indicate that the only difference between the data under the signal plus noise hypothesis and the noise only hypothesis is the power level. Consequently, it is expected that the optimal detector would perform power or energy estimates on each data set. This is what we find in the likelihood ratio detector.

11.2 Likelihood Ratio Detector

Likelihood Ratio Function

The likelihood ratio test is found from (9.5) and (5) to be

$$\Lambda(\mathbf{Y}) = \left(\prod_{m=1}^{M} \frac{\sigma_m^2}{\sigma_{S,m}^2 + \sigma_m^2} \right) \exp\left[\frac{1}{2} \sum_{m=1}^{M} \frac{\mathbf{Y}_m^{*T} \mathbf{Y}_m}{\sigma_m^2} \left(1 - \frac{\sigma_m^2}{\sigma_{S,m}^2 + \sigma_m^2} \right) \right] \underset{H_0}{\overset{H_1}{\gtrless}} \Lambda_T, \quad (6a)$$

for a given data set where the sum in (6a) is over M frequency channels. Note that those terms in which there is no signal component, i.e. $\sigma^2_{S,m} = 0$, vanish from the sum. Since the signature for many targets consists of only a sparse set of line components, the sum in (6a) frequently reduces to sums over just a few frequency channels. For targets emitting a strong characteristic tonal, the sum can be reduced to a single term. Since the signal component frequencies are usually known only to within a range of frequencies, multiple detectors must be configured to search that range. This can be done automatically by computer but is often performed by visual screening of spectral data by operators.

The power weightings in the spectral sum in (6a) are unknown. While the noise powers are known, i.e. can be estimated, at best only the relative magnitudes of the signal powers and not the absolute levels are known. However, since in (6a) all the power weightings are positive and the sum increases monotonically with the data, an equivalent statistic is given by the log likelihood ratio statistic

$$\lambda(\mathbf{Y}) = \sum_{k=1}^{K} \frac{\mathbf{Y}_{m_k}^{*T} \mathbf{Y}_{m_k}}{\sigma_{m_k}^2} \underset{H_0}{\overset{H_1}{\gtrless}} \lambda_T. \quad (6b)$$

This detector is a normalized sum of spectral powers in which the normalization is by the estimated noise power in each frequency channel. The noise estimate can be obtained by averaging the analysis band levels during periods when no signal is present. This averaging is generally performed over a time sequence of samples but may also include averaging over adjacent frequency channels. The K terms in the sum are for the m_k frequency channels that are hypothesized to contain the signal components. The problem with this detector is that it weights the weak frequency channels disproportionately heavily. However, for the cases considered in the following, i.e. a single channel; equal signal-to-noise ratios in all channels; and different but weak signal-to-noise ratios in all channels, this problem doesn't occur.

Likelihood Ratio Statistics

For the general case, the statistics of the energy detector in (6b) are complicated but they reduce to the well-known Chi-squared distribution for several operationally important cases. These reductions and their applications to sonar problems are demonstrated in the following. In all instances the conditional means and variances of the energy detector are found from (6b) to be

$$E_1(\lambda(\mathbf{Y})) = 2N\sum_{k=1}^{K}\left(\frac{\sigma_{S,k}^2}{\sigma_k^2}+1\right), \qquad E_0(\lambda(\mathbf{Y})) = 2NK, \tag{7a,b}$$

$$\mathrm{Var}_1(\lambda(\mathbf{Y})) = 4N\sum_{k=1}^{K}\left(\frac{\sigma_{S,k}^2}{\sigma_k^2}+1\right), \qquad \mathrm{Var}_0(\lambda(\mathbf{Y})) = 4NK, \tag{7c,d}$$

where the subscripts 1,0 denote the hypotheses H_1 and H_0, and the K frequency channels, hypothesized to contain the target signature components, have been sequentially indexed with index k. The derivation of these results is somewhat tedious but straightforward. The detection index from (9.10) is found to be

$$d = NK\left(\frac{1}{K}\sum_{k=1}^{K}\frac{\sigma_{S,k}^2}{\sigma_k^2}\right)^2, \tag{7e}$$

where the term in parentheses is the average in-water signal-to-noise ratio over all frequency channels .

The log likelihood ratio detector in (6b) is the sum of magnitude squared zero mean Gaussian random variables. For those cases in which these variables have equal variances, the detector output is Chi-squared distributed[3] with 2NK degrees-of-freedom, i.e. the total number of independent terms in the sequence of data windows. This occurs in all instances for the noise only output since each term in the energy detector is normalized by its expected power, resulting in unit variance for each term, and the expected total output value is 2NK by (7b). Therefore, for the noise only case the energy detector output is always distributed in accordance with the Chi-squared distribution

$$p(\lambda|H_0) = \frac{1}{2^{NK}\Gamma(NK)}\lambda^{(NK-1)}\exp[-\lambda/2]. \tag{8}$$

[3]Abramowitz and Stegun, (26.4).

For the signal plus noise case the underlying Gaussian random variables in the energy detector in (6b) remain zero mean but the variances are, in general, not all equal since they are dependent upon the signal-to-noise ratios in the separate frequency channels as indicated in (7a). Consequently, for the general signal-plus-noise case the detector output is not Chi-squared distributed and, unfortunately, a simple distribution for the general case is not available. However, for three operationally important cases the statistics of the detector reduce to the well-known Chi-squared distribution. These cases are discussed in the following and include the cases of: a single frequency channel; signal-to-noise equal in all frequency channels; signal-to-noise unequal but low in all channels.

ROC Curves - Single Frequency Channel

If the frequency channels are "unganged" and detections are based solely upon the levels found in each channel separately, the variances of each term in the detector sum are equal and the detector output for both the signal plus noise and noise only cases are Chi-squared distributed

$$p(\lambda|H_1) = \frac{1}{2^N\left(\sigma_S^2/\sigma^2+1\right)^N\Gamma(N)}\lambda^{(N-1)}\exp\left[-\lambda/2\left(\sigma_S^2/\sigma^2+1\right)\right], \tag{9a}$$

$$p(\lambda|H_0) = \frac{1}{2^N\Gamma(N)}\lambda^{(N-1)}\exp[-\lambda/2], \tag{9b}$$

where the number of degrees-of-freedom 2N is twice the number of spectral time samples in the frequency channel.

The ROC curves for the single channel case can be immediately determined from (9.8) and (9) to be

$$P_D = \frac{1}{2^N\left(\sigma_S^2/\sigma^2+1\right)^N\Gamma(N)}\int_{\lambda_T}^{\infty}\lambda^{(N-1)}\exp\left[-\lambda/2\left(\sigma_S^2/\sigma^2+1\right)\right]d\lambda, \tag{10a}$$

$$P_{FA} = \frac{1}{2^N\Gamma(N)}\int_{\lambda_T}^{\infty}\lambda^{(N-1)}\exp[-\lambda/2]d\lambda, \tag{10b}$$

where the lower limit on the integrals is the log likelihood threshold in (6b). For long term integrations, i.e. N >> 1, those expressions reduce to[4]

$$P_D = \frac{1}{\sqrt{2\pi}}\int_{z_D}^{\infty}dz\,\exp\left(-z^2/2\right), \quad z_D = \frac{\lambda_T/\left(\sigma_S^2/\sigma^2+1\right)}{2\sqrt{N}} - \sqrt{N}, \tag{11a}$$

[4] Abramowitz and Stegun, (26.4.11).

$$P_{FA} = \frac{1}{\sqrt{2\pi}} \int_{z_{FA}}^{\infty} dz \exp\left(-z^2/2\right), \qquad z_{FA} = \frac{\lambda_T}{2\sqrt{N}} - \sqrt{N}. \tag{11b}$$

The probability of detection expression in (11a) can be further simplified by approximating the lower limit on the probability of detection integral

$$\left(\frac{1}{\sigma_S^2/\sigma^2+1}\right)\left[\left(\frac{\lambda_T}{2\sqrt{N}} - \sqrt{N}\right) - \sqrt{N}\left(\frac{\sigma_S^2}{\sigma^2}\right)\right] \cong \left[\left(\frac{\lambda_T}{2\sqrt{N}} - \sqrt{N}\right) - \sqrt{N}\left(\frac{\sigma_S^2}{\sigma^2}\right)\right]. \tag{11c}$$

The validity of this approximation rests on the notion that, for large degrees-of-freedom parameter, i.e. time-bandwidth product, the required signal-to-noise ratio is low. (It is shown below that about 120 samples, i.e. 240 degrees-of-freedom, is the minimum number of samples needed to qualify for the large time-bandwidth approximation for many operationally important scenarios.) With this approximation, the lead term in the right member of (11c) is the same as the lower limit on the probability of false alarm integral and the last term is the square root of the detection index, i.e. from (7e) we obtain the detection index expression

$$d = N\left(\frac{\sigma_S^2}{\sigma^2}\right)^2. \tag{11d}$$

We note that the ROC equations defined by (11) are identical in form to the active known signal ROC expressions in (10.12). Consequently, Figures 10.2a and 10.2b can be used to determine the required detection index for the single channel passive detection case. However, as discussed in the following, the detection indices for the active known signal and for the passive unknown signal relate to the in-water signal-to-noise ratio differently and, also, the detector thresholds for both cases relate to the ROC curves differently.

Example 3: To achieve the system operating point, $P_D = 50\%$ and $P_{FA} = 0.01\%$, during the search for tonals in Example 1, the required detection index, by Figure 10.2b, is 14. Also, for the chosen system operating point, the lower limit in (11b) is about 3.71902[5] which yields the threshold $\lambda_T = 321.5$. Since by (7) the mean and standard deviation, for the noise only hypothesis, are 240 and 21.9, respectively, it is seen that the threshold is required to be higher than the mean by 3.7 standard deviations to achieve the required false alarm rate. ♦

The large time-bandwidth approximations in (11) coupled with the ROC curves in Figure 10.2 are very useful approximations but they are not a complete solution to the problem. The detection index in (11d) indicates that the detector output signal-to-noise ratio grows with the number of degrees-of-freedom, i.e. in

---

[5] Abramowitz and Stegun, Table 26.6.

proportion to the data time-bandwidth product, which suggests that, in principal, even the weakest signal would be detected if enough data were accumulated. However, in practice the signal does not remain in the same analysis band indefinitely and weak signals can drift in and out of an analysis band without being detected. To cope with these situations, shorter integration times must be chosen to match the data window duration to the time that the signal is expected to be in the analysis band. However, if the integration time required to match the signal is less than about 120 spectral samples, the large time-bandwidth approximations in (11) are not appropriate and the exact relations in (10) must be used. By means of those relations, a set of ROC curves, shown in Figure 2, is developed by noting that for the given operating point, P_D and P_{FA}, and number of samples N, the probability of false

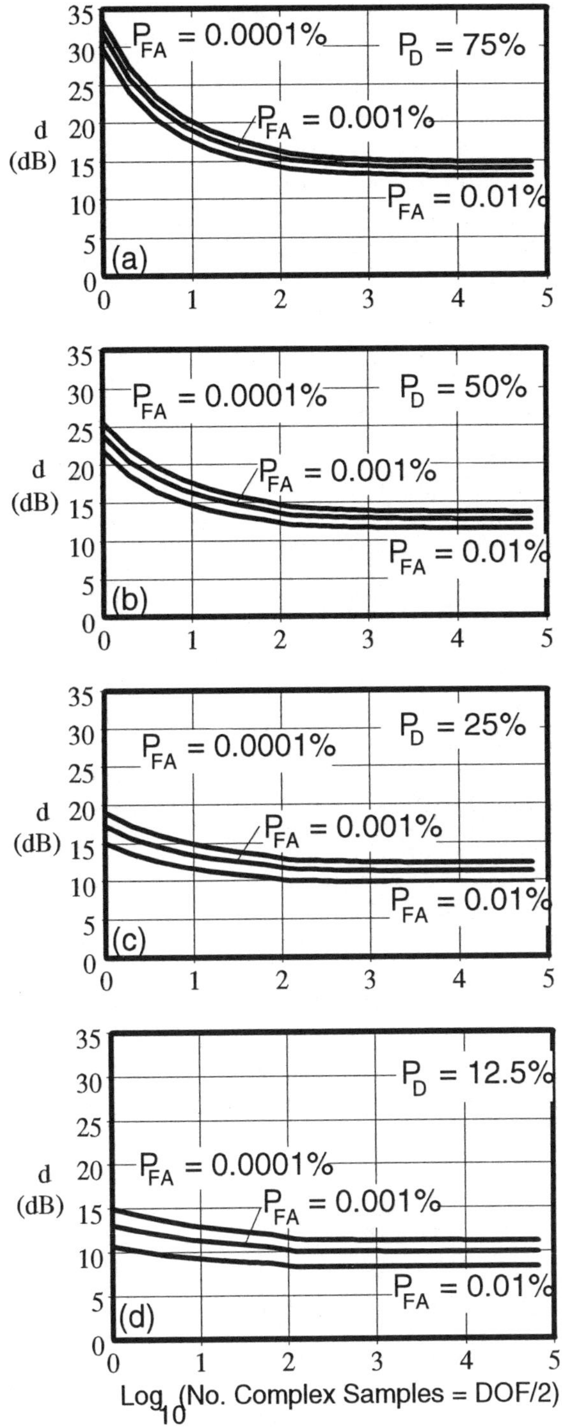

Figure 2 Passive Detection Index versus Number of Complex Samples and Degrees-of-Freedom (DOF)

alarm expression in (10b) determines the required detector threshold λ_T and the probability of detection expression in (10a) determines the required in-water signal-to-noise ratio and, hence, the detection index by (11d). Sets of such ROC curves showing the required detection indices versus degrees-of-freedom are presented in Figure 2 for selected values of detection and false alarm probabilities.

Example 4: To achieve the system operating point, P_D = 50% and P_{FA} = 0.01%, during the search for tonals in Example 1, but using only 10 spectral samples, i.e. 20 degrees-of-freedom, the required detection index, by Figure 2, is 15 dB or, equivalently, 32 as compared to 14 for the analogous large time-bandwidth case in Examples 1 and 3. The detector threshold, for 10 spectral samples is found to be[6] $\lambda_T = 52$ from (10b). Since by (7) the noise only mean and standard deviation are 20 and 6.3, respectively, we see that the threshold is required to be higher than the mean by 5 standard deviations to achieve the required false alarm rate. ♦

Example 5: To achieve the system operating point, P_D = 50% and P_{FA} = 0.01%, during the search for the transient in Example 2, using only 2 degrees-of-freedom, i.e. 1 spectral sample, the required detection index, by Figure 2, is 22 dB or, equivalently, 158.5. The detector threshold, for 2 degrees-of-freedom, is found to be[7] $\lambda_T = 18.5$ from (10b). Since by (7) the noise only mean and standard deviation are 2 and 2, respectively, we see that the threshold is required to be higher than the mean noise by 8.25 standard deviations to achieve the required false alarm rate. ♦

Robertson[8] has developed a general set of ROC curves for the case of a sine wave in Gaussian noise for a wide range of the degrees-of-freedom parameter. These curves can be applied to the case at hand in so far as the received target tonals can be modeled as sine waves. The results obtained from the Robertson curves are approximately the same as those obtained from Figure 2.

Detection Threshold (DT) - Single Frequency Channel

The detection threshold is the minimum received in-water signal-to-noise ratio that is required to achieve the desired detection performance level. The received in-water signal-to-noise ratio is the ratio of the signal and noise intensities

$$\mathrm{SNR_R} = 10\mathrm{Log}\left(\frac{\int\limits_{\text{signal band}} d\omega P(\omega)}{\int\limits_{\text{noise band}} d\omega N_0(\omega)} \right), \tag{12}$$

where the terms “signal band” and “noise band” denote the frequency bands for signal and noise and they are assumed to fill their respective bands. Since the noise is broad band, the noise band is the same as the analysis band but for signals the

[6] Abramowitz and Stegun, Table 26.7.
[7] Abramowitz and Stegun, Table 26.7.
[8] Robertson, G. H., “Operating Characteristics for a Linear Detector of CW Signals in Narrow-Band Gaussian Noise,” Bell System Technical Journal, **46**, 4, 1967.

signal band is the same as the analysis band only for broad band signals and is contained by the analysis band for narrow band signals. For broad band signals the choice of analysis band has no impact on the in-water signal-to-noise ratio measure, assuming an analysis band less than or equal to the signal band and constant spectral densities over the analysis band, but for narrow band signals the choice of analysis band can have a significant impact on the in-water signal-to-noise ratio measure.

For broad band signals the detection threshold for an analysis band from (12) can be written

$$DT = 10\text{Log}\left(\frac{P_m \Delta f}{N_0 \Delta f}\right) = 10\text{Log}\left(\frac{\sigma_S^2}{\sigma^2}\right), \tag{13a}$$

where P_m is the minimum spectral density for detection, N_0 is the noise spectral density, and both P_m and N_0 are assumed to be constant over the analysis band which is less than or equal to the signal bandwidth. From the detection index expression in (11d) the broad band detection threshold in (13a) can be written

$$DT = 10\text{Log}\left(\sqrt{\frac{d}{T\Delta f}}\right) = 5\text{Log}(d) - 5\text{Log}(T\Delta f). \tag{13b}$$

Equation (13b) shows that, for a given detection index, the detection threshold decreases as the time-bandwidth of the data increases. The implication is that for optimal detection performance, i.e. requiring the lowest in-water signal-to-noise ratio for detection, the analysis bandwidth must be matched to the signal bandwidth and the total integration time must be as long as possible. Matching the analysis bandwidth to the signal bandwidth includes a maximum amount of signal energy and limits the amount of noise to the in-band noise.

It might be thought that increasing the analysis bandwidth to include more signal energy would also include more noise energy leaving the in-water signal-to-noise ratio within the analysis band constant, i.e. from (12), for constant signal and noise spectra, the received signal-to-noise ratio $SNR = P/N_0$ is constant. Moreover, it might be thought that this would also leave the detection performance unchanged. However, while the in-water received signal-to-noise ratio does remains constant, the detection performance does not because the required minimum signal-to-noise ratio, i.e. detection threshold, decreases. The decrease occurs because increasing the analysis bandwidth to match the signal bandwidth shortens the Fourier transform data windows which produces a greater number of independent data samples summed in the detector. From the minimum signal-to-noise ratio defined in (13b) it can be found that the minimum signal-to-noise ratio decreases with increases in the degrees-of-freedom at the rate $(\delta SNR_{min})/SNR_{min} = -0.5(\delta N/N)$. Consequently, the signal excess, $SE = SNR_R - DT$, increases under this scenario because the detection threshold decreases. Since the received in-water signal-to-noise ratio within the analysis band is constant, the reduction in the minimum signal-to-noise ratio increases signal excess and improves detection performance.

Example 6: For the transient detection in Examples 2 and 5, the detection index was found to be d = 158.5 with a single sample "integration time" of 0.002 seconds and an analysis band of 500 Hz. The detection threshold for this case is

$$DT = 5\mathrm{Log}(158.5) - 5\mathrm{Log}(0.002\,\mathrm{sec} \times 500\mathrm{Hz}) = 11\ \mathrm{dB}\,, \tag{13c}$$

which shows that the detection of short transients requires a relatively strong signal, i.e. one that would be visually evident on a time plot, before a detection can be made at the required false alarm rate. This type of detection is most appropriate over short ranges as can occur during submarine-to-submarine encounters. ♦

For narrow band signals the detection index from (11d) can be written

$$d = T\Delta f\left(\frac{\sigma_S^2}{N_0 \Delta f}\right)^2, \tag{13d}$$

where it is assumed that the signal spectrum is concentrated in a single line component with intensity σ_S^2 within the analysis bandwidth. Formally, the analysis for narrow band signals could follow that for broadband signals by defining the in-water signal-to-noise ratio to be $\sigma_S^2/N_0\Delta f$. However, this is unsatisfactory because it requires the in-water signal-to-noise ratio to be related to the analysis band which can vary over processing channels. Instead, the reference noise band is taken be 1 Hz which might be more or less than the analysis bandwidth. With this definition the in-water signal-to-noise ratio for detection is independent of analysis bandwidth and the detection threshold becomes

$$DT = 10\mathrm{Log}\left(\frac{\sigma_S^2}{N_0 \times 1\mathrm{Hz}}\right) = 10\mathrm{Log}\left(\sqrt{\frac{d\Delta f}{T \times (1\mathrm{Hz})^2}}\right) = 5\mathrm{Log}(d) - 5\mathrm{Log}\left(\frac{T}{\Delta f}\right). \tag{13e}$$

In the last term of (13e) the implicit factor of $(1\ \mathrm{Hz})^2$ requires that the signal duration T be measured in seconds and the analysis band be measured in Hz. To optimize detection performance, (13e) shows that the detection threshold can be decreased by decreasing the analysis bandwidth and by increasing the integration time as much as possible. The decrease in analysis bandwidth decreases the noise power in the band but, because the signal is overcontained by the analysis band, the signal power remains constant and the signal-to-noise ratio within the analysis band increases. From the signal-to-noise ratio defined in (13d) within parentheses, it is found that the relative increase in signal-to-noise ratio within the analysis band is equal to the relative decrease in bandwidth, i.e. $(\delta SNR)/SNR = -\delta\Delta f/\Delta f$. Moreover, from (13e) the minimum signal-to-noise ratio for detection decreases with decreases in relative bandwidth, i.e. $(\delta SNR_{min})/SNR_{min} = 0.5(\delta\Delta f/\Delta f)$. Since both the received in-water signal-to-noise ratio and detection threshold are defined in terms of a 1 Hz noise band, the reduction in minimum signal-to-noise required for detection increases signal excess and improves detection performance.

Example 7: For the tonal search in Examples 1, 3 and 4, the detection index was found to be d = 14 for 20 minutes integration and d = 32 for 100 seconds integration both with an analysis bandwidth of 0.1 Hz. The detection thresholds for these cases are

$$\text{20 minute integration:} \quad DT = 5\text{Log}(14) - 5\text{Log}(1200/0.1) = -14.7 \text{ dB}, \qquad (13f)$$

$$\text{100 second integration:} \quad DT = 5\text{Log}(32) - 5\text{Log}(100/0.1) = -7.5 \text{ dB}. \qquad (13g)$$

In these examples the longer processing time affords the luxury of a 7.2 dB lower detection index. For both cases, though, the signals would not likely be visible on a time plot of 1 Hz analysis band data. They are "pulled out of the noise" by decreasing the analysis band width and increasing the degrees-of-freedom though long term integration.♦

The detection threshold calculations in (13) implicitly assume that the detector is "matched" to the signal, i.e. that the signal is continuously integrated throughout its duration. However, this will not always be achieved because the signal can fade in amplitude during the integration process or the integration process may be prematurely terminated while the signal is still strong. Similarly, the signal can drift in both frequency and receive beam number and cause it to transition from one beam-band detector to another. For such situations in which the signal is in-and-out of the detector window, the detector will not accumulate the maximum signal energy resulting in less than optimal detection results. There are two ways the analysis beam-band can be mismatched to the signal causing degradation to the detector performance: 1. The signal remains in the detector window but the window duration is not matched to the signal duration; 2. The signal drifts into another beam-band causing it to transition out of the detector window which results in a mismatch between the window duration and signal duration. Both cases are considered in the following.

1. For the case in which the signal fades prior to the termination of the detection integration, i.e. $T_S \leq T$, excess noise is accumulated. In contrast, for the case in which the detector integration prematurely terminates, i.e. $T \leq T_S$, insufficient signal energy is accumulated. For the respective cases the detection index in (11d) can be written

$$T_S \leq T: \quad d = T\Delta f\left(\frac{\sigma_S^2 T_S}{\sigma^2 T}\right)^2 \rightarrow 10\text{Log}\left(\frac{\sigma_S^2}{\sigma^2}\right) = DT_{eff} = DT_O + 5\text{Log}\left(\frac{T}{T_S}\right), \qquad (14a)$$

$$T \leq T_S: \quad d = T\Delta f\left(\frac{\sigma_S^2}{\sigma^2}\right)^2 \rightarrow 10\text{Log}\left(\frac{\sigma_S^2}{\sigma^2}\right) = DT_{eff} = DT_O + 5\text{Log}\left(\frac{T_S}{T}\right). \qquad (14b)$$

In the last expressions, the quantity DT_{eff} is the effective detection threshold given in terms of the operational detection threshold $DT_O = 5\text{Log}(d/T_S\Delta f)$. We see that, while the system is configured to operate at detection threshold DT_O, in fact, it operates at the greater detection threshold Dt_{eff}. This has the effect of lowering the probability of detection for the system but does not change the false alarm rate.

Since the correction terms for both cases increase the detection threshold, i.e. are positive, we can write (14a,b) as

$$DT_{eff} = DT_O + 5\left|\log\left(\frac{T}{T_S}\right)\right|, \qquad DT_O = 5\mathrm{Log}\left(\frac{d}{T_S \Delta f}\right), \tag{14c}$$

where T is the duration of the detector integration and T_S is the signal duration.

Example 8: If, for the narrow band tonal detection in Examples 1, 3, 4 and 6, the mismatch in signal duration time and processing time was, say, a factor of two (or a half) the detection penalty would be

$$\text{Detection Threshold Increase} = 5\left|\mathrm{Log}(2)\right| = 5\left|\mathrm{Log}(0.5)\right| = 1.5 \text{ dB}. \tag{14d}$$

For the case, $T/T_S = 2$, the processor integrates twice as long as the signal is in the analysis band whereas for the case, $T/T_S = 0.5$, the processor integrates only half of the time that the signal is in the analysis band. In both cases the detection threshold would effectively be raised by 1.5 dB. This has the effect of decreasing the FOM by 1.5 dB. If we model the FOM as a simple power law, i.e. $FOM = -10\mathrm{Log}(1/r^{\alpha})$, we find by differentiation that $\Delta r/r = \Delta FOM/10\alpha$. For the decrease in FOM of 1.5 dB, we arrive at the loss expression for detection range, $\Delta r/r = -0.15/\alpha$.. This result indicates that the cost of the processor mismatch, within a cylindrical spreading regime, would be a decrease in maximum detection range of 15% while, for a spherical spreading regime, the cost would be a decrease in maximum detection range of 7.5%.♦

2. For the case in which the signal drifts into another beam-band causing it to transition out of the detector window, i.e. $T_T \le T$ and $T_T \le T_S$ where T_T is the transition time and, say, $T \le T_S$, we see that after time T_T we have both conditions of excess noise energy accumulation and insufficient signal energy accumulation. For this case the detection index, from (11d), can be written

$$d = T\Delta f\left(\frac{\sigma_S^2 T_T}{\sigma^2 T}\right)^2 \rightarrow 10\mathrm{Log}\left(\frac{\sigma_S^2}{\sigma^2}\right) = DT_{eff} = DT_O + 5\log\left(\frac{T}{T_T}\right) + 5\mathrm{Log}\left(\frac{T_S}{T_T}\right). \tag{14e}$$

In the last expressions, the quantity DT_{eff} is, again, the effective detection threshold given in terms of the operational detection threshold $DT_O = 5\mathrm{Log}(d/T_S\Delta f)$ and, while the system is configured to operate at detection threshold DT_O, in fact, it operates at the greater detection threshold DT_{eff}. The first correction term $5\mathrm{Log}(T/T_T)$ arises because excess noise is accumulated in the detector due to the extended window duration. The second term $5\mathrm{Log}(T_S/T_T)$ arises because insufficient signal energy is accumulated due to the signal passing into an adjacent analysis band. As for the mismatches in Case 1, both effects lower the probability of detection for the system but do not change the false alarm rate.

Example 9: If for the narrow band tonal detection in Example 8, the signal occupied only half of the detection window $T_T = T/2$ and the signal passed into the adjacent analysis band half way through its duration $T_T = T_S/2$ the detection penalty would be

$$\text{Detection Threshold Increase} = 5\text{Log}(2) + 5\text{Log}(2) = 3 \text{ dB}. \qquad (14f)$$

This has the effect of decreasing the FOM by 3 dB and, using the simple FOM model in Example 8, the cost of the processor mismatch, within a cylindrical spreading regime, would be a decrease in maximum detection range of 30% while, for a spherical spreading regime, the cost would be a decrease in maximum detection range of 15%. This example bears a similarity to the fading signal problem in Example 8 but with twice the loss. They differ because the problem in Example 8 was due to a doubling of the noise energy while in this example the problem is due to both a doubling of the noise energy and a halving of the signal energy. ♦

The importance of adaptive processing to ensure that the detection window is matched to the received signal is apparent from the above discussions. This is a difficult problem but, for stationary surveillance systems with earth fixed beams, and, for mobile systems with azimuthally fixed beams and compensation for own-ship Doppler on all beams, the adaptation is somewhat simplified since it must compensate just for the target movements from beam-to-beam and for target Doppler shifts between analysis bands.

Multiple Frequency Channels - Equal Signal-to-Noise Ratios

For the case of equal signal-to-noise ratios in all frequency channels the variances of all terms in the detector output are equal and the detector output is Chi-squared distributed with 2NK degrees-of-freedom

$$p(\lambda|H_1) = \frac{1}{2^{NK}\left(\sigma_S^2/\sigma^2+1\right)^{NK}\Gamma(NK)}\lambda^{NK-1}\exp\left[-\lambda/2\left(\sigma_S^2/\sigma^2+1\right)\right], \qquad (15a)$$

$$p(\lambda|H_0) = \frac{1}{2^{NK}\Gamma(NK)}\lambda^{NK-1}\exp[-\lambda/2]. \qquad (15b)$$

These relations are identical in form to the single frequency channel relations in (9) with the number of samples increased to NK, i.e. the number of samples is dependent on the K frequency channels as well as the N time samples. Accordingly, the single frequency channel case and multiple frequency channel, with equal signal-to-noise ratios, are equivalent and the results for the former case can be directly extended to the latter case.

Multiple Frequency Channels - Unequal Signal-to-Noise Ratios

For the case of unequal signal-to-noise ratios in some or all of the frequency channels the statistics of the energy detector output under the signal plus noise hypothesis become more complicated. For each channel the output remains Chi-squared distributed

$$p(\lambda_k|H_1) = \frac{1}{2^N\left(\sigma_{S,k}^2/\sigma_k^2+1\right)^N\Gamma(N)}\lambda_k^{N-1}\exp\left[-\lambda/2\left(\sigma_{S,k}^2/\sigma_k^2+1\right)\right], \quad (16a)$$

where the subscript k (k = 1,2,···,K) denotes the detector output for the kth frequency channel. Since the total detector output is the sum of all channels and all channels are independent, the probability distribution of the total detector output is given by the inverse Fourier transform of the product of characteristic functions for each channel where the characteristic function for the kth channel is

$$\begin{aligned}\Phi_k(\omega|H_1) &= \int_0^\infty p(\lambda_k|H_1)\exp(j\omega\lambda_k)d\lambda_k \\ &= \frac{1}{(-2j)^N\left(\sigma_{S,k}^2/\sigma_k^2+1\right)^N\left(\omega+\dfrac{j/2}{\sigma_{S,k}^2/\sigma_k^2+1}\right)^N}. \end{aligned} \quad (16b)$$

Consequently, the characteristic function for the multiple channel detector output can easily be written

$$\Phi(\lambda|H_1) = \frac{1}{(-2j)^{NK}\prod_{k=1}^{K}\left(\sigma_{S,k}^2/\sigma_k^2+1\right)^N\prod_{k=1}^{K}\left(\omega+\dfrac{j/2}{\sigma_{S,k}^2/\sigma_k^2+1}\right)^N}. \quad (16c)$$

The inverse Fourier transform of the this function gives the distribution function for the multiple channel detector output with arbitrary signal-to-noise ratios but, for the general case, the integration does not lead to a simple closed form representation for the probability density function. However, for the important case of low signal-to-noise ratios, $\sigma^2{}_{S,k}/\sigma^2{}_k \ll 1$, such a representation does exist.

From the approximations

$$\prod_{k=1}^{K}\left(\sigma_{S,k}^2/\sigma_k^2+1\right)^N \cong \left(1+\langle SNR\rangle\right)^{NK}, \quad (17a)$$

$$\prod_{k=1}^{K}\left(\omega+\frac{j/2}{\sigma_{S,k}^2/\sigma_k^2+1}\right)^N \cong \left(\omega+\frac{j/2}{\langle SNR\rangle+1}\right)^{NK}, \quad (17b)$$

where the average signal-to-noise ratio over all frequency channels is given by

$$\langle SNR\rangle = \frac{1}{K}\sum_{k=1}^{K}\frac{\sigma_{S,k}^2}{\sigma_k^2}, \quad (17c)$$

the multiple channel characteristic function in (16c) becomes

$$\Phi(\lambda|H_1) = \frac{1}{(-2j)^{NK}(\langle SNR\rangle + 1)^{NK}\left(\omega + \frac{j/2}{\langle SNR\rangle + 1}\right)^{NK}}. \tag{17d}$$

Comparison with the single channel characteristic function in (16b) shows that the statistics of the multiple frequency channel output for the case of unequal signal-to-noise ratios reduces, for the case of low signal-to-noise ratios, to a Chi-squared distribution. Accordingly, the performance of the detector for unequal but low signal-to-noise ratios is the same as for the multiple frequency channel case with equal signal-to-noise ratios but with the constant signal-to-noise ratio parameter replaced by the average signal-to-noise ratio over frequency channels. The restriction to low signal-to-noise ratios means that, practically speaking, this result is valid only for large numbers of degrees-of-freedom.

11.3 System P_D and P_{FA}

Operational spectral grams consist of sequences of detector spectral beam-band outputs that exhibit "line" components or broad band energy components. Both such components can "track" through a sequence of adjacent analysis beam-bands. These data can be tracked automatically by computer algorithms and are commonly presented to operators as spectral waterfall grams on paper or electronic displays. The operators perform visual inspection of the displayed grams and search for sequences of detections within "threat" bands. Of special importance are harmonic relationships among detected line components and special harmonic markers can be overlaid on the data to aid in their identification.

The detection performance of the operators and/or computer algorithms, based upon sequences of detections, are, in general, improved over the detection performance outlined in the preceding because decisions are made based upon additional information as compared to the detection performance in the preceding sections which are based upon a single detector output within the spectral gram. To relate the performance for the single detector output and the sequence of gram data we note that sequences of the events "missed detection" and "correct dismissal" are independent events. For example, given that a sequence of previous detector outputs were missed detections or were correct dismissals gives no information that the next output will be either a missed detection or a correct dismissed because a correct detection or false alarm is always imminent. Since these dismissal events for the accumulated gram data is just the intersection of the individual dismissal events, this observation gives the results

$$1 - P_{OD} = (1 - P_D)^{NK}, \tag{18a}$$

$$1 - P_{OFA} = (1 - P_{FA})^{NK}, \tag{18b}$$

where the terms P_{OD} and P_{OFA} are the operational probabilities of detection and false alarm for the sequence of N detector outputs over K frequency bands and P_D and P_{FA} are the probabilities of detection and false alarm for each single detector output.

These relations relate the single output probabilities of detection and false alarm that are required to achieve the operational probabilities of detection and false alarm which are based upon a sequence of spectral gram data. Solving (18) gives

$$P_D = 1-\left(1-P_{OD}\right)^{1/NK}, \tag{19a}$$

$$P_{FA} = 1-\left(1-P_{OFA}\right)^{1/NK}. \tag{19b}$$

These relations relate the operationally important operating point for a sequence of data to the ROC operating point for the single detector output.

Example 10: If the desired operational performance is P_{OD}= 99.9999% and P_{OFA} = 0.01%, for NK = 100, the required single datum performance is P_D = 12.9% and P_{FA} = 0.0001%. The detection index for this case can be estimated from Figure 2 to be approximately 15 dB.♦

For passive systems there is an interdependence between the probabilities of detection and false alarm analogous to that for active systems. For example we can write the relations

$$P_{OFA} + NK \xrightarrow{(19b)} P_{FA} \xrightarrow{(10b)} \lambda_T, \tag{20a}$$

$$P_D + \lambda_T \xrightarrow{(10a)} DT + A_O \to FOM. \tag{20b}$$

The first relation asserts that choosing an operational probability of false alarm and total number of detection cells, then with (19b) the single cell probability of false alarm is determined. Similarly, this single cell probability of false alarm and (10b) determine the detector threshold. Moving to the second relation we find that choosing a probability of detection, with the previously determined detector threshold, (10a) determines the detection threshold which, with the operations area, in turn, determines the FOM either from (2.3) or (2.4c). (Analogous to the active detection case this operations area concept encompasses the geometric area, the propagation conditions, the noise background and likely targets to be found.) If the FOM as determined by the process in (20b) does not meet or exceed the system FOM, then the system will not achieve its operating point within the operations area. To determine the discrepancy in the operating point, we run (20b) backwards,

$$FOM + A_O \to DT + z_T \xrightarrow{(10a)} P_D. \tag{20c}$$

The system FOM with the chosen operations area leads to the detection threshold with which, in conjunction with the detector threshold from (20a), (10a) leads to the probability of detection. This probability of detection with the probability of false alarm define the operating point within the operations area. Since this probability of detection applies primarily at the outer limits of the operations area and since it is never graphically evident without extensive testing, the resulting operating point for the system is generally accepted as an operational trade-off.

CHAPTER 12
LOCALIZATION OF TARGETS

Once a target has been detected, it can be localized[1] for monitoring and prosecution. Localization of a target is achieved by means of parameters acquired from detection data. The resulting localization estimates, e.g. target position and velocity, become elements of the state space description of the target. The particular detection parameters acquired by a given system are dependent upon the characteristics of that system but, in general terms, for active systems they include Doppler compression and time of detection of the target echo while for passive systems they include frequency of intercepted radiation and the rate of change of frequency. The detection parameters for both active and passive systems also include the receive beam or receive beam-set on which the target was detected. All of the detection parameters can be arrayed in a measurement vector for which, in principle, the covariance matrix is known from the system design, and the target state vector and its covariance determined therefrom. The target state vector usually includes range, bearing, velocity and, sometimes, depth of target. While the bearing estimates are usually obtained from the bearing of the beam holding the target or a beam weighted interpolation of the bearings of the beam-set holding the target, estimation procedures for the other state space parameters are not as straight forward. In the following we discuss methods for estimating each of those state space parameters and their variances from detection data.

12.1 Localization Using Active Detection Data

The echo received from a target is shown in (7.41) to have the form

$$s_R(t) = s_a\left(\alpha\left(t - \frac{r_{ST}}{c} - \frac{r_{TR}}{c}\right)\right), \tag{1a}$$

where s_a is the transmitted analytic signal, r_{ST} and r_{TR} are the ranges from, respectively, source to target and target to receiver, c is the effective sound speed over those paths which, here, is assumed to be the same over both paths and α denotes the Doppler compression on the signal. The Doppler compression is given by

$$\alpha = \left(\frac{1+v_T/c_T}{1-v_S/c_S}\right)\left(\frac{1+v_R/c_R}{1-v_T/c_T}\right), \tag{1b}$$

---

[1] Target localization and target tracking, discussed in Chapter 13, are related concepts. We distinguish between them in the following way. Target localization is performed by transforming detection parameters from one or more detections into a single state space estimate. Target tracking is performed by iteratively transforming detection parameters from a sequence of detections into a sequence of state space estimates. The sequence of state space estimates from tracking procedures lie on a smoothed track as compared to a sequence of state space estimates obtained from target localization procedures. The smoothed track estimate is a more accurate estimate of target disposition and can be extrapolated and interpolated.

where the quantities v and c denote, respectively, a platform speed and the sound speed at that platform. The subscripts on those parameters denote the source, target and receive platforms in an obvious way and, for each case, the platform speed is the closing speed. For a collocated source and receiver and for platform speeds that are small compared to the sound speed, the platform movements will be small compared to the nominal range r, i.e. $r \cong r_{ST} \cong r_{TR}$, and (1) reduces to

$$s_R(t) = s_a\left(\alpha\left(t - \frac{2r}{c}\right)\right), \qquad \alpha \cong 1 + \frac{2v_S}{c_S} + \frac{2v_T}{c_T}. \tag{2a,b}$$

The Doppler compression and time of arrival measurements are obtained during matched filter detection and, in both (1) and (2), attenuation due to geometric spreading and absorption is unimportant and has been ignored.

Matched Filter Detection

The matched filter was derived in Chapter 10 and is given as the correlator in (10.16). In the derivation of (10.16) it was assumed that the signal component of the received data was Doppler compressed and time delayed by, respectively, α and t_d. The matched filter replica signal was assumed to be similarly compressed and delayed, i.e. to have the form $s_a(\alpha(t - t_d))$, thereby aligning the replica with the received signal and maximizing the matched filter output. In fact, the Doppler compression and time delay of the received signal are unknown and the search parameters of the matched filter are adjusted to effect a search over a Doppler band of frequencies and a range interval. Accordingly, the received signal is more correctly given by $s_a(\alpha'(t - t_d'))$ where α' and t_d' are the unknown Doppler compression and time of arrival parameters. We substitute this received signal into (10.16b) with the variable change $t \rightarrow t + t_d$ to obtain

$$\lambda(y|\alpha, t_d) = \left| \int_0^T s_a^*(\alpha'(t - (t_d' - t_d)))s_a(\alpha t)dt \right|. \tag{3}$$

In this form, the maximum output is seen to be achieved for the Doppler search parameter $\alpha = \alpha'$ and time delay search parameter $t_d = t_d'$ which are the Doppler compression and time delay of the received signal. When the matched filter replica signal is aligned with the received signal, the magnitude of the matched filter output is the energy of the analytic signal which is approximately E_a, i.e. excluding small Doppler compression effects. This calculation "compresses" the signal into a single time. We note that, while the time delay of the matched filter output corresponds to the time t_d, since the matched filter must accumulate the entire echo to make this calculation, the matched filter output is not in real-time but is delayed an amount equal to the signal duration. Given that the matched filter produces the Doppler compression and time delay of the signal, we must relate those parameters to the range rate and range of the target. The range rate estimate is given directly from (1b) or (2b) but the range estimate requires a little more discussion.

Range Estimation

The fundamental relation from which target range is determined is

$$r = \frac{ct_d}{2}, \tag{4}$$

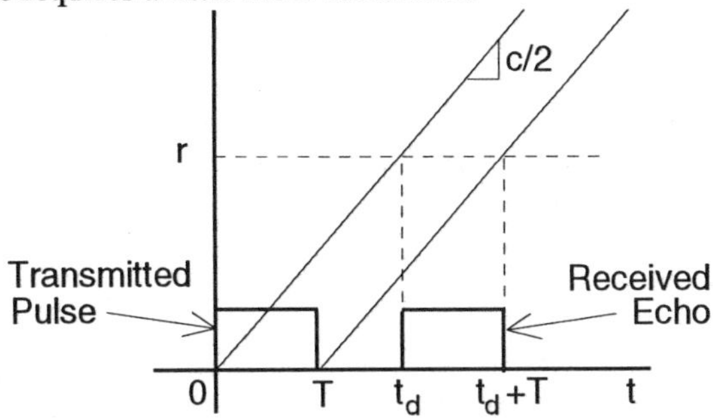

Figure 1 Echo Structure and Target Range versus Echo Reception Time for a Single Transmitted Pulse and a Point Target

where r is the range to the target, c is the sound speed and t_d is the estimated time of detection of the target at the receiver. A plot of this relation is shown in Figure 1 for a single pulse of duration T and a point target at range r. During the reception period t_d to $t_d + T$ the echo is received from the target having the same form as the transmitted signal. The leading edge of the received signal is the time marker for t_d.

If the target in Figure 1 were spatially extended, as are all real targets, the received echo would have an appearance as suggested by Figure 2. In this figure several features are apparent. As the transmitted pulse sweeps over the target it initially scatters from the point of first contact and then scatters from all ensonified target surfaces beyond that point up to the leading edge of the pulse. The echo can build during this portion of the transmission and reach a maximum level when either the entire target is ensonified, i.e. the spatial extent of the pulse is longer than the range extent of the target, or when the entire transmitted pulse is scattered by the target, i.e. the spatial extent of the pulse is shorter than the range extent of the target. As the trailing edge of the pulse sweeps over the target, the echo amplitude can decrease in reverse order of the initial amplitude increase. Of course, echoes from real submarine targets usually exhibit more level variability than this simple model suggests. However, if the "target" were a region on the ocean bottom from which the echo was backscattered reverberation or if it were a reverberant surface patch in a convergence zone, the received reverberation could approximate this simple model. In any event, the range extent from which scattering is obtained is $r'-r = c\tau/2$, where $\tau = t_d'-t_d$ as indicated in Figure 2. The quantity $c\tau$ is the range extent of the target for the long pulse case and is the range extent of the pulse for the short

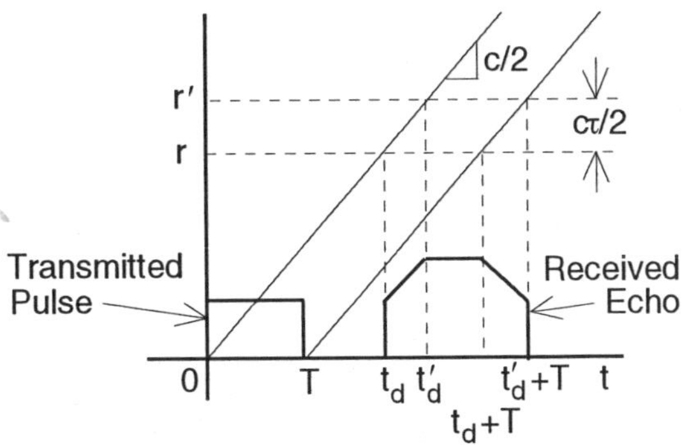

Figure 2 Echo Structure and Target Range versus Echo Reception Time for a Single Transmitted Pulse and Extended Target

pulse case. The total duration of the echo would be $T+\tau = t_d'+T-t_d$. The duration of the possibly maximum level is $T-\tau = t_d+T-t_d'$ and the "shoulders" are each of duration τ. For the long pulse case $T+\tau \leq 2T$ and for the short pulse case $T+\tau \geq 2T$. We show the long pulse case in Figure 2. As for the case of the point target in Figure 1, the time t_d measured by the matched filter for the extended target corresponds to the leading edge of the received signal reflected from the point of first contact with the target and is an appropriate value for the range estimate in (4). However, for the extended target we generally find additional detector threshold crossings corresponding to the time interval from beyond t_d which come from reflection points beyond the point of first contact that can be larger in amplitude than the initial return. Because of this it is common practice to estimate the time of arrival of the peak value or a centroid value of the matched filter output.

Range Rate and Range Errors

Relations for the uncertainty in the range rate and range estimates can be found from the differentials of (2b) and (4)

$$\delta\alpha = \frac{2}{c_T}\delta v_T, \qquad \delta r = \frac{c}{2}\delta t_d, \tag{5a,b}$$

where we assume that the sound speeds are known. If the sound speeds are not known it's easy to include uncertainties for those variables as well. Squaring and averaging (5) we obtain the variance relations

$$\sigma_\alpha^2 = \frac{4\sigma_{v_T}^2}{c_T^2}, \qquad \frac{\sigma_r^2}{r^2} = \frac{\sigma_{t_d}^2}{t_d^2}, \tag{6a,b}$$

where it is assumed that the errors in the target range rate and time of arrival are zero mean. These errors are due to the Doppler and time resolution of the matched filter output. To determine these values we turn to the ambiguity function.

Ambiguity Function

The matched filter expression in (3) generates a surface above the (α,t_d) plane, as those search parameters are varied, and the search values of interest are the values corresponding to the peak of the surface. However, because noise perturbs the surface, it's clear that the shape of the unperturbed surface can strongly determine the accuracy of the Doppler compression and time delay estimates. Ideally, the surface height would be a maximum at the correct Doppler compression and time delay coordinates and zero elsewhere. Unfortunately, as we show in the following, this ideal cannot be achieved.

The matched filter output function, when normalized by the energy, is called the ambiguity function and the corresponding surface points have values between zero and unity. It is so-called because the (α,t_d) values for those surface points having the same elevation are all equally valid target parameters. A convenient expression for the ambiguity surface for both analysis and implementation can be obtained by replacing the signal in (3) by its modulated carrier form to give

$$\chi(\alpha,t_d)=\frac{1}{E_a}\left|\int_0^T m^*(\alpha'(t-(t'_d-t_d)))m(\alpha t)\exp(j(\alpha-\alpha')t)dt\right|. \tag{7}$$

Since for narrow band signals the modulation is relatively unaffected by the Doppler compression, (7) for narrow band sonar systems has the form

$$\chi(\alpha,t_d)=\frac{1}{E_a}\left|\int_0^T m^*(t-(t'_d-t_d))m(t)\exp(j(\alpha-\alpha')t)dt\right|. \tag{8a}$$

We can express this in the frequency domain as

$$\chi(\alpha,t_d)=\frac{1}{2\pi E_a}\left|\int_{-\infty}^{\infty} M^*(\omega)M(\omega+\alpha-\alpha')\exp(-j\omega(t'_d-t_d))d\omega\right|, \tag{8b}$$

$$m(t)=\frac{1}{2\pi}\int_{-\infty}^{\infty} M(\omega)\exp(-j\omega t)d\omega, \quad M(t)=\int_{-\infty}^{\infty} m(t)\exp(j\omega t)dt, \tag{9a,b}$$

where (8a) leads to (8b) via the Fourier transform relations in (9).

To illustrate a specific case, we consider the ambiguity surface of a pulsed signal with Gaussian amplitude modulation and linear frequency modulation represented by

$$s_a(t)=\exp\left(-\frac{\beta^2(t-T/2)^2}{T^2}\right)\exp j\left[\omega_0 t+\mu(t-T/2)^2+\phi\right], \quad 0\le t\le T \tag{10}$$

where ω_0 is the angular carrier frequency, ϕ is an arbitrary phase and the quantities β and μ are positive constants that control, respectively, the amplitude and frequency modulation and T is the duration of the signal. The instantaneous angular frequency of the signal, which is given by the time derivative of the phase, is found to be $\omega = \omega_0+2\mu(t-T/2)$. This shows that 2μ is the rate of change of the instantaneous angular frequency and the band of frequencies swept is $W = \mu T/\pi$. The amplitude modulation is imposed to reduce the power surge to the transducer elements and the frequency modulation increases the bandwidth of the signal. If β and/or μ vanish, the signal becomes, respectively, a "boxcar" modulated pulse and/or a narrow band tonal. Without loss of generality, we assume that the signal has undergone no Doppler compression and has zero time delay since their effects are just shifts over the (α,t_d) plane. Accordingly, we substitute (10) into (8a) with α' and t' both zero and obtain, after some straight forward computations, the ambiguity function

$$\chi(\alpha, t_d) = \frac{1}{E_a}\left[\int_{-T/2}^{T/2} \exp\left(-\frac{2\beta^2 t^2}{T^2}\right)dt\right]\exp\left[-\frac{1}{2}\left(\frac{\alpha^2 T^2}{4\beta^2} - \frac{\mu T^2 \alpha t_d}{\beta^2} + \frac{m^2\beta^2 t_d^2}{T^2}\right)\right], \quad (11a)$$

$$m^2 = 1 + \frac{\mu^2 T^4}{\beta^4}, \quad (11b)$$

where m is called the compression ratio. The integral in this expression is just the energy of the signal and it cancels out of the expression. The remaining function is seen to peak at the origin of coordinates. Of course, if the signal in (10) were time delayed and Doppler compressed, (11) would just be shifted by the time delay and Doppler compression with minor rescaling of the amplitude.

Examination of (11) shows the ambiguity surface to have elliptical horizontal cross sections. The loci of these cross sections are given by

$$C^2 = -2\ln\chi(\alpha, t_d) = \frac{\alpha^2 T^2}{4\beta^2} - \frac{\mu T^2 \alpha t_d}{\beta^2} + \frac{m^2\beta^2 t_d^2}{T^2}, \quad (11c)$$

where C^2 is a constant that defines the cross section level. For the $C^2 = 1$ contour in (11c), i.e. the contour level at height equal to $\exp(-1/2)$, the half-widths of the ambiguity surface are

$$\alpha = \frac{2\beta}{T}, \qquad t_d = \frac{T}{m\beta}. \quad (12a,b)$$

The widths of the ambiguity surface are good measures of the accuracy of the Doppler compression and time delay estimates in the presence of noise which are required for (6). We see that, while the width along the Doppler axis is relatively bandwidth independent - it depends only on signal duration - the width along the time delay axis decreases with increasing signal bandwidth, i.e. m increases with band width which is why m is called the compression ratio. As β increases, the signal is more concentrated around the time $t = T/2$ and the signal duration is effectively decreased. This improves the time delay estimate but causes the Doppler resolution to decrease.

The slope of the ridge of the ambiguity surface can be found by rotating the coordinate system such that (11c) has the canonical elliptical form. This is conveniently done by first transforming the variables to the dimensionless variables

$$u = \frac{\alpha T}{2}, \qquad v = \frac{t_d}{T}, \quad (13a,b)$$

which reduces (11c) to

$$C^2 = \frac{u^2}{\beta^2} - 2\frac{\mu T^2}{\beta^2}uv + m^2\beta^2 v^2. \quad (13c)$$

Rotating coordinates in (13c) via the transformation

$$\begin{aligned} u &= \zeta\cos\theta - \eta\sin\theta, \\ v &= \zeta\sin\theta + \eta\cos\theta, \end{aligned} \tag{14a}$$

and setting the coefficient of the $\zeta\eta$ product term to zero, we obtain the canonical form

$$C^2 = A\zeta^2 + B\eta^2, \tag{14b}$$

$$2A = \left(\beta^{-2} + m^2\beta^2\right)\left[1 - \sqrt{1 - \frac{4}{\left(\beta^{-2} + m^2\beta^2\right)^2}}\right] \xrightarrow{\mu T^2 \to \infty} \frac{2}{m^2\beta^2}, \tag{15a}$$

$$2B = \left(\beta^{-2} + m^2\beta^2\right)\left[1 + \sqrt{1 - \frac{4}{\left(\beta^{-2} + m^2\beta^2\right)^2}}\right] \xrightarrow{\mu T^2 \to \infty} 2m^2\beta^2 . \tag{15b}$$

The angle of axis rotation of the ambiguity surface ridge is given by

$$\tan\theta = \left[\frac{\sqrt{\mu^2 T^4 + 4} - \mu T^2}{\sqrt{\mu^2 T^4 + 4} + \mu T^2}\right]^{1/2} \xrightarrow{\mu T^2 \to \infty} \frac{1}{\mu T^2}, \tag{15c}$$

where θ is the angle between the u axis and the slope of the ambiguity surface ridge. Consequently, the equation of the slope of the ambiguity surface ridge in the (u,v) coordinate system is $v = (1/\mu T^2)u$. Transforming this to (α, t_d) coordinates, using (13a,b) we obtain

$$\alpha = 2\mu t_d . \tag{16}$$

We see that the angle of rotation of the ambiguity ridge is the rate of change of the instantaneous frequency.

Example 1: Ambiguity surface contours at the exp(−1/2) level are shown in Figure 3 for a 10 second duration pulse with Gaussian amplitude modulation and for both linear frequency modulation and a narrow band tonal. Figures 3a and 3b are for the case of weak amplitude modulation with parameter $\beta = 1$. For this case the narrow band tonal ambiguity surfaces show a large ambiguity in time delay and very little ambiguity in Doppler. In comparison, Figures 3c and 3d, which are for the case of strong amplitude modulation with parameter $\beta = 2$, show reduced ambiguity in time delay and increased ambiguity in Doppler. Figures 3a and 3c are for the relatively narrow band linear frequency modulated signal with parameter $\mu = 0.1$ and the slope of the ambiguity surface ridges are, accordingly, $2\mu = 0.2$. In comparison, Figures 3b and 3d, which are for a more broad band linear frequency modulated signal with parameter $\mu = 0.2$, have slopes of $2\mu = 0.4$. Examination of the ambiguity surface widths along the coordinate axes shows that they are consistent with (12a,b). ♦

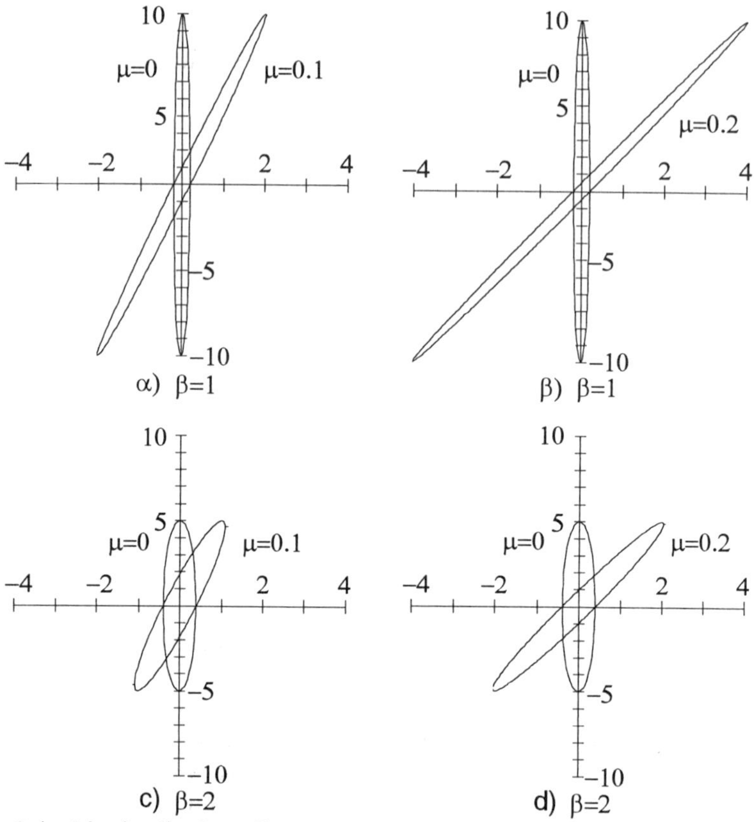

Figure 3 Ambiguity Surface Contours for a Signal of Duration T = 10 seconds with Gaussian Amplitude Modulation $\exp(-\beta^2 t^2/T^2)$ and Linear Frequency Modulation $\exp(j\mu t^2)$. The Bandwidths, which vary μ, are : 0, 1/π and 2/π Hz.

To determine certain general properties of the ambiguity surface shape we begin by Fourier transforming the squared ambiguity function. Shifting the coordinates to the peak in (8a), i.e. we let $\beta = \alpha - \alpha'$ and $\tau = t_d' - t_d$, with infinite integration limits the Fourier transform of the square of that function is found to be

$$V(t,\omega) = \frac{1}{2\pi E_a^2}\int_{-\infty}^{\infty} d\beta \int_{-\infty}^{\infty} d\tau \chi^2(\beta,\tau)\exp(j(\omega\tau - \beta t))$$

$$= \frac{1}{2\pi E_a^2}\int_{-\infty}^{\infty} dt_1 m(t_1)\int_{-\infty}^{\infty} dt_2 m^*(t_2)\int_{-\infty}^{\infty} d\tau m^*(t_1-\tau)m(t_2-\tau)\exp(j\omega\tau)$$

$$\times \int_{-\infty}^{\infty} d\beta \exp(j\beta(t_1 - t_2 - t)) . \quad (17a)$$

Since the integral over β is the Dirac delta function $2\pi\delta(t_1 - t_2 - t)$, the integral over t_2 is immediately performed to yield

$$V(t,\omega) = \frac{1}{E_a^2} \int_{-\infty}^{\infty} dt_1 m_a(t_1) m_a^*(t_1 - t) \int_{-\infty}^{\infty} d\tau m_a^*(t_1 - \tau) m_a(t_1 - t - \tau) \exp(j\omega\tau)$$

$$= \frac{1}{E_a^2} \left| \int_{-\infty}^{\infty} dt_1 m_a^*(t_1 - t) m_a(t_1) \exp(j\omega t_1) \right|^2 , \qquad (17b)$$

where the variable change $\tau \rightarrow t_1 - \tau$ leads to the second equality. Comparing (8a) and (17b) we see that the Fourier transform of the squared ambiguity surface is just the squared ambiguity surface.

$$V(t,\omega) = \chi^2(\omega,t) . \qquad (17c)$$

Moreover, we see that

$$\chi^2(-\omega,-t) = \chi^2(\omega,t) , \qquad (17d)$$

$$V(0,0) = \frac{1}{E_a^2} \left| \int_{-\infty}^{\infty} dt_1 \left| m_a(t_1) \right|^2 \right|^2 = 1 , \qquad (17e)$$

which shows that the magnitude of the ambiguity surface is axially symmetrical about its peak and the volume under the magnitude of the ambiguity surface is unity. Unity volume for all signals asserts that there is a conservation of ambiguity which means that if the ambiguity is reduced for a certain range of (α, t_d) parameters then the ambiguity must increase for other ranges of (α, t_d) parameters. The challenge to signal designers is to concentrate the ambiguity within those regions where measurements are to be made.

The "sharpness" of the peak of the ambiguity surface can be assessed from the second derivatives of (8). For the centered functions these are found to be

$$\frac{\partial^2 \chi(0,0)}{\partial \beta^2} = -\frac{1}{E_a} \int_{-\infty}^{\infty} t^2 \left| m_a(t) \right|^2 dt = -\sigma_t^2 , \qquad (18a)$$

$$\frac{\partial^2 \chi(0,0)}{\partial \tau^2} = -\frac{1}{2\pi E_a} \int_{-\infty}^{\infty} \omega^2 \left| M_a(\omega) \right|^2 d\omega = -\sigma_\omega^2 , \qquad (18b)$$

where σ_t and σ_ω are the "root mean square" duration and bandwidth of the signal. These derivatives are necessarily negative, because the peak is convex upward, and show that the peak of the surface is sharper when the magnitude of the signal duration and bandwidth are larger. These relations also indicate that the width of the peak in the Doppler dimension is determined by the rms duration of the signal and the width of the peak in the time delay dimension is determined by the rms spectral width of the signal. So, in general terms, we can infer that to achieve good Doppler resolution we need a long signal, and to achieve good range resolution we need a broad band signal. However, because the volume under the ambiguity surface is conserved, there are limits to the trade-off between Doppler resolution and time delay resolution. To determine the limits of the trade-off we turn next to the uncertainty principle for signals.

If an acoustic signal could be designed such that its ambiguity surface had the bulk of the ambiguity surface volume concentrated over some small region of the (α, t_d) plane, this signal would be very effective for detection performance. Unfortunately, the uncertainty principle for signals asserts that there is a limit to how concentrated a signal can be over the ambiguity plane. To show this we start by forming the positive function

$$\frac{1}{2\pi E}\int_0^{\infty}\left|x\omega M_a(\omega)+\frac{dM_a(\omega)}{d\omega}\right|^2 d\omega \geq 0, \tag{19}$$

where x is some parameter. Expanding this function, it is seen to have the quadratic form in x

$$ax^2+bx+c\geq 0, \tag{20a}$$

with real coefficients

$$a=\frac{1}{2\pi E_a}\int_0^{\infty}\omega^2\left|M_a(\omega)\right|^2 d\omega=\sigma_\omega^2, \tag{20b}$$

$$\begin{aligned} b &= \frac{1}{2\pi E_a}\int_0^{\infty}\omega\frac{d}{d\omega}\left|M_a(\omega)\right|^2 d\omega \\ &= \frac{\omega}{2\pi E_a}\left|M_a(\omega)\right|^2\Big|_0^{\infty} - \frac{1}{2\pi E_a}\int_0^{\infty}\left|M_a(\omega)\right|^2 d\omega \\ &= -1, \end{aligned} \tag{20c}$$

$$\begin{aligned} c &= \frac{1}{2\pi E_a}\int_0^{\infty}\left|\frac{d}{d\omega}\int_0^{\infty}m_a(t)\exp(-j\omega t)dt\right|^2 d\omega = \frac{1}{2\pi E_a}\int_0^{\infty}\left|\int_0^{\infty}-jtm_a(t)\exp(-j\omega t)dt\right|^2 \\ &= \frac{1}{2\pi E_a}\int_0^{\infty}\int_0^{\infty}dt_1dt_2m_a(t_1)m_a^*(t_2)\int_0^{\infty}d\omega\exp(-j\omega(t_1-t_2)) \\ &= \frac{1}{E_a}\int_0^{\infty}dt\,t^2\left|m_a(t)\right|^2=\sigma_t^2. \end{aligned} \tag{20d}$$

Since the quadratic form in (20a) is positive, there can be values of x that satisfy that relation only if

$$b^2-4ac\leq 0. \tag{21}$$

Substituting the coefficients from (20) gives the signal uncertainty relation.

$$\sigma_\omega^2 \sigma_t^2 \geq \frac{1}{4}. \tag{22}$$

This relation asserts that the width of the ambiguity function along the time delay axis σ_t and the width along the Doppler axis σ_ω are constrained such that their product has a lower bound. This implies that there are constraints on how accurately the time delay and Doppler compression on a signal can be measured.

An alternative way of looking at the ambiguity surface volume is to consider horizontal slices through the surface having areas A(h) where h is the elevation above the (α, t_d) plane. The volume under the ambiguity surface can be expressed in terms of these areas as

$$V = \frac{1}{E_a} \int_0^1 A(h)dh = \frac{A(h_{eff})}{E_a} = 1, \tag{23}$$

where $A(h_{eff})$ is the effective cross sectional area of the ambiguity surface. We see that this area is equal to the energy of the signal for all signals but that the elevation of the effective cross sectional area is signal dependent. This is another way of viewing the principle of conservation of ambiguity. An example of this principle can be seen in the ambiguity surface cross sections in Figure 3.

Multiple Transmissions

In the preceding we have considered procedures for measuring Doppler compression and time of arrival for the case of a single transmitted signal. Frequently, though, it is advantageous to transmit a sequence of signals during one transmission cycle. The echoes from these transmissions can be processed in one of two ways. The sequence can be viewed as a single signal and the echoes processed as such to yield a single ambiguity surface. Or, the echoes can be processed individually and their separate ambiguity surfaces combined in post-processing procedures. The former procedure is one way signal designers achieve flexibility in plying their trade. For example, we have seen in Figure 3 that the linear frequency modulated signal distributes its ambiguity over an inclined ridge. Consequently, for a single signal, an accurate estimate of either Doppler compression or time delay can be produced only if one of these variables is previously known which, of course, is not the usual case. However, if two linear frequency modulated signals are transmitted, one with an upsweep in frequency and one with a down sweep in frequency, i.e. the so-called "housetop" transmission, it is easy to see that the ambiguity surface of the signals processed jointly will be concentrated within the region of the correct Doppler compression and time delay. Other interesting signal combinations also produce useful results. Since all of the preceding discussions apply to these cases, they are not pursued here any further.

For the case of echoes from multiple signals that are individually processed with results combined in post-processing, there are four general cases of interest;

1. same signals, same band,
2. same signals, different bands,
3. different signals, same band,
4. different signals, different bands.

An example of the first case is a sequence of two or more identical broad band frequency modulated signals. A single matched filter searches the band for echoes and each echo produces ambiguity surfaces that are nominally separated in time by the interval between the signal transmissions. However, if the target has a non-zero range rate, the time interval between the ambiguity surfaces will be perturbed from the nominal interval. This is a measurable quantity if the interval between transmissions is large enough such that the range change of the target during the interval is large compared to the range resolution of the broad band signals. The result is a signal combination that provides both a good range rate estimate and a good range estimate simultaneously. An example of the second case is a sequence of narrow band tonals that are transmitted at different frequencies. Such a signal serves two purposes. It provides accurate estimates of the target range rate and it is insensitive to multi-path interference. This insensitivity arises because, while echoes from any one of the narrow band tonals can be destructively canceled due to multi-path, it is highly unlikely that all tonals will be simultaneously canceled. To implement this procedure we need separate matched filters processing each tonal band with the matched filter outputs time aligned and combined using some type of summation algorithm. An example of the third case is a housetop transmission as discussed above. For this case two matched filters search the same frequency band for echoes and produce separate ambiguity surfaces for the upsweep and downsweep frequency modulated signals. The surfaces are time aligned and combined using some type of product algorithm. For the fourth case, the signals can be processed independently or the received ambiguity functions can be time aligned and combined by various techniques.

Reverberation can be a serious problem when individually processing echoes that are in the same frequency band, i.e. cases 1 and 3 above. The echoes of the following waveforms arrive with and are obscured by the reverberation from the leading waveforms. When this problem becomes severe, cases 2 and 4 waveforms must be used. The penalty for this choice, though, is, generally, a reduction in individual waveform bandwidth, due to the constraints of the total available system bandwidth, and, consequently, a reduction in range resolution. These problems can be mitigated in a convergence zone environment by restricting the duration of the waveforms. A discuss of this problem in given in Appendix 10-A.

It might be thought that separate matched filter processing of echoes with post-processing combinatorics is not an efficient approach to detection and localization. In terms of maximizing estimation accuracy, this is correct. However, in terms of minimizing processing cost, these procedures are more economical than joint matched filtering because of the reduction in time and bandwidth requirements.

12.2 Localization Using Passive Detection Data

Passive systems have certain advantages over active systems in localizing targets. They can link detections to appropriate targets more accurately than active systems through the use of target signature analysis and they generally have receiving apertures that afford high bearing resolution. Also, they are covert which means that targets are less likely to perform evasive maneuvers during the localization procedures. While it is true that passive systems have no direct measure of range, in contrast to active systems, they can infer target range through use of wide receiving apertures and platform movements. The principal disadvantage that passive systems have, as compared to active systems, is that target localization generally requires more data and more processing time than for an active system. And, of course, target and propagation conditions must remain favorable during this period.

Range Estimation from Wave Front Curvature

Target range can be estimated from measurements of the curvature of the wave front at the receiving array. This technique requires the receiving aperture to be wide enough to measure the wave front curvature or, equivalently, the target to be close enough to produce a curvature in the wave front at the receiver. These measurements are typically made using a towed array, a hull mounted flank array or separate receive arrays which might include a towed array.

For a target at bearing θ, the wave front curvature at the receive array can be estimated by correlating the received signal from the forward half of the array with the received signal from the aft half of the array and by measuring the target bearings at the two apertures. From Figure 4 we obtain the trigonometric relations

$$R_1 \sin\theta_1 = R_2 \sin\theta_2 = R \sin\theta \,, \tag{24a}$$

$$R_1 \cos\theta_1 = R \cos\theta - L/4 \,, \tag{24b}$$

$$R_2 \cos\theta_2 = R \cos\theta + L/4 \,, \tag{24c}$$

$$\theta_1 = \theta + \alpha_1 \,, \qquad \theta_2 = \theta - \alpha_2 \,. \tag{24d,e}$$

Combining, we obtain

$$R_2 - R_1 = R \sin\theta \left(\frac{1}{\sin\theta_2} - \frac{1}{\sin\theta_1} \right) = c\tau \,, \tag{25a}$$

$$\theta = (\theta_1 + \theta_2 + \alpha_1 - \alpha_2)/2 \,. \tag{25b}$$

Solving for the range R gives

$$\begin{aligned} R &= c\tau \left(\frac{\sin\theta_1 \sin\theta_2}{(\sin\theta_1 - \sin\theta_2)\sin((\theta_1 + \theta_2 + \alpha_1 - \alpha_2)/2)} \right) \\ &\cong c\tau \left(\frac{\sin\theta_1 \sin\theta_2}{(\sin\theta_1 - \sin\theta_2)\sin((\theta_1 + \theta_2)/2)} \right) . \end{aligned} \tag{26}$$

The approximate result is obtained by noting that, since $\sin\theta_1 - \sin\theta_2$ and $\alpha_1 - \alpha_2$ are small quantities, only the term that is first order in these quantities need be retained.

An approximate expression for the relative error in this estimate can be obtained by differentiating the range,equation to give

$$\frac{\delta R}{R} = \frac{\delta\tau}{\tau} + \delta\theta_1\left(\cot\theta_1 - \frac{\cos\theta_1}{\sin\theta_1 - \sin\theta_2} - \frac{1}{2}\cot\left(\frac{\theta_1+\theta_2}{2}\right)\right)$$
$$+\delta\theta_2\left(\cot\theta_2 + \frac{\cos\theta_2}{\sin\theta_1 - \sin\theta_2} - \frac{1}{2}\cot\left(\frac{\theta_1+\theta_2}{2}\right)\right). \quad (27a)$$

where δR, $\delta\tau$, $\delta\theta_1$ and $\delta\theta_2$ are the errors in the indicated variable. Assuming these errors are zero mean and statistically independent we obtain the variance relation

$$\frac{\sigma_R^2}{R^2} = \frac{\sigma_\tau^2}{\tau^2} + \sigma_{\theta_1}^2\left(\frac{1}{2}\cot(\theta) - \frac{1}{\alpha_1+\alpha_2}\right)^2 + \sigma_{\theta_2}^2\left(\frac{1}{2}\cot(\theta) + \frac{1}{\alpha_1+\alpha_2}\right)^2, \quad (27b)$$

where the relations in (24) were used and only first order terms retained. It is apparent that, for end-fire received signals at $\theta = 0$, the error due to the time delay estimate is minimum, since τ is a maximum, but the error due to the array aperture is unbounded, since the cotangent is unbounded and $\alpha_1 + \alpha_2$ vanishes. Similarly, for broadside received signals at $\theta = \pi/2$, the error due to the time delay estimate is unbounded, since τ vanishes, while the error due to the aperture is minimum, since the cotangent vanishes and $\alpha_1 + \alpha_2$ is a maximum. Therefore, it appears that our best range estimation accuracy is obtained for signals received near quartering aspects. For this case (27b) reduces to

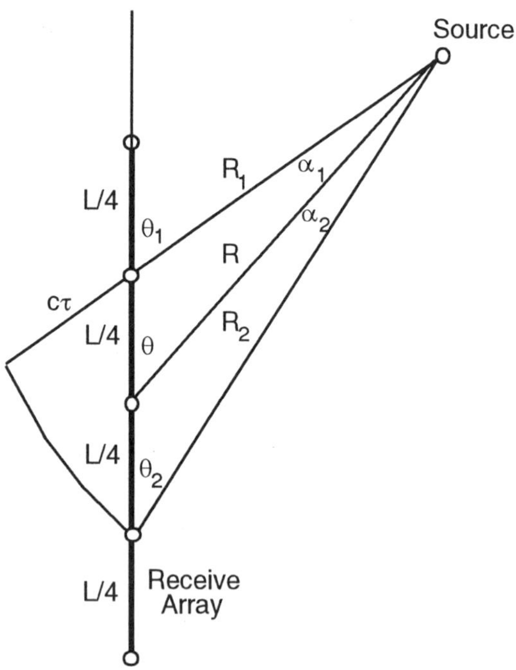

Figure 4 Passive Range Estimation using Wave Front Curvature for a Horizontal Array

$$\frac{\sigma_R^2}{R^2} = \frac{\sigma_\tau^2}{\tau^2} + \frac{2\sigma_\theta^2}{(\alpha_1+\alpha_2)^2}, \quad (27c)$$

where the error in both angles is σ_θ. The angle subtended by the array is approximately $\alpha_1+\alpha_2 \cong (L/2)\sin(\theta)/R$ which, for quartering aspect, gives us the error relation

$$\frac{\sigma_R^2}{R^2} = \frac{\sigma_\tau^2}{\tau^2} + 16\left(\frac{R}{L}\right)^2 \sigma_\theta^2. \tag{27d}$$

Example 2: For a 200 m towed array receiving acoustic energy at 45° in a 100 Hz band, we have $\delta\tau$ = 1/100 Hz = 0.01 sec and τ = (200 m/2)/($\sqrt{2}\times$1500 m/sec) $\cong$ 0.05 sec. Assuming the acoustic energy is at nominally 1000 Hz with wavelengths of about 1.5 m, the beam width of each subaperture would be approximately 2.8(1.5 m)/(π(100 m)) = .013 radians. If the error in the look direction of each subaperture is a quarter of a beam width, i.e. 0.00325 radians, then for a target at, say, 50 array lengths, i.e. 10 km, the relative range error would be about 68% for a single look. Since the standard deviation decreases as $\sqrt{N}$, N = number of measurements, then 100 measurements would yield an estimate with relative error of about 7%.♦

The application of this passive ranging technique using towed array data will have reduced errors because of the long aperture length of the towed array but, because of the uncertainty in towed array element positions due to array wander and tilt, accurate estimates of array position is required to achieve optimal results.

A more rigorous approach for determining range and depth of a target is the matched field processing method[2]. Matched field processing performs matched filter processing to match data from a vertical array to the data that would be obtained from a source target at a given range and depth. The ambiguity function for the match is computed over both range and depth and the ambiguity function peak defines the estimate of the target position. Two problems that are encountered with the matched field processing method are:

1. The matched field processing method relies strongly on a knowledge of the acoustic propagation in the medium including the acoustics of the ocean floor. This will never be known completely and attempts at improving the knowledge base are met with increasing cost and diminishing returns.

2. The matched field processing method requires a large number of computations to produce the ambiguity surface. Furthermore, the number of computations increases rapidly as the knowledge base concerning the medium increases.

<u>Target Position and Velocity from Bearings Only Measurements</u>

The determination of target position and velocity, which is more generally known as target motion analysis (TMA), can be performed passively for the case of

---

[2] Baggeroer, A.B., Kuperman, W.A. and Schmidt, H., "Matched Field Processing: Source Localization in Correlated Noise as an Optimum Parameter Estimation Problem," J. Acoust. Soc. Am., **83**, 571-587, February, 1988.

bearings-only measurements or for the case of bearings and Doppler measurements. We assume that the target is traveling on a constant track with velocities v_x and v_y in, respectively, the x and y directions as indicated in Figure 5. Target positions and velocity can be estimated from bearings only measurements by performing one or more ownship course changes during the measurement period. Target bearings are seen in Figure 5 to be related to target states for all legs of ownship track by the expression,

$$\beta(t)+\theta(t)=\tan^{-1}\left(\frac{x(t)}{y(t)}\right), \tag{28}$$

where $\beta(t)$ is the bearing of the target with respect to ownship, $\theta(t)$ is ownship course and x(t) and y(t) are the coordinates of the target with respect to the ownship coordinate system. The relative target coordinates on the first leg of the ownship track are given by

$$\begin{aligned} x(t) &= x_0 + v_x t - vt\sin(\theta(t)), \qquad 0<t<T \\ y(t) &= y_0 + v_y t - vt\cos(\theta(t)), \qquad 0<t<T \end{aligned} \tag{29a}$$

where x_0 and y_0 are the initial coordinates of the target at t = 0, i.e. when the target was initially detected, and where T is the duration of the first leg. The relative target coordinates on the second leg of ownship track are given by

$$\begin{aligned} x(t) &= \left[x_0 + v_x T - vT\sin(\theta(T))\right] + v_x(t-T) - v(t-T)\sin(\theta(t)), \qquad T<t \\ y(t) &= \left[y_0 + v_y T - vT\cos(\theta(T))\right] + v_y(t-T) - v(t-T)\cos(\theta(t)), \qquad T<t \end{aligned} \tag{29b}$$

as indicated in Figure 5. Of course, similar expressions can be written for successive legs of ownship track. Using more legs would give increased accuracy to our estimate, but it is sufficient to consider only two legs to illustrate the principal.

For each bearing measurement, we substitute (29a) or (29b), depending upon the leg from which the bearing and course measurements came, into (28). This gives a set of equations for the unknown initial target coordinates x_0 and y_0 and unknown velocity v_x and v_y. Since there are four unknowns, it is sufficient to have four measurements for their estimation. However, if we have many more measurements a more accurate solution can be gained by performing a least mean square fit to the data. For the case of N measurements we form the sum of squared errors

$$e^2=\sum_{k=1}^{N}\left(\beta_k+\theta_k-\tan^{-1}\left(\frac{x_k(x_0,y_0,v_x,v_y)}{y_k(x_0,y_0,v_x,v_y)}\right)\right)^2, \tag{30}$$

where β_k and θ_k are measurements and the arctangent term is computed from (29a) and (29b). Performing the following partial differentiations with respect to the unknown parameters

$$\frac{\partial e^2(x_0, y_0, v_x, v_y)}{\partial x_0} = 0, \qquad \frac{\partial e^2(x_0, y_0, v_x, v_y)}{\partial v_x} = 0,$$
$$\frac{\partial e^2(x_0, y_0, v_x, v_y)}{\partial y_0} = 0, \qquad \frac{\partial e^2(x_0, y_0, v_x, v_y)}{\partial v_y} = 0, \tag{31}$$

gives a system of four nonlinear equations to solve for the unknown parameters. This can be done by Newton's iterative method.

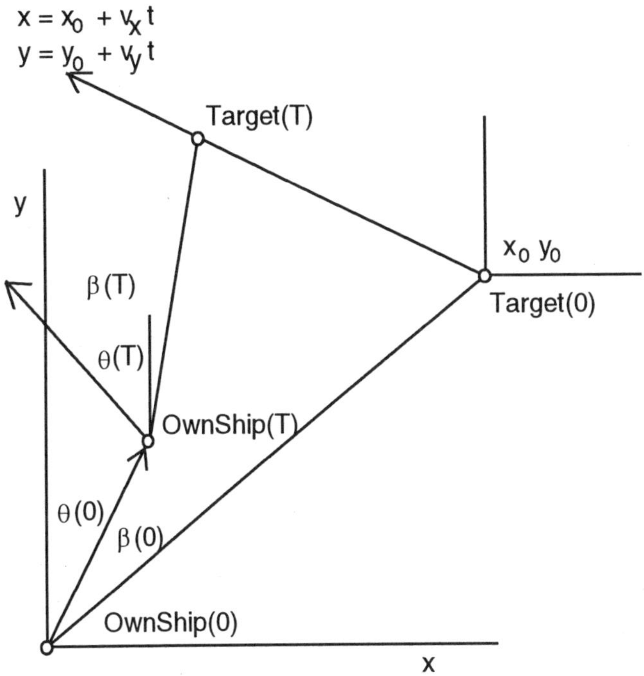

Figure 5 Passive TMA - Bearings Only

Target Range from Bearing and Doppler Data

For the case of bearings and Doppler measurements the target range can be estimated without an ownship course change. The observed frequency from the source is related to the target range rate by

$$f_{obs}(t) = f_s\left(1 - \frac{\dot{r}(t)}{c}\right). \tag{32a}$$

Differentiation gives

$$\dot{f}_{obs}(t) = -f_s\left(\frac{\ddot{r}}{c}\right). \tag{32b}$$

Assuming the velocities of the target and own-ship are constant, we have

$$\left(r\dot{\beta}\right)^2 + \dot{r}^2 = \text{constant}\,, \tag{33a}$$

where β is the bearing to the target. Differentiating gives

$$\ddot{r}(t) = -r(\dot{\beta})^2\,, \tag{33b}$$

and substitution into (32b) gives the range estimation expression

$$r = \frac{c\dot{f}_{obs}}{\dot{\beta}^2 f_{obs}}\,, \tag{34}$$

where the approximation $f_{obs} \cong f_s$ was adopted. This result can give estimates of target range on a measurement by measurement basis. However, since these measurements are noisy and they should be improved by linear regression techniques.

CHAPTER 13
TARGET TRACKING

Once a target has been detected, it can be tracked for monitoring and prosecution. In general terms, a tracking algorithm estimates target positions and velocities iteratively over time and fits a track to those estimates. This track estimate can be interpolated to give estimates of the target motion at intermediate track points and extrapolated to give estimates of target position after the time of the last detection. By projecting target motion beyond the last detection, ASW forces and ASW weapons can be directed to the position where the target is most likely to be found at the time of interception. In an environment of sporadic detections, track estimates and track projections are our best estimate of the present and future target positions as the most recent detection slips steadily into the past. Of course, an operator could form an "eyeball" estimate of both the target track and the current position from a simple plot of a sequence of detection localizations, as described in Chapter 12, and this is sometimes done. However, a target tracking algorithm provides a smoothed track estimate and provides these estimates automatically. The data throughput rate is much greater and the accuracy of the estimates, being based upon an objective motion model, is more consistent.

There are several tracking algorithms in common use and everyone has their favorite. In the following, we discuss the simple α-β tracker and the more elegant Kalman tracker. Both are predictor-corrector algorithms. In such algorithms a motion model predicts the next target position based upon all the available data. Then, upon receipt of the next detection, the predicted target position is corrected to reflect all the available data, including the new detection, to give an updated position estimate. The process is then repeated. The primary difference between the α-β tracker and the Kalman tracker is that the tracker coefficients are ad hoc for the former and satisfy an optimality criterion for the latter. More specifically, the Kalman coefficients minimize the error in the estimated track and are dependent upon the assumed maneuvering capabilities of the target, the data collection system and the track history.

The foregoing gives a general overview of how tracking is performed in an environment in which detections can be unambiguously associated with particular targets. This condition prevails when the targets are few in number and are separated, e.g. either spatially or temporally. In the event, though, that two or more targets are operating in close proximity to each other, or one or more targets are operating in a clutter environment in which many false targets are detected, the detections can no longer be unambiguously associated with a particular target and we must resort to multi-target tracking algorithms. For such cases, the data are sorted to produce the most likely target scene. We discuss in the following a new data association algorithm in which possible target scenes are ranked in probability. The highest probability target scene is the chosen operational view of the environment but, as new data are collected, the relative probabilities of alternative target scenes may change which would change the operational view of the environment. We also discuss a new approach to determining track confidence scores which provide measures of track quality.

13.1 The Predictor-Corrector Tracker

The predictor-corrector tracker is based upon a target motion model and a measurement equation. If we denote the target state vector at time t_k as $\mathbf{x}_k$, which generally consists of position and velocity information, the target dynamics can be represented most simply by the target motion equation

$$\mathbf{x}_k = \Phi_{k,k-1}\mathbf{x}_{k-1} + \mathbf{w}_{k-1}, \tag{1a}$$

where the matrix $\Phi_{k,k-1}$ is the state transition matrix. The additive term $\mathbf{w}_{k-1}$ is the modeling noise which is assumed to be zero mean Gaussian with the known covariance

$$\mathbf{Q}_{k-1} = \mathbf{E}\left(\mathbf{w}_{k-1}\mathbf{w}_{k-1}^T\right), \tag{1b}$$

where $\mathbf{E}$ denotes expectation. Similarly, if we denote the vector of data received from the target at t_k as $\mathbf{d}_k$, which can consist of measurements such as range, bearing and range rate, and denote the mapping between the target state vector and the data by the measurement relationship $\mathbf{h}_k$, the data and target state are related by the measurement equation

$$\mathbf{d}_k = \mathbf{h}_k\left(\mathbf{x}_k\right) + \mathbf{v}_k. \tag{1c}$$

In this expression, the measurement relationship $\mathbf{h}_k$ is generally nonlinear and the additive measurement noise term $\mathbf{v}_k$ arises because the measurement relationship is either not known exactly or is approximated. The additive noise is generally referred to as the measurement residual. The statistics of the measurement residual are assumed to be Gaussian distributed with zero mean and the known covariance matrix

$$\mathbf{R}_k = \mathbf{E}\left(\mathbf{v}_k\mathbf{v}_k^T\right). \tag{1d}$$

More general statements of the tracking problem are possible[1] and discussions of these formulations can be found in the references.

We determine an estimate of the target state vector, which we denote as $\hat{\mathbf{x}}_k$ where the caret signifies estimate. The estimate is assumed to satisfy the motion and measurement equations

$$\hat{\mathbf{x}}_k' = \Phi_{k,k-1}\hat{\mathbf{x}}_{k-1}, \tag{2a}$$

$$\mathbf{d}_k = \mathbf{h}_k\left(\hat{\mathbf{x}}_k'\right) + \mathbf{e}_k. \tag{2b}$$

In the target motion equation, the updated target state estimate is referred to as the predicted target state and is denoted by a prime. Of course, the state transition matrix in (1a) and (2a) does not, in general, produce the next actual target state

---

[1] Gelb, A. (ed.), *Applied Optimal Estimation,* The MIT Press, Cambridge, Mass., 1974.

because the transition matrix will not generally represent the motion of the target as ordered by the submarine captain. Similarly, the measurement mapping of the predicted state onto the collected data will not, in general, be correct because of errors in the predicted state and because of uncertainties in the mapping and the zero mean Gaussian distributed vector $\mathbf{e}_k$ denotes those errors. These errors are unavoidable but are mitigated with each new detection.

We combine the motion equation and measurement equation from (2a,b) to produce an updated estimate of the target state that can be expressed in the predictor-corrector form

$$\hat{\mathbf{x}}_k = \hat{\mathbf{x}}'_k + \mathbf{K}_k[\mathbf{d}_k - \mathbf{h}_k(\hat{\mathbf{x}}'_k)]. \tag{2c}$$

Then, given a previous state estimate of the target at time t_{k-1}, the state estimate at time t_k is given by the predicted state, obtained from (2a), corrected by the weighted measurement residual, obtained from (2b). The weighting coefficient, denoted $\mathbf{K}_k$, is taken to be time dependent, as is the case for the Kalman tracker, but $\mathbf{K}_k$ can be time independent for the α-β tracker.

The α-β Tracker

We consider the simple example of a bearings-only tracker for passive sonar systems for which only bearings measurements are available. The state vector is chosen to have the components of target bearing θ_k and target bearing rate which is its time derivative. The estimated state vector and the predicted estimated state vector from (2a) are

$$\hat{\mathbf{x}}_k = \begin{bmatrix} \hat{\theta}_k \\ \hat{\dot{\theta}}_k \end{bmatrix}, \qquad \begin{bmatrix} \hat{\theta}'_k \\ \hat{\dot{\theta}}'_k \end{bmatrix} = \begin{bmatrix} 1 & T_k \\ 0 & 1 \end{bmatrix} \begin{bmatrix} \hat{\theta}_{k-1} \\ \hat{\dot{\theta}}_{k-1} \end{bmatrix}. \tag{3a,b}$$

The first factor in the right member of (3b) is the state transition matrix $\Phi_{k,k-1}$ and T_k is the time to the last detection. The measurement equation from (2b) is given by

$$\vartheta_k = [1 \quad 0] \begin{bmatrix} \hat{\theta}'_k \\ \hat{\dot{\theta}}'_k \end{bmatrix} + e_k, \tag{3c}$$

where ϑ_k is the measured bearing to the target and the first factor in the right member is the measurement matrix $\mathbf{h}_k$ which, in this model, is linear. Combining (3b) and (3c) we obtain the α-β tracker as in (2c)

$$\begin{bmatrix} \hat{\theta}_k \\ \hat{\dot{\theta}}_k \end{bmatrix} = \begin{bmatrix} \hat{\theta}'_k \\ \hat{\dot{\theta}}'_k \end{bmatrix} + \begin{bmatrix} \alpha \\ \beta \end{bmatrix} \left[\vartheta_k - [1 \quad 0] \begin{bmatrix} \hat{\theta}'_k \\ \hat{\dot{\theta}}'_k \end{bmatrix} \right]. \tag{3d}$$

The elements of the vector $[\alpha,\beta]^T$ are the tracker weighting coefficients. Expanding (3d), we obtain the update equations for the target bearing and bearing rate

$$\hat{\theta}_k = \left(\hat{\theta}_{k-1} + T_k\hat{\dot{\theta}}_{k-1}\right) + \alpha\left[\vartheta_k - \left(\hat{\theta}_{k-1} + T_k\hat{\dot{\theta}}_{k-1}\right)\right], \tag{4a}$$

$$\hat{\dot{\theta}}_k = \hat{\dot{\theta}}_{k-1} + \beta\left[\vartheta_k - \left(\hat{\theta}_{k-1} + T_k\hat{\dot{\theta}}_{k-1}\right)\right]. \tag{4b}$$

We see in (4) that, if the predicted measurement is less than the actual measurement, weighted errors are added to the predicted target bearing and the predicted bearing rate to yield the updated states. Conversely, if the predicted measurement were greater than the actual measurement we subtract the weighted errors. In both cases, of course, α and β are positive. The actual values of α and β are chosen to give the desired tracker response. For small values of α and β, the added corrections are small and the updated state responds slowly to the data giving a smooth track. Conversely, for large values of α and β, the tracker responds more quickly to the data but the track estimate is more irregular.

The α-β tracker is an effective tracker and one that is easy to implement. The coefficients can be fixed or chosen anew at each update of the tracker. For example, when the bearing rate is large, the coefficients can be chosen large so that the tracker responds quickly to bearing changes and, when the bearing rate is small, the coefficients can be chosen small to yield stable estimates of bearing. However, it is unsettling that the coefficients are chosen somewhat arbitrarily for all scenarios. In contrast, the Kalman tracker applies dynamic time dependent weightings to the measurement residuals and does so in an optimal manner.

The Kalman Tracker

The weights on the Kalman tracker are determined by minimizing the mean squared difference between the actual track and the estimated track where the mean squared difference is given by the covariance matrix

$$\mathbf{P}_k = E\left[\left(\mathbf{x}_k - \hat{\mathbf{x}}_k\right)\left(\mathbf{x}_k - \hat{\mathbf{x}}_k\right)^T\right]. \tag{5}$$

Minimizing this expression gives the matrix of Kalman tracker coefficients and a brief derivation of this calculation is given in Appendix 13-A. In the following, we examine the results of those calculations. More extensive discussions are given in the references.

Starting from an initial target state estimate $\hat{\mathbf{x}}_0$ with error covariance $\mathbf{P}_0$ in (5) we predict the next estimates by means of the transition matrix $\Phi_{k,k-1}$

$$\hat{\mathbf{x}}'_k = \Phi_{k,k-1}\hat{\mathbf{x}}_{k-1}, \tag{6a}$$

$$\mathbf{P}'_k = \Phi_{k,k-1}\mathbf{P}_{k-1}\Phi^T_{k,k-1} + \mathbf{Q}_{k-1}, \tag{6b}$$

where k = 1 for the initial calculations, $\mathbf{P}_k'$ is the predicted error covariance matrix and the modeling noise $\mathbf{Q}_{k-1}$ is known from (1b).

The Kalman gain matrix is then computed from the predicted error covariance and measurement relation $\mathbf{h}_k$

$$\mathbf{K}_k = \mathbf{P}'_k \mathbf{H}_k^T \left(\mathbf{H}_k \mathbf{P}'_k \mathbf{H}_k^T + \mathbf{R}_k \right)^{-1} , \tag{6c}$$

$$\mathbf{H}_k = \left. \frac{\partial \mathbf{h}_k(\mathbf{x})}{\partial \mathbf{x}} \right|_{\mathbf{x}=\hat{\mathbf{x}}'_k} . \tag{6d}$$

In these relations $\mathbf{R}_k$ is the covariance of the measurement residual in (1d) and $\mathbf{H}_k$ is the Jacobian of the measurement relation evaluated at the predicted target state, i.e. $\mathbf{H}_k$ is the linearized measurement relation.

The target state and error covariance are then updated with the next measurement to obtain

$$\hat{\mathbf{x}}_k = \hat{\mathbf{x}}'_k + \mathbf{K}_k \left[\mathbf{d}_k - \mathbf{h}_k(\hat{\mathbf{x}}'_k) \right] , \tag{6e}$$

$$\mathbf{P}_k = \left[\mathrm{I} - \mathbf{K}_k \mathbf{H}_k \right] \mathbf{P}'_k , \tag{6f}$$

where (6e) is the same as (2c). The process is then repeated for $k = 2,3,\cdots$. The modeling errors $\mathbf{Q}_{k-1}$ and noise residual errors $\mathbf{R}_k$ are, of course, presumed to be known in these calculations and these quantities can change as the iteration progresses.

As examples of the Kalman tracker we consider two important cases for ASW sonar: a fixed active system and a fixed passive system with two coordinated receiving arrays. For both systems, the initial state vector and initial error covariance matrix are chosen to be given by

$$\hat{\mathbf{x}}_0 = \begin{bmatrix} \hat{x}_0 \\ \hat{y}_0 \\ \hat{\dot{x}}_0 \\ \hat{\dot{y}}_0 \end{bmatrix} , \qquad \mathbf{P}_0 = \begin{bmatrix} \sigma_0^2 & 0 & 0 & 0 \\ 0 & \sigma_0^2 & 0 & 0 \\ 0 & 0 & \upsilon_0^2 & 0 \\ 0 & 0 & 0 & \upsilon_0^2 \end{bmatrix} , \tag{7a,b}$$

where the state vector consists of Cartesian x-y position elements and x-y velocity elements. We make the simplifying assumption that the initial errors in the positions and velocities are independent and also assume that the x-y position variances are equal and the x-y velocity variances are equal. For linear motion, the transition matrix is written

$$\Phi_{k,k-1} = \begin{bmatrix} 1 & 0 & T_k & 0 \\ 0 & 1 & 0 & T_k \\ 0 & 0 & 1 & 0 \\ 0 & 0 & 0 & 1 \end{bmatrix} , \tag{7c}$$

where T_k is the time increment between the k and k−1 target states and is more fully discussed below. Under these assumptions, the predicted state and error covariance for $k = 1$ are found from (6) to be

$$\hat{\mathbf{x}}_1' = \begin{bmatrix} \hat{x}_0 + T_1\hat{\dot{x}}_0 \\ \hat{y}_0 + T_1\hat{\dot{y}}_0 \\ \hat{\dot{x}}_0 \\ \hat{\dot{y}}_0 \end{bmatrix}, \quad \mathbf{P}_1' = \begin{bmatrix} \sigma_0^2 + T_1^2\upsilon_0^2 & 0 & T_1\upsilon_0^2 & 0 \\ 0 & \sigma_0^2 + T_1^2\upsilon_0^2 & 0 & T_1\upsilon_0^2 \\ T_1\upsilon_0^2 & 0 & \upsilon_0^2 & 0 \\ 0 & T_1\upsilon_0^2 & 0 & \upsilon_0^2 \end{bmatrix} + \mathbf{Q}_0 . \quad (7d,e)$$

The motion model noise covariance $\mathbf{Q}_{k-1}$ is assumed to be diagonal matrix that contains small variance terms to ensure that the tracker does not ignore the data as discussed in Appendix 13-A. At the first iteration we see that the target is predicted to move on a straight line and the track position errors and track velocity errors are each predicted to remain independent. However, the position errors and velocity errors are predicted to become correlated. We also see that the predicted position errors have grown while the predicted velocity errors remain constant.

Kalman Tracking for a Fixed Active System:

For the case of a fixed active system, as indicated in Figure 1, the initial target state is obtain from the range, bearing and range rate provided by the initial detection. The initial position error covariance is determined from the range resolution of the signal Δr and the azimuthal resolution of the receiving beam $\Delta\theta$. The range resolution is found from the matched filter time delay resolution and the initial velocity error covariance is determined from the matched filter Doppler resolution.

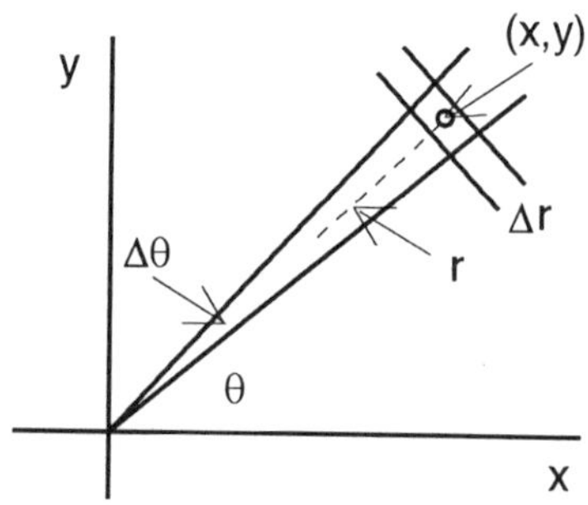

Figure 1 Coordinates for Fixed Active Sonar System with Position Error

The components of the data vector are taken to be the time delay between the receipt of a detection and the start of the transmission, denoted τ_k, the beam bearing θ_k and the Doppler shift on the received signal σ_k. The relation between the Doppler shift and the Doppler compression is $\sigma_k = \alpha_k - 1$. The time delay and Doppler shift are obtained from the Doppler compensated matched filter in the detection system and the bearing is found from the array beam on which the signal is detected. More accuracy can be obtained if the bearing measurement is refined by a weighted sum of detections over several beams. The data vector and the mapping between the target state coordinates and the data vector are

$$\mathbf{d}_k = \begin{bmatrix} \tau_k \\ \theta_k \\ \sigma_k \end{bmatrix}, \qquad \mathbf{h}(\mathbf{x}) = \begin{bmatrix} \frac{2}{c}\sqrt{\left(x^2 + y^2\right)} \\ \tan^{-1}(y/x) \\ \frac{2}{c}\frac{x\dot{x} + y\dot{y}}{\sqrt{\left(x^2 + y^2\right)}} \end{bmatrix}, \qquad (8a,b)$$

where x and y are the Cartesian coordinates of the target and c is the sound speed. The Jacobian of the measurement relation is found from (6d) to be

$$\mathbf{H}_k = \begin{bmatrix} 2\hat{x}'_k / c\hat{r}'_k & 2\hat{y}'_k / c\hat{r}'_k & 0 & 0 \\ -\hat{y}'_k / \hat{r}'^2_k & \hat{x}'_k / \hat{r}'^2_k & 0 & 0 \\ 0 & 0 & 2\hat{x}'_k / c\hat{r}'_k & 2\hat{y}'_k / c\hat{r}'_k \end{bmatrix}. \tag{8c}$$

The measurement residual covariance is assumed to be given by

$$\mathbf{R}_k = \begin{bmatrix} \sigma_\tau^2 & 0 & 0 \\ 0 & \sigma_\theta^2 & 0 \\ 0 & 0 & \sigma_\alpha^2 \end{bmatrix}, \tag{8d}$$

where the elements are the squared resolutions.

With these results the Kalman gain matrix is readily found from (6c) and the updated target state and covariance matrix are readily found from (6e) and (6f). The tracker can then be iteratively updated for $k = 2,3,\cdots$. Since the algebra is straightforward but tedious, these calculations are best performed by machine based algorithms and are omitted here.

The time increment in the transition matrix in (7c) for the fixed active system projects the previous target state, at time t_{k-1}, to the time at which the signal is reflected from the target, i.e. the state times t_k correspond to the positions of the target at the times of echo generation and not echo reception. Denoting the time of the last detection as t', the transmission was initiated at time $t' - \tau_k$ and the updated state estimate of the target is computed at time $t_k = t' - \tau_k/2$. To determine the state at the current time, it must be extrapolated. The time increment in the transition matrix is $T_k = t_k - t_{k-1}$. Note that the time increment is found from the detection data but it is not part of the data vector. Alternatively, the problem can be redefined to include T_k in the data vector.

Kalman Tracking for a Fixed Passive System:

For the case of a fixed passive system with two coordinated arrays, as indicated in Figure 2, the signals from the two arrays are correlated with Doppler compensation to determine the differential time delay τ_k and differential Doppler shift σ_k of the signals at the two arrays. These differential values and the bearings $\theta_{1,k}$ and $\theta_{2,k}$ of the target from the arrays are the elements of the data vector. More accuracy can be obtained if the bearing measurements are refined by a weighted sum of detections over several beams. The data vector and the mapping between the target state coordinates and the data vector are given by

$$\mathbf{d}_k = \begin{bmatrix} \tau_k \\ \sigma_k \\ \theta_{1,k} \\ \theta_{2,k} \end{bmatrix}, \qquad \mathbf{h}(\mathbf{x}) = \begin{bmatrix} \frac{2}{c}(r_1 - r_2) \\ \frac{2}{c}(\dot{r}_1 - \dot{r}_2) \\ \theta_1 \\ \theta_2 \end{bmatrix}, \tag{9a,b}$$

where

$$r_1 = \sqrt{r^2 + (d/2)^2 + 2r(d/2)\cos\theta}\,, \qquad r_2 = \sqrt{r^2 + (d/2)^2 - 2r(d/2)\cos\theta}\,,$$

$$r = \sqrt{x^2 + y^2}\,, \qquad \theta = \tan^{-1}(y/x),$$

$$\theta_1 = \tan^{-1}\left(\frac{y}{x + d/2}\right), \qquad \theta_2 = \tan^{-1}\left(\frac{y}{x - d/2}\right).$$

In these relations, the quantities x and y are the Cartesian coordinates of the target with respect to the point midway between the arrays, r and θ are the corresponding cylindrical target coordinates, d is the spacing between the arrays and c is the sound speed.

The target position is readily seen to lie on an hyperbola where the arrays are located at the foci and the bearings to the target intersect to give the target position on one branch of the hyperbola. This is indicated in Figure 2 where the hyperbola is imagined to pass through the target position at x,y.

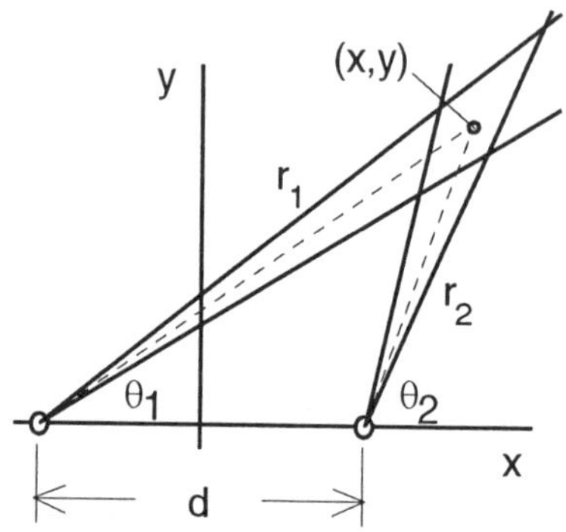

Figure 2 Coordinates for Fixed Passive System

The Jacobian of the measurement relationship in (9b) gives the linear measurement relation in (6d) and the covariance of the measurement residuals is similar to the covariance for the active case in (8d) except that it is a 4×4 diagonal matrix in which the elements are the squared resolutions. From these relations, the Kalman gain matrix in (6c) is readily computed. Then the updated target state and error covariance in (6e) and (6f) are readily obtained. The tracker can then be iteratively updated for $k = 2,3,\cdots$. Since the algebra is straightforward but tedious, these calculations are best performed by machine based algorithms and are omitted here.

The time increment in the transition matrix in (7c) projects the previous target state, at time t_{k-1}, to the time at which the detected data signal is radiated by the target. Denoting the time of the last detection as t′, the radiation was emitted at $t' - r_k/c$, where r_k is the range to the target from one of the arrays, and the updated state estimate of the target is at time $t_k = t' - r_k/c$. The time increment in the transition matrix is $T_k = t_k - t_{k-1}$. Note, however, that the time increment is dependent upon r_k which is not known until the current state is estimated. Accordingly, we must approximate this quantity. One simple approach determines the range for the detection directly from the beam intersection for the detection. This makes T_k data dependent but not part of the data vector. Alternatively, the problem can be redefined to include T_k in the data vector.

13.2 Multi-Target Multi-Sensor Tracking

Multiple target scenarios have always been important for naval missions such as forward force projection and convoy operations. A major increase, though, in the importance of multiple sensor detection systems, as fleet ASW assets as well as surveillance assets become integrated, has increased the need for multi-target multi-sensor tracking capabilities. As a result, such tracking systems have been developed to maximize the usefulness of all available data and to provide a comprehensive and standardized view of the environment. Generally speaking, such a view, even one that may not be entirely correct but that is correctable, is far better than to suffer the "fog of war" confusion brought on by conflicting views concerning the disposition of the enemy.

In scenarios in which multiple targets and/or multiple sensor detection systems are operating, it is not always clear which detection goes with which target but in all cases the detection data must be associated with appropriate targets prior to sending the data to the tracker. Ambiguities in data association can arise for many reasons. For active systems, multiple target detections from one or more sensor systems can be localized to regions of space and time and can be easily associated with the correct target if they are spatially and/or temporally separated. However, active detections from targets that are in close proximity to each other and maneuvering cannot always be correctly partitioned. For passive systems, multiple target detections from a single sensor system can be localized to beam-bands and can be easily associated with the correct target if they are azimuthally and/or spectrally separated. However, passive detections from targets that are in the same beam-band cannot always be correctly partitioned. Similarly, target detections from different active or different passive sensor systems, especially widely spaced and non-synchronously sampled sensor systems, may be incorrectly associated because of unknown Doppler shifts on the data and because of uncertainties in the beam directions. These problems are further compounded for active systems by clutter detections and for passive systems by surface traffic. The problems of data association become even greater when active and passive data are to be jointly associated.

As a result of these complexities, we are required to adopt data association strategies that will be effective for all scenarios. This can be done by forming all possible data associations leading to all possible target scenes that are supported by the detection data. Each target scene is scored and ranked with the highest ranked target scene being the operational view of the environment. As the situation develops, the scores for the target scenes will change and the operational view of the environment will continuously evolve provided that the top ranked target scene remains on top. However, if the relative scores of the target scenes change, the operational view of the environment can undergo a paradigm shift as the operational target scene is replaced by a lower ranked target scene, e.g. a single track in the former view may be resolved into two tracks in the latter view. There are various methodologies for partitioning the data into target scenes and for scoring the target scenes. Some of these methods are discussed in the references. In the following we examine an alternative approach in which we also obtain probabilistic confidence scores for the tracks within each target scene. The confidence scores measure how well tracks conform to motion models which thereby indicate the probabilities that targets of interest underlie the detections. This provides a ready means for rejecting erratic tracks which would include clutter tracks.

Multi-Target Scene Generation

Each target scene that is generated is, from a statistical viewpoint, an hypothesis concerning the disposition of targets. In the immediately following, we consider the hypotheses that are generated by a single scanning sensor system. For such a system we denote the detection data as d_{jk}, (j=1,2,3,...; k = 1,2,3,...), where j denotes the scan number and k denotes the detection within the scan. All possible hypotheses are formed in accordance with the tree structure indicated in Figure 3 which depicts the case of two scans each receiving two detections. The initial hypothesis in the tree is denoted, $h_0 = (\phi)$, where ϕ denotes the empty set. As such, the initial hypothesis is empty of detections and has unity probability. This asserts that, prior to any data collection, the operator is in a state of total ignorance concerning target dispositions but is absolutely certain that the ignorance is valid or, equivalently, is completely certain that nothing is known regarding the target scene. Upon receipt of the first detection d_{11} the initial hypothesis evolves into two target scene hypotheses, denoted (FA) and (d_{11}). In the first of these hypotheses, the detection is attributed to a false alarm while in the second it is attributed to a newly detected target for which, at this point, only one detection is available. Upon receipt of the next detection d_{12} the false alarm hypothesis evolves into two new hypotheses, denoted (FA:FA) and (FA:d_{12}), and the new-target hypothesis evolves into two new hypotheses, denoted (d_{11}:FA) and (d_{11}:d_{12}). Each succeeding hypothesis in the data association chain is similarly formed where, in each case, a plus sign "+" denotes a joining of detection data to form a track and a colon ":" denotes a separator between tracks. In the general hypothesis generation procedure, for each existing hypothesis, each new detection is first associated as a false alarm. It is then associated with the most recent detection in each target track, unless that detection and the new detection are from the same scan, and is lastly associated as a new-target detection. This procedure is repeated with each existing hypothesis. It is clear that a blind application of this procedure leads to an explosive growth in the number of hypotheses. To ensure that the data processing system is not overwhelmed by the computational load of maintaining a large number of hypotheses and to facilitate real-time processing of the data, weak hypotheses must be "pruned" on an ongoing basis to eliminate improbable target scenes. Fortunately, most associations will be found to be improbable and the hypotheses containing these associations can be immediately eliminated. When an hypothesis is pruned, that limb of the tree ceases to grow but the growth of the remaining limbs continues as before. The risk incurred in eliminating low probability hypotheses is the possibility of prematurely eliminating the "truth" hypothesis because it temporarily carried a low score. This possibility is viewed as an event having low probability of occurrence and having low cost consequences when it does occur because one of the retained hypotheses will always be very similar to the pruned truth hypothesis.

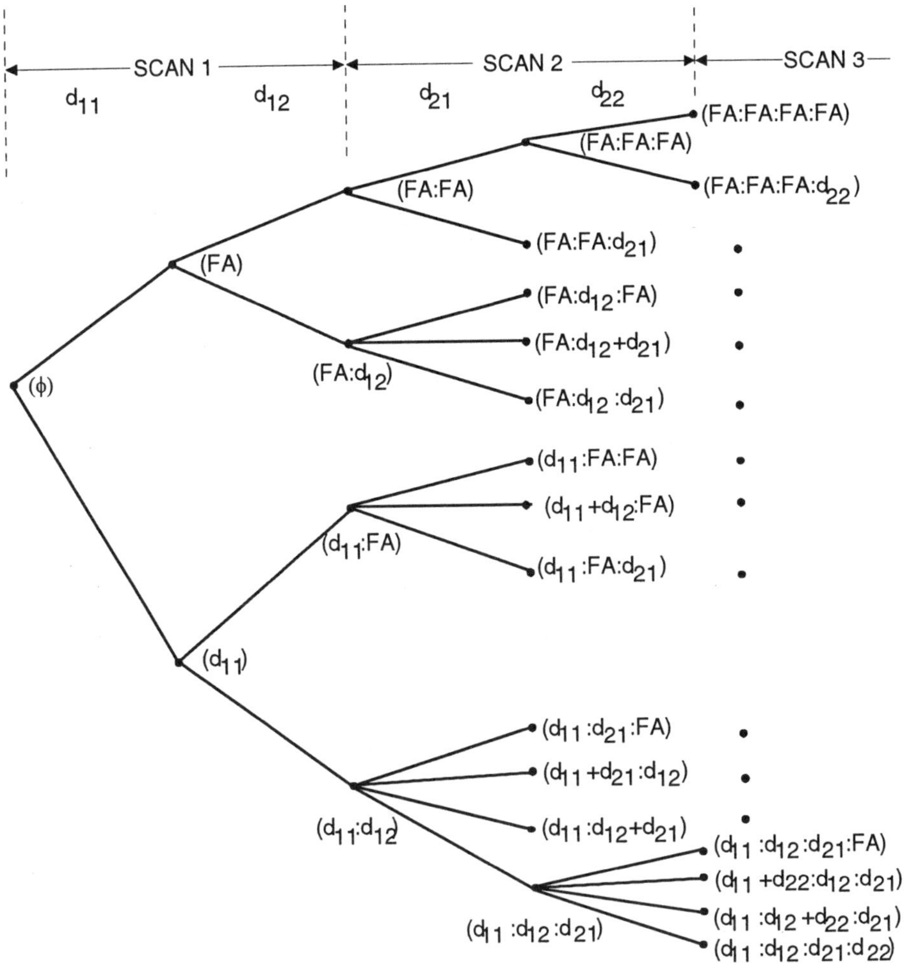

Figure 3 Data Association Tree Structure for a Single Scanning System

The hypothesis generation process depicted in Figure 3 is a "blind" data association process for a single scanning system. It is blind in the sense that it forms all possible target scene hypotheses supported by the data with the exception that detections from any one scan are assumed to not associate. However, for some modes of operation, detection data can be received from random azimuths. For systems that collect detection data using random search patterns, it is possible that the same target could be redetected within a scan period. Similarly, for systems that receive detections from other sensor systems, the same target could be redetected within the same scan period. For these systems a totally blind association process must necessarily be performed. For example, in the association process described above, the detections d_{11} and d_{12} would be associated to form the added target scene hypothesis, $(d_{11}+d_{12})$ even though these data were acquired on the same scan. An important point to note is that, this hypothesis could be technically formed for a single scanning system since it would be immediately eliminated as a zero probability hypothesis. It is merely uneconomical to include the formation of this hypothesis for a single scanning system.

Target Scene Hypothesis Probabilities

When a new detection is acquired, each existing hypothesis generates up to $N_T + 2$ new hypotheses, where N_T denotes the number of tracks in the hypothesis. The root hypothesis is termed the parent hypothesis and the generated hypotheses are termed the child hypotheses (Earlier hypotheses on the tree are referred to as ancestral and later hypotheses are referred to as descendants). The transitions produced by the detection d along any set of branches from the parent hypothesis h_q to the child hypotheses $h_{q'}$ is denoted

$$h_q \xrightarrow{d} h_{q'} . \tag{10a}$$

In Figure 3, for example, we see the transitions

$$\begin{aligned}(d_{11}) &\xrightarrow{d_{12}} (d_{11}\!:\!FA),\\ (d_{11}) &\xrightarrow{d_{12}} (d_{11}\!:\!d_{12}).\end{aligned} \tag{10b}$$

In these expressions, the indices denoting the evolution from the parent index q to the child indices q′ are symbolic because the processes of hypothesis generation and pruning adapt to the tactical situation and cannot be predicted. Consequently, no general formula for the indices is possible. In practice this is unimportant since, in a practical algorithm, the hypotheses are generated recursively and specific index numbers need not be defined or called out. The important point is that the index q of the parent hypothesis denotes one of several parent hypotheses and the index q′ ranges over the indices of the child hypotheses that sprout from each parent node.

The relationship between the hypotheses and their tracks and the relationship between the tracks and their associated detections are expressed, respectively, by the set intersection relations

$$h_q = \bigcap_{j_q} T_{j_q} , \tag{11a}$$

$$T_n = \bigcap_{j_n} T'_{j_n} . \tag{11b}$$

These relations represent the event of an hypothesized target scene as what is common to its constituent tracks, each denoted T, and, similarly, the event of a track is represented as what is common to its track elements, i.e. the detections denoted T′. This is necessary since any one track can exist in several hypotheses and any one detection can exist in several tracks. It is what is common to these elements that define the whole. It is also important to note that these relations assert that hypotheses, tracks and track elements are events and not merely collections of data.

We also note that, while every hypothesis contains all detections, each hypothesis partitions the detections differently to form tracks. If we consider the set of all tracks T so formed, the indices j_q denote the subset of all tracks that are associated with hypothesis h_q. Similarly, the track element events T′ are drawn from the set of all detections and the track element indices j_n denote the subset of all detections that are included in track T_n. Since, analogous to hypotheses, track events and track element events are developed recursively and evolve unpredictably, the track indices and data association indices are symbolic.

Note a distinction here. The track element event T′ is not the event of detection. It is the event that the detection came from the target underlying the track. In symbols, the track element event is denoted T′ = [target|d] whereas the event of detection is denoted [d|target].

The operational target scene hypothesis at each stage of data association is the one having the greatest probability. The hypothesis probabilities are determined iteratively via a Bayesian recursion relation. These probabilities are conditioned on the received data, the previous hypothesis and other sources of information that includes historical and intelligence sources. The latter sources are included in the information set, denoted I, which is explicitly called out in all calculations.

To develop the Bayesian recursion relation, form the conditional probability relations

$$\begin{aligned} P(h_{q'}, h_q, d|I) &= P(h_{q'}, h_q|d, I)P(d|I), \\ &= P(h_{q'}, d|h_q, I)P(h_q|I)\,. \end{aligned} \tag{12a}$$

Rearranging the right members gives the relation

$$P(h_{q'}, h_q|d, I) = \frac{P(h_{q'}, d|h_q, I)P(h_q|I)}{P(d|I)}\,. \tag{12b}$$

Summing over all parent hypotheses gives the relation

$$P(h_{q'}|I') = \frac{\sum_q P(h_{q'}, d|h_q, I)P(h_q|I)}{P(d|I)}\,, \tag{12c}$$

where I' = d∪I denotes the updated information set, which is the union of the detection event and the old information set. In this summation we used the marginal probability relation

$$\sum_q P(h_{q'}, h_q) = P(h_{q'})\,. \tag{12d}$$

Since the probability $P(h_{q'},d|h_q,I)$ in (12c) vanishes for h_q not parental to $h_{q'}$ we can simplify that relation to

$$P(h_{q'}|I') = \frac{P(h_{q'}, d|h_q, I)P(h_q|I)}{P(d|I)}. \tag{13a}$$

This gives a Bayesian recursion relation for the hypothesis probabilities in which the probabilities of the new hypotheses $P(h_{q'}|I')$ are in the form of updates from the probability of the parent hypothesis $P(h_q|I)$. This is the desired form but a more useful expression of it can be found. Summing over the child hypotheses in (13a) we obtain

$$\sum_{q'} P(h_{q'}|I') = P(h_q|d, I), \tag{13b}$$

which indicates that the total probability of the children of h_q is a "corrected" probability, i.e. it is the probability of the parent corrected after receiving the last detection. Actually, since the total probability of the children hypotheses must be just the probability of the parent hypothesis, the parent hypothesis must be independent of the present detection. As we progress this will be found to be the case. Substituting (13b) into (13a) gives an expression for the normalizing constant $P(d|I)$ which, upon resubstitution into (13a), gives the result

$$P(h_{q'}|I') = \left[\frac{P(h_{q'}, d|h_q, I)}{P(d|h_q, I)}\right] P(h_q|d, I). \tag{14a}$$

Performing the decomposition

$$P(h_{q'}, d|h_q, I) = P(h_{q'}|d, h_q, I)P(d|h_q, I), \tag{14b}$$

(14a) becomes

$$P(h_{q'}|I') = P(h_{q'}|d, h_q, I)P(h_q|d, I). \tag{14c}$$

This is nearly in the recursive form for the hypothesis probability updates that we want but we need to express the last factor in terms of the probability of the parent hypothesis. If, in fact, as asserted above, the parent hypothesis supports all follow-on data we would obtain from (14c) the desired recursive form

$$P(h_{q'}|I') = P(h_{q'}|d, h_q, I)P(h_q|I). \tag{14d}$$

This will be found to be the case but for now we carry the excess baggage.

To determine the last factor of (14c) we form the conditional probabilities

$$\begin{aligned} P(h_q, d|I) &= P(h_q|d, I)P(d|I), \\ &= P(d|h_q, I)P(h_q|I). \end{aligned} \tag{15a}$$

The right members give

$$P(h_q|d, I) = \frac{P(d|h_q, I)P(h_q|I)}{P(d|I)}. \tag{15b}$$

Noting that the sum over all parent hypotheses gives unity,

$$\sum_q P(h_q|d, I) = 1\,, \tag{15c}$$

we obtain another expression for the normalizing constant P(d|I) in (15b). Substituting this result into (15b) gives

$$P(h_q|d, I) = \frac{P(d|h_q, I)P(h_q|I)}{\sum_q P(d|h_q, I)P(h_q|I)}\,. \tag{16a}$$

Equation (16a) represents the corrected parent hypothesis probability in terms of the previous parent hypothesis probability. It corrects the parent hypothesis probability using the probability of receiving the detection, given the target scene, but before the data has been associated in the update. If the data shows no preferences, i.e. if it supports all target scenes equally well

$$P(d|h_q, I) = \text{constant}\,, \tag{16b}$$

the corrected probability reduces to the old parent probability

$$P(h_q|d, I) = P(h_q|I)\,. \tag{16c}$$

We demonstrate that (16b) is, in fact, the case in (23) and that (16c) follows.

Prior to this demonstration, we collect the results in (14c) and (16a) and summarize the hypothesis probability equations

$$P(h_{q'}|I') = P(h_{q'}|d, h_q, I)P(h_q|d, I)\,, \tag{17a}$$

$$P(h_q|d, I) = \frac{P(d|h_q, I)P(h_q|I)}{\sum_q P(d|h_q, I)P(h_q|I)}\,. \tag{17b}$$

These equations express the Bayesian recursion probability update in terms of, what might be called, a "look-ahead" probability and a "look-back" probability. The first factor of the right member of (17a) is the look-ahead probability. It performs a look-ahead in the sense that it uses the received data, the parent hypothesis and other information, as available, to determine a first-cut at the probability of the child hypotheses. The second factor of (17a) is the look-back probability which is defined in (17b). It performs a look-back in the sense that it redetermines the probability of the parent hypothesis in the light of the received data to determine how well the received data justifies the parent hypothesis. Together, the product of the look-ahead and look-back probabilities give the probability of the child hypothesis. The look-ahead factor is determined below in Section A and the look-back factor is shown to be just the parent hypothesis probability in Section B.

A. Look-Ahead Factor: $P(h_{q'}|d,h_q,I)$:

The look-ahead factor $P(h_{q'}|d,h_q,I)$ in (17a) is determined from the events that the new detection is a false alarm, an element of the updated track or a new-target which are, respectively, represented by the events

$$h_{q'} = \begin{cases} FA \cap \left(\bigcap_{j_q} T_{j_q} \right) \\ (T'_{j_{q'}} \cap T_{j_{q'}}) \cap \left(\bigcap_{j_q \neq j_{q'}} T_{j_q} \right), \\ NT \cap \left(\bigcap_{j_q} T_{j_q} \right) \end{cases} \tag{18}$$

where the appropriate equation is selected. Since track events within an hypothesis are statistically independent, i.e. knowing that any one of the tracks exists does not shed light on the existence of the other tracks, the probabilities of the hypotheses in (18) can be expressed

$$P(h_{q'}|d,h_q,I) = P(FA)P(\bigcap_{j_q} T_{j_q}|d,h_q,I) = P(FA)\,, \tag{19a}$$

$$\begin{aligned} P(h_{q'}|d,h_q,I) &= P\left((T'_{j_{q'}} \bigcap T_{j_{q'}}) \left(\bigcap_{j_q \neq j_{q'}} T_{j_q} \right) |d,h_q,I \right), \\ &= P(T'_{j_{q'}}|d,h_q,I)P(\bigcap_{j_q} T_{j_q}|d,h_q,I) \\ &= P(T'_{j_{q'}}|d,h_q,I)\,, \end{aligned} \tag{19b}$$

$$P(h_{q'}|d,h_q,I) = P(NT)P(\bigcap_{j_q} T_{j_q}|d,h_q,I) = P(NT)\,. \tag{19c}$$

In each expression, the last equalities are obtained from the observation that every track in the parent hypothesis is a certainty, given that hypothesis, and has unity probability.

We can develop a relationship between the probabilities in (19). In the recursive process of building hypotheses, the child hypotheses are each formed from the parent hypothesis by asserting the new detection to be a false alarm, by linking it to one of the tracks in the parent hypothesis or by assuming it to be a new-target detection Since this exhausts all possibilities, one of these events must be true and the probability of their union is unity

$$P(FA \bigcup_{j_{q'}} T'_{j_{q'}} \bigcup NT) = 1 . \tag{20a}$$

Since the events in (20a) are mutually exclusive, we have the probability relationship

$$P(FA) + \sum_{j_{q'}} P(T'_{j_{q'}} | d, h_q, I) + P(NT) = 1 . \tag{20b}$$

The probability that the new detection is an element of one of the tracks in the parent hypothesis is dependent upon how well the new detection "fits" that track. Since these probabilities are independently determined and the probability of false alarm is known, the new-target probability is computed from (20b).

To determine how well a detection fits a track, we define, in Appendix 13-B, events that a detection associates with existing tracks. These association events are not equivalent to the events that a detection is an element of and belongs to a target track. This is so because a detection can be associated with all target tracks but a detection is an element of, at most, only one target track, i.e. the association events are not mutually exclusive but the events of belonging are mutually exclusive. Further, we find it necessary to define two types of association events. The first type are termed events of association and are denoted A_j, and the second type are termed reduced events of association and are denoted B_j. The events of association A_j include all detections that associate with the tracks and the reduced events of association B_j include all detections that associate with the tracks - excluding all false alarms and new-target detections. By extending a theorem by Feller concerning event unions in (13.9-B), we obtain the following relation between the probability that a detection is an element of a target track and the probabilities of the reduced events of association

$$P(T'_n) = S_1(n) - \frac{S_2(n)}{2} + \frac{S_3(n)}{3} - \ \dots \ \pm \frac{S_N(n)}{N} , \tag{21a}$$

$$\begin{aligned}
S_1(n) &= \sum_{i} P(B_i)\delta_{n,i} \\
S_2(n) &= \sum_{i,j} P(B_i)P(B_j)(\delta_{n,i} + \delta_{n,j}) \\
S_3(n) &= \sum_{i,j,k} P(B_i)P(B_j)P(B_k)(\delta_{n,i} + \delta_{n,j} + \delta_{n,k}) \\
&\vdots \\
S_N(n) &= P(B_1)P(B_2)P(B_3)\cdots P(B_N)
\end{aligned} \tag{21b}$$

where there are N tracks in the hypothesis. The relationship between the probability of the reduced events of association and the probability of the events of association are given in (13.2d-B) as

$$P\left(B_j\right) = P\left(A_j\right)\left(1 - P(FA)\right)\left(1 - P(NT)\right) , \tag{22a}$$

where the probability of new-target is obtained from (13.2f-B)

$$P(NT) = \frac{\left[1-P\left(\bigcup_{j_{q'}} A_{j_{q'}}\right)\right][1-P(FA)]}{1-P\left(\bigcup_{j_{q'}} A_{j_{q'}}\right)[1-P(FA)]}, \tag{22b}$$

and the probabilities of the events of association are obtained from (13.13-B)

$$P(A) = \frac{1}{2^{N/2}\Gamma(N/2)} \int_{\hat{\lambda}}^{\infty} \lambda^{(\frac{N-2}{2})} e^{-\lambda/2} d\lambda\,. \tag{22c}$$

The lower limit in (22c) $\hat{\lambda}$ is the distance between the measurement vector obtained from the current detection and the measurement vector expected to be found on the current detection and is defined in (13.13-B). These probabilities can be computed using approximations or using a look-up table.

B. Look-Back Factor $P(h_q|d,I)$

The factor $P(d|h_q,I)$ in the look-back factor in (17b) is the probability of getting a detection given that the target scene is described by hypothesis h_q. This quantity is related to the probability of detection. For example, for the case of a one track hypothesis, we have $P(d|T,I)$ which has the appearance of the probability of detection but it is not. The probability of detection is the probability of getting a threshold crossing given that there is an underlying target. The probability of interest here is the probability of getting a threshold crossing given a target track within the target scene. In this case the detection d is the event of being a false alarm or of being a threshold crossing due to an underlying target. Denoting the event of a threshold crossing as D, it can be a false alarm, or linked with an existing target or be a new-target. In any case the event is given by

$$[d|h_q,I] = \left[FA \bigcup \left(\bigcup_{j_q}\left(D \bigcap T_{j_q}\right)\right) \bigcup (D \bigcap NT) | h_q, I\right]. \tag{23a}$$

The probability of the event is

$$\begin{aligned} P(d|h_q,I) &= P_{FA} + P_D \sum_i P(T_i) + P_D P_{NT} \\ &= 1-(1-P_D)(1-P_{FA}), \end{aligned} \tag{23b}$$

where (20b) was used. This expression asserts that the probability of getting detection d in the target scene h_q is the probability of not missing a target detection and not missing a false alarm. If the probability of false alarm vanished, this expression reduces to the probability of detection. In all cases it is a constant and independent of the parent hypothesis. This may be puzzling since an hypothesis with many tracks might be thought to have a higher probability of producing a detection than an hypothesis with few tracks. The puzzle is resolved when it is considered that no time window is assumed. An hypothesis with many tracks might require less time to produce a detection than an hypothesis with few tracks, but the probability of each hypothesis producing a detection is the same. Since the hypotheses are updated only when a detection is received, instead of at fixed intervals of time, the above look-back probability is the appropriate quantity. Since (23b) is equivalent to (16b), the update hypothesis probability reduces to (14d).

In summary, the updated hypothesis probabilities are

$$P(h_{q'}|I') = \begin{Bmatrix} P(FA) \\ P(T^{\prime}_{Jq'}|d, h_q, I) \\ P(NT) \end{Bmatrix} P(h_q|I), \tag{24}$$

where the appropriate term is chosen depending upon whether the detection is being associated as a false alarm, element of a target track or new-target. In these relations, P(FA) is the system false alarm probability and P(NT) is the probability of a new-target given by (22b).

As the hypothesis generation process evolves, the hypothesis probabilities tend to decrease as the total probability "mass" is spread over an increasing number of hypotheses. This is proper and has no impact on the choice of the operational target scene. This is true even when hypotheses are pruned and even if the probabilities of the retained hypotheses are renormalized to sum to unity. This can be implicitly seen in the definition of an hypothesis in (11a) which shows that the hypothesis probability is the product of track probabilities - as the number of tracks increase or their probabilities decrease the hypothesis probability decreases. In fact, track probabilities decrease as they accumulate data, as can be determined from (11b), i.e. $P(T_n) = \prod P(T^{\prime}_{Jn})$. This is the proper measure of the probability of getting a particular track from all the detections received but it provides no measure of the quality of the track, i.e. the probability that it represents an underlying target. For track quality measures we turn to track confidence scores discussed next.

13.3 Track Confidence Scores

Track confidence scores are required to provide the operator with discriminants for rapidly distinguishing reliable tracks from questionable tracks. The availability of these scores expedite system operations and provide a means of ranking target tracks in terms of their operational importance. To define the confidence scores for a track, we assume that we have a track that has been updated N times and define the data set X_N to be the collection of data used to update the track. Some of the elements of the data set will be detections and some will be missed detections. Defining T_N and N_N to be, respectively, the event that a target underlies the N data and the event that no target underlies the N data, the confidence scores are defined

$$\text{confidence a target is present} = P(T_N \mid X_N, I)\,, \tag{25a}$$

$$\text{confidence no target is present} = P(N_N \mid X_N, I)\,. \tag{25b}$$

These scores are probabilities conditioned upon the track data and other apriori information. The other information might be historical in nature or obtained from off-board sensors or intelligence reports.

The track confidence scores are operational performance measures and are related to the system performance measures of probability of detection and probability of false alarm. In a general sense, these system performance measures have the forms

$$\text{Probability of Detection} = P(X_N \mid T_N, I)\,, \tag{26a}$$

$$\text{Probability False Alarm} = P(X_N \mid N_N, I)\,. \tag{26b}$$

The track confidence scores are seen to be, effectively, inversions of the system performance probabilities and represent real-time operational performance as opposed to statistical engineering performance.

The confidence scores can be integrated into the multi-target multi-sensor tracking algorithm in the following manner. We append to the track events in (11a) the events that the track represents either an underlying target or no underlying target

$$h_q = \bigcap_{j_q}\left[T_{j_q} \bigcap (T_N \bigcup N_N \mid X_N, I)\right]. \tag{27a}$$

This modified hypothesis event consists of the common element of all the tracks in the target scene in which each track is acknowledged to represent either a target or no target. Since the latter event is the certain event, i.e. $C = T_N \cup N_N$, appending it to each track has no impact on any of the preceding analysis. However, once the tracks and target scenes with their probabilities are generated, the probabilities of the track events become

$$P(T_{j_q}) = P(T_{j_q})(P(T_N \mid X_N, I) + P(N_N \mid X_N, I))\,, \tag{27b}$$

where the last factor in parenthesis is unity. This result shows that the confidence scores can be embedded in the track probabilities. The confidence score for an underlying target is also referred to as the probability of target and, since the confidence scores sum to unity, it is sufficient to know just this probability. Algorithms for computing the probability of target are developed in Section 14.2.

APPENDIX 13-A: KALMAN TRACKER[1]

The Kalman tracker is a least mean square algorithm for estimating target tracks. It is based upon models of target motion and of the detection system which are, respectively,

$$\mathbf{x}_k = \Phi_{k,k-1}\mathbf{x}_{k-1} + \mathbf{w}_{k-1}, \tag{1a}$$

$$\mathbf{d}_k = \mathbf{h}_k(\mathbf{x}_k) + \mathbf{v}_k. \tag{1b}$$

In these relations $\mathbf{x}_k$ is the state vector of the target and $\mathbf{d}_k$ is the measurement vector. The state vector consists of components that represent the target coordinates which might be expressed in Cartesian coordinates, e.g. $(x, y, \dot{x}, \dot{y})$. The measurement vector consists of components that represent detection data which might be expressed in cylindrical polar coordinates, e.g. $(r, \theta, \dot{r})$. Of course, other coordinate systems are possible and the state vector could include, say, acceleration just as the data vector could, in principle, include classification or other information. The motion model is expressed as a linear relationship in which the transition matrix $\Phi_{k,k-1}$ maps the previous state into the current state. The measurement model is expressed as a nonlinear mapping $\mathbf{h}_k$ between the target state and the measurements. We have chosen a general nonlinear measurement model because we are interested in mappings between state vectors expressed in Cartesian coordinates and data vectors expressed in cylindrical polar coordinates. The linear motion model was chosen for its simplicity but it is, essentially, a flat earth motion model. This is adequate when the target does not move far between detections but, for large state changes between detections or when predictions are required that span a significant part of the earth's surface, the linear motion model must be extended to a nonlinear spherical earth model. We won't do this but it can be done using methods described in the following. The additive terms $\mathbf{w}_k$ and $\mathbf{v}_k$ represent modeling noise and are assumed to be zero mean Gaussian noise with covariance matrices

$$\mathbf{Q}_k = \mathbf{E}\left(\mathbf{w}_k \mathbf{w}_k^T\right), \tag{1c}$$

$$\mathbf{R}_k = \mathbf{E}\left(\mathbf{v}_k \mathbf{v}_k^T\right), \tag{1d}$$

where $\mathbf{E}$ denotes expectation. These modeling noise statistics are assumed to be known and noises are assumed to be independent of each other and independent of the state vector.

The actual state of the target is unknown but the Kalman tracker provides an estimate $\hat{\mathbf{x}}_k$ of the track that is unbiased and that has minimum mean square error with the track, i.e.

$$\mathbf{E}(\hat{\mathbf{x}}_k - \mathbf{x}_k) = 0, \tag{2a}$$

---

[1] Gelb, A. (ed.), *Applied Optimal Estimation*, The MIT Press, Cambridge, Mass., 1974.

$$E\left[(\hat{\mathbf{x}}_k - \mathbf{x}_k)_T(\hat{\mathbf{x}}_k - \mathbf{x}_k)\right] = \text{minimum}. \tag{2b}$$

Actually, assuming that a track estimate can be found that produces a minimum mean square error, we have for that track estimate

$$\frac{\partial}{\partial \hat{\mathbf{x}}_k} E\left[(\hat{\mathbf{x}}_k - \mathbf{x}_k)^T(\hat{\mathbf{x}}_k - \mathbf{x}_k)\right] = 2E(\hat{\mathbf{x}}_k - \mathbf{x}_k) = 0, \tag{2c}$$

which shows that this track estimate is unbiased. Consequently, if we can find the minimum mean square error solution both conditions will be met. The mean square error condition is used to determine the gain matrix of the Kalman tracker.

The Kalman tracker has a predictor-corrector form in which the predicted estimate of the target state is given by

$$\hat{\mathbf{x}}'_k = \Phi_{k,k-1}\hat{\mathbf{x}}_{k-1}. \tag{3a}$$

The target state update is

$$\hat{\mathbf{x}}_k = \hat{\mathbf{x}}'_k + \mathbf{K}_k\left[\mathbf{d}_k - \mathbf{h}_k(\hat{\mathbf{x}}'_k)\right], \tag{3b}$$

where the second term provides the correction to the predicted state. In the update equation, the state estimate at k−1 is dependent upon all the data received up to that time but the predicted state does not have the benefit of the latest measurement. This is provided by the correction term in which the quantity $\mathbf{K}_k$ is the Kalman gain matrix and the quantity in braces is the measurement residual

$$\mathbf{e}_k = \mathbf{d}_k - \mathbf{h}_k(\hat{\mathbf{x}}'_k). \tag{3c}$$

The residual is the difference between the measurement received and the predicted measurement.

To implement the update equation, the Kalman gain matrix needs to be determined. Denoting the track estimation error and projected track estimation error, respectively, as

$$\tilde{\mathbf{x}}_k = \hat{\mathbf{x}}_k - \mathbf{x}_k, \tag{4a}$$

$$\tilde{\mathbf{x}}'_k = \hat{\mathbf{x}}'_k - \mathbf{x}_k, \tag{4b}$$

we obtain an expression for these errors from the difference between (1) and (3)

$$\tilde{\mathbf{x}}_k = \tilde{\mathbf{x}}'_k + \mathbf{K}_k\left[\mathbf{h}_k(\mathbf{x}_k) - \mathbf{h}_k(\hat{\mathbf{x}}'_k)\right] + \mathbf{K}_k\mathbf{v}_k. \tag{5a}$$

This can be put in a more computationally convenient form by expanding the measurement mapping around the predicted target state estimate

$$\mathbf{h}_k(\mathbf{x}) = \mathbf{h}_k(\hat{\mathbf{x}}'_k) + \mathbf{H}_k(\mathbf{x} - \hat{\mathbf{x}}'_k), \tag{5b}$$

where $\mathbf{H}_k$ is the Jacobian matrix

$$\mathbf{H}_k = \left.\frac{\partial \mathbf{h}_k(\mathbf{x})}{\partial \mathbf{x}}\right|_{\mathbf{x}=\hat{\mathbf{x}}'_k} . \tag{5c}$$

Substituting (5b) and (5c) into (5a) with (4b) we obtain

$$\tilde{\mathbf{x}}_k = \left[\mathrm{I} - \mathbf{K}_k\mathbf{H}_k\right]\tilde{\mathbf{x}}'_{k-1} + \mathbf{K}_k\mathbf{v}_k . \tag{5d}$$

The covariance of the track estimation error is immediately found form (5d) to be

$$\begin{aligned}\mathbf{P}_k &= \mathbf{E}\left[\tilde{\mathbf{x}}_k\tilde{\mathbf{x}}_k^T\right] \\ &= \left[\mathrm{I} - \mathbf{K}_k\mathbf{H}_k\right]\mathbf{P}'_k\left[\mathrm{I} - \mathbf{K}_k\mathbf{H}_k\right]^T + \mathbf{K}_k\mathbf{R}_k\mathbf{K}_k^T ,\end{aligned} \tag{6a}$$

where the predicted error covariance is given by

$$\mathbf{P}'_k = \mathbf{E}\left[\tilde{\mathbf{x}}'_k\tilde{\mathbf{x}}'^T_k\right] . \tag{6b}$$

The predicted error covariance at k can be expressed in terms of the error covariance at k−1 by substituting (4b), (3a) and (1a) into (6b) to obtain

$$\mathbf{P}'_k = \Phi_{k,k-1}\mathbf{P}_{k-1}\Phi_{k,k-1}^T + \mathbf{Q}_{k-1} . \tag{6c}$$

This gives an iterative prediction formula for the covariance matrix. Both error covariance and predicted error covariance matrices are both seen to be symmetric and positive definite.

We can expand the error covariance relation in (6a) to obtain

$$\mathbf{P}_k = \mathbf{P}'_k - \mathbf{K}_k\mathbf{H}_k\mathbf{P}'_k - \mathbf{P}'^T_k\mathbf{H}_k^T\mathbf{K}_k^T + \mathbf{K}_k\left(\mathbf{H}_k\mathbf{P}'_k\mathbf{H}_k^T + \mathbf{R}_k\right)\mathbf{K}_k^T , \tag{7a}$$

which has the form of a quadratic in $\mathbf{K}_k$. Noting that the term in parentheses is symmetric and positive definite, there is a matrix S_k such that

$$\mathbf{S}_k\mathbf{S}_k^T = \mathbf{H}_k\mathbf{P}'_k\mathbf{H}_k^T + \mathbf{R}_k . \tag{7b}$$

We won't need to compute this matrix but we can use the form in (7b) to complete the square of (7a) and determine a matrix $\mathbf{A}_k$ such that error covariance can be written in the form

$$\mathbf{P}_k = \mathbf{P}'_k + \left(\mathbf{K}_k\mathbf{S}_k - \mathbf{A}_k\right)\left(\mathbf{K}_k\mathbf{S}_k - \mathbf{A}_k\right)^T - \mathbf{A}_k\mathbf{A}_k^T . \tag{7c}$$

Expanding this gives

$$\mathbf{P}_k = \mathbf{P}'_k - (\mathbf{K}_k\mathbf{S}_k)\mathbf{A}_k^T - \mathbf{A}_k(\mathbf{K}_k\mathbf{S}_k)^T + \mathbf{K}_k\left(\mathbf{S}_k\mathbf{S}_k^T\right)\mathbf{K}_k^T, \tag{7d}$$

and comparison with (7a) shows that we must have

$$\mathbf{S}_k\mathbf{A}_k^T = \mathbf{H}_k\mathbf{P}'_k, \tag{8a}$$

$$\mathbf{A}_k\mathbf{S}_k^T = \mathbf{P}'_k\mathbf{H}_k^T. \tag{8b}$$

These are equivalent expressions which provide the relationship

$$\mathbf{A}_k = \mathbf{P}_k'^T\mathbf{H}_k^T\mathbf{S}_k^{-T}, \tag{8c}$$

where the superscript −T means inverse of the transpose or transpose of the inverse - the order being immaterial. Substituting into (7c) gives

$$\mathbf{P}_k = \mathbf{P}'_k + \left(\mathbf{K}_k\mathbf{S}_k - \mathbf{P}'_k\mathbf{H}_k^T\mathbf{S}_k^{-T}\right)\left(\mathbf{K}_k\mathbf{S}_k - \mathbf{P}'_k\mathbf{H}_k^T\mathbf{S}_k^{-T}\right)^T - \mathbf{P}'_k\mathbf{H}_k^T\mathbf{S}_k^{-T}\left(\mathbf{P}'_k\mathbf{H}_k^T\mathbf{S}_k^{-T}\right)^T. \tag{9a}$$

From this relationship we can determine the Kalman gain matrix such that the condition in (2b) is satisfied. That condition can be written as

$$\mathbf{E}\left[(\hat{\mathbf{x}}_k - \mathbf{x}_k)_T(\hat{\mathbf{x}}_k - \mathbf{x}_k)\right] = \mathrm{Trace}\mathbf{P}_k = \mathrm{minimum}. \tag{9b}$$

Since the first and last terms in (9a) are independent of the gain matrix and the trace of the middle term is positive, we achieve a minimum by requiring

$$\mathbf{K}_k = \mathbf{P}'_k\mathbf{H}_k^T\mathbf{S}_k^{-T}\mathbf{S}_k^{-1} = \mathbf{P}'_k\mathbf{H}_k^T\left(\mathbf{S}_k\mathbf{S}_k^T\right)^{-1} = \mathbf{P}'_k\mathbf{H}_k^T\left(\mathbf{H}_k\mathbf{P}'_k\mathbf{H}_k^T + \mathbf{R}_k\right)^{-1}. \tag{10a}$$

This is the Kalman gain matrix. Substituting the Kalman gain matrix back into (7a) gives the update equation for the state covariance

$$\mathbf{P}_k = [\mathrm{I} - \mathbf{K}_k\mathbf{H}_k]\mathbf{P}'_k. \tag{10b}$$

The principal results of the preceding can be summarized as follows. Starting from an initial target state estimate $\hat{\mathbf{x}}_0$ with error covariance $\mathbf{P}_0$, we iterate to produce updates to the target state estimate and error covariance. At k−1 with target state estimate $\hat{\mathbf{x}}_{k-1}$ and error covariance $\mathbf{P}_{k-1}$, we predict the next estimates by means of the transition matrix

$$\hat{\mathbf{x}}'_k = \Phi_{k,k-1}\hat{\mathbf{x}}_{k-1}, \tag{11a}$$

$$\mathbf{P}'_k = \Phi_{k,k-1}\mathbf{P}_{k-1}\Phi^T_{k,k-1} + \mathbf{Q}_{k-1}. \tag{11b}$$

The Kalman gain matrix is then computed from the predicted error covariance and measurement relation

$$\mathbf{K}_k = \mathbf{P}'_k\mathbf{H}^T_k\left(\mathbf{H}_k\mathbf{P}'_k\mathbf{H}^T_k + \mathbf{R}_k\right)^{-1}. \tag{11c}$$

The target state and error covariance are then updated with the next measurement to obtain

$$\hat{\mathbf{x}}_k = \hat{\mathbf{x}}'_k + \mathbf{K}_k\left[\mathbf{d}_k - \mathbf{h}_k(\hat{\mathbf{x}}'_k)\right], \tag{11d}$$

$$\mathbf{P}_k = \left[\mathrm{I} - \mathbf{K}_k\mathbf{H}_k\right]\mathbf{P}'_k. \tag{11e}$$

The process is then repeated. The modeling errors $\mathbf{Q}_{k-1}$ and $\mathbf{R}_k$ are, of course, presumed to be known in these calculations and can change as the iteration progresses. A principal function of the motion modeling noise $\mathbf{Q}_{k-1}$ is to ensure that the tracker does not become indifferent to the data. If the error covariance were to vanish at some point, then, without the motion modeling noise, the predicted error covariance in (11b) and the Kalman gain matrix in (11c) would also vanish. If this occurs, it follows from (11d) and (11e) that the tracker would ignore the data and rely solely on the predicted track states and that the error covariance would remain zero continuously. With the addition of the modeling noise covariance in (11b), this condition is prevented and the tracker continues to respond to new data.

APPENDIX 13-B: TRACK ELEMENT PROBABILITIES

The probabilities of the updated target scene hypotheses are given by the probabilities that the new detection is either a false alarm, an element of an existing track or a new-target detection. From (13.19), this is expressed as

$$P(h_{q'}|d, h_q, I) = \begin{cases} P(FA) \\ P(T'_{j_{q'}}|d, h_q, I), \\ P(NT) \end{cases} \tag{1a}$$

where, from (13.20),

$$P(FA) + \sum_{j_{q'}} P(T'_{j_{q'}}|d, h_q, I) + P(NT) = 1. \tag{1b}$$

Since the probability of false alarm is known, we must determine the probabilities that a detection is either an element of an existing track or an element of a new-target track. The former are determined from probabilities of association between the new detection and existing tracks and the latter is determined from (1b).

Postponing, temporarily, a definition of what is meant by “event of association”, we examine some of the properties that these events must have. One important distinction between the event that a detection is an element of an existing track and the event that a detection associates with an existing track is that the former events are mutually exclusive while the latter events are not, i.e. a new detection can be an element of at most one track but it can be associated with all tracks. Since a new detection must necessarily be a false alarm, an element of one of the existing tracks or a new-target detection and, also, it must associate with all existing tracks, we can express these events in the form

$$FA \bigcup \left(\bigcup_{j_{q'}} T'_{j_{q'}} \right) \bigcup NT = FA \bigcup \left(\bigcup_{j_{q'}} A_{j_{q'}} \right) \bigcup NT, \tag{2a}$$

where A_j are the association events. Both members of (2a) are the certain event and, while all events in the left member are mutually exclusive, the association events are not mutually exclusive with any events in the right member and we cannot write term-by-term equalities that would give us the probabilities of the track element events in terms of the track association events. However, we can write the relation

$$\bigcup_{j_{q'}} T'_{j_{q'}} = \bigcup_{j_{q'}} A_{j_{q'}} - \bigcup_{j_{q'}} A_{j_{q'}} \bigcap (FA \bigcup NT), \tag{2b}$$

where the second term of the right member removes those false alarms and new-target detections that associate with the tracks. With elementary manipulations we can rewrite (2b) as

$$\bigcup_{j_{q'}} T'_{j_{q'}} = \bigcup_{j_{q'}} \left(A_{j_{q'}} \cap FA^c \cap NT^c \right) = \bigcup_{j_{q'}} B_{j_{q'}} . \tag{2c}$$

In this form we have equality between the unions of track element events and the events B_j which are just reduced association events, i.e. the association events from which false alarm and new target detections that associate with existing tracks have been excluded. Since the B_j events are not mutually exclusive, we still cannot write term-by-term equalities in (2c) but, by means of a theorem by Feller concerning unions of events, we can determine the probabilities of track element events in terms of the probabilities of the reduced association events. This is done in Section 13.1-B.

Given those determinations, our problem is solved if we can determine the probabilities of the reduced association events. Toward this end we see from the definition of the reduced association events in (2c) that

$$P\left(B_{j_{q'}}\right) = P\left(A_{j_{q'}}\right)\left(1 - P(FA)\right)\left(1 - P(NT)\right), \tag{2d}$$

where independence between each association event and the event that the detection is neither a false alarm nor a new target is assumed. Consequently, the probabilities of the reduced association events are obtained from the association events and, also, from the probabilities of false alarm and new-target detections. The probabilities of association are determined in Section 13.2-B, the false alarm probability is known and the probability of new-target detection is found from (1b). In fact, from (1b) and (2b) we infer

$$P(FA) + P\left(\bigcup_{j_{q'}} A_{j_{q'}}\right)\left(1 - P(FA)\right)\left(1 - P(NT)\right) + P(NT) = 1, \tag{2e}$$

which immediately gives the probability of new-target detection

$$P(NT) = \frac{\left[1 - P\left(\bigcup_{j_{q'}} A_{j_{q'}}\right)\right]\left[1 - P(FA)\right]}{1 - P\left(\bigcup_{j_{q'}} A_{j_{q'}}\right)\left[1 - P(FA)\right]}. \tag{2f}$$

A number of limits implied by (2f) are of interest. If the track association probability vanishes, i.e. $P(\cup A) = 0$, then the detection must be either a new-target or a false alarm, i.e. $P(NT) = 1 - P(FA)$. If the track association probability is unity, i.e. $P(\cup A) = 1$, then a new-target detection is unlikely, i.e. $P(NT) = 0$. If the environment is especially noisy, i.e. $P(FA) = 1$, then, however well the detection associates with an existing track, it is unlikely to be a new-target, i.e. $P(NT) = 0$. Further, if the environment is especially quiet, i.e. P(FA) = 0, then the detection is necessarily from an existing target or a new-target, i.e. either P(NT) = 1 or $P(\cup A) = 1$. With a little reflection we see that each of these cases describes the situation as we should expect it. In sum, we see that the preceding discussions, coupled with the results in the following sections, completes the solution to our problem of determining the probabilities that a detection is an element of an existing track or from a new target.

13.1-B: Probabilities of Track Elements

The events of association are dependent upon the received detection, the parent hypothesis tracks and other information sources. The most important information source for our purposes is the target motion model for each track in the parent hypothesis but classification information can also prove to be a powerful discriminant. Detections that are consistent with the target motion model for a particular track have a high probability of being associated with that target as opposed to detections that imply, say, that the target has veered markedly from its track. Detections that carry classification information that is consistent with a target classification also have a high probability of being associated with that target as opposed to detections that imply that the target is a different class of target. Other information that might assert the presence of a target in a particular region can raise the probabilities of associations of weak detections made in that region. Similarly, detections falling only within a specified signal-to-noise range might be associated with tracks to discriminate between weak false alarms and strong clutter returns. The important point for this discussion is that the association events and their probabilities can include all significant elements of the problem. However, for our purposes in the following, the association probabilities will be determined solely from target motion model considerations.

To determine the relationship between the events that a detection is an element of one of many tracks and the events that the detection is associated with several tracks, we proceed by initially examining the case of two tracks. It is assumed that the association probabilities between the detection and each of the tracks have been determined. For a discussion of the determination of these probabilities, see Section 13.2-B. For the case of two tracks, where B_1 and B_2 are the reduced events of association with targets 1 and 2, the certain event can be written

$$C = (B_1 \cap B_2^c) \cup (B_1^c \cap B_2) \cup (B_1 \cap B_2) \cup (B_1^c \cap B_2^c), \tag{3a}$$

where the superscript "c" denotes complement. The first term is the reduced event of association with target 1 and not target 2; the second term is the reduced event of association with target 2 and not target 1; the third term is the reduced event of association with both targets; and the fourth term is the reduced event of association with neither target. The first three terms are the event of the detection being an element of either target track 1 or 2 and the fourth term is the event of the detection being either a new-target or false alarm. Resolving as in (2c) we obtain

$$T_1' \bigcup T_2' = (B_1 \bigcap B_2^c) \bigcup (B_2 \bigcap B_1^c) \bigcup (B_1 \bigcap B_2)\,. \tag{3b}$$

The left member is the union of two mutually exclusive events and to put the right member in the same form we need to draw upon additional information sources to resolve the ambiguity when the detection associates with both tracks 1 and 2. Denote the information that tips the decision toward targets 1 and 2 as, respectively, I_1 and I_2, where their union is the whole of available information, $I_1 \cup I_2 = C$, and their intersection vanishes, $I_1 \cap I_2 = 0$, i.e. all ambiguous information is excluded. Forming the union of this information with (3b) and expanding we obtain

$$T_1' \bigcup T_2' = \left\{(B_1 \bigcap B_2^c) \bigcup \left[(B_1 \bigcap B_2) \bigcap I_1\right]\right\} \bigcup \left\{(B_1^c \bigcap B_2) \bigcup \left[(B_1 \bigcap B_2) \bigcap I_2\right]\right\}. \tag{3c}$$

Equation (3c) can be resolved into the events of the detection being an element of tracks T_1 and T_2.

$$T_1' = (B_1 \bigcap B_2^c) \bigcup \left[(B_1 \bigcap B_2) \bigcap I_1\right], \tag{4a}$$

$$T_2' = (B_1^c \bigcap B_2) \bigcup \left[(B_1 \bigcap B_2) \bigcap I_2\right]. \tag{4b}$$

These events satisfy the relations required by the track element events

$$T_1' \bigcup T_2' = B_1 \bigcup B_2\,, \quad T_1' \bigcap T_2' = 0\,. \tag{5a,b}$$

The probabilities of the events in (4) can be written

$$P(T_1') = P(B_1) - P(B_1)P(B_2)P(I_2)\,, \tag{6a}$$

$$P(T_2') = P(B_2) - P(B_1)P(B_2)P(I_1)\,, \tag{6b}$$

where it is assumed that the reduced association events are statistically independent. Since on average the additional information is expected to tilt the decision equally toward targets 1 and 2, we have the probabilities: $P(I_1) = 1/2$ and $P(I_2) = 1/2$. This is equivalent to assuming that the additional information is obtained from a fair coin toss. Of course, another set of fractions can be chosen - but they must sum to unity. The probabilities of the reduced association events are obtained from (2d) and the probability of a new-target is obtained from (2h). Both require the probabilities of association which are obtained from Section 13.2-B.

For the case of three tracks, the above process is modified by the injection of additional information regarding the resolution of tracks 1 vs. 2, 1 vs. 3 and 2 vs. 3, i.e. $I_{1,3} \cup I_{1,2} \cup I_{2,3} = C$, and we assume that these information sets are unbiased with probabilities of 1/3 each, i.e. a fair coin toss of a three-sided coin. We omit this analysis and its extensions to greater numbers of targets because the algebra becomes increasingly tedious. Instead, for the case of unbiased information, we can develop formulas for the track probabilities by extending a theorem by Feller[1] concerning the union of N events.

Consider the union of N reduced association events

$$U = \bigcup_{i=1}^{N} B_i \, , \tag{7}$$

where the left member corresponds to our union of track element events in (2c) and the right member is our union of reduced association events. The probability of the union is found from Feller's theorem which we reproduce in summary form.

Form the N probability sums

$$\begin{aligned} S_1 &= \sum_i P(B_i), \\ S_2 &= \sum_{i,j} P(B_i \cap B_j), \\ S_3 &= \sum_{i,j,k} P(B_i \cap B_j \cap B_k), \\ &\vdots \end{aligned} \tag{8a}$$

where $i < j < k < \cdots \le N$. This index convention ensures that each combination of event intersections occurs only once in each sum and the last sum in the sequence is $S_N = P(B_1 \cap B_2 \cap \cdots \cap B_N)$. In the nth sum we are combining the N association events taken n at a time. Accordingly, the total number of terms in the sum S_n is given by the binomial coefficient $\binom{N}{n}$.

Feller proves that the probability of the union in (7) is given by

$$P(U) = S_1 - S_2 + S_3 - S_4 + \ \ldots \ \pm S_N \, . \tag{8b}$$

To show this in brief terms, we assume that a detection d falls in exactly n of the N reduced association events. For the left member of (7), the probability of U, conditioned on d, is the probability that d falls in at least one of the association events and its probability is

$$P(U|d) = P(d) \, . \tag{8c}$$

For the right member of (7), without loss of generality, we number the reduced association events such that d falls in the first n reduced association events. The probability of the right member, conditioned on d, is obtained by adding in and

[1] Feller, William, *An Introduction to Probability Theory and Its Applications*, 3rd ed., **1**, pp. 98-100, Wiley, 1968.

subtracting out probability "masses" to avoid double counting, i.e. add the sum of probabilities of events B_i, given in S_1; minus the sum of probabilities of the intersecting events $B_i \cap B_j$, given in S_2; plus the sum of probabilities of the intersecting events $B_i \cap B_j \cap B_k$; given in S_3; etc. The first sum is seen to be: $S_1 = nP(d)$, because $P(B_i|d) = P(d)$ for the first n terms in the sum and vanish for the remaining terms. By similar arguments, the second sum is seen to be given by $S_2 = \binom{n}{2} P(d)$ and the higher order sums follow. Combining terms as in (8b), the probability of the right member of (7) is[2]

$$\left[\binom{n}{1} - \binom{n}{2} + \cdots \pm \binom{n}{n}\right] P(d) = P(d)\,. \tag{8d}$$

This shows that both sides of (8b) are equal for the arbitrary detection d and, consequently, they are equal for all detections which proves (8b).

We extend Feller's theorem by defining the new probability sums

$$\begin{aligned} S_1(p) &= \sum_i P(B_i)(\delta_{p,i}), \\ S_2(p) &= \sum_{i,j} P(B_i \cap B_j)(\delta_{p,i} + \delta_{p,j}), \\ S_3(p) &= \sum_{i,j,k} P(B_i \cap B_j \cap B_k)(\delta_{p,i} + \delta_{p,j} + \delta_{p,k}), \\ &\vdots \end{aligned} \tag{9a}$$

where, as before, $i < j < k < \cdots \leq N$ and the symbol $\delta_{p,q}$ denotes the Kroneker delta function which is unity for $p = q$ and zero otherwise. The argument of the sums, denoted p, is taken to be the number of the track in the target scene hypothesis and the last sum is, again, $S_N(p) = NP(B_1 \cap B_2 \cap \cdots \cap B_N)$. This represents a decomposition of Feller's sums in which

$$\sum_{p=1}^{N} S_n(p) = nS_n\,, \tag{9b}$$

where the S_n are Feller's sums in (8a). The probability that a detection is an element of a track is then given by

$$P(T'_p) = S_1(p) - \frac{S_2(p)}{2} + \frac{S_3(p)}{3} - \ \ldots \ \pm \frac{S_N(p)}{N}\,. \tag{9c}$$

---

[2] Abramowitz and Stegun, (3.1.7).

The normalization factors represent ambiguity resolutions via fair coin tosses. To show that this result properly represents the track element probabilities, sum (9c) over all tracks. From (9b) we obtain Feller's series in (8b). Comparison of (9) with (6) for the case of 2 tracks shows that these results are equivalent as do similar comparisons for 3 and more tracks.

13.2-B: Probabilities of Association

The event that a detection is associated with a target is the event that the detection is in the target's proximity. From (13.3c-A) and (13.1b-A) the measurement residual of a detection with a target is

$$\mathbf{e}_k = \mathbf{d}_k - \mathbf{h}_k(\hat{\mathbf{x}}'_k) = \mathbf{h}_k(\mathbf{x}_k) - \mathbf{h}_k(\hat{\mathbf{x}}'_k) + \mathbf{v}_k, \tag{10a}$$

where $\mathbf{d}_k$ denotes the detection data, $\mathbf{h}_k$ maps the predicted target state and target state onto an expected detection and k denotes a sampling time. Expanding the measurement relation, as in (13.5-A), and computing the mean and covariance matrix we obtain

$$\mathbf{E}(\mathbf{e}_k) = 0, \tag{10b}$$

$$\mathbf{E}\left(\mathbf{e}_k\mathbf{e}_k^T\right) = \mathbf{H}_k\mathbf{P}_k\mathbf{H}_k^T + \mathbf{R}_k, \tag{10c}$$

$$\mathbf{P}_k = \mathbf{E}\left[(\mathbf{x}_k - \hat{\mathbf{x}}'_k)(\mathbf{x}_k - \hat{\mathbf{x}}'_k)_T\right], \qquad \mathbf{R}_k = \mathbf{E}\left(\mathbf{v}_k\mathbf{v}_k^T\right). \tag{10d,e}$$

Since the measurement residual is assumed to be Gaussian distributed, its probability density function, from 8.9-A, is

$$\mathrm{p}(\mathbf{e}_k) = \frac{\exp\left[-\frac{1}{2}\mathbf{e}_k^T\left(\mathbf{H}_k\mathbf{P}_k\mathbf{H}_k^T + \mathbf{R}\right)^{-1}\mathbf{e}_k\right]}{(2\pi)^{N/2}|\mathbf{H}_k\mathbf{P}_k\mathbf{H}_k^T + \mathbf{R}_k|^{1/2}}, \tag{11}$$

where N is the number of elements in the measurement vector $\mathbf{d}_k$ and $|\cdot|$ denotes determinant.

The measurement residual is the difference between the actual measurement and the expected measurement. Therefore, for a given a detection having measurement residual $\hat{\mathbf{e}}_k$, the strength of the association of the measurement with the target varies as the likelihood of the measurement. Based upon this notion, we define the probability of association to be the probability that any other measurement would have a lesser likelihood. This can be written

$$\mathrm{P(A)} = \mathrm{P}[\mathrm{p}(\mathbf{e}_k) \le \mathrm{p}(\hat{\mathbf{e}}_k)]. \tag{12a}$$

Equivalently, we can express this notion in terms of the log-likelihood statistic

$$P(A) = P\left(\lambda \geq \hat{\lambda}\right), \quad \lambda = \mathbf{e}_k^T\left(\mathbf{H}_k\mathbf{P}_k\mathbf{H}_k^T + \mathbf{R}_k\right)^{-1}\mathbf{e}_k . \tag{12b,c}$$

The log-likelihood statistic is chi-squared distributed and has unit variance with N degrees of freedom

$$p(\lambda) = \frac{\lambda^{(\frac{N-2}{2})} e^{-\lambda/2}}{2^{N/2}\Gamma(N/2)}, \tag{12d}$$

where N is the number of data elements in the data vector $\mathbf{d}_k$.

Combining (12b,c) and (12d) gives the expression for the probability of association

$$P(A) = \frac{1}{2^{N/2}\Gamma(N/2)} \int_{\hat{\lambda}}^{\infty} \lambda^{(\frac{N-2}{2})} e^{-\lambda/2} d\lambda , \tag{13a}$$

$$\hat{\lambda}_k = \hat{\mathbf{e}}_k^T(\mathbf{H}_k\mathbf{P}_k\mathbf{H}_k^T + \mathbf{R}_k)^{-1}\hat{\mathbf{e}}_k . \tag{13b}$$

For practical calculations, values of this result can be obtained from approximations or from look-up tables.

CHAPTER 14
OPERATIONAL PERFORMANCE

ASW forces are employed to secure the use of the sea by friendly forces and/or to deny the use of the sea to opposing forces. Both can be achieved simultaneously resulting in total control of the sea but a lesser degree of sea control will usually be the case. Secure use of the sea can be achieved without clearing the sea of all enemy forces as long as they are sufficiently restrained from interfering with friendly force operations, e.g. enemy submarines may unobtrusively transit the area submerged. Similarly, the denial of the sea to opposing forces can be achieved without establishing secure use of the sea as long as the enemy is capable of returning in force to restrict friendly force use of the area. Accordingly, the operational objectives of sea use and sea denial can be met either separately or jointly.

Sea use operations include battle group and convoy protection. In the age of long range cruise missiles, battle group protection requires that an area of sea, approximately 500 km radius around the battle group, be secured and that during battle group transit this secure area is required to move with the battle group at a speed of about 15 knots or more. Tactical ASW forces can readily maintain coverage during battle group transit but shore based surveillance forces and SURTASS require prepositioning of assets and hand-off procedures. While force projection against an enemy's infrastructure is the primary objective of the battle group, which takes precedence over ASW operations, a large portion of the battle group assets are given over to ASW operations, which makes the infliction of maximum damage on enemy submarine forces a strong secondary objective. Convoy protection requires a secure area of 100 km radius around the convoy, moving at a speed of about 10 knots during transit, to ward off short range missile and torpedo attacks. Over the length of the journey, the convoy may be handed off from one ASW support group to another each taking turns performing escort services within its region of ocean. Alternative to a moving secure area, convoy protection can be performed by the establishment of a 200 km wide corridor through which the convoy passes.

Sea denial operations include barrier operations, which are generally established at choke points, port ingress/egress operations, conducted outside enemy or friendly ports, and area denial operations. Choke points of historical importance include: the Greenland-Iceland-United Kingdom (GIUK) gap; the Skaggerak between Norway and Denmark; Sea of Japan entrances such as the Kunashir Passage and Straits of Tsushima; the Straits of Malacca between the Malayan Peninsula and Indonesia; the Bosphorus, Sicilian Channel and Straits of Gibraltar in the Mediterranean; and the Yucatan Channel and Straits of Florida in the Caribbean. Port ingress and egress operations can monitor deployments of enemy forces or can be used to delouse Allied SSBNs of trailing enemy SSNs as they depart from home port and return to home port. Area denial operations can be mounted by shore based SOSUS/SURTASS and airborne P3 operations to force enemy submarines farther from Allied shore lines or to remain outside areas through which a battle group or convoy is to pass.

These concepts of operations for ASW systems are based upon strategic concepts that pertain, specifically, to the operational level of strategy[1]. The operational level of strategy is concerned with the implementation of strategy and battle plans through the control of tactical units which, if effectively performed, results in an outcome that is more than just the sum of tactical unit actions. The operational level acts as the central nervous system of the war effort and is the level at which command, control, communications and information (C^3I), as well as logistics, are established and maintained. It administrates the battle and adapts battle plans to the course of the conflict. Operational concepts for tactical units specify the roles that these forces are intended to play within the overall theater of operations and operational performance indicates how well these forces can be expected to perform. Operational performance is a technical measure of performance and, while it establishes a baseline for tactical performance, the outcome of a conflict between two proactive and reactive opponents will be determined by many factors, not the least of which are, surprise, courage, will and training.

Two measures for estimating operational performance for ASW sonar are discussed in the following. These are:

1. the "probability of target detectability", i.e. the probability of detection of a target within the operations area over a period of time; and

2. the "probability of detected target", i.e. the probability that a target track estimate is in fact due to an underlying target.

These two performance measures can help in establishing guidelines, in the first case, for target search patterns and, in the second case, target prosecution scenarios. They can be adapted to provide tactical performance measures and/or tactical performance aids by the inclusion of specific tactical data and by their inclusion in tactical processing algorithms but are here used to assess what-if scenarios that lead to technical operational performance estimates. Our viewpoint is from the perspective of active sonar but analogous results for passive sonar can be similarly developed.

14.1 Probability of Target Detectability

The probability of target detectability is an extension of the engineering concept of probability of detection. However, whereas engineering probability of detection applies to a single interrogation of a single detection cell and is conditioned upon a target present within that cell, the probability of target detectability applies to a sequence of interrogations of an operational area and is conditioned upon a target present within that area and accounts for the operational deployment of the ASW platform and the maneuverability of the target.

---

[1] The operational level of strategy falls in a hierarchy of strategic concepts that include: grand strategy, military strategy, theater strategy, operations and tactics.

General Relationships

A target in assumed to be within the ASW sonar operations area and ASW forces are tasked with locating it. More than one sonar unit may be involved and the sonar coverage may either be sufficient to interrogate the entire operations area without maneuvering or the sonar coverage "footprint" may be less than the operations area and maneuvers, such as "sprint and drift" or "continuous search", may be required to cover the entire area. In all cases the probability of target detectability is defined to be the probability of detecting the target within the operations area in N interrogations

$$P_D(N) = P(\text{ target detection in N interrogations}) . \tag{1a}$$

An interrogation is considered to be active sonar transmission and is a sampling of the acoustic environment during an interval of time. In all cases the probabilities are conditioned upon a target present in the operations area but, for notational convenience, the repeated statements of this condition are suppressed.

An equivalent expression for target detectability is:

$$P_D(N) = 1 - P(\text{ N misses in N interrogations}) . \tag{1b}$$

We denote the events "miss" and "detection" on the nth interrogation as, respectively, m_n and d_n. In this terminology miss and detection assert, respectively, no target detection and target detection - as opposed to no threshold crossing and threshold crossing, i.e. the misses include all false alarms. The compound event of all misses in N interrogations is the event

$$M_N = \bigcap_{j=1}^{N} m_j , \tag{2}$$

where the intersection of misses in all interrogations asserts that all interrogations produced misses. With this notation the probability of target detectability is

$$P_D(N) = 1 - P(M_N) , \tag{3a}$$

which can be written

$$P_D(N) = 1 - \prod_{n=1}^{N} P(m_n | M_{n-1}) = 1 - \prod_{n=1}^{N} \left[1 - P(d_n | M_{n-1})\right] . \tag{3b}$$

In this expression, M_0 is the certain event, i.e. prior to an interrogation of the area a miss of the target is a certainty, and the particular detection cells yielding the detections depend upon the tactical scenario and are unknown at the operational level which is our concern.

Since the event of detection of the target during the nth sampling implies that a target is present we can define the event

$$T_n = [\text{target within the nth search area}], \tag{4a}$$

where the search area is the area ensonified. Expanding the last probability term in (3b) we obtain

$$P(d_n | M_{n-1}) = P(d_n, T_n | M_{n-1}) = P(d_n | T_n, M_{n-1}) P(T_n | M_{n-1}). \tag{4b}$$

The first equality holds because the event of a detection d_n implies that a target is present, i.e. T_n. Since the term $P(d_n|T_n,M_{n-1})$ is conditioned upon the presence of a target in the detection cell, it is independent of whatever happened on the previous n−1 interrogations and the factor is just the engineering probability of detection P_D. Collecting results gives

$$P_D(N) = 1 - \prod_{n=1}^{N} \left[1 - P_D P(T_n | M_{n-1})\right], \tag{5}$$

where P_D is assumed to be constant for all target detection cells and to be the design detection probability. This is not strictly correct because P_D depends upon the received signal-to-noise ratio which varies across detection cells. However, since the design detection probability is the minimum detection probability, the approximation yields a conservative estimate of target detectability which is appropriate for the purposes of determining operational performance. The term $P(T_n|M_{n-1})$ is the probability of target availability during the nth interrogation and it is determined by the scenario.

Example applications of the general result in (5) are discussed for the scenarios of stationary and mobile active ASW systems that are operating in duct, bottom bounce (BB) and convergence zone (CZ) environments. These environments are characterized by circular regions of sonar coverage that are defined by a maximum range. For the duct environment, sonar coverage within the region is complete but for the BB and CZ environments inner circular regions of coverage are excluded because acoustic energy does not penetrate those regions. In all cases it is found that target mobility can enhance the system detection performance and the cases of fast targets and fixed targets are compared which serve to bound the performance for intermediate speed targets. Moreover, for mobile ASW systems, detection performance can also be enhanced through system mobility and estimates of the impact of search procedure for mobile systems is discussed.

Stationary ASW Systems

Stationary ASW sonar systems include sonobuoy fields, SOSUS or, during short periods of time, mobile sonar systems, e.g. tactical systems or SURTASS. These systems, being fixed in area coverage, must rely on the target coming to the system to be detected. Consequently, it comes as no surprise that target mobility is important to detection performance. Fast targets enter and leave the sonar footprint more frequently than slow targets and offer more detection opportunities. Since the target is in and out of the ensonified search area, the presence of a fast target within the search region is independent of the misses received on previous interrogations and the probability of target availability for fast targets reduces to the a priori probability that the target is in the search area, i.e. for

$$\text{fast targets:} \qquad P(T_n|M_{n-1}) = P(T_n) = \frac{A_S}{A_o}, \tag{6a}$$

where A_S is the ensonified search area, A_O is the operations area and the right member asserts total ignorance regarding target position.

In the limit of a fixed target, detection is possible only if the target is located within the sonar footprint and previous misses become a clue as to the whereabouts of the target. From Bayes' theorem, the probability of target availability can be written in the form

$$P(T_n|M_{n-1}) = \frac{P(M_{n-1}|T_n)P(T_n)}{P(M_{n-1})}, \tag{6b}$$

where the factor, $P(M_{n-1}|T_n)$, is the probability of receiving n−1 misses on the target, given that the target is in the search area, and where the factor $P(T_n)$ is the a priori probability of the target being in the search area. The denominator of (6b) can be expanded

$$P(M_{n-1}) = P(M_{n-1}|T_n)P(T_n) + P\left(M_{n-1}|T_n^{\,c}\right)P\left(T_n^{\,c}\right), \tag{6c}$$

where the superscript "c" denotes the complement of the event T_n which is the event that no target is present in the search region. Each quantity in (6c) has been determined above. Collecting results gives the probability of target availability expression for

$$\text{fixed targets:} \qquad P(T_n|M_{n-1}) = \frac{(1-P_D)^{n-1}(A_s/A_o)}{(1-P_D)^{n-1}(A_s/A_o)+(1-A_s/A_o)}. \tag{6d}$$

Equations (6a) and (6d) show that on the initial transmission, n = 1, the probabilities of target availability for the fast and fixed targets are the same but as misses are accumulated the probability of availability for fixed targets decreases to zero whereas that probability for fast targets remains constant.

Example 1: For the sake of illustrative examples, stationary active systems having one CZ coverage are considered and their performance in duct, BB and CZ propagation environments against fast and fixed targets are compared. The engineering probabilities of detection for each environment are set by the operating point on the receiver operating characteristic curves and typical values are specified.

$$\text{Duct:} \quad P_D = 0.5; \quad \frac{A_{duct}}{A_{ops}} = 1, \quad (7a)$$

$$\text{BB:} \quad P_D = 0.5; \quad \frac{A_{BB}}{A_{ops}} = \frac{1}{2}, \quad (7b)$$

$$\text{CZ:} \quad P_D = 0.9; \quad \frac{A_{cz}}{A_{ops}} = \frac{1}{5}, \quad (7c)$$

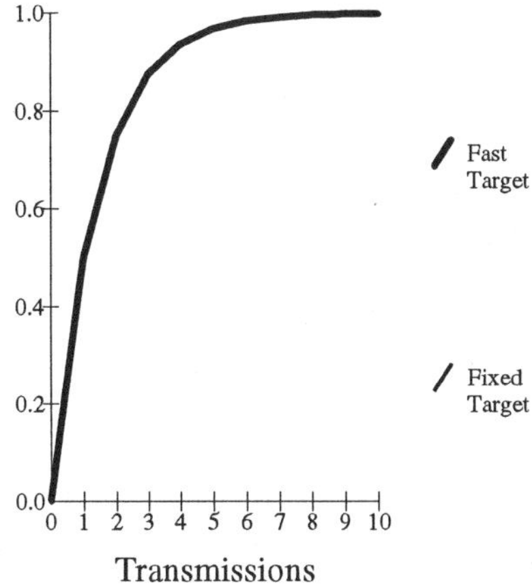

Figure 1a $P_D(N)$ for Duct

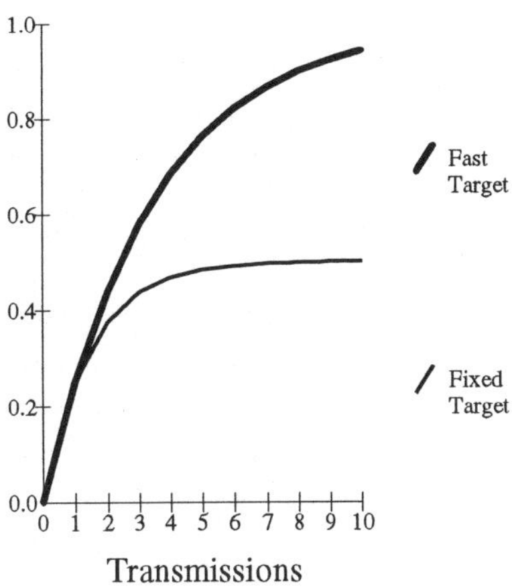

Figure 1b $P_D(N)$ for BB

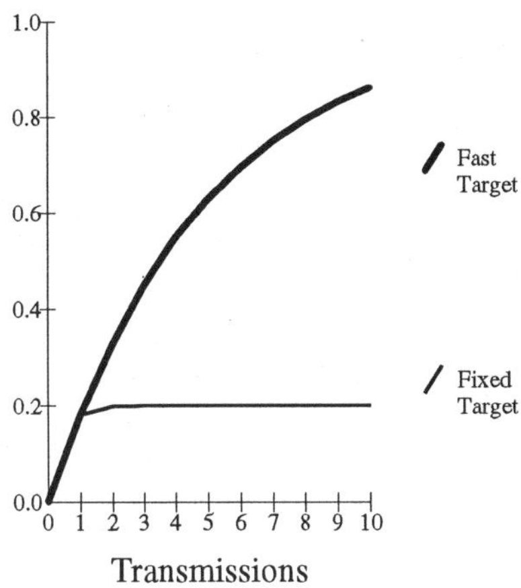

Figure 1c $P_D(N)$ for CZ

Figure 1 $P_D(N)$ for Stationary ASW Systems

The probabilities of target availability are given in (6a) and (6d) and assumed values for the probabilities of the target being in the ensonified areas are specified in (7). Plots of (5) for these values are graphed in Figure 1. For duct environments, the superposition of the detection performance curves for fast and fixed targets, as shown in Figure 1a, occurs because both targets are similarly ensonified on each transmission. This also causes the rapid increase in probability of detectability to near certainty within a few transmissions. Such is the benevolence of duct environments. For BB and CZ environments, the detection performance for both target types increases more slowly and the performance against fixed targets is markedly degraded as is apparent in Figures 1b and 1c. These curves suggest that the detection performance for fast targets in BB and CZ environments requires, respectively, about 2 and 4 as many transmissions as required for comparable performance in duct environments. Moreover, they indicate that, while the initial rise in detection probability for fixed targets parallels that for fast targets, detection performance against fixed targets is degraded and eventually approximates the probability of target present in the ensonified region which is caused by an accumulation of misses implying that the target is not in the ensonified area. Even more pessimistically, for targets that react to sonar transmissions and evade ensonification by avoiding ensonified regions, either by exiting the operations area, by surfacing, or by skillful maneuvering, the detection performance is expected to be further degraded from that indicated in Figure 1.♦

Mobile ASW Systems

Mobile ASW systems search for submarine targets by sequentially repositioning their sensors or by sweeping out areas of the ocean while transiting. These systems include tactical systems and SURTASS. Tactical helicopter dipping sonar systems can cover about 1000 square kilometers of ocean within about four hours by repositioning after several transmissions. Active and passive systems aboard surface ships can sweep out about 2000 square kilometers of ocean per hour in an aggressive search and SURTASS can sweep out about 3000 square kilometers per hour. Similarly to stationary ASW systems, the detection performance of mobile ASW systems can be enhanced by target mobility, which brings the target to the ASW system, but, additionally, mobile systems can enhance detection performance by seeking out the target.

Target detectability for mobile ASW systems is given by the general expression in (5) but we must rewrite this expression in terms of factors over search areas as well as transmissions. In this rewrite of (5) it is necessary to distinguish between missed detections within the current search area and missed detections within previous search areas. For the current kth search area (5) becomes

$$P_D(N) = 1 - \prod_{n=1}^{N_1}\left[1 - P_D P\left(T_{n,1} | M_{n-1}\right)\right] \cdots$$
$$\cdots \prod_{n=1}^{N_{k-1}}\left[1 - P_D P\left(T_{n,k-1} | M_{n-1+N_{k-2}+\cdots+N_1}\right)\right] \cdots$$
$$\cdots \prod_{n=1}^{q}\left[1 - P_D P\left(T_{n,k} | M_{n-1+N_{k-1}+\cdots+N_1}\right)\right], \qquad (8)$$
$$N = N_1 + \cdots + N_{k-1} + q$$

where N_{k-1} denotes the number of transmissions in the (k−1)th area, q is the current number of transmissions in the kth area and $T_{n,k}$ is the event of target present on the nth transmission in the kth area. The event of missed detections from (2) can be written

$$M_{n-1+N_{k-1}+\cdots+N_1} = \left(\bigcap_{j=1}^{n-1} m_{j,k}\right) \cap \left(\bigcap_{j=1}^{N_{k-1}} m_{j,k-1}\right) \cap \cdots \cap \left(\bigcap_{j=1}^{N_1} m_{j,1}\right), \tag{9a}$$

which can be abbreviated as

$$M_{n-1+N_{k-1}+\cdots+N_1} = M_{n-1,k} \cap M_{N_{k-1}+\cdots+N_1}, \tag{9b}$$

where

$$M_{n-1,k} = \left(\bigcap_{j=1}^{n-1} m_{j,k}\right), \quad M_{N_{k-1}+\cdots+N_1} = \left(\bigcap_{j=1}^{N_{k-1}} m_{j,k-1}\right) \cap \cdots \cap \left(\bigcap_{j=1}^{N_1} m_{j,1}\right). \tag{9c,d}$$

With these abbreviations, (8) can be written in the form

$$\begin{aligned} P_D(N) = 1 - \prod_{n=1}^{N_1}\left[1 - P_D P\left(T_{n,1} | M_{n-1,1}\right)\right] \cdots \\ \cdots \prod_{n=1}^{N_{k-1}}\left[1 - P_D P\left(T_{n,k-1} | M_{n-1,k-1} \cap M_{N_{k-2}+\cdots+N_1}\right)\right] \cdots \\ \cdots \prod_{n=1}^{q}\left[1 - P_D P\left(T_{n,k} | M_{n-1,k} \cap M_{N_{k-1}+\cdots+N_1}\right)\right]. \\ N = N_1 + \cdots + N_{k-1} + q \end{aligned} \tag{10}$$

The probability of target availability for cases of fast and fixed targets are considered and for each of these cases we consider an idealistic and an optimistic scenario.

Fast Targets:

For the case of fast targets, the probability of target availability is independent of misses in the current search area

$$P\left(T_{n,k} | M_{n-1,k} \cap M_{N_{k-1}+\cdots+N_1}\right) = P\left(T_{n,k} | M_{N_{k-1}+\cdots+N_1}\right). \tag{11a}$$

Since the cumulative misses within previous search areas implies that the target was not present, an idealistic scenario can be developed from this expression in which the event of misses in all previous search areas is assumed to mean that the target was and is not present in those areas. In this case the probability of the target being present in the current kth search area is just the ratio of that search area to the total operations area reduced by the previously searched areas. Hence, we obtain

$$P\left(T_{n,k}|M_{N_{k-1}+\cdots+N_1}\right)=\frac{A_s(k)}{\left(A_o-\sum_{j=1}^{k-1}A_s(j)\right)}=\frac{1}{K-(k-1)}. \tag{11b}$$

In the latter equality the ensonified search areas are assumed to be equal in size and to, collectively, just cover the operations area, i.e. $A_O = KA_S$. This expression asserts that within the initial search area k = 1 nothing is known of the target position but, with the accumulation of misses within each successive search area, the probability evolves and finally asserts total confidence that the target is in the last search area. The idealism of this scenario can only be assured by an exhaustive probing of each search area and a sweep search strategy that ensures that the target cannot escape and will be flushed.

A less demanding scenario for fast targets is the case in which the event of target present is independent of all previous misses which reduces (11a) to the a priori probability of target availability

$$P\left(T_{n,k}|M_{n-1,k}\cap M_{N_{k-1}+\cdots+N_1}\right)=P\left(T_{n,k}\right). \tag{11c}$$

Being independent of the sonar data, the a priori probability must be set using other sources of information which can include historical and intelligence data. An optimistic scenario having lesser detection performance than the idealistic scenario can be developed from this expression by assuming no knowledge regarding the position of the target throughout the search procedure. This result is given by

$$P\left(T_{n,k}\right)=\frac{A_s(k)}{A_o}=\frac{1}{K}, \tag{11d}$$

where the last equality is obtained by assuming equal sized areas that, collectively, just cover the operations area, i.e. $A_O = KA_S$. These idealistic and optimistic scenarios for the fast target case are graphed in Figure 2.

Fixed Targets:

For the case of fixed targets, the event of target present is dependent upon misses in previous search areas and, also, upon misses in the current search area but the dependence on each set of misses is different. Considering the recent misses first and proceeding as in (6d), we obtain

$$P\left(T_{n,k}|M_{n-1,k}\cap M_{N_{k-1}+\cdots+N_1}\right)=$$

$$\frac{(1-P_D)^{n-1}P\left(T_{n,k}|M_{N_{k-1}+\cdots+N_1}\right)}{(1-P_D)^{n-1}P\left(T_{n,k}|M_{N_{k-1}+\cdots+N_1}\right)+\left(1-P\left(T_{n,k}|M_{N_{k-1}+\cdots+N_1}\right)\right)}. \tag{12a}$$

Assuming an idealistic sweep of the operations area by the ASW system, as for the fast target case in (11b), an idealistic scenario can be developed for fixed targets by substituting that equation into (12a).

Alternatively, the probability of target availability, conditioned upon misses in the previous search areas, can be similarly expanded by Bayes' theorem to obtain

$$P\left(T_{n,k}|M_{N_{k-1}+\cdots+N_1}\right)$$
$$= \frac{P\left(M_{N_{k-1}+\cdots+N_1}|T_{n,k}\right)P\left(T_{n,k}\right)}{P\left(M_{N_{k-1}+\cdots+N_1}|T_{n,k}\right)P\left(T_{n,k}\right)+P\left(M_{N_{k-1}+\cdots+N_1}|T^c_{n,k}\right)\left(1-P\left(T_{n,k}\right)\right)}$$
$$= \frac{P\left(T_{n,k}\right)}{P\left(T_{n,k}\right)+P\left(M_{N_{k-1}+\cdots+N_1}|T^c_{n,k}\right)\left(1-P\left(T_{n,k}\right)\right)}, \quad (12b)$$

where the last equality holds because the probability of misses in the previous k−1 search areas is unity if the target is in the current kth search area. The event $T^c{}_{n,k}$ is the complement of $T_{n,k}$ and is the condition that the target is not in the current search area. This effectively asserts that the target is in one of the previous search areas or in one of the areas yet to be searched. An optimistic scenario can be developed from this result by substituting (11d) into (12b) and by expanding the remaining term to obtain

$$P\left(M_{N_{k-1}+\cdots+N_1}|T^c{}_{k,n}\right) = \frac{P\left(M_{N_{k-1}+\cdots+N_1}\cap\left(T_{<k}\cup T_{>k}\right)\right)}{P\left(T^c_{n,k}\right)}$$
$$= \frac{P\left(M_{N_{k-1}+\cdots+N_1}\right)P\left(T_{<k}\right)+P\left(M_{N_{k-1}+\cdots+N_1}\right)P\left(T_{>k}\right)}{P\left(T^c_{n,k}\right)}$$
$$= \frac{\left(1-P_D\right)^{N_T}\dfrac{(k-1)}{K}+\left(1-\dfrac{k}{K}\right)}{\left(1-\dfrac{1}{K}\right)}. \quad (12c)$$

In the first equality the notation $T_{<k}$ and $T_{>k}$ denote that the target is not in the kth search area, i.e. that it is in either the previously searched areas or in one of the areas to be searched. In the second equality we assumed that the misses in the previously searched areas are independent of which area the target is in. In the last equality we assumed that the number of transmissions in each search area is the same and is denoted N_T and that the search areas are all of equal size and collectively just cover the operations area, i.e. $A_O = KA_S$.

Example 2: Scenarios for mobile ASW systems are developed by extending the examples for stationary ASW systems. An operations area four times greater than that assumed for the stationary case was arbitrarily chosen, which is alternatively assumed to be a duct, BB and CZ environment. The engineering probabilities of detection are as given in (7) and, under the assumption that the search areas just cover the operations area, the numbers of search areas K are chosen such that, with reference to (7), $K = 4A_O/A_S$. These values and the number of transmissions per area are given in (13). Comparisons between fast and fixed targets for both idealistic and optimistic scenarios in duct, BB and CZ propagation environments are presented in Figure 2.

Duct: $K = 4$, $N_T = 10$, (13a)

BB: $K = 8$, $N_T = 5$, (13b)

CZ: $K = 20$, $N_T = 2$, (13c)

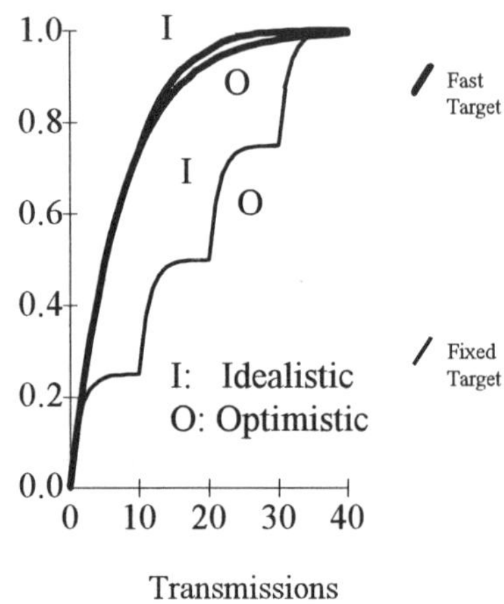

Figure 2a $P_D(N)$ for Ducts

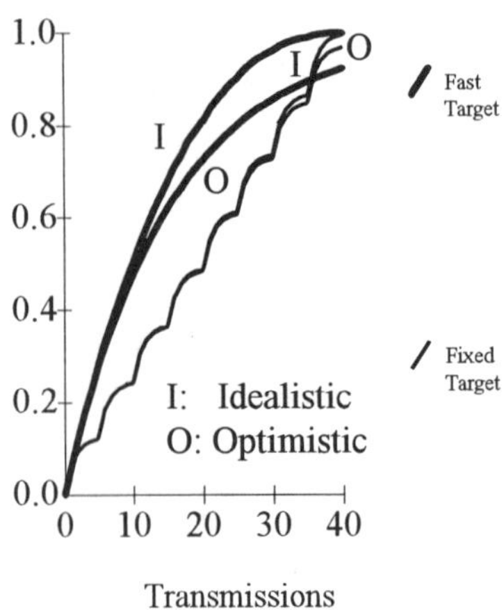

Figure 2b $P_D(N)$ for BB

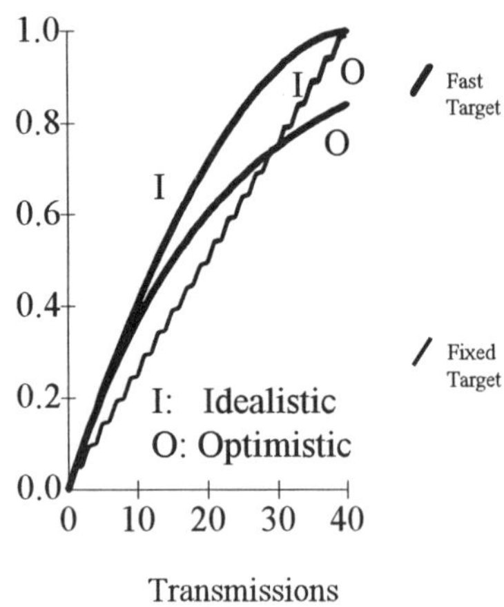

Figure 2c $P_D(N)$ for CZ

Figure 2 $P_D(N)$ for Mobile Systems

An overview of Figure 2 indicates that, for all cases, the performance of the idealistic scenario exceeds that of the optimistic scenario. Further, detection performance against fast targets is generally better than against fixed targets initially because fast targets present detection opportunities in all search areas. However, eventually detection performance against fixed targets meets or exceeds that of fast targets. Moreover, while the idealistic case leads to virtually certain detection in all cases, the differences between the idealistic and the optimistic scenarios are not large over extended time intervals. The largest differences are for fast targets in CZ environments because the smaller the ensonified area, the less time a fast target spends in the area which reduces its probability of detectability as compared to a fixed target. For fixed targets the differences between idealistic and optimistic scenarios are negligible because both cases are essentially equivalent, i.e. the target does not slip away and both scenarios rely on attrition. The figures also show that detection performance by mobile systems against fixed targets is substantially improved over detection performance by stationary systems against fixed targets, as indicated in Figure 1, because the system is able to seek out the target.

The results graphed in Figure 2 are informative but they are not realistic for several reasons. For realistic scenarios, two important modifications would include, on the one hand, the assumption that the ensonified areas overlap and do not just cover the operations area, and, on the other hand, the assumption that the entire operations area may not be thoroughly searched. For real systems the ensonified areas necessarily overlap, e.g. think of covering the operations area with nonoverlapping CZ rings, which will require more search areas to completely cover the operations area. Consequently, while the performance indicated in Figure 2 may be achieved in a more realistic scenario, it will most certainly require more time, more search areas and more transmissions than indicated in Figure 2. Furthermore, for real searches against hostile targets, a systematic search may not be possible and the entire operations area may not be searched with a sufficient number of transmissions to detect the target. This would result in detection performance that is less than that indicated in Figure 2. Moreover, extending the discussion into the realm of pessimistic scenarios, the target may react to the ASW sonar transmissions and attempt to evade ensonified areas by exiting the operations area or by surfacing or, worse, may aggressively hunt the ASW forces, forcing their fallback to a more cautious posture. All of these possibilities would reduce detection performance from that indicated in Figure 2. Because of these "real world" possibilities, the most effective search scenarios will generally involve an integrated system of multiple sensor systems in which target scenes are produced from shared data and the search pattern adapts to the evolving scene. ♦

14.2 Probability of Detected Target

In an on-going search of the operations area, detections received by ASW sonar systems are used to develop a target scene. At each interrogation of the environment, the received detections may be from previously detected targets or may be from new targets. To sort through the detection ambiguities, all detections are tested to determine if they can be associated with previously detected targets, in which case they are used to update estimates of those target tracks, and otherwise are viewed as new target detections and used to initiate a new track. From the resulting target tracks and their extrapolations, information is obtained that can aid in target classification, threat assessment and in coordinating ASW assets during target prosecutions. However, the complexity of the detection data can lead to incorrect data associations resulting in false tracks in addition to valid tracks. Because extra crew effort is fatiguing, communication link capacities are limited, machinery is subject to wear and tear, and munitions are hard to replace; it is critical to expend ASW assets only on valid threats. Consequently, criteria are established for recognizing valid target tracks as an aid in determining effective prosecution scenarios. Probability of detected target is one such criterion and is the probability that a target underlies a track estimate based upon the track data and upon constraints of what constitutes valid target motion.

General Relationships

Considering a sequence of detection data that have been linked by the tracker to produce a target track estimate, the issue is whether or not the data represent an underlying target or a false track that can mislead the ASW effort. The sequence of detection data that make up the track has the form

$$X_N = \left[x_0, x_1, \ldots, x_N\right], \tag{14a}$$

$$x_n = \begin{cases} 1 & \text{"hit"} \\ 0 & \text{"miss"} \end{cases} \tag{14b}$$

where the index denotes the nth interrogation time. The interrogation times are usually periodic but this is not a requirement. The tracker links the hits to form a track, and, in the place of the misses, the tracker either extrapolates or interpolates to cover the missed detections.

The hits in the sequence of detection data can arise from three possible scenarios. They may be hits on a submarine target exclusively; they may be hits on a submarine target with additional false alarm hits, due to reverberation and/or noise, that have been linked erroneously with the target to form a track; or they may be reverberation and/or noise hits only. For the first two cases, a target underlies the detection data, and the hypothesized event that represents these cases of target detected is denoted

$$T_N = \left[t_0, t_1, \ldots, t_N\right], \tag{14c}$$

where the subevent t_n denotes an hypothesized target underlying the data on the nth detection. The hypothesized subevents may be in error and, if a marked divergence between the detection data and the track of a creditable target occur, the probability of the event, i.e. its validity, decays and the hypothesis is discarded. The hypothesized event of no target detected is denoted

$$N_N = \left[n_0, n_1, \ldots, n_N\right], \tag{14d}$$

where the subevent n_n hypothesizes that there is no target underlying the data on the nth detection. The credibility of these two sets of hypotheses can be discriminated by comparing their respective probabilities conditioned upon the detection data.

The probability of target detected is defined to be

$$P(T_N | X_N) = P(\text{Target} \mid \text{All Detection Data}), \tag{14e}$$

and the probability of no-target detected is defined to be

$$P(N_N | X_N) = P(\text{No - Target} \mid \text{All Detection Data}). \tag{14f}$$

As a consequence of these definitions

$$P(T_N | X_N) + P(N_N | X_N) = 1, \tag{14g}$$

because the events target and no-target are mutually exclusive. These measures of confidence have been previously discussed in Section 13.3 in connection with the multi-target multi-sensor tracker and can be viewed as an integral part of that tracker.

Consideration of the joint probability of target detected and the most recent detection datum $P(T_N, x_N | X_{N-1})$ and its complement $P(N_N, x_N | X_{N-1})$ lead immediately to the expanded forms

$$P(T_N | X_N) = \frac{P(x_N | T_N, X_{N-1}) P(T_N | X_{N-1})}{P(x_N | X_{N-1})}, \tag{15a}$$

$$P(N_N | X_N) = \frac{P(x_N | N_N, X_{N-1}) P(N_N | X_{N-1})}{P(x_N | X_{N-1})}. \tag{15b}$$

These are the sought after probabilities in (14) and each term of these expressions is determined in the immediately following.

The first factors in the right members of (15) are the probabilities of receiving the current detection datum conditioned upon the two events: a target (or alternatively no target) underlying the detection data; and the previously received detection data. Since the current presence of a target t_N or absence of a target n_N is sufficient to determine the probability of the current detection datum, these events are independent of the previously received detection data and, if x_N is a hit, they are just the probabilities of, respectively, detection and false alarm and, if x_N is a miss, are the probabilities of, respectively, missed detection and correct dismissal.

The second factors in (15) can be expanded to yield

$$P(T_N | X_{N-1}) = P(t_N | T_{N-1}, X_{N-1}) P(T_{N-1} | X_{N-1}), \tag{16a}$$

$$P(N_N | X_{N-1}) = P(n_N | N_{N-1}, X_{N-1}) P(N_{N-1} | X_{N-1}). \tag{16b}$$

The first factors in the right members represent the events of target detected and no target detected during the current period and are dependent upon the corresponding events during the previous detection periods. Since they are independent of the previously received detection data, they reduce to the probabilities $P(t_N|T_{N-1})$ and $P(n_N|N_{N-1})$. These are the motion models for the case of a target and no target. The last factors in the right members of (16) are just the probabilities of target detected and no target detected at the end of the previous detection period and are the basis of an iterative solution.

The denominators of (15) are the probability of receiving the current detection datum conditioned upon the previously received detection data. This term can be expanded to form

$$P(x_N|X_{N-1}) = P(x_N|T_N, X_{N-1})P(T_N|X_{N-1}) + P(x_N|N_N, X_{N-1})P(N_N|X_{N-1}), \quad (16c)$$

in which each quantity was determined in the immediately preceding.

Collecting terms, (15) can be written in two possible ways. If x_N is a hit, (15) becomes

$$P(T_N|X_N) = \frac{P_D P(t_N|T_{N-1})P(T_{N-1}|X_{N-1})}{P_D P(t_N|T_{N-1})P(T_{N-1}|X_{N-1}) + P_{FA} P(n_N|N_{N-1})P(N_{N-1}|X_{N-1})}, \quad (17a)$$

$$P(N_N|X_N) = \frac{P_{FA} P(n_N|N_{N-1})P(N_{N-1}|X_{N-1})}{P_D P(t_N|T_{N-1})P(T_{N-1}|X_{N-1}) + P_{FA} P(n_N|N_{N-1})P(N_{N-1}|X_{N-1})}, \quad (17b)$$

but, if x_N is a miss, (15) becomes

$$P(T_N|X_N) = \frac{(1-P_D)P(t_N|T_{N-1})P(T_{N-1}|X_{N-1})}{(1-P_D)P(t_N|T_{N-1})P(T_{N-1}|X_{N-1}) + (1-P_{FA})P(n_N|N_{N-1})P(N_{N-1}|X_{N-1})}, \quad (17c)$$

$$P(N_N|X_N) = \frac{(1-P_{FA})P(n_N|N_{N-1})P(N_{N-1}|X_{N-1})}{(1-P_D)P(t_N|T_{N-1})P(T_{N-1}|X_{N-1}) + (1-P_{FA})P(n_N|N_{N-1})P(N_{N-1}|X_{N-1})}. \quad (17d)$$

Equations (17) are the general results from which special cases will be examined. These expressions provide an iterative formulation for determining the current probabilities of target detected and no target detected based upon track data.

For the initial hit, (17) becomes

$$P(T_0|X_0) = \frac{P_D P(t_0)}{P_D P(t_0) + P_{FA} P(n_0)}, \quad (18a)$$

$$P(N_0|X_0) = \frac{P_{FA} P(n_0)}{P_D P(t_0) + P_{FA} P(n_0)}. \quad (18b)$$

To arrive at these results it is noted that the events T_{-1}, N_{-1} and X_{-1} are empty and, from (16), the probabilities $P(T_{-1} \mid X_{-1})$ and $P(N_{-1} \mid X_{-1})$ are unity. The probabilities $P(t_0)$ and $P(n_0)$ are a priori probabilities, i.e. before any acoustic data is collected, of a target or no target being in any detection cell during the first interrogation. The value of $P(t_0)$ specifies the probability of finding a target in each detection cell and is based upon all that is known about the detection environment, which includes historical data, current intelligence data and other sources of data and the parameters of the detection system. Being mutually exclusive, the a priori probabilities satisfy the relation

$$P(t_0) + P(n_0) = 1, \tag{18c}$$

which determines the a priori probability of no target present. The case in which the initial event is a miss, i.e. $x_0 = 0$, is not relevant to the ASW problem and is not pursued.

For the first and higher order iterations, the target motion model term, i.e. the probabilities $P(t_N|T_{N-1})$, are determined from the characteristics of target motion and these quantities play an important role in identifying valid track trajectories. Whereas for the initial detection, the target motion model term reduces to the a priori probability of target present, for the first iteration it is conditioned upon the presence of the target at the location of the first linked detection. Assuming random target motion the target motion model for the first iteration is

$$P(t_1|T_0) = \iint_{\text{Cell}} d\theta dR\left[Rg(R,t)\right], \tag{19a}$$

where R is the range to the position of the linked detection: t is the time required for the target to travel to that new position; and g(R,t) is the probability density function for target motion. The integration in (19a) is carried out over the current detection cell. Since at the first iteration a track estimate is unavailable, unless one is obtained externally, the density function in (19a) is for a target having unknown initial direction of travel and the lack of dependence of the density function on the angular coordinate reflects that condition. In contrast, the target motion model for the second and higher iterations, for which a track estimate is assured, is given by

$$P(t_N|T_{N-1}) = \iint_{\text{Cell}} d\theta dR\left[R\hat{g}(R,\theta,t)\right], \tag{19b}$$

where R,θ,t are the coordinates of the current target position, as determined by the tracker, relative to the last target position and track. The function $\hat{g}(R,\theta,t)$ is the probability density function for the current target position in which the initial direction of the target is known.

The probability of no target present, $P(n_N|N_{N-1})$, for the case of the initial detection, reduced to the a priori probability of no target present, $P(n_0) = (1-P(t_0))$. For all higher iterations this function has the similar form

$$P(n_N|N_{N-1}) = 1 - P(t_N), \tag{19c}$$

which follows from the identity, $P(t_N, n_N|N_{N-1}) = 1$. The probability $P(t_N)$ is the a priori probability of target present during the Nth interrogation period. It is chosen, as for the initial detection, but it can change as the search progresses and as targets enter and leave the operations area. The a priori probabilities provide information to the algorithm regarding the environment beyond that given by the acoustic data.

Example 3: Consider a target traversing a continuous random walk track at a constant but unknown speed. The target speed is arbitrary but it is expected to be the target's maximum quiet speed. Moreover, assume that the target has been detected by active sonar with the detection history indicated in Figure 3a. After each detection, the target position density function is assumed to be a modified Washburn[2] density function in which the target speed is uniformly distributed between minimum and maximum speeds which are given, respectively, by V_m and V_M. Additionally assume the Washburn target turn rate parameter is small compared to the interrogation frequency, i.e. $\alpha t \ll 1$ where α is the target turn rate and t is the time between transmissions. Under this assumption, for an interrogation period of, say, 2 minutes, the turn rate on average must be small compared to 30 turns per hour with, say, no more than 6 turns per hour being an acceptable parameter. The assumption of a small turn rate is required if meaningful tracks are to be generated and is consistent with the maneuvers of patrolling targets and also for targets that are attempting to evade and escape, i.e. the turn rate of evasive targets is restricted by its need to put distance between itself and the hunter. The assumed transmission interval of 2 minutes corresponds to a maximum detection range of about 30 to 40 nm for which the additional duty cycle time allows for reverberation to decay and for processing time.

For simplicity, the ASW system is assumed to transmit single CW pulses omnidirectionally and to receive directionally and, additionally, it is assumed to perform energy detection and to process only range-bearing information and to process no Doppler information. The ROC operating point is assumed to be $P_D = 0.5$ and $P_{FA} = 0.0001$. The assumed probability of detection corresponds to a duct environment and the assumed probability of false alarm corresponds to approximately one false alarm per transmission assuming the detection cells are about 1/4 nm^2 on average. This average detection cell resolution can be approximately achieved by 1 sec CW pulses and 5° receive beams. This assumed ASW system is especially simple and does not correspond to any operational system but it serves to illustrate the following simple example of probability of target detected.

Following the initial system detection, the modified Washburn density function for unknown target direction, to first order in the turn rate parameter, i.e. $\alpha t \ll 1$, is given by the expression[3]

---

[2] See Appendix 14-A for a discussion of the Washburn probability density functions.

[3] See Appendix 14-B for a discussion of the Washburn probability density functions with random target speed.

$$g(R,t) = \frac{1}{2\pi R(V_M - V_m)t}\left[\left(U\left(\frac{R}{V_m t}-1\right) - U\left(\frac{R}{V_M t}-1\right)\right)(1-\alpha t) + \alpha t\left(\sin^{-1}\left(\left[\frac{R}{V_m t}\right]\right) - \sin^{-1}\left(\left[\frac{R}{V_M t}\right]\right)\right)\right], \quad (20a)$$

where U(·) is the unit step function

$$U(x) = \begin{cases} 1, & x \geq 0 \\ 0, & x < 0 \end{cases} \quad (20b)$$

and the range function is defined by

$$\left[\frac{R}{Vt}\right] = \begin{cases} \frac{R}{Vt}, & 0 \leq R \leq Vt \\ 1. & Vt \leq R \leq \infty \end{cases} \quad (20c)$$

From (20a) it can be seen that the probability of target position is nonzero only within the circle $R \leq V_M t$ and is concentrated within the ring of radii $V_m t \leq R \leq V_M t$.

For the first and higher iterations of the probability of target detected algorithm, the modified Washburn density function for known initial target direction, to first order in the turn rate parameter, i.e. $\alpha t << 1$, is[4]

$$\hat{g}(R,\theta,t) = \frac{1}{2\pi R\cos\theta(V_M - V_m)t}\left[2\pi(1-\alpha t)\left(U\left(\frac{R}{V_m t}-1\right) - U\left(\frac{R}{V_M t}-1\right)\right)\delta(\tan\theta) + (\alpha t)\ln\left(\frac{1-\left[\frac{R}{V_M t}\right]\cos\theta}{1-\left[\frac{R}{V_m t}\right]\cos\theta}\right)\right], \quad (20d)$$

where the $\theta = 0$ coordinate is oriented along the path of target motion. The probability is seen to be nonzero only within the circle of radius $R \leq V_M t$ and to be concentrated in the Delta function ridge positioned along the $\theta = 0$ axis in the range interval $V_m t \leq R \leq V_M t$.

Probability of detected target graphs are shown in Figures 3b, 3c and 3d for three cases of a priori probability of target present as indicated. The three cases correspond, respectively, to a target known to be within an area of radius 7.5 nm, 15 nm and 30 nm, where P_0 = (detection cell resolution)/(area of containment). The purpose of the

[4] See Appendix 14-B.

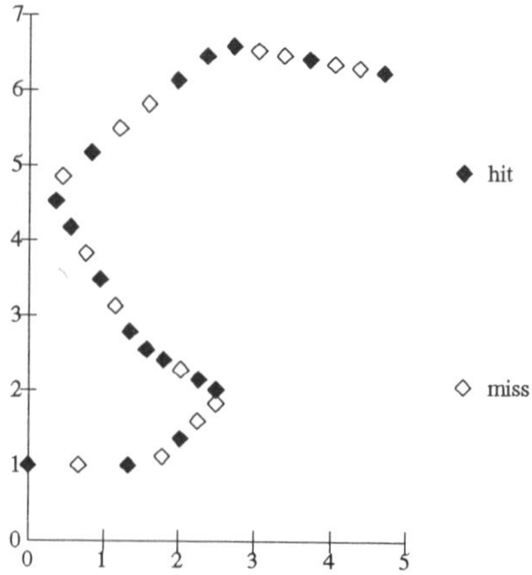

Figure 3a Target Track and Detection History

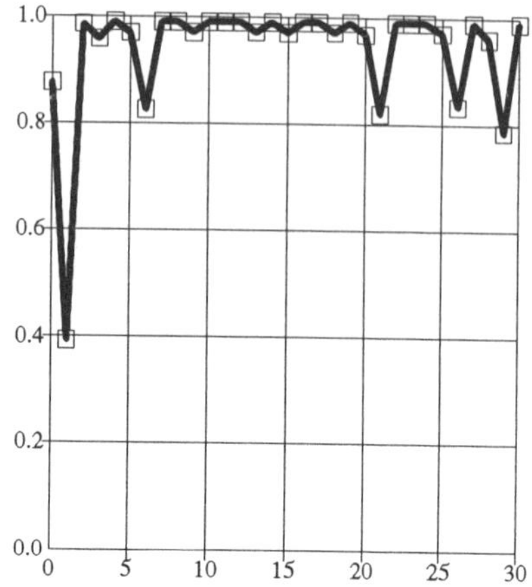

Figure 3b Probability of Detected Target ($P_0 = 1.4\times10^{-3}$, R = 7.5 nm)

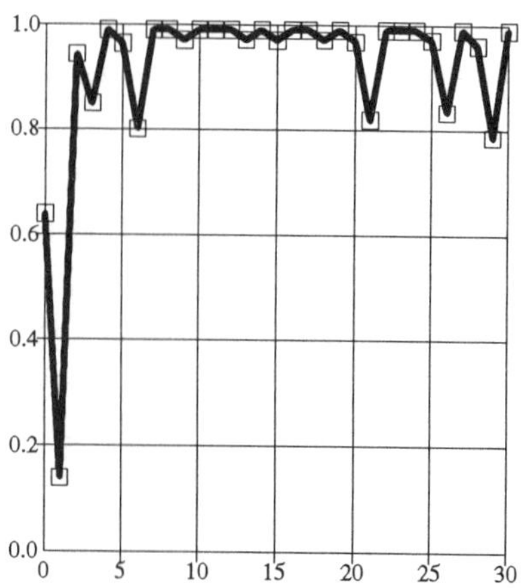

Figure 3c Probability of Detected Target ($P_0 = 3.5\times10^{-4}$, R = 15 nm)

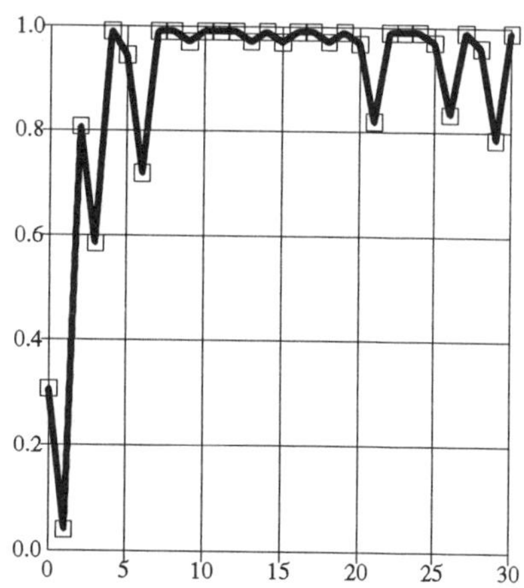

Figure 3d Probability of Detected Target ($P_0 = 8.8\times10^{-5}$, R = 30 nm)

Figure 3 Probability of Detected Target

comparison is to demonstrate that the particular choice of a priori probability is unimportant to the final results. As indicated in Figure 3, a priori information that places the target in a more constrained region raises the probability of target detected for the initial detections within that region. However, as more detections on the target are obtained, the weight of the detection data information overwhelms the a priori information and, for the example, after about three detections the probability of target detected is insensitive to the initial conditions.

A point-by-point examination of the detection history suggests important results concerning missed detections and target turns. It is found that after the initial target acquisition period and the probability of target detected has become independent of the initial conditions:

1. single misses cause insignificant reductions in probability of target detected but two sequential misses cause a 20% drop in probability of target detected;

2. under the Washburn model, which anticipates a target track following a continuous random walk, target turns cause insignificant reductions in probability of target detected.

Whereas it required about three detections to acquire the target with a high degree of confidence, point one indicates that a sequence of three misses would result in a markedly reduced confidence that the target was still being held. Also, while the Washburn model anticipates a turning target and does not strongly penalize detections that deviate from straight line motion, as indicated in point two, another motion model, appropriate for a target transiting in a straight line, would produce significant reductions in probability of target detected at each turn.

Since, in the absence of a priori information regarding the target, prosecution scenarios can require about three detections for high confidence pursuits, target prosecution in BB and CZ environments, in which only sporadic detection is the norm, can be expected to be difficult at best. In general, the target prosecution problem can be mitigated by either a system of distributed sensors that might produce relatively short range detections or by high quality a priori information that might come from previous target holdings on other ASW detection systems. ♦

APPENDIX 14-A: WASHBURN PDF FOR CONSTANT TARGET SPEED

A target is presumed to have been detected at some point within the area of sonar coverage and, prior to its next detection, it is presumed to have moved in a continuous random walk pattern. Consequently, the position of the target after the first detection, and prior to the subsequent detection, is a random process that evolves in time. The time dependent distribution function of the target position is required to determine the statistics of the target motion. Two cases of this function have been derived by Washburn[1] and a brief derivation of those results is given in the following. More details can be found in the original paper. The first case considers target motion in which the initial target direction is unknown to the observer and the second case considers target motion in which the initial target direction is known to the observer. The first of these cases corresponds to an initial target detection from which no estimate of the target track is available. The second case corresponds to the most recent detection in a sequence of detections from which a track estimate has been obtained. In both cases the speed of the target is presumed to be constant and to be known to the observer. The condition of known target speed is relaxed in Appendix 14-B.

14.1-A: Initial Target Direction Unknown

The time dependent position of the target is given by X_t and Y_t where the point of last detection defines the origin of coordinates in both space and time. During target maneuvers, the target travels course legs of lengths D_n and directions θ_n, n = 1,2,3,..., at the speed V. After time t the target will have performed N_t turns which satisfies the relation

$$\sum_{n=1}^{N_t} D_n \le Vt \le \sum_{n=1}^{N_t+1} D_n \,, \tag{1}$$

i.e. N_t is the number of turns completed at time t. Following Washburn, the density function for the target position at time t is denoted

$$g_n(x,y,t) = f_{X_t,Y_t \mid N_t}(x,y \mid n) \,, \tag{2}$$

where the left member is an abbreviated representation of the right member and the right member asserts the existence of the conditional density function of target position after n turns, in proper probabilistic functional notation. Washburn has developed an iterative process to determine all nth order density functions.

The zeroth order density function corresponds to the target motion prior to the first turn. It can be immediately written

[1] Washburn, Alan, "Probability Density of a Moving Particle," Operations Research, Vol. 17, No. 5, September-October, 1969.

$$g_0(x,y,t) = \left[\frac{1}{2\pi\,(Vt)^2}\right]\delta\left(\sqrt{\frac{(x^2+y^2)}{(Vt)^2}}-1\right), \tag{3}$$

in which δ denotes the Dirac delta function and the probability is concentrated on a ring at range Vt. This is to be expected since the initial target direction is unknown. The integral of this function over the x-y plane, and all nth order position density functions to be determined, can be shown to be unity as, of course, must be the case for probability density functions. To determine the higher order density functions, Washburn's iterative process is followed.

The nth order density function, for n = 1,2,3,⋯, can be written as the marginal density function

$$g_n(x,y,t) = \int_0^{Vt} du \int_0^{2\pi} d\theta f_{X_t,Y_t,D_1,\theta_1 \mid N_t}(x,y,u,\theta \mid n), \tag{4a}$$

in which the integrand is the joint density between the final position of the target and the position at the end of the first leg where D_1 and θ_1 denote the length and direction of the first leg. Expanding the integrand in terms of conditional probabilities gives

$$g_n(x,y,t) = \int_0^{Vt} du \int_0^{2\pi} d\theta f_{X_t,Y_t \mid D_1,\theta_1,N_t}(x,y \mid u,\theta,n)\, f_{D_1,\theta_1 \mid N_t}(u,\theta \mid n). \tag{4b}$$

The first factor in the integral is just the density function for the target position; but it is now conditioned, not only on the number of legs N_t but also on the event of the first leg. The effects of this additional conditioning are two: since the event of the first leg is a given, the number of random course legs is reduced to n–1 which means this factor is related to g_{n-1}; and the origin for the distribution of the n–1 random course legs is displaced to the end of the first leg. Mathematically this can be expressed as

$$f_{X_t,Y_t \mid D_1,\theta_1,N_t}(x,y \mid u,\theta,n) = g_{n-1}(x-u\cos\theta,\, y-u\sin\theta,\, t-u/V). \tag{5}$$

Equation (5) is the basis of an iterative result in which the nth order position density function is expressed as a functional of the (n–1)st order density function for all the course legs following the first leg. Of course the first leg is also a random event and has the distribution represented by the second factor in the integral in (4b). The integration over both factors gives the average (n–1)st order position density function for all first legs which yields the desired result of the nth order position density function.

The second factor in the integral in (4b) can be rewritten, using Bayes' formula, as

$$f_{D_1,\theta_1 \mid N_t}(u,\theta \mid n) = \frac{P(N_t = n \mid D_1 = u,\, \theta_1 = \theta)}{P(N_t = n)} f_{D_1,\theta_1}(u,\theta). \tag{6a}$$

Taking each factor in (6a) in succession, the probability that n turns have been executed, given that the first leg has been completed, is the probability that n−1 turns have been completed during the period of time after the first turn, i.e.

$$P(N_t = n \mid D_1 = u, \theta_1 = \theta) = P(N_{t-u/V} = n-1), \tag{6b}$$

where the number of turns is assumed to satisfy a Poisson distribution

$$P(N_t = n) = \left[\frac{(\alpha t)^n}{n!}\right]\exp(-\alpha t). \tag{6c}$$

The parameter α in (6c) is the turn rate parameter with units of number of turns per unit time. For the last factor in (6a), the leg length and course direction are assumed to be independent random variables with the leg length being exponentially distributed, and having turn rate parameter α/V, and the course direction being uniformly distributed over the full circle

$$f_{D_1,\theta_1}(u,\theta) = \frac{1}{2\pi}\left[\frac{\alpha}{V}\exp\left(-\frac{\alpha u}{V}\right)\right]. \tag{6d}$$

The turn rate parameter α/V has units of number of turns per unit distance and its reciprocal V/α is the mean length of a course leg. Combining results reduces (6a) to the form

$$f_{D_1,\theta_1 \mid N_t}(u,\theta \mid n) = \frac{n}{2\pi(Vt)^2}\left(1-\frac{u}{Vt}\right)^{n-1}, \tag{6e}$$

and (4) becomes

$$g_n(x,y,t) = \frac{n}{2\pi Vt}\int_0^{Vt} du \int_0^{2\pi} d\theta\, g_{n-1}(x-u\cos\theta, y-u\sin\theta, t-u/V)\left(1-\frac{u}{Vt}\right)^{n-1}. \tag{7}$$

This is a key result that can be iterated to generate all nth order position density functions.

The first order position density function is obtained from the zeroth order function given in (3). With some manipulation the combined expressions give

$$g_1(r,t) = \frac{1}{(2\pi Vt)^2}\int_0^{Vt} dv \int_0^{2\pi} d\theta\,\delta\left(\sqrt{\frac{r^2+v^2-2rv\cos\theta}{(1-v)^2}}-1\right)\frac{1}{(1-v)^2}, \tag{8a}$$

$$r = \frac{R}{Vt}, \quad R = \sqrt{x^2+y^2}, \quad v = \frac{u}{Vt}.$$

This result is found to be independent of angle.

From the Delta function property

$$\int \delta(f(x))g(x)dx = g(x_0) / f'(x_0), \quad f(x_0) = 0, \tag{8b}$$

where the integration interval contains x_0 but no other zeros of the function f(·), the integration over v in (8a) is performed to give[2]

$$g_1(r,t) = \frac{1}{(2\pi Vt)^2}\int_0^{2\pi}\frac{d\theta}{1 - r\cos\theta} = \frac{1}{2\pi(Vt)^2}\frac{1}{\sqrt{1-r^2}}, \tag{8c}$$

where g(r,t) is zero for r > 1. This result shows that after one course turn the probability of target position is still greatest near the circle of maximum range, R = Vt, but the probability of finding the target within the interior of the circle is greater than zero in contrast to the zeroth order position density. It is found that each succeeding position density function shifts the more probable positions from the circle of maximum range toward intermediate ranges.

It is readily shown that iterating (7) produces the general result

$$g_n(r,t) = \frac{1}{2\pi(Vt)^2}n\left(1-r^2\right)^{\frac{n}{2}-1}, \tag{9}$$

where g(r,t) is zero for r > 1. This general result is verified by showing that it satisfies (7)

$$g_n(r,t) = \frac{n(n-1)}{(2\pi Vt)^2}\int_0^{v_0}dv\int_0^{2\pi}d\theta\left[1 - \frac{r^2 + v^2 - 2rv\cos\theta}{(1-v)^2}\right]^{\frac{n-3}{2}}(1-v)^{n-3}. \tag{10a}$$

The integration limit v_0 is the normalized length of the first leg and is less than unity since the first leg does not terminate on the circle of maximum range. To determine v_0, we observe that (9) implies

$$\frac{r^2 + v^2 - 2rv\cos\theta}{(1-v)^2} < 1 \quad \rightarrow \quad v < v_0 = \frac{1-r^2}{2(1 - r\cos\theta)}. \tag{10b,c}$$

Manipulating (10a) gives

$$g_n(r,t) = \frac{n(n-1)}{(2\pi Vt)^2}\left(1-r^2\right)^{\frac{n-3}{2}}\int_0^{2\pi}d\theta\int_0^{v_0}dv\left[1 - \frac{v}{v_0}\right]^{\frac{n-3}{2}}$$

---

[2] Gradshteyn and Ryzhik, (3.613-1).

$$= \frac{n}{(2\pi Vt)^2}\left(1-r^2\right)^{\frac{n-1}{2}} \int_0^{2\pi} \frac{d\theta}{1-r\cos\theta}, \tag{10d}$$

which immediately gives the general result[3] in (9) as was to be proved.

Each of the nth order position density functions gives the probability of target position at time t assuming that n turns have been completed. To determine the probability of target position at time t, the probability of the number of turns completed at time t must be included. This is obtained from the expected value of all nth order density functions

$$g(x,y,t) = \sum_{n=0}^{\infty} g_n(x,y,t) P(N_t = n). \tag{11a}$$

Substituting from (3), (6c) and (9) gives

$$g(r,t) = \frac{\exp(-\alpha t)}{2\pi (Vt)^2}\left[\delta(r-1) + \sum_{n=1}^{\infty} n\left(1-r^2\right)^{\frac{n}{2}-1}\frac{(\alpha t)^n}{n!}\right]. \tag{11b}$$

The sum can be immediately evaluated[4] to give

$$g(r,t) = \frac{\exp(-\alpha t)}{2\pi (Vt)^2}\left[\delta(r-1) + \frac{\alpha t}{\sqrt{1-r^2}}\exp\left(\sqrt{1-r^2}\,\alpha t\right)\right], \tag{12}$$

where g(r,t) is zero for $r > 1$. This is Washburn's probability density function for the position of a particle moving on a continuous random walk. It can be shown to integrate to unity over the plane as, of course, it must.

14.2-A: Initial Target Direction Known

The target is assumed to be traveling on a continuous random walk after the most recent detection and prior to a subsequent detection but, contrary to the case in which the initial target direction is unknown, the direction of the first leg of the random walk is assumed to be known. This situation arises when a target track estimate is available which can be obtained from previous detections, a track estimate from another platform, intelligence data, etc. The initial direction can be taken, without loss of generality, to be in the $\theta = 0$ direction, i.e. along the x-axis of the coordinate system erected locally at the last detection position. The position density function for this case can be determined from the preceding results.

---

[3] Gradshteyn and Ryzhik, (3.613-1).
[4] Abramowitz and Stegun, (4.2.1).

Prior to the target's first turn, the target position is known and has the position density function

$$f_{X_t,Y_t|N_t=0}(x,y|n=0) = \frac{1}{(Vt)^2}\delta\left(\frac{x}{Vt}-1\right)\delta\left(\frac{y}{Vt}\right), \tag{13a}$$

which asserts that the target is at position (Vt, 0) at time t. Since the probability that no turns occur during the time interval (0,t) is given by $P(N_t = 0) = \exp(-\alpha t)$, the probability density function for finding the target at position (Vt,0) is

$$f_{X_t,Y_t}(x,y) = \frac{\exp(-\alpha t)}{(Vt)^2}\delta\left(\frac{x}{Vt}-1\right)\delta\left(\frac{y}{Vt}\right). \tag{13b}$$

If, in fact, the target turned at time τ, where $0 \le \tau \le t$, then the target position at time t would be $(V\tau + x, y)$ where x and y are random variables. The distribution of these random variables is given by the position density function derived in (12), after being displaced to the position of the first turn. In mathematical terms, after the target's first turn at time τ the distribution of target positions is given by

$$f_{X_t,Y_t|N_\tau=1}(x,y|n=1) = g(x-V\tau, y, t-\tau). \tag{14a}$$

Since τ is unknown operationally, all possible values of τ must be considered. Subdividing the interval (0,t) into N subintervals, the probability of the event that one target turn occurred in the subinterval τ_n to $\tau_n + \Delta\tau_n$ is $P(N_{\tau_n+\Delta\tau_n} = 1) - P(N_{\tau_n} = 1)$ where $P(\cdot)$ is the Poisson density function in (6c) and $n = 1,2,3,\cdots,N$. Since the events of the first turn occurring in each interval are mutually exclusive, their probabilities are added to give the probability of one turn in the interval (0,t). Moreover, the probability density function of finding the target at position (x,y) after the first turn is the average value of the density in (14a)

$$f_{X_t,Y_t}(x,y) = \lim_{\Delta\tau \to 0}\sum_{n=1}^{N}\Delta\tau\left[\frac{P(N_{\tau_n+\Delta\tau_n}=1)-P(N_{\tau_n}=1)}{\Delta\tau}\right]g(x-Vt, y, t-\tau), \tag{14b}$$

which in the limit becomes

$$f_{X_t,Y_t}(x,y) = \alpha\int_0^t d\tau \exp(-\alpha\tau) g(x-V\tau, y, t-\tau). \tag{14c}$$

Since the first turn either occurs or does not occur, the probabilities add to give the general probability density function for the target position

$$\hat{g}(x,y,t) = \frac{\exp(-\alpha t)}{(Vt)^2}\delta\left(\frac{x}{Vt}-1\right)\delta\left(\frac{y}{Vt}\right) + \alpha\int_0^t d\tau \exp(-\alpha\tau) g(x-V\tau, y, t-\tau), \tag{15}$$

where g(·) is given by (12). Substituting from (12), the integral in (15) becomes

$$I = \frac{\alpha\, t\exp(-\alpha t)}{2\pi (Vt)^2}\left[\int_0^{W_0} dw\, \frac{\delta(\rho-1)}{(1-w)^2} + \int_0^{W_0} dw\, \frac{\alpha\, t\exp\left[\alpha t(1-w)\sqrt{1-\rho^2}\right]}{(1-w)\sqrt{1-\rho^2}}\right], \qquad (16a)$$

$$\rho = \frac{(x - V\tau)^2 + y^2}{V^2(t-\tau)^2} = \frac{r^2 + w^2 - 2xw/Vt}{(1-w)^2} \le 1,$$

$$w = \frac{\tau}{t}, \qquad W_0 = \frac{1+r}{2}, \qquad r = \frac{\sqrt{x^2+y^2}}{Vt}.$$

The first integral in (16a) can be evaluated using the Delta function property in (8b) to yield

$$I_1 = \frac{1}{(1 - x/Vt)}, \qquad (16b)$$

for $r \le 1$ and zero otherwise. The second integral in (16a) can be evaluated by making the substitutions

$$(1-w)\sqrt{1-\rho^2} = \sqrt{2\left(1-\frac{x}{Vt}\right)}\sqrt{w_0 - w}, \qquad w_0 - w = w_0 v^2, \qquad (17a,b)$$

to yield

$$I_2 = \frac{\exp\left(\alpha t\sqrt{1-r^2}\right) - 1}{\left(1-\dfrac{x}{Vt}\right)}. \qquad (17c)$$

Combining results gives the probability density function for the target position

$$\hat{g}(x,y,t) = \frac{\exp(-\alpha t)}{(Vt)^2}\delta\left(\frac{x}{Vt}-1\right)\delta\left(\frac{y}{Vt}\right) + \frac{\alpha t\exp\left(\alpha t\left[1-\sqrt{1-r^2}\right]\right)}{2\pi (Vt)^2\left(1-\dfrac{x}{Vt}\right)}, \qquad (18a)$$

where

$$r = \frac{\sqrt{x^2+y^2}}{Vt}, \qquad x = (Vt)r\cos(\theta). \qquad (18b,c)$$

APPENDIX 14-B: WASHBURN PDF FOR UNKNOWN TARGET SPEED

The Washburn probability density functions, derived in Appendix 14-A, give the distribution of target positions for case of a target traversing a continuous random walk. Two density functions were derived for the cases of unknown initial target direction and known initial target direction. In both cases the target speed was assumed to be constant and known. A constant and known target speed is appropriate for the case in which the target evades at maximum speed, which is assumed to be known, or some other known speed. However, for cases in which the target is assumed to evade at a constant speed for which only an estimate of the target speed is available, it is a random variable. In the following, modified Washburn probability density functions are derived in which target speed is assumed to be constant but to be uniformly distributed between a minimum and maximum value.

The Washburn density functions are repeated below for convenience, from (14.12-A) and (14.18-A), in which the conditioning on target speed is made explicit:

1. unknown initial direction

$$g(r,t|V)=\frac{\exp(-\alpha t)}{2\pi(Vt)^2}\left[\delta(r-1)+\frac{\alpha t}{\sqrt{1-r^2}}\exp\left(\sqrt{1-r^2}\,\alpha t\right)\right], \tag{1a}$$

2. known initial direction

$$\hat{g}(r,\theta,t|V)=\frac{\exp(-\alpha t)}{(Vt)^2}\delta\left(\frac{x}{Vt}-1\right)\delta\left(\frac{y}{Vt}\right)+\frac{\alpha t\exp\left(\alpha t\left[1-\sqrt{1-r^2}\right]\right)}{2\pi(Vt)^2\left(1-\frac{x}{Vt}\right)}, \tag{1b}$$

$$r=\frac{\sqrt{x^2+y^2}}{Vt}=\frac{R}{Vt},\quad x=R\cos(\theta),\quad y=R\sin(\theta).$$

In both cases the density functions are zero for $r>1$.

Assuming the target speed is uniformly disributed between a minimum and maximum value, i.e. V_m and V_M respectively,

$$f_V(V)=\frac{1}{(V_M-V_m)},\quad V_m\le V\le V_M$$
$$=0,\quad \text{elsewhere} \tag{2}$$

the Washburn density functions become

$$g(R,t)=\int_{V_m}^{V_M}dVf_V(V)g\left(\frac{R}{Vt},t|V\right),\quad \hat{g}(R,\theta,t)=\int_{V_m}^{V_M}dVf_V(V)\hat{g}\left(\frac{R}{Vt},\theta,t|V\right). \tag{3a,b}$$

Each of these integrals is evaluated in the following.

14.1-B: Initial Target Direction Unknown

With the variable change, V = R/rt, (3a) becomes

$$g(R,t) = \int_{\frac{R}{V_M t}}^{\frac{R}{V_m t}} dr \left[\frac{(R/t) f_V(R/rt)}{r^2}\right] g\left(r, t \middle| \frac{R}{Vt}\right), \tag{4a}$$

where the term in square brackets is the density function for r

$$f_r(r) = \frac{R}{r^2 (V_M - V_m) t}. \quad 0 \le \frac{R}{V_M t} \le r \le \frac{R}{V_m t} \le 1 \tag{4b}$$

As R ranges from zero to infinity, r achieves the limits zero to unity because of the constraints on the support for the density function of V. Collecting results, the density function in (4) becomes

$$g(R,t) = \frac{\exp(-\alpha t)}{2\pi R (V_M - V_m) t} \int_{\left[\frac{R}{V_M t}\right]}^{\left[\frac{R}{V_m t}\right]} dr \left[\delta(r-1) + \frac{(\alpha t)\exp\left(\alpha t \sqrt{1-r^2}\right)}{\sqrt{1-r^2}}\right], \tag{5a}$$

where the bracket in the integration limit denotes

$$\left[\frac{R}{Vt}\right] = \begin{cases} \frac{R}{Vt}, & 0 \le R \le Vt \\ 1. & Vt \le R \le \infty \end{cases} \tag{5b}$$

The first integral in (5a) is

$$I_1 = U\left(\frac{R}{V_m t} - 1\right) - U\left(\frac{R}{V_M t} - 1\right), \tag{5c}$$

where U(·) denotes the unit step function

$$U(x) = \begin{cases} 1, & x \ge 0 \\ 0. & x < 0 \end{cases} \tag{5d}$$

The second integral in (5a) is

$$I_2 = \alpha t \int_{\left[\frac{R}{V_M t}\right]}^{\left[\frac{R}{V_m t}\right]} dr \frac{\exp\left(\alpha t \sqrt{1-r^2}\right)}{\sqrt{1-r^2}}, \tag{5e}$$

which does not have an elementary closed form solution but, by expanding the exponential function in a Taylor series and making the variable change $x = r^2$, it can be expressed in the infinite series form

$$I_2 = \frac{1}{2}\sum_{n=0}^{\infty}\frac{(\alpha t)^n}{n!}\left[B_{\left[\frac{R}{V_m t}\right]^2}\left(\frac{1}{2},\frac{n+1}{2}\right) - B_{\left[\frac{R}{V_M t}\right]^2}\left(\frac{1}{2},\frac{n+1}{2}\right)\right], \tag{5f}$$

where $B_X(a,b)$ is the incomplete Beta function[1].

The advantage of the series form is that approximations for small values of αt can easily be obtained. For such values of αt the target turn rate is small compared to the target detection frequency which must be the case if a meaningful target track estimate is to be achieved. Moreover, since the choice of a high turn rate by an evasive target would lengthen the target time in the surveillance area and delay its escape, the low turn frequency assumption is appropriate. The probability density function for the target position under the low turn rate assumption is, to first order in αt,

$$g(R,t) = \frac{1}{2\pi R(V_M - V_m)t}\left[\left(U\left(\frac{R}{V_m t}-1\right) - U\left(\frac{R}{V_M t}-1\right)\right)(1-\alpha t) + \alpha t\left(\sin^{-1}\left(\left[\frac{R}{V_m t}\right]\right) - \sin^{-1}\left(\left[\frac{R}{V_M t}\right]\right)\right)\right]. \tag{6}$$

It can be shown that this approximation to the target position density function satisfies the requirement

$$\int_0^{2\pi} d\theta \int_0^{\infty} dR R g(R,t) = 1. \tag{7}$$

14.2-B: Initial Target Direction Known

Proceeding as in Section 12.1-B (3b) can be written

$$\hat{g}(R,\theta,t) = \frac{\exp(-\alpha t)}{2\pi R(V_M - V_m)t}\int_{\left[\frac{R}{V_M t}\right]}^{\left[\frac{R}{V_m t}\right]} dr\left[2\pi\delta(r\cos\theta - 1)\delta(r\sin\theta) + \frac{(\alpha t)\exp\left(\alpha t\sqrt{1-r^2}\right)}{1-r\cos\theta}\right]. \tag{8}$$

From the delta function property in (14.8b-A) the first integral of (8) is determined to be

[1] Abramowitz and Stegun,(6.6.1).

$$I_1 = \begin{cases} \dfrac{2\pi}{\cos\theta}\,\delta(\tan\theta), & V_m t \le R \le V_M t \\ 0. & \text{elsewhere} \end{cases} \tag{9}$$

The second integral has no elementary closed form solution but an approximate integration to first order in αt can be determined as in Section 14.1-B giving

$$\hat{g}(R,\theta,t) = \frac{1}{2\pi R\cos\theta\left(V_M - V_m\right)t}\left[\left(\frac{R}{V_m t} - 1\right)2\pi(1-\alpha t)\left(U - U\left(\frac{R}{V_M t} - 1\right)\right)\delta(\tan\theta) + (\alpha t)\ln\left(\frac{1-\left[\dfrac{R}{V_M t}\right]\cos\theta}{1-\left[\dfrac{R}{V_m t}\right]\cos\theta}\right)\right], \tag{10}$$

It can be shown that this approximation to the target position density function satisfies the requirement

$$\int_0^{2\pi} d\theta \int_0^{\infty} dR R\hat{g}(R,\theta,t) = 1\,. \tag{11}$$

Bibliography

Abramowitz, M. and I.A. Stegun, *Hndbk of Mathematical Functions*, Nat. Bur. Stan., Dept. Comm., 1972.
Bar-Shalom, Y. and T.E. Fortmann, *Tracking and Data Association*, Academic Press, Boston, MA, 1988.
Beckmann, P. and A. Spizzichino, *Scattering of EM Waves/Rough Surfaces*, Artech, Norwood, MA, 1987.
Beyer, R.T., *Nonlinear Acoustics*, Naval Ships System Command, Dept. of Navy, 1974.
Blackman, S.S., *Multiple-Target Tracking with Radar Applications*, Artech, Denham, MA, 1986.
Bleistein, N., *Mathematical Methods for Wave Phenomena*, Academic Press, Orlando, FL., 1984.
Brekhovskikh, L.M., Lieberman, D. ed., Beyer, R.T. tr., *Waves in Layered Media*, Academic Press, 1960.
Burdic, W.S., *Underwater Acoustic System Analysis*, Prentice-Hall, Englewood Cliffs, NJ, 1984.
Caruthers, J.W., *Fundamentals of Marine Acoustics*, Elsevier, Amsterdam, 1977.
Clay, S.C. and H. Medwin, *Acoustical Oceanography*, John Wiley, New York, 1977.
Coates, R.F.W., *Underwater Acoustic Systems*, John Wiley, New York, 1989.
DeSanto, J.A. ed., *Ocean Acoustics*, Topics in Current Physics, **8**, Springer-Verlag, Berlin, 1979.
Elmore, W.C. and M.A. Heald, *Physics of Waves*, Dover, New York, 1985.
Flatté, S.M., ed., *Sound Transmission through a Fluctuating Ocean*, Cambridge U. Press, Cambridge, 1979.
Franks, L.E., *Signal Theory*, Prentice-Hall, Englewood Cliffs, NJ, 1969.
Gelb, Arthur, ed., *Applied Optimal Estimation*, M.I.T. Press, Cambridge, MA, 1974.
Gradshteyn, I.S. and I.M. Ryzhik, *Table of Integrals, Series and Products*, Academic Press, New York, 1980.
Horton, C.W. Sr., *Signal Processing of Underwater Acoustic Waves*, U.S. Gov. Printing Office, 1969.
Horton, J.W. , *Fundamentals of SONAR*, United States Naval Institute, Annapolis, MD, 1957.
Hunt, F.V., *Electroacoustics*, Harvard University Press, Cambridge, MA, 1954.
Jenkins, G.M. and D.G. Watts, *Spectral Analysis & its Applications*, Holden-Day, San Francisco, CA, 1968.
Keller, J.B. ed. et.al., *Wave Propagation and Underwater Acoustics*, LNP, **70**, Springer-Verlag, Berlin, 1977.
Kennedy, R.S., *Fading Dispersive Communication Channels*, Wiley-Interscience, New York, 1969.
Kinsler, L.E. and A.R. Frey, *Fundamentals of Acoustics*, John Wiley, New York, 1962.
Lamb, Sir H., *Hydrodynamics*, Dover, New York, 1945.
Lamb, Sir H., *The Dynamical Theory of Sound*, Dover, New York, 1960.
Lindsay, R.B., *Mechanical Radiation*, McGraw-Hill, New York, 1960.
Love, A.E.H., *A Treatise on the Mathematical Theory of Elasticity*, Dover, New York, 1944.
Miller, D. and J. Jordan, *Modern Submarine Warfare*, Military Press, New York, 1987.
Minkler, G. and J. Minkler, *CFAR*, Magellon Book Co., Baltimore, MD, 1990.
Monin, A.S., et. al., Lumley, J.J. ed.., *Variability of the Oceans*, Wiley-Interscience, New York, 1974.
Morse, P.M., *Vibration and Sound*, McGraw-Hill, New York, 1948.
Morse, P.M., and H. Feshbach, *Methods of Theoretical Physics*, **I, II**, McGraw-Hill, New York, 1953.
Morse, P.M. and K.U. Ingard, *Theoretical Acoustics*, McGraw-Hill, New York, 1968.
Nguyen, H.P., *Submarine Detection from Space*, Naval Institute Press, Annapolis, MD, 1993.
Officer, C.B., *Introduction. to the Theory of Sound Transmission*, McGraw-Hill, New York, 1958.
Ogilvy, J.A., *Theory of Wave Scattering from Rough Surfaces*, Adam Hilger, New York, 1991.
Operations Analysis Study Group, *Naval Operations Analysis*, Naval Institute Press, Annapolis, MD, 1977.
Phillips, O.M., *The Dynamics of the Upper Ocean*, Cambridge University Press, Cambridge, 1980.
Physics of Sound in the Sea, Sum. Tech. Rpt. Division 6, **8**, Nat. Def. Res. Council, Dept. Navy, 1968.
Pickard, G.L., *Descriptive Physical Oceanography*, Pergamon Press, Oxford, 1975.
Princ. and Appl. of Underwater Sound, Sum. Tech. Rpt. Div. 6, **7**, Nat. Def. Res. Council, Dept. Navy, 1968.
Rayleigh, J.W.S., *The Theory of Sound*, **I, II**, Dover, New York, 1945.
Ross D., *Mechanics of Underwater Noise*, Pergamon Press, New York, 1976.
Selin, I., *Detection Theory*, Princeton University. Press, Princeton NJ, 1965.
Sibul, L.H., *Adaptive Signal Processing*, IEEE Press, 345 East 47 St., New York, 1987.
Steinberg, B.D., *Principles of Aperture and Array System Design*, Wiley-Interscience, New York, 1976.
Sorenson, H.W., *Kalman Filtering: Theory and Application*, IEEE Press, 345 East 47 St., New York., 1985.
Tolstoy, I. and Clay, C.S., *Ocean Acoustics*, McGraw-Hill, New York, 1966.
Tucker, D.G. and B.K. Gazey, *Applied Underwater Acoustics*, Pergamon Press, Oxford, 1966.
Urick, R.J., *Principles of Underwater Sound*, McGraw-Hill, New York, 1975.
Urick, R.J., *Sound Propagation in the Sea*, Defense Advanced Research Projects Agency, 1979.
Vakman, D.E., *Sophisticated Signals & Uncertainty Principle in Radar*, Springer-Verlag, New York, 1968.
Van Trees, H.L., *Detection, Estimation, and Modulation Theory*, **I, II, III**, John Wiley, New York, 1968.
Whalen, A.D., *Detections of Signals in Noise*, Academic Press, New York, 1971.
Woodward, P.M., Probability and Information Theory with App. to Radar, Artech, Denham, MA, 1980.
Ziomek, L.J., *Underwater Acoustics, A Linear Systems Approach*, Academic Press, Orlando, FL, 1985.

INDEX

Book Ordering Information

Copies of this book can be ordered from:

Sage Publishing Company
P.O. Box 460394
Escondido, CA 92046

Order Form

Book Price	$50.00
Postage/Handling	2.00
Sum	52.00
CA Resident's Sales Tax	3.88
Total per Book	
Quantity	
Total Sale	
Discount (for 2+ books) 10% of Total Sale	
Remittance	

Inquire regarding volume discounts.

Author Information

Dr. Grace has been a sonar scientist for over 30 years. Throughout his career he has worked in government and university laboratories and in small research and development companies. Much of this work was performed under contract with the U.S. Navy on anti-submarine warfare sonar systems. Additionally, he has taught sonar engineering courses and other courses at the university level. For the past 10 years, he has been an independent consultant and much of his efforts during this period were in support of the U.S. Navy Low Frequency Aperture Program. This work included research and development and at-sea testing of the LFA system. More recently, his efforts have been focused on freelance writing. He currently resides in the Elfin Forest area of Southern California with his wife, Alice Courteau.